Planar Microwave Sensors

Planar Microwave Sensors

Ferran Martín
Paris Vélez
Jonathan Muñoz-Enano
Lijuan Su
Universitat Autònoma de Barcelona

Registered Office(s)
John Wiley & Sons, Inc., 111 River Street, Hoboken, NJ 07030, USA

Editorial Office
111 River Street, Hoboken, NJ 07030, USA

For details of our global editorial offices, customer services, and more information about Wiley products visit us at www.wiley.com.

Wiley also publishes its books in a variety of electronic formats and by print-on-demand. Some content that appears in standard print versions of this book may not be available in other formats.

Library of Congress Cataloging-in-Publication Data
Names: Martín, Ferran, 1965- author. | Vélez, Paris, author. |
  Muñoz-Enano, Jonathan, author. | Su, Lijuan, author.
Title: Planar microwave sensors / Ferran Martín, Paris Vélez, Jonathan
  Muñoz-Enano, Lijuan Su.
Description: Hoboken, New Jersey : Wiley-IEEE Press, [2023]
Identifiers: LCCN 2022017876 (print) | LCCN 2022017877 (ebook) | ISBN
  9781119811039 (cloth) | ISBN 9781119811046 (adobe pdf) | ISBN
  9781119811053 (epub)
Subjects: LCSH: Microwave detectors.
Classification: LCC TK7876 .M22 2022 (print) | LCC TK7876 (ebook) | DDC
  621.381/3–dc23/eng/20220622
LC record available at https://lccn.loc.gov/2022017876
LC ebook record available at https://lccn.loc.gov/2022017877

Cover Design: Wiley
Cover Image: Courtesy of Ferran Martín, Paris Vélez, Jonathan Muñoz-Enano, Lijuan Su

Set in 9.5/12.5pt STIXTwoText by Straive, Pondicherry, India

To Anna, Alba, and Arnau
To Pedro, Ana, Julia, Miguel, Arantxa, and Irai
To Claudio, Loli, Adrian, Antonia, and Raquel
To Gerard and Oscar

And to the memory of Prof. Tatsuo Itoh, a Mentor and a Guide, awarded the Honoris Causa *Doctorate by the Universitat Autònoma de Barcelona in 2015, among many other distinctions and honors.*

Contents

Preface

Perseverance, humility and courage are necessary qualities in a good scientist, more than wisdom, curiosity, and intelligence. Generosity is probably the highest virtue of a great person.

Someone, somewhere, sometime

In today's society, there is an increasing need to sense multitude of variables of different types. Data collection from the environment, or, in general, from a certain system, is a fundamental requirement in order to gain insight on the state of such system, and thus take appropriate decisions and actions, either through human intervention or autonomously, when necessary. The subject of analysis of the system can be as diverse as large-scale (macro) systems, medium-sized systems, or microsystems. Examples include the space, the atmosphere, a forest, a city, a crop field, a civil infrastructure, a factory, industrial machinery, the home, a specific indoor/outdoor area, daily objects, persons, animals, food, a biological sample, etc. In certain cases, retrieving information of the system is the main, and sometimes unique, objective, without any further active action that modifies the system, or represents any kind of control over it. For example, weather forecasting is based on the measurement of ambient variables such as temperature, atmospheric pressure, relative humidity, wind velocity, etc., and the information provided by weather forecasters (inferred from such environmental data using complex meteorological models) is very useful for citizens and administrations for obvious reasons (in particular, when extreme meteorological conditions are predicted). However, obtaining these environmental variables has an informative intention (i.e. the weather forecast), exclusively. There are other macro systems, for instance, a region in the Earth susceptible to seismic action, where the sensed data (seismic variables, in this particular case) are not used to generate preventive actions (earthquakes cannot be avoided) but to make useful predictions (that avoid major catastrophes or, at least, protect the population).

However, in most systems, sensing is necessary as a first step to generate actions that modify, or perturb, it according to certain requirements. For example, any motion control system, present in many industrial scenarios (e.g. elevators, conveyor belts, and servomotors), is equipped with sensors that measure motion variables, such as position, velocity, etc., and such variables are used by system actuators to generate correcting actions, if needed. Another clear example is the autonomous and intelligent vehicle, where a set of sensors of different types continuously collects data, which are used not only to assist the driver (if it is present) but also to autonomously take decisions (the so-called unmanned vehicle is another paradigmatic example). In healthcare, smart systems able to monitor vital constants, to measure variables of medical interest, e.g. glucose content in blood, or to detect unexpected body movements (e.g. in disabled or elderly people), indicative of potential dangerous events (such as lipothymia or ictus), are of the highest interest. Naturally, such systems need sensors to collect these data, but the main relevant and distinctive feature of smart healthcare systems is their ability (not always present) to generate specific actions from the retrieved data. For example, in the event of a sudden increase of glucose in blood above a certain threshold, detected by a dedicated real-time glucose-monitoring sensor, an alert indicative of hyperglycemia should be activated, in order to prevent the patient from dramatic irreversible effects. Alternatively, in hospital environment, a smart system should be able to automatically inoculate insulin to the patient, in a controlled way, in order to compensate for the excess of sugar in blood.

The three examples reported in the previous paragraph (motion control, the autonomous vehicle, and smart systems for healthcare) are representative of three sectors where system intelligence is penetrating significantly and are thereby experiencing a considerable (digital) transformation, namely, the Smart Industry (intimately related to the fourth industrial revolution, also called Industry 4.0), the Intelligent Car, and the Smart Healthcare. Nevertheless, there are many other sectors, which are nowadays in the process of transformation toward the digital world (or Smart World), including agriculture, packaging, food industry, city management and sustainability, and civil engineering (e.g. structural health monitoring), to cite some of them. Thus, terms such as Smart Agriculture, Smart Packaging, Smart Cities, or Ambient Intelligence, among others, are becoming progressively more familiar within our society.

To make the Smart World concept a reality, or, at least, to achieve further levels of intelligence within the above-cited fields, efforts at different levels are needed. At system level, enabling technologies for the so-called Internet of Things (IoT), e.g. radiofrequency identification (RFID), near-field communications (NFC), wireless sensor networks (WSN), energy harvesting, cloud computing, big data analysis, communication protocols, and embedded systems, is the subject of an intensive research activity, and the progress in such technologies is fundamental to envisage a future interconnected and intelligent world. Nevertheless, one key factor for the deployment of IoT and related applications is the recent implantation of the fifth generation of mobile networks (5G), with higher capacity, connectivity, and broader bandwidths (among other advantages) as compared to 4G. At device level, the key components in smart sensing systems are sensors. The research activity in the sensors domain has experienced a very significant growth in recent years. There are many types of sensors, exploiting different technologies, e.g. optical sensors, acoustic sensors, and magnetic sensors, but the sensors that are expected to play a key role in future smart systems are microwave sensors, and particularly planar sensors, the subject of this book. Their low cost, small size, and low profile, as well as the possibility of sensor implementation in flexible and organic substrates (by either subtractive or additive processes) are important attributes of planar microwave sensors. Other important aspects of microwave technologies are low-cost generation and detection systems, microwave interaction with the materials at different scales (i.e. through the near field or the far field), wave propagation (and penetration) in many different types of materials, and system functionality in hostile and harsh environments, e.g. with pollution and dirtiness (encountered in many industrial scenarios), or under adverse meteorological conditions. Planar microwave sensors can be implemented in combination with other technologies, such as microfluidics, micromachining, lab-on-a-chip, textiles, etc., and, inherently, exhibit a potentially wireless connectivity. Moreover, the sensor substrate can integrate the associated sensor electronics, needed for signal generation, post-processing, and (eventually) communication purposes, representing a reduction in system complexity and cost. Versatility is another relevant characteristic of planar microwave sensors. Despite the fact that such devices are (canonically) permittivity sensors, able to detect changes in the dielectric properties of the immediate environment, it is possible to use these sensors to determine material composition or to detect defects or anomalies in samples or targets, in both cases related to permittivity changes. Nevertheless, it is also possible to use planar microwave sensors to measure many other parameters, including physical and chemical variables (e.g. temperature, humidity, motion, concentration of certain substances in samples, and gas detection) or to perform biological analysis (e.g. bacterial growth and presence of certain analytes in biosamples). In some cases, smart materials (i.e. functional materials with dielectric properties highly sensitive to the measurand), reagents (i.e. chemical agents able to activate a chemical reaction), or bioreceptors (i.e. biological elements that bind to a specific analyte) are needed in order to boost up sensor sensitivity, a key performance parameter. Finally, let us mention that by using biodegradable substrates and organic inks, planar microwave sensors are potential candidates for the implementation of "green" sensing systems (for instance, RFID sensors with battery-free and chipless sensing tags). Nevertheless, there are many challenging aspects, mainly related to performance degradation inherent to the use of eco-friendly materials, that should be addressed to make "green sensing" a reality.

The previous advantageous characteristics of planar microwave sensors explain the huge efforts dedicated to their research and optimization in recent years and justify the publication of the present book. Though a planar microwave sensor in a real scenario is composed of three main blocks (the electromagnetic module, including the sensitive element, the electronics module, responsible for signal generation and post-processing, and the communication module), and, eventually, of a mechanical part, this book is essentially focused on the electromagnetic, or microwave, block. Nevertheless, in some of the reported sensor implementations (proof-of-concept demonstrators), part of the electronics is also included (especially, when this is necessary to retrieve and visualize the sensing data). The book tries to emphasize the underlying physics behind the considered sensing mechanisms and provides useful guidelines for sensor design, inferred from detailed analysis of the structures under study, mainly devoted to sensitivity optimization. Most of the reported sensors in the book are focused on permittivity measurements, material characterization, and motion sensing, and mainly correspond to recent contributions by the authors. Nevertheless, there are many other prototypes (especially in the subtopic of RFID sensors) proposed by other researchers working in the field. Let us also mention that by using functional materials, many of the ideas presented in this book can be extrapolated to the implementation of sensors devoted to the measurement of a wide variety of physical, chemical, and biological/medical variables.

The book contains as many chapters as considered sensing working principles, plus an additional introductory chapter (Chapter 1) and a concluding chapter (Chapter 8). Thus, Chapter 1 is an introduction to the topic of the book, planar microwave sensors. However, a general overview of other microwave (mainly nonplanar and remote) sensors, extended also to other sensing technologies (such as optics, acoustics, magnetic, mechanical, and electric), is carried out in the chapter. The chapter proposes a general and useful categorization scheme of microwave sensors, where it is shown that planar sensors belong to the category of non-remote sensors, in contrast to the remote counterparts, mainly RADAR sensors and radiometers. Special attention is dedicated to the classification of planar microwave sensors, the subject of interest in the book. Probably the most useful categorization criterion for such sensors is the one that obeys to their working principle (the classification adopted in this book). Chapter 1 ends with a comparison of planar microwave sensors with other sensors based on microwaves and other technologies.

Chapter 2 is devoted to frequency-variation sensors, probably those planar microwave sensors that constitute the main subject of research in the field to date. These sensors are based on resonant structures, with resonance frequency (and magnitude) sensitive to the dielectric properties of the surrounding environment. The chapter presents a set of (planar) semi-lumped resonators useful for sensing and includes a sensitivity analysis, based on the circuit models describing the sensor, that links the sensitivity to the characteristics of the sensor material and circuit parameters. It is shown that these sensors are useful for the dielectric characterization of solid and liquid samples. Moreover, it is demonstrated that, by considering sensor arrays and multifrequency sensing, it is possible to resolve the dielectric properties of the sample, with potential application to the detection of anomalies in organic tissues (e.g. malignant or cancerous cells). The chapter also presents a multifrequency method for the selective determination of components in complex mixtures (e.g. liquid solutions), which is based on the dispersive behavior of the permittivity of the different constitutive elements of the composite.

The working principle discussed and studied in Chapter 3 is phase variation. Such sensors are single-frequency devices that can be implemented by means of transmission-mode or reflective-mode structures. Moreover, phase-variation sensors can operate either differentially or as single-ended devices. The phase of the transmission or the reflection coefficient is the canonical output variable, sensitive to the dielectric properties of the medium in contact (or in proximity) with the sensitive element, typically a transmission line section, or a planar resonator. It is shown that, by exploiting the highly dispersive behavior of artificial lines, high sensitivities in phase-variation sensors based on such artificial lines can be achieved. Several examples representative of the specific approaches are included in the chapter. Nevertheless, probably the most notable contribution of Chapter 3 concerns the introduction of a sensor strategy useful to optimize the sensitivity without the need to increase the area of the sensing region (the usual procedure in phase-variation sensors). The technique applies to reflective-mode structures, and it is based on cascading a step-impedance line (with quarter-wavelength sections) to the sensing element, typically an open-ended half- or quarter-wavelength line section, or a planar resonator tuned to the operating frequency. An exhaustive analysis, useful to predict the sensitivity in these highly sensitive reflective-mode sensors, is carried out in Chapter 3. Although most phase-variation sensors are devoted to permittivity measurements or dielectric characterization of solids and liquids, examples of motion sensors based on phase variation are also included in Chapter 3.

Chapter 4 is dedicated to coupling-modulation sensors, i.e. planar microwave sensors where the (variable) electromagnetic coupling between a transmission line (or a transmission line-based structure) and a planar resonant element (or a set of resonators) is the physical sensing mechanism. Since such electromagnetic coupling depends on the relative position and orientation between the line and resonant element/s, it follows that the canonical application of coupling-modulation sensors is the measurement of displacements and velocities (both linear and angular). The output variable in such sensors is usually the magnitude of the transmission, or reflection, coefficient at the operating frequency (such sensors are typically single-frequency devices). Sensors based on electromagnetic symmetry properties, particularly symmetry truncation, and sensors based on line-to-resonator proximity are presented in the chapter. Special attention is focused on the so-called electromagnetic encoders (linear and rotary), since such sensors may have application in industrial scenarios where motion control is required and where their optical counterparts (optical encoders) may experience difficulties related to the presence of pollution, grease, contaminants, etc. Nevertheless, the chapter reports also an example of a coupling-modulation sensor devoted to material characterization.

In Chapter 5, frequency splitting, generated in a transmission line (or transmission line-based structure) symmetrically loaded with a pair of identical resonators when such resonators are asymmetrically loaded, or perturbed, is the considered working principle. These sensors are not true differential sensors, but they exhibit similar characteristics, in particular, they are robust against potential cross-sensitivities caused by ambient factors (e.g. temperature and humidity). The chapter highlights that inter-resonator coupling tends to degrade the sensitivity, especially in the limit of small perturbations. Thus, several methods reported to circumvent this limitative aspect are presented, discussed, and experimentally validated in Chapter 5.

Although differential sensing is not exactly a working principle, but rather a strategy to eliminate the undesired effects of potential common-mode stimuli, a dedicated chapter (Chapter 6) is focused on differential-mode sensors. Nevertheless, some examples of differential sensors are included in previous chapters (e.g. in Chapter 3, where differential phase-variation sensors are reported, since the physical sensing mechanism in such sensors is phase variation). True differential sensors should consist at least of two independent and noninteracting (uncoupled) sensor units (sensor pair). Each unit can operate independently as a single-ended sensor, but in differential sensors, the input and output variables are the incremental variables of their single-ended constitutive counterparts. In most sensors in Chapter 6, the output variable is the cross-mode transmission, or reflection, coefficient, which is, intrinsically, a differential variable, whereas the input variable is usually the differential permittivity between the sample under test and a reference sample. The chapter shows that such sensors can be used as highly sensitive comparators, as well. The reported examples include differential-mode sensors for solid and liquid characterization. It is remarkable that the high sensitivity and resolution of some of the sensors included in the chapter and able to discriminate, e.g. concentrations of electrolytes in aqueous solutions as small as 0.25 g/L. The chapter demonstrates also the possibility to determine the total electrolyte concentration in animal (horse) urine, of potential interest as a real-time monitoring and prescreening method, useful to detect certain pathologies related to an excess, or defect, of electrolytes.

Chapter 7 is devoted to the topic of RFID sensors, justified by the increasing demand of distributed sensors wirelessly connected to the central unit (the reader in the framework of RFID) within the context of the IoT and the Smart World. The key elements in RFID sensing systems are the sensor tags, equipped with sensing functionality and, usually, with identification capability. The chapter reports a classification scheme of RFID sensors where the main distinction is between chipped- and chipless-RFID sensors. The former are not true planar sensors, as far as they include an integrated circuit where the identification code is stored. Nevertheless, in the so-called electromagnetic chipped-RFID sensors, sensing is performed by means of the planar antenna of the tag, eventually altered to increase the sensitivity, rather than by means of a dedicated electronic sensor module (as the so-called electronic chipped-RFID sensors use). By contrast, chipless-RFID sensor tags are fully planar, and, consequently, such sensors have been the subject of a further consideration in Chapter 7, despite the fact that their performance cannot compete against the one of chip-based tags. The chapter succinctly reviews several functional (or smart) materials typically used in chipless-RFID sensors (as coating films) in order to sensitively measure several physical or chemical variables (e.g. temperature, or humidity, to cite some variables sensed in several of the reported prototypes). The chapter includes a list of potential applications of RFID sensors in fields as diverse as Smart Agriculture, Smart Packaging, Smart Healthcare, Structural Health Monitoring, Smart Cities, Smart Cars, Space, etc. It ends by presenting some commercial RFID sensors, by pointing out the main limitative aspects of RFID sensors (primarily concerning chipless-RFID tags) and by pointing out some challenging issues for future investigation. The main efforts should be focused on the topic of chipless-RFID sensing tags, since their limited performance is one of the bottlenecks toward the full deployment of the IoT.

Finally, Chapter 8 summarizes the different sensing approaches considered in the previous chapters in the form of a comparative analysis and points out the main concluding remarks.

It is the authors' hope that the present book constitutes a reference manuscript in the topic of planar microwave sensors. The authors have done their best to generate a useful product of practical interest for the academy and the industry, specially for engineers, students, and researchers involved in sensing technologies, in RF/microwave engineering, in IoT systems, and, in general, in any transversal field where planar microwave sensors are expected to play a fundamental role in next future.

Ferran Martín
Paris Vélez
Jonathan Muñoz-Enano
Lijuan Su
Bellaterra (Cerdanyola del Vallès), Barcelona
February 2022

Acknowledgments

This book is mainly the result of the research activity by the authors in the field of planar microwave sensors carried out in recent years. However, it also includes many ideas, sensing devices, and their practical applications (extracted from the available literature) that belong to others. Thus, the book has been written under the perspective, viewpoint, and experience of the authors, but the objective has been to provide a wide overview on the topic and an up-to-date state-of-the-art. It is impossible to enumerate all the people whose ideas have contributed to make this book a reality, but all the authors of the cited literature in the different chapters should be felt part of the list to which the authors are in debt. Nevertheless, special thanks are given to some researchers with whom the authors have had a close and fruitful collaboration in the field of planar microwave sensors in recent years, including Dr. Katia Grenier (CNRS-LAAS, Toulouse, France), Prof. David Dubuc (University of Toulouse and CNRS-LAAS, Toulouse, France), Dr. Amir Ebrahimi (RMIT University, Melbourne, Australia), Dr. Ali Karami-Horestani (Gdansk University of Technology, Poland), Dr. Marta Gil-Barba (Universidad Politécnica de Madrid, Spain), Dr. Ignacio Gil (Universitat Politècnica de Catalunya, Spain), and Dr. Rosa Villa (IMB-CNM-CSIC, Spain). Dr. Paris Vélez and Jonathan Muñoz-Enano explicitly express their gratitude to Dr. Katia Grenier and Prof. David Dubuc for accepting them to be part of their Group at CNRS-LAAS during their respective research stays (both in the field of microfluidic sensors). Dr. Paris Vélez acknowledges also Dr. Rosa Villa and Dr. Ignacio Gil for accepting him in their respective Groups during two consecutive research stays, one devoted to microfluidic sensors and the other one to textile-based sensors. Prof. Ferran Martín would also like to mention the numerous and productive discussions on electromagnetism topics related to planar microwave sensors (in part included in the book) with Prof. Francisco Medina and his Team at the Universidad de Sevilla, Spain.

The authors are also very grateful to the members of their Group and Research Transfer Centre (CIMITEC), ascribed to the Departament d'Enginyeria Electrònica at the Universitat Autònoma de Barcelona, Spain. Among them, special thanks are given to Dr. Ferran Paredes and Dr. Cristian Herrojo, since some of the results related to electromagnetic encoders included in the book have been mostly generated by them (and also by Dr. Jordi Naqui and Dr. Javier Mata-Contreras, past members of CIMITEC). Jan Coromina and Pau Casacuberta are also included in the acknowledgment list, since part of their PhD research is devoted to planar microwave sensors, and hence related to the topics of the book. The authors are also in debt to Dr. Gerard Sisó and Javier Hellín, who have fabricated many of the reported prototypes, to Anna Cedenilla, for handling the permissions of reproduced figures, and to Marta Mora, head of the Administrative Staff of the Department.

Several funding institutions, research agencies, and companies have supported the research activity carried out by the authors and reported in the book. Among them, special thanks are given to the Ministry of Science and Innovation and National Research Agency (MCIN/AEI 10.13039/501100011033), Spain, and ERDF European Union, for the financial support through the projects TEC-2016-75650-R, RTC-2017-6303-7, PID2019-103904RB-I00, and PDC2021 121085-I00 (European Union Next Generation EU/PRTR), to the AGAUR Research Agency, Catalonia Government, for their support through the projects 2014SGR-157, 2017SGR-1159, and 2014LLAV00046, and to the European Space Agency (ESA) for the contract 4000111799/14/NL/SC. The authors have had several collaborations, or are in close contact, with companies in the field of sensors and related topics, including EMXYS, Hohner Automáticos, García Carrión, and ZIP BCN Solutions, among others. Let us also mention the support received by the authors for their professional development and progress. This support includes ICREA (Institució Catalana de Recerca i Estudis Avançats), a Catalan Institution that has awarded Prof. Ferran Martín (calls 2008, 2013, and 2018), the TECNIOSPRING Program (ACCIÓ, Catalonia Government, and Horizon 2020 Marie Sklodowska-Curie Funds), that has awarded Dr. Paris Vélez through the project TECSPR15-1-0050, and the Juan de la Cierva Program (Ministry of Science and Innovation, Spain), for the grants IJCI-2017-31339 and IJC-2019-040786-I given to Dr. Paris Vélez and Dr. Lijuan Su, respectively. Let us also include in the list the FI-Grant (Secretaria d'Universitats i Recerca, Government of Catalonia, and European Social Fund) awarded to Jonathan Muñoz-Enano for the realization of the PhD.

Finally, the authors would like to express their most sincere gratitude to their respective families for creating the necessary atmosphere to write a complex manuscript like this, for being so patient during many times the authors have had an almost exclusive dedication to the preparation of the book, and for their continuous and unconditional support. This book also belongs to them. Thank you very much!

Ferran Martín
Paris Vélez
Jonathan Muñoz-Enano
Lijuan Su

About the Authors

Ferran Martín received BS degree in Physics from the Universitat Autònoma de Barcelona (UAB), Spain, in 1988 and PhD degree in 1992. He is a full professor of Electronics in the Electronics Engineering Department, UAB, since 2007. He is the head of the *Microwave Engineering, Metamaterials and Antennas* (GEMMA) Group, director of *CIMITEC* (a Research and Technology Transfer Centre, ascribed to UAB), and past head of the Electronics Engineering Department (period 2015–2021). His research interests include microwave circuits and sensors, metamaterials, and radiofrequency identification (RFID). Ferran Martín has organized several international events related to metamaterials and related topics, including Workshops at the *IEEE International Microwave Symposium* (years 2005 and 2007) and *European Microwave Conference* (2009, 2015, 2017, and 2018), and the *Fifth International Congress on Advanced Electromagnetic Materials in Microwaves and Optics* (Metamaterials 2011), where he acted as Chair of the Local Organizing Committee. He has acted as a guest editor for several special issues on metamaterials and sensors in different international journals, and he is the associated editor of *IET Microwaves Antennas and Propagation* and *Sensors*.

Ferran Martín has authored and co-authored over 650 technical conferences, letters, journal papers, and book chapters. Professor Martín is the co-author of the book on metamaterials entitled *Metamaterials with Negative Parameters: Theory, Design and Microwave Applications* (John Wiley & Sons Inc.), author of the book *Artificial Transmission Lines for RF and Microwave Applications* (John Wiley & Sons Inc.), co-editor of the book *Balanced Microwave Filters* (Wiley/IEEE Press), co-author of the book *Time-Domain Signature Barcodes for Chipless-RFID and Sensing Applications* (Springer), and co-author of the book *Planar Microwave Sensors* (Wiley/IEEE Press). Ferran Martín has generated 22 PhDs; he has filed several patents on metamaterials and other topics related to microwave engineering; and he has headed dozens of research projects and development contracts with companies. Among his distinctions, Ferran Martín has received the 2006 Duran Farell Prize for Technological Research; he holds the *Parc de Recerca UAB – Santander* Technology Transfer Chair; and he has been the recipient of three ICREA ACADEMIA Awards (calls 2008, 2013, and 2018). He is Fellow of the IEEE (since 2012) and Fellow of the IET (since 2016).

Paris Vélez (Senior Member, IEEE) was born in Barcelona, Spain, in 1982. He received the degree in telecommunications engineering, specializing in electronics, the degree in electronics engineering, and the PhD degree in electrical engineering from the Universitat Autònoma de Barcelona, Spain, in 2008, 2010, and 2014, respectively. His PhD thesis concerned common-mode suppressed differential microwave circuits based on metamaterial concepts and semi-lumped resonators. During the PhD degree, the Spanish government awarded him a Pre-Doctoral Teaching and Research Fellowship from 2011 to 2014. From 2015 to 2017, he was involved in subjects related to metamaterial-based sensors for fluidic detection and characterization at LAAS-CNRS through a

TECNIOSpring Fellowship cofounded by the Marie Curie Program. From 2018 to 2020, he has worked in the miniaturization and optimization of passive RF/microwave circuits and sensors based on metamaterials through the Juan de la Cierva Fellowship. At present, he is a postdoctoral researcher in CIMITEC, Universitat Autonoma de Barcelona, Spain. His current research interests include the optimization and miniaturization of planar sensors, microfluidic sensors, and textile-based sensors. He is a reviewer for *IEEE Transactions on Microwave Theory and Techniques* and many other journals.

Jonathan Muñoz-Enano was born in Mollet del Vallès (Barcelona), Spain, in 1994. He received the bachelor's degree in Electronic Telecommunications Engineering in 2016 and the master's degree in Telecommunications Engineering in 2018, both at the Universitat Autònoma de Barcelona (UAB), Spain. In July 2022, he obtained PhD degree within the doctoral program in Electronics Engineering and Telecommunication in the same university, with the thesis entitled "Highly sensitive planar microwave sensors for dielectric characterization of solids, liquids, and biosamples." Actually, he is working as a postdoctoral research in CIMITEC, Universitat Autònoma de Barcelona, Spain.

Lijuan Su was born in Qianjiang, Hubei, China in 1983. She received the BS degree in communication engineering and the MS degree in circuits and systems from the Wuhan University of Technology, Wuhan, China, in 2005 and 2013, respectively, and the PhD degree in electronic engineering from the Universitat Autònoma de Barcelona, Barcelona, Spain, in 2017. From November 2017 to December 2019, she worked as a postdoctoral researcher with the Flexible Electronics Research Center, Huazhong University of Science and Technology, Wuhan, China. She is currently a postdoctoral researcher in CIMITEC, Universitat Autònoma de Barcelona, Spain. Her current research interests focus on the development of novel microwave sensors with improved performance for biosensors, dielectric characterization of solids and liquids, defect detection, industrial processes, and so on.

List of Acronyms

2-CSR	2-turn complementary spiral resonator
2-SR	2-turn spiral resonator
A/D	analog to digital
AC	alternating current
AIDC	automatic identification and data capture
AM	amplitude modulation
ASK	amplitude shift keying
BAP	battery-assisted passive
BC-SRR	broadside-coupled split-ring resonator
BST	barium strontium titanate
CdS	cadmium sulfide
CNT	carbon nanotube
CPW	coplanar waveguide
CRLH	composite right-left handed
CSRR	complementary split-ring resonator
CST	computer simulation technology
CWT	continuous wavelet transform
DB-DGS	dumbbell-shaped defect ground structure
DC	direct current
DGS	defect ground structure
DI	deionized
DPF	density per frequency
DPL	density per length
DPS	density per surface
EIW	electro-inductive wave
ELC	electric-LC
EPC	electronic product code
FM-CW	frequency modulation continuous wave
FoM	figure of merit
GPR	ground penetrating radar
GO	graphene oxide
HF	high frequency
HFSS	high frequency simulation software
HIoT	Health Internet of Things
IC	integrated circuit
ID	identification

IDC	interdigital capacitor
IEEE	Institute of Electrical and Electronics Engineers
IEC	International Electrotechnical Commission
IoE	Internet of Everything
IoMT	Internet of Medical Things
IoT	Internet of Things
ISE	ion selective electrode
ISM	industrial, scientific, and medical
ISO	International Organization for Standardization
LASER	light amplification by stimulated emission of radiation
LED	light emitting diode
LIDAR	light detection and ranging
LUT	liquid under test
MCSRR	modified complementary split-ring resonator
MIW	magneto inductive wave
MLC	magnetic-LC
MRM	monostatic radar module
MUT	material under test
NFC	near-field communications
NIST	National Institute of Standards and Technology
NRW	Nicolson–Ross–Weir
OCSRR	open complementary split-ring resonator
OSRR	open split-ring resonator
PCB	printed circuit board
PDMS	polydimethysiloxane
PEEK	polyether ether ketone
PEDOT:PSS	poly(3,4-ethylenedioxythiophene):poly(styrenesulfonate)
PEN	polyethylene naphthalate
PET	polyethylene terephthalate
PEUT	polyetherurethane
PHEMA	poly 2-hydroxyethylmethacrylate
PIE	pulse interval encoding
PIR	passive infrared
PLA	polylactic acid
PMMA	polymethyl methacrylate
PPR	pulse per revolution
Pt-rGO	platinum-reduced graphene oxide
PVA	polyvinyl alcohol
QR	quick response
R/T	reflection/transmission
RADAR	radio detection and ranging
RCS	radar cross section
rGO	reduced graphene oxide
REF	reference
RF	radiofrequency
RFID	radiofrequency identification
SAW	surface acoustic wave

SHM	structural health monitoring
SIR	step-impedance resonator
SISS	step-impedance shunt stub
SIW	substrate integrated waveguide
SMD	surface-mount device
SNR	signal-to-noise ratio
SONAR	sound navigation and ranging
SRR	split-ring resonator
S-SRR	S-shaped split-ring resonator
SWIPT	simultaneous wireless information and power transfer
TDR	time domain reflectometry
TFT	thin-film transistor
UHF	ultra high frequency
UV	ultra violet
UWB	ultra wideband
VCO	voltage-controlled oscillator
VNA	vector network analyzer
WL	Weiner Lower
WU	Weiner Upper
WPT	wireless power transfer
WSN	wireless sensor network

1

Introduction to Planar Microwave Sensors

Within today's paradigm of Internet of Things (IoT) [1–6], or, more generally, Internet of Everything (IoE) [7], and the advent of the Fourth Industrial Revolution (also known as Industry 4.0) [8–12], there has been an increasing demand for low-cost sensors, able to continuously monitoring a multiplicity of parameters, including environmental, physical, chemical, biological, or medical variables, among many others, of interest in diverse scenarios. This trend will continue with the progressive implantation of the 5th generation of mobile networks, designated as 5G, which is expected to satisfy the large-scale connectivity requirements that today's modern society demands (an interesting example is the autonomous and connected vehicle, necessarily equipped with hundreds of sensors). Sensors and sensor networks are key components in the so-called "Smart World" (or digital world), understood as an interconnected world equipped with sensors, actuators, displays, and computational elements, where the ultimate goals are to enhance the quality of life of citizens, to ease the action of policy decision-makers, and to contribute to the sustainability (maintenance and protection) of the overall environment [13]. Within this framework, Smart Cities, Ambient Intelligence, Smart Healthcare, Smart Agriculture, Smart Industry, etc., are specific areas that have experienced a significant transformation toward the digital world in recent years. Key enablers, including not only IoT and 5G (and related technologies), but also big data, artificial intelligence, cloud-computing, etc., have contributed to such progress. However, "Intelligence" is still in its infancy in most of the cited areas, and further technological efforts at different levels, and within different fields, are needed in order to make the "Smart World" concept a reality. Sensors constitute one of such fields, where a significant research activity is going on, particularly in the topic of planar microwave sensors, the subject of this book.

Microwave sensors implemented in planar technology are expected to play an increasingly important role within the IoT and the "Smart World," since these sensors can potentially satisfy many demanding requirements and provide technical solutions to several challenging aspects. Planar microwave sensors exhibit low cost, small size, and low profile, and can be implemented in many different types of substrates (by means of both subtractive and additive processes), including flexible and biodegradable substrates. Thus, implementing conformal sensors [14, 15] or environmentally friendly ("green") sensors [16, 17] operating at microwaves is feasible. It is also possible to fabricate planar microwave sensors in combination with other technologies, such as microfluidics [18, 19], micromachining [20], lab-on-a-chip [21–23], textiles (fabric) [24–26] and, thereby opening the sensing possibilities and application scenarios (e.g. liquid sensing, and wearables). Moreover, microwaves are very sensitive to the properties of the materials to which they interact. This means that, based on permittivity measurements, microwave sensors are very useful for material characterization and analysis. Defect detection, determination of material composition, biosensing, and chemical sensing, among others, are potential applications of planar microwave sensors, as it has been demonstrated in the available literature (and as it will be shown throughout this book and references included therein). Monitoring ambient variables, such as temperature and humidity, or other physical variables (e.g. pressure) is also possible by means of planar microwave sensors. In these cases, functional films (or "smart" films, a usual designation), with ambient-dependent dielectric properties, are typically used [27–30]. Many efforts have also been dedicated to the implementation of planar microwave sensors for measuring spatial variables (of interest, e.g. in motion control applications), where the key advantageous aspect, as compared with other competing technologies (e.g. optics), is their major robustness for operation in hostile and harsh environments [31]. Microwaves exhibit further interesting properties for sensing, such as low-cost generation and detection systems, interaction with the materials at different scales (i.e. through the near field or the far field), and wireless connectivity. This last aspect is critical for the implementation of IoT systems based on low-cost wireless sensor networks (WSN), implemented, for example, by means of radiofrequency identification (RFID) sensor nodes [27–37].

This chapter is an introduction to planar microwave sensors, with the aim of pointing out the main sensor types, their advantages and limitations as compared with other sensing approaches, and their categorization, or classification. Nevertheless, there are other sensors operating at radiofrequency (RF) and microwave frequeny, which will be succinctly reviewed in this chapter. A general overview of different sensing technologies, besides microwaves, as well as the relevant sensor characteristics indicative of their overall performance, are also included in the next section of the present chapter.

1.1 Sensor Performance Indicators, Classification Criteria, and General Overview of Sensing Technologies

Probably the most general, or less restrictive, definition of "sensor" is as follows: *a sensor is a device able to provide information of a certain variable.* Many types of variables (usually referred to as measurand) can be inferred, or measured, by means of sensors, including physical variables (e.g. temperature, velocity, pressure, and moisture), chemical variables (e.g. gas composition or detection), biological parameters (e.g. glucose or electrolyte content in blood), and medical health indicators (e.g. arterial pressure, or cholesterol level), among others. According to the general definition provided above, the scope of the term includes sensors whose complexity is low, or insignificant. For instance, measuring the height of a person, or the distance between opposite walls in a room, is as simple as using a length gauge, and visual inspection. Such "simple sensor" does not utilize any transduction mechanism. However, there are more sophisticated, accurate, and reliable options for measuring lengths and displacements, including optical, ultrasound, and microwave systems, among others, where the measurand (length in the considered case) is determined after at least one transduction process. In such process, energy conversion from one form to another is involved. For example, in an ultrasound distance-measurement sensor (usually referred to as SONAR – sound navigation and ranging – in aquatic navigation systems), where the time lapse between the emitted and received ultrasound pulse (echo) determines the distance to the target, the acoustic energy is converted to electric energy, and the measuring distance is shown, e.g. in a display, after several signal processing steps. The sensors of interest in this book are those that respond to a certain variable (stimulus) with an electrical signal (typically after one or more transduction mechanisms). Such signal can be monitored in a computer, processed, or transmitted by wires or wirelessly. In such sensors, the input variable is thus the measurand (or stimulus), whereas the output variable is an electric, or electronic, signal, in the form of a direct current (DC) or alternate current (AC), voltage, charge, etc., and described by means of the amplitude (or magnitude), frequency, phase, digital code, etc., of such signal.

1.1.1 Performance Indicators

The main relevant parameters indicative of sensor performance, and their definitions, are as follows [38]:

- Linearity: capability of the sensor to provide a linear response. The linearity is determined by the deviation of the transfer function, or dependence of the output signal with the input signal, from a straight line.
- Sensitivity: variation of the output signal in response to a variation in the input signal (or variable). In nonlinear sensors, the sensitivity is not constant, and it is given by the derivative of the output signal with respect to the input signal, or slope of the transfer function.
- Resolution: the smallest variation of the input signal that can be reliably detected. In analog sensors, the resolution is intimately related to the noise level, whereas in digital sensors, it is given by the span corresponding to a bit (for instance, in angular displacement sensors based on optical or electromagnetic rotary encoders, the angular resolution is determined by the number of pulses per revolution, PPR, as $360°/PPR$).
- Input and output dynamic range: range of values of the input variable that the sensor can measure, and generated span in the output variable. The input and output dynamic ranges are related by the transfer function of the sensor. The dynamic range is usually limited by the floor noise and by saturation effects.
- Accuracy: degree of closeness to the true value of the measurand. It is the opposite concept of error, or difference between the measured and actual value of the magnitude or variable of interest.
- Repeatability and precision: sensor's ability to provide identical outputs for the same input. Within the context of sensors, precision and repeatability are synonymous. Precision is related to the standard deviation of measured data, but not to the proximity of the mean value to the true value. Thus, precision and accuracy should not be confused. A sensor might be very precise and exhibit a limited accuracy, or vice versa.
- Selectivity: capability of the sensor to selectively determining the value of a certain parameter of interest, in a multivariable measurement or in the presence of uncontrollable stimuli (e.g. caused by variations in ambient factors). For example, there are sensors where the input variable is the concentration of a certain substance in a mixture (e.g. a liquid solution with various solute substances). If the different substances have similar influence on the output variable, the selectivity is necessarily poor. Sensor selectivity is related to the existence of cross-sensitivities between the output variable and undesired stimuli. Lack of selectivity is the cause of sensor measuring errors or inaccuracies. Many sensors exhibit certain cross-sensitivities to environmental variables, such as temperature and humidity, or even atmospheric pressure. In such cases, it is convenient to perform the measurements in a controllable environment. Alternatively, differential sensors, to be defined later, robust against cross-sensitivities (since environmental parameters are seen as common-mode stimuli), can be used to mitigate the effects of ambient factors.
- Bandwidth: determines how fast the sensor responds to changes in the input signal. Fast-response sensors, or, equivalently, low-decay-time sensors, exhibit wide frequency bandwidths.
- Noise: disturbance in the output (generally) electrical signal of the sensor. The sources of noise can be diverse (thermal noise, shoot noise, flicker noise, etc.).

- Stability: degree to which sensor characteristics remain constant over time. Changes in sensor characteristics, or drifts, can be due to aging, wearing or degradation effects in certain components, and/or changes in the signal-to-noise ratio, etc.
- Robustness: ability of the sensor to reliably working under adverse conditions or extreme stimulus.
- Operating lifetime: time interval where the sensor is able to provide its functionality. In battery-assisted sensors (e.g. in certain WSNs), the lifetime is determined by power consumption. Nevertheless, there are several factors determining sensor lifetime, including the operational environment, etc.

1.1.2 Sensors' Classification Criteria

Sensors can be classified according to various criteria [39]. For example, depending on whether they need an external energy source for operation, or not, sensors can be divided in two categories: active and passive. Most sensors are active, in particular, most microwave sensors are based on the interaction of microwave signals with the material under test (MUT), analyte, or target, and therefore they need a microwave signal generator.[1] An exception is the radiometer, a sensitive microwave receiver able to detect thermal radiation of objects. Passive sensors, also called self-generating sensors, generate their own signals and do not require any external source of energy, i.e. passive sensors "listen to what is happening" (examples include thermocouples, piezoelectric sensors, photodiode sensors, and radiometers).

Another binary categorization of sensors concerns the nature of the output signal: analog or digital. Analog sensors generate an analog signal in response to what they sense, contrary to digital sensors. Typically, digital sensors include an analog-to-digital (A/D) converter. An example of a digital sensor is a chip-based electronic RFID sensor (to be discussed in Chapter 7), where the measured data is sent to the reader in digital form. Nevertheless, it should be mentioned that digital sensors might refer also to sensing devices where the output signal is a set of pulses, encoding the measured data. Canonical examples of such pulse-based sensors are the optical and electromagnetic rotary encoders [40, 41], useful for determining angular positions and velocities (the latter are discussed in detail in Chapter 4).

Sensors can also be divided between contact and contactless sensors [42–45]. In the former, the MUT, target, or analyte must be in physical contact with the sensitive element of the sensor. Contactless sensors, by contrast, are able to retrieve the measured data without physical contact. Many optical, microwave, and ultrasound sensors are contactless. For example, the measuring-distance ultrasound sensor mentioned before is an example of a contactless sensor. Radio detection and ranging (RADAR) sensors belong also to the category of noncontact sensors. In some cases, contact between the sensitive element and the target is not needed, but proximity is necessary. For example, in electromagnetic rotary encoders [41] (see Chapter 4), the rotating element, designated as rotor, is not in contact with the sensitive element, the stator. However, proximity between the rotor and the stator is required in order to detect the metallic inclusions printed, or etched, in the rotor (through near-field coupling). Such inclusions in motion generate the necessary pulses to infer the angular

---

1 It should be clarified that, in Chapter 7, certain types of RFID sensors have been classified as passive in the sense that such sensing systems, including the interrogator (reader) and the tag (the sensitive element), use tags that do not need a battery. In such passive tags, the required energy is obtained from the ambient (through harvesting) or from the interrogation signal. However, the tags, in general, are not able to provide their own signals and need a source of energy for operation. Clearly, the whole RFID sensing system, including not only the sensing tag but also the interrogator, is active, according to the definition given in the text (since the tags provide the measured data in response to an interrogation signal, generated by the reader, an active component).

position and velocity of the rotor (see further details in Chapter 4). Contactless sensors should not be confused with wireless sensors [34, 35, 46–51]. The most relevant characteristic of wireless sensors is that the sensor element (usually referred to as sensor node in WSNs) sends the measured data to a central unit, located at a certain distance, through a wireless link. Thus, the measured data is wirelessly, or contactless, forwarded to the central unit. However, the sensing mechanism might require contact with the stimulus. For example, chipless-RFID permittivity sensors devoted to the analysis of concrete samples are reported in Chapter 7 [52]. The communication between the sensor and the central unit (the reader, in the terminology of RFID technology) proceeds wirelessly, through the far field, but the samples (concrete) must be in contact with the sensitive element of the sensor (a delay line in the reported example). Thus, such device can be classified as a wireless contact sensor. Note that contact sensors not necessarily must be in contact with a certain sample or target to perform their function. For example, a sensor devoted to ambient temperature measurements, e.g. using the temperature-dependent dielectric constant of certain materials such as ferroelectrics or certain polyamides, provides the temperature information corresponding to the position where such sensitive material is located. Thus, the sensor provides the temperature of the "local" environment to which it is "in contact." However, the temperature of specific targets (objects) can be inferred remotely (i.e. contactless) by means of infrared temperature sensors, able to detect the infrared energy emitted by all materials [53, 54].

There are sensors that provide their measured data relative to a reference value, whereas other sensors react to the stimulus on an absolute scale. In the former case, the sensors are designated as relative, or incremental-type, sensors, whereas absolute sensor is the term coined for the second class. Concerning temperature measurements, there are two clear examples of sensors corresponding to each type. Thermistors are absolute sensors that utilize temperature-dependent resistors and are used in many industrial applications [55]. Thermocouples are relative sensors able to provide the temperature difference between two points [56]. Thermocouples exploit the Seebeck effect, or electromotive force generated between two points of an electrically conducting material when there is a temperature difference, or gradient, between them. The generated voltage depends on the specific conductive material. Therefore, by using two metallic wires of dissimilar materials connected at a junction (the temperature measuring junction), and leaving the two other extremes (terminations) of the wires unconnected but bonded to a region with uniform (reference) temperature, it follows that a temperature difference between the junction and the extremes of the wires will generate a voltage difference between such terminations. Sometimes, relative sensors are used to prevent from measuring errors caused by undesired cross-sensitivities to common-mode stimulus, typically caused by variations in ambient factors (temperature or humidity).[2] In certain applications, rather than measuring an absolute value of a variable, the interest is to compare a magnitude with a reference, or a sample (the MUT) with a reference (REF) sample. In this case, the sensor can be designated as comparator. One potential application of comparators is defect detection, where the analyzed (potentially defected) samples are compared with a reference [57]. Several examples of relative sensors, also called differential-mode sensors, and comparators are later reported in this book.

The previous classification schemes of sensors are binary, and, given a sensing device, it is relatively easy to discern to which category it belongs for either case. Sensors can be classified according to other (more complex) criteria. For example, sensors can be grouped as a function of their field of application. Examples include (but are not limited to) fields as diverse as agriculture, automotive,

---

2 Variations in temperature or humidity with position at the typical scales of the sensors (with sizes below few centimeters in many cases) are not expected, and therefore such variables are seen as common-mode stimuli by relative sensors.

space, health/medical, civil engineering, industrial engineering, energy/power, environment/meteorology, manufacturing, marine, military, scientific instrumentation, information/telecommunications, and domestic appliances.

The variable under measurement can also be considered for sensor classification. In this case, due to the innumerable list of potential measurands, several groups and subgroups (and even a further division) can be identified [39]. Thus, three main groups of sensors classified attending to the nature of the measurand can be identified: biological, chemical, and physical sensors. However, sometimes, it is difficult to determine to which group a certain variable, or sensor, belongs (in particular, there is some indeterminacy between biological and chemical sensors, because many biological variables, or substances, are also of chemical nature). Examples of chemical sensors include gas sensors, or electrolyte-concentration sensors based on electrochemical techniques, among others. Nevertheless, these latter sensors can be considered biosensors, as well, provided they are applied to the measurement of electrolyte content, e.g. in blood or urine, i.e. bio-samples. Most sensors belong to the group of physical sensors. In this case, several subcategories, such as acoustic, electric, magnetic, mechanical, optical, and thermal, can be distinguished, and many sensor types can be, in turn, identified within each subgroup. For example, mechanical sensors include displacement, position and velocity sensors, force/strain sensors, pressure sensors, etc. Measurands such as electric current, voltage, permittivity, conductivity, and many others can be inferred by means of electric-type sensors. Finally, examples of thermal sensors are temperature and thermal conductivity sensors, to cite some of them.

Sensors can be classified according to the involved sensing mechanism, exploited effect, or conversion phenomena. In this case, there are also three main groups, i.e. physical sensors (again the dominant one in terms of subcategories), chemical sensors, and biosensors. As a representative example, a thermocouple is a thermoelectric physical sensor where sensing is based on thermal-to-electric energy conversion. Further types of physical sensors categorized according to the involved physical phenomenon are photoelectric, piezoelectric, or magnetoelectric sensors, among many others. Electrochemical processes and chemical transformation are the basis of most chemical sensors, whereas biosensors typically exploit biochemical transformation.

There are additional sensor classification schemes, not discussed in this chapter (see [39] for a complete overview of such categorizations). However, probably the most useful classification of sensors is the one given by the dominant technology involved in sensing. In this regard, optical, magnetic, acoustic (including ultrasound), mechanical, electrical, and microwave (or, more generally, electromagnetic) sensors, among others, can be identified as sensor types that clearly exploit a specific technology. It is not the aim of this chapter to provide a detailed overview of such types of sensors but to briefly discuss some aspects of interest and to point out some representative examples of some of the cited groups (see next Section 1.1.3). Microwave sensors, the main subject of this book, are discussed separately in a dedicated unit (Section 1.2).

1.1.3 Sensing Technologies

In this section, the most relevant sensing technologies, as pointed out in the preceding paragraph, are briefly overviewed. For a more in-depth study, the authors recommend various general books and handbooks on sensors and sensing technologies (e.g. [38, 58–61]). Additional literature relative to the specific considered technologies is also given in the next sections.

1.1.3.1 Optical Sensors

Probably, optical sensors are those dominating the general sensor market. The reason is that optical technology offers a number of advantages, combined with limited cost, small size, and a broad range of applications in sensing [62–69]. In general, optical sensors are highly sensitive and accurate, and are relatively small. The small dimensions are related to the fact that the required optical source and detector (typically a light emitting diode – LED – and a phototransistor, respectively) are semiconductor devices, intrinsically small. Moreover, such components are relatively cheap, and therefore optical sensors are quite competitive in terms of cost. The limitative aspect comes from the fact that optical beams are not able to penetrate certain substances, as RF or microwave signals do, and consequently optical sensors cannot be applied in certain applications, such as subcutaneous or implant sensors, among others. Moreover, optical sensors are less robust than other sensors, e.g. magnetic sensors or microwave sensors, against harsh environments (i.e. subjected to extreme temperatures or radiation), or ambient conditions subjected to pollution or dirtiness, as those encountered in many industrial applications. Nevertheless, despite these limitations, optical sensors are probably the strongest competitors to microwave sensors, at least in certain applications. For example, optical rotary encoders typically exhibit very high resolutions by virtue of the achievable number of pulses per revolution (several thousands in many prototypes) [40, 70, 71]. By contrast, the microwave counterparts [41, 72–75] cannot provide such high number of pulses per revolution, since electromagnetic signals in the microwave range (with guided wavelengths of the order of few centimeters) are not able to detect inclusions in the sub-millimeter range.

Optical sensors are useful in many daily applications, and they are integral part of many commonly used devices and systems, such as computers, heavy machinery, elevators, copy machines, alarm systems, and ambient light systems. Optical sensors of different types are able to provide measurements of variables as diverse as position and velocity, vibrations, temperature, pressure, chemical and biological components, and strain. The optical rotary encoder, referred to before, is a canonical example of an optical sensor, useful to accurately measuring angular displacements and velocities. Encoders are used in elevators, spatial vehicles, servomechanisms, etc. Like many other optical sensors, optical encoders use a light emitter device (usually an LED or a LASER – light amplification by stimulated emission of radiation) and a detector (a photodiode or a phototransistor, typically), and the operating principle is based on the transmission and interruption of light. Specifically, in optical encoders, the apertures present in the rotor element (which must be positioned between the emitter and the detector) are transparent to light. Thus, the light beam can be detected when one of such apertures crosses the optical path between the emitter and the detector, thereby generating a pulse (see Chapter 4 for further details, specifically Fig. 4.29). These optical sensors, where the emitter sends the light beam to the detector (located in a separate position), are designated as through-beam sensors. In other optical sensors, light transmission/reflection is the considered sensing mechanism, i.e. the emitted beam is reflected back to the detector by an object or a target (retroreflective sensors). For example, distances to (or velocities of) a target can be accurately measured by means of LIDAR (light detection and ranging) systems, where a pulsed (typically infrared) LASER is used to determine the distance/velocity of the target from the time delay in the reflected signal [76, 77]. The operating principle of the LIDAR is identical to that of the RADAR or SONAR, the unique difference being that RADAR systems utilize radio waves (including also microwaves and even higher frequencies within the electromagnetic spectrum), and SONARs are based on acoustic waves, rather than optical signals. As compared with RADAR and SONAR sensors, LIDAR sensors are more accurate and fast. Moreover, LIDARs are extremely selective, since LASER beams are very directive. By contrast, as mentioned, light is not able to penetrate materials as radio or acoustic waves do. For that reason, RADARs or SONARs are necessary in certain applications where the space between the target and the emitter contains elements or materials opaque to light propagation (e.g. geo-localization). LIDARs have applications in many fields, including topography, astronomy, geology, physics of the atmosphere, traffic control, and imaging, among others.

Optical sensors are also very interesting for the analysis of biological samples [78, 79]. Indeed, optical sensors probably constitute the most common type of electromagnetic biosensors.[3] The main reason is the highly achievable sensitivity and selectivity for the detection of analytes as diverse as toxins, drugs, antibodies, proteins, tumor cells, viruses, etc. Moreover, as compared with other analytical techniques, optical biosensors enable the direct, real-time, and label-free.[4] (in some cases) detection of several types of biological components. Cost and size are also interesting characteristics of optical biosensors that make them very competitive against other biosensor types (e.g. electrochemical, piezoelectric, and magnetic). A bio-recognition (also called bio-receptor) sensing element and an optical transducer are constitutive parts of most optical biosensors (Fig. 1.1). The bio-recognition elements are enzymes, antibodies, antigen, nucleic acids, receptors, cells, tissues, etc., which interact with the optical field, providing information of the analyte, bound to the bio-receptor. Concerning the transduction mechanism, surface plasmon resonance, evanescent wave fluorescence, or optical waveguide interferometry exploit the evanescent field in order to detect the interaction between the bio-recognition element and the analyte. Nevertheless, there are other optical transducers, such as photonic crystals, and optical resonators, that can be used for optical biosensing. It is not the aim of this section to provide an exhaustive overview of optical biosensors. Nevertheless, the interested reader can find information in various surveys on the topic [80–88].

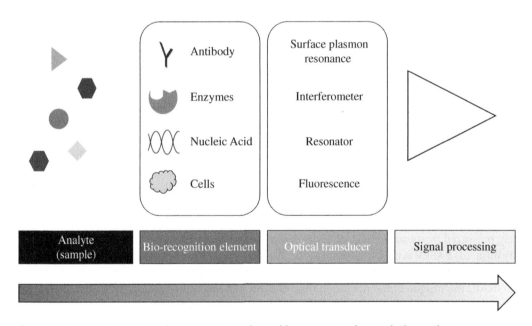

Figure 1.1 Sketch of an optical biosensors based on a bio-receptor and an optical transducer.

---

3 Nevertheless, the biosensor market is dominated by electrochemical sensors.

4 Label-free biosensors use the intrinsic properties of the analytes in order to detect their presence. However, there are biological analytes that are difficult to detect based on their intrinsic properties. In these cases, biosensors require labels in order to enhance the sensitivity (label-assisted sensors).

1.1.3.2 Magnetic Sensors

Similar to optical sensors, magnetic sensors can be strong competitors of microwave sensors in certain applications [89–95]. Magnetic sensors are devices able to detect magnetic fields, generated either by a magnet or by a varying current. These sensors can be useful in various applications, but there are a variety of devices focused on the measurement of linear and angular displacements, as well as proximity, based on magnetic sensors. Magnetic sensors use a wide variety of physical effects, but the most common sensors are coils [96], reed switches [97], Hall-effect sensors [98–100], and magnetoresistive sensors [101–103] (Fig. 1.2). Coils are the simplest form of magnetic sensors, able to detect variations in a magnetic field. In brief, approaching or moving away a magnet to a coil modifies the magnetic flux density in the coil, and, as a result, an electromotive force and current are induced in the coil. A reed switch is a sensor consisting of two metal pieces (reeds), typically made of a magnetic material (e.g. nickel), extending from the left and right side of a glass tube, with a gap in the central overlapping region. When an external magnetic field is applied, the reeds are magnetized, and the overlapping parts come into contact due to their mutual attraction. As a result, the switch turns on. Hall-effect sensors are able to detect static magnetic fields, contrary to coil-based sensors. These sensors are based on the Hall effect, or the voltage generation in a semiconductor or metallic material layer when a magnetic field is applied perpendicularly to a current flowing in that material. This voltage (Hall voltage) is a result of the electromotive force that arises

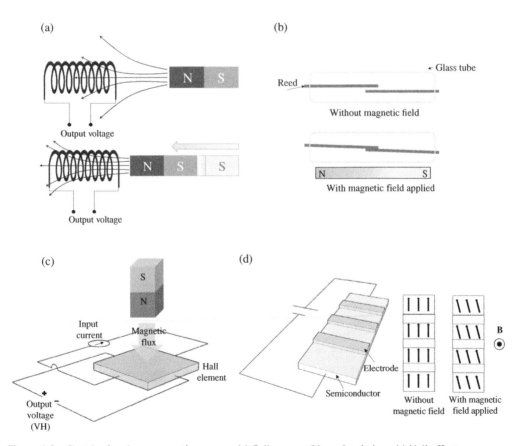

Figure 1.2 Sketch of various magnetic sensors. (a) Coil sensor; (b) reed switches; (c) Hall effect sensor; (d) magnetoresistive sensor.

orthogonally to both the current and the magnetic field. Hall-effect devices can be used as current sensors, angle sensors, proximity sensors, etc. Finally, magnetoresistive sensors base their operating principle on the variation in the resistance of certain magnetic materials, such as iron, nickel, or cobalt, caused by changes in a magnetic field. There are various types of magnetoresistive sensors, but the most sensitive are those based on the tunnel magnetoresistance effect, a quantum mechanical phenomenon (see further details in [104, 105]). Other magnetic sensors exploiting different physical effects can be found in [89–95].

1.1.3.3 Acoustic Sensors

Among acoustic sensors [106–108], acoustic sound sensors, or microphones, are well known for years. Such sensors are based on an acoustic-to-electric transducer that utilizes a thin piece of a material, called diaphragm, which vibrates when hit by sound waves. Such vibration is converted to an electrical signal through a capacitance variation. However, the devices that have generated the greatest interest among acoustic sensors are surface acoustic wave (SAW) sensors [109–115] (Fig. 1.3). Their working principle is the modulation of a surface acoustic wave caused by the physical phenomenon, or variable, under measurement. SAW sensors consist of a piezoelectric substrate with an input and an output interdigitated transducer, and a delay line in between, where the acoustic wave propagates. The input transducer transforms an electrical signal to a surface acoustic wave by virtue of the piezoelectric effect, whereas the output transducer converts the acoustic wave back to an electrical signal. Sensing relies on the fact that the output transducer captures any change caused in the mechanical wave, through their electrical response. Since the characteristics of the acoustic waves depend on the properties of the surface of the substrate, it follows that sensing any phenomenon, or variable, altering such properties is possible. Pressure, strain, torque, and temperature are variables that can be inherently measured by means of SAW sensors, since they alter the length along the surface of the device. Obviously, sensing films coated to the surface can also be used in order to enhance sensor sensitivity to certain variables. In particular, by using such films, the functionality of SAW sensors can be extended to the detection of chemical and biological species, or gases, or to the measurement of humidity, viscosity, etc. (see [114] and references therein). SAW sensors may exhibit performances superior to those of microwave sensors in certain applications, but their cost is usually higher. Chipless-RFID sensors based on SAW technology have also been proposed [116–119]. Sound sensors devoted to distance measurements and ranging, known as SONAR, are also acoustic sensors. Such sensors are of especial interest in subaquatic zones (underwater) due to the very good propagation of sound waves in water (and are common equipment in submarines and ships).

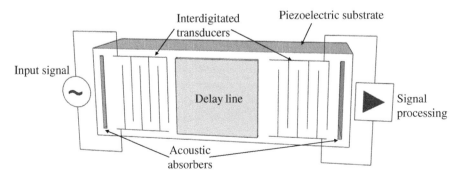

Figure 1.3 Sketch of a SAW sensor.

1.1.3.4 Mechanical Sensors

Mechanical sensors are sometimes defined as measuring devices sensitive to changes in mechanical properties [120, 121]. Nevertheless, there are many sensors of this type that utilize, e.g. optics, microwaves, or magnetism, rather than mechanics, as the main technology for sensing. For example, optical or microwave rotary encoders are sensitive to angular displacement and velocity (i.e. mechanical motion), but such sensors cannot be classified as mechanical sensors. However, there are sensors where a mechanical effect, such as vibration, is the fundamental mechanism. This is the case of the so-called micro-cantilevers [122, 123], i.e. mechanical sensors useful for ultrasensitive mass detection [124, 125] (Fig. 1.4). In such sensors, the bending or the frequency shift of an oscillating cantilever is the specific detection mechanism. Typically, the shift in resonance frequency is used to determine the mass attached to the cantilever. Other measurands that can be detected with micro-cantilevers are acceleration and force. In these cases, cantilever bending is exploited for sensing. Other mechanical sensors are based on the piezoresistive and piezoelectric effects. The former is the variation in the resistance of certain materials when they are subjected to strain or mechanical deformation. This effect can be canonically applied to the implementation of pressure and force sensors [126, 127]. The piezoelectric effect is the generation of an electric field when the material (a crystal, such as quartz) is subjected to stress. Piezoelectric sensors are useful for measuring pressure, force, acceleration, and temperature, among others [128, 129]. In these sensors, metallic electrodes are typically attached to the piezoelectric material in order to transduce the mechanical variable to an electrical one. Note that SAW sensors are also based on piezoelectric materials. Despite the fact that such sensors have been considered to belong to the class of acoustic sensors, sound waves are mechanical waves. Consequently, SAW sensors are, indeed, mechanical sensors, as well.

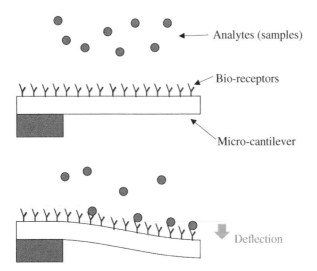

Figure 1.4 Sketch of a micro-cantilever sensor for mass or analyte sensing. When the analyte is settled in the receptor (e.g. a bio-receptor), the suspended mass of the cantilever changes, thereby causing bending. In the static mode of operation, the transduction mechanism is cantilever deflection, which can be retrieved by means of the angle of the reflected beam in response to an optical stimulus. In the dynamic operational mode, the shift in the resonance frequency of the cantilever, caused by the increment of suspended mass, is the mechanism used for sensing.

1.1.3.5 Electric Sensors

It could be accepted that electric sensors are devices devoted to the measurement of electrical variables, such as voltage or current. However, within the context of the present categorization scheme, based on the main involved sensing technology, electric sensors are devices that exploit a transduction mechanism of electrical nature. Capacitive, inductive, and resistive sensors are examples of electrical sensors, able to detect changes in capacitance, inductance, and resistance, respectively [130]. If such variables are inferred by means of electrical methods, the device can be considered a true electric sensor. However, there are other technologies, e.g. microwaves, which can be used for measuring such variables (for instance, a capacitance change can be detected by a shift in the resonance frequency of a microwave oscillator). In this case, the sensor cannot be categorized as an electric sensor. In electric capacitive sensors [131, 132], the capacitance can be measured from the frequency variation of an oscillator circuit (operating at low frequencies), or by means of the attenuation generated in a harmonic signal. It is also possible to infer the capacitance value by applying a harmonic signal to a capacitive divider, where one of the capacitances is the one under measurement, whereas the other one has a well-known value. By measuring the output voltage in the contact point between both capacitances, the unknown capacitance can be found from the ratio of capacitances, which equals the ratio of the output/input signal amplitudes. The potentiometer is a well-known example of a resistive sensor, able to convert a linear or an angular displacement into an output voltage by moving a sliding contact along the surface of a resistive element [133] (Fig. 1.5). Finally, inductive sensors are typically based on coils. Such coils have been discussed in the paragraph dedicated to magnetic sensors, since they actually exploit magnetism for sensing. However, as far as the output signal is typically an electric current, it is reasonable to consider that these sensors are electric sensors, as well.

1.2 Microwave Sensors

This section provides a general overview of microwave sensors. The aim is to briefly discuss the main microwave sensor types, their categorization, and their advantages and drawbacks as compared with other sensors. A section is dedicated to planar microwave sensors, the subject of this book. Such section is an introductory excerpt, since the next chapters will deal exhaustively with such sensors. Nevertheless, the classification of planar microwave sensors, according to various

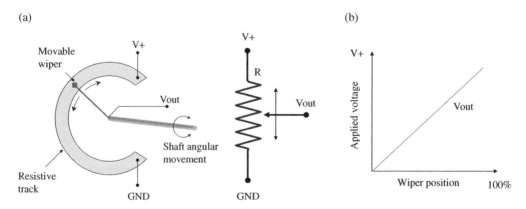

Figure 1.5 Sketch of a potentiometer used as angle sensor (a) and working principle (b).

criteria, is treated in detail in Section 1.3, whereas Section 1.4 provides a qualitative comparison of planar microwave sensors with other (nonplanar) microwave sensors and other sensing technologies.

In the broadest sense of the term, microwave sensors utilize devices and electromagnetic fields in the range between hundreds of MHz up to several THz. That is, such "broad" designation includes millimeter-wave and THz sensors, since their working principles are very similar to those of the microwave counterparts, and, in general, different from those governing optical sensors. The scope of the term may also include sensors operating at the ultrahigh frequency (UHF) band (300 MHz–1 GHz) since the limit between that band and microwaves is very diffuse. Nevertheless, sensor performance, cost, size, and other characteristics may significantly vary with the operating frequency. It should also be mentioned that the planar sensors discussed throughout this book operate at frequencies mainly circumscribed within the UHF band and the lower sub-bands of the microwave spectrum (and, indeed, in the sensor categorization by frequency of operation, to be discussed later, the distinction between THz, millimeter-wave, microwave, and UHF sensors is adopted).

There are two main types of microwave sensors: those based on RADAR technology or, more generally, on remote sensing, and those devoted to the analysis of material properties and measurement of physical, chemical, or biological variables based on the in situ interaction of microwaves with matter (the sensors subject of this book belong to the second class). The next sections succinctly review these sensor types, with special emphasis on the latter category.

1.2.1 Remote Sensing: RADARs and Radiometers

Remote sensors gather information of an object or a phenomenon without physical contact with it, and, typically, the sensing element is located at significant distances from the target[5] [134–137]. Remote sensing can be applied in different fields, including Earth observation, land surveying, atmosphere analysis, agriculture, geology, geography, space, meteorology, oceanography, traffic control, and many others. Remote sensors can be divided in two groups: active and passive. In active remote sensing, the sensor includes a source of electromagnetic radiation (a transmitter) that illuminates the target and a detector (receiver) that collects the signal reflected back from the target. The so-called RADAR sensor belongs to this class [138–142]. In passive remote sensing, the sensor detects the electromagnetic energy emitted by an object or scattered by it.

Microwave radiometers are an example of passive remote sensors, able to detect microwave radiation emitted by matter (radiometers are self-generating sensors, as indicated before) [143, 144]. The architecture of a microwave radiometer consists of an antenna, the front-end, composed of a mixer and a low noise amplifier, and a back-end for signal processing at intermediate frequencies. Down conversion to a lower (intermediate) frequency (heterodyning) is often used in order to ease signal amplification and signal processing. Nevertheless, low noise amplifiers are becoming available at higher frequencies, i.e. up to 100 GHz, making heterodyne techniques obsolete. Radiometers are mainly used for water vapor and temperature observations in weather and climate monitoring, but there are other applications, such as remote sensing of Earth's ocean and land surfaces (i.e. gathering ocean temperature and wind speed, ice characteristics, soil and vegetation properties, etc.) [145–148].

RADAR sensors, as active devices generating their own electromagnetic signal, use the time delay between such emitted signal and the signal reflected back by the target (collected in the receiver) in order to determine its location, velocity, and direction. RADAR sensors are useful for localization purposes and are of special interest in geology, archeology, and civil engineering, among other

---

5 For example, remote sensors located in satellites or aircrafts are applied to detect and identify objects on Earth.

sectors, due to their capability to detect underground buried targets. The so-called ground-penetrating RADAR (GPR) is a clear example of an active remote sensor useful in such mentioned fields (e.g. for the detection of buried pipes and mines, and measurement of ice thickness in frozen areas) [149–151]. Traffic control, anti-collision systems, and automotive driving assistance are additional well-known applications of RADAR (and LIDAR) sensors [142, 152–156]. Nevertheless, RADAR sensors are also of interest for microwave imaging [138, 157–160]. In this case, the working principle is the unique and characteristic interaction between the different targets and the incident electromagnetic radiation (described by the spectral response).

There are different types of RADAR sensors. In pulsed RADARs, either monostatic or bistatic,[6] the distance to the target is determined by the transit time a short pulse emitted by the transmitter antenna needs to reach the receiver after being reflected back by the target. This simple form of RADAR has limitations, especially for short-range distance measurements, where very short pulses are needed (the accuracy of the distance measurement is inversely related to the length of the pulse). A different form of distance measurement using RADAR exploits frequency modulation. In these systems, referred to as frequency-modulation continuous-wave (FM-CW) RADAR, the frequency of the transmitted signal is continuously varied over time [161]. Since such signal needs a finite time to return back to the receiver, it follows that the frequency of such signal is different to the one of the signal being broadcasted by the transmitter. By comparing the frequency difference between both signals, the distance to the target can be obtained. FM-CW RADARs can operate at relatively low frequencies, and their cost is typically low. These systems have been used as altimeters, and have found application in those industrial scenarios where dust and pollution prevent from the use of infrared and optical sensors. For the measurement of the relative velocity of the target with regard to the RADAR sensor, it is possible to use pulsed or FM-CW RADAR systems. It suffices to infer the distance at two different instants of time. However, for quasi-instantaneous speed measurements, Doppler RADARs are preferred [162]. In such RADAR, the transmitter sends an unmodulated continuous wave (CW). The velocity is determined from the frequency shift of the reflected signal, caused by the relative velocity between the target and the sensor (Doppler effect).[7] Doppler RADARs are also simple and low-cost systems that have found applications in diverse scenarios, in particular, vehicle speed control. Such systems are also useful in environments subjected to dirt or pollution. Techniques for improving the performance of RADAR systems, including impulse compression techniques, ultrawide band (UWB) RADAR, etc., have been proposed [140, 163, 164], but are out of the scope of this book. Figure 1.6 depicts the basic architecture corresponding to an FM-CW RADAR and (CW) Doppler RADAR system.

The frequency of electromagnetic waves used in RADAR depends on the specific application, but the frequency range is comprised between the UHF band (or even lower frequencies in certain cases) and the millimeter-wave band. For example, GPRs operate at low frequencies (typically below 2–3 GHz) because such frequencies penetrate more efficiently in Earth. As frequency increases, the attenuation caused by atmosphere (due to absorption) also increases, but high-frequency RADARs are prone to generate tightly focused beams with relatively small-sized high-gain antennas. Thus, medium-range RADARs, such as airborne RADARs, operate between the L and the K_a bands, but many short-range targeting RADARs, e.g. automotive RADARs and collision avoidance RADARs, operate in the millimeter-wave band.

Let us mention, to end this section, that RADAR technology has also been applied to the implementation of presence and motion detectors [165–169]. As compared with their infrared

---

6 Monostatic RADARs use the same antenna for transmission and reception, whereas in bistatic RADARs, signal transmission and reception is performed by two different antennas.
7 Nevertheless, there are also pulsed Doppler RADARs and FM-CW Doppler RADARs.

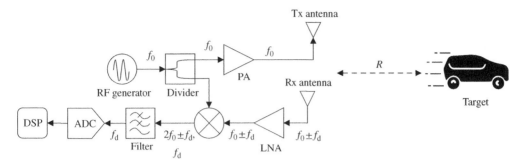

Figure 1.6 Basic architecture of a CW Doppler RADAR devoted to the determination of the velocity of a moving target, which causes a Doppler shift of $\pm f_d$. The frequency of the signal generator is f_0. The Doppler frequency f_d, necessary to infer the target velocity, is obtained by filtering the higher frequency component, $2f_0 \pm f_d$, at the output port of the mixer. The PA and LNA blocks correspond to a power amplifier and low-noise amplifier, respectively. The target speed is determined after an analog-to-digital conversion (ADC) and digital signal processing (DSP) stages. The basic architecture of a FM-CW RADAR, useful to provide the range (distance) of the target, is essentially the same. The unique difference is that a VCO is needed in order to generate the frequency-varying (FM-CW) signal injected to the power divider. In this case, the distance of the target is determined also by the incremental frequency at the input ports of the mixer, caused by the time delay of the transmitted signal.

counterparts (the passive infrared –PIR– detector, a pyroelectric sensor able to detect different levels of infrared radiation), RADAR presence and motion detectors are not affected by ambient temperature and exhibit a wider detection range (though their cost is higher and they are more power consuming than PIRs). Further details on RADAR sensors and radiometers can be found in the literature [134–144].

1.2.2 Sensors for *In Situ* Measurement of Physical Parameters and Material Properties: Non-remote Sensors

RADAR sensors and radiometers are contactless and free space (*ex situ*) sensors, where the sample under test, or target under study, is located at significant distances from the sensing element (sensing is performed remotely). This section deals with another type of microwave sensors, i.e. those that require proximity, not necessarily contact, with the sample or material under analysis. In such in situ sensors, the electromagnetic field interacts with the sample, which modifies certain measurable characteristics of the sensing element (e.g. a resonance frequency, magnitude, phase, or quality factor). Note that wireless and RFID sensors (to be studied in Chapter 7) are proximity sensors, despite the fact that the measured data is wirelessly sent to a central unit or reader (in some cases located at substantial distances from the sample). The reason is that in these devices, the sensitive element (a sensing tag in the nomenclature of RFID sensors) must be in proximity, or even in contact, with the target. By contrast, in remote sensing, the entire sensing function is carried out by a system distantly separated from the target.

Microwave sensors are of special interest for the dielectric characterization of samples or analytes, including solids, liquids, gases, chemical components or composites, bio-samples, etc. By virtue of the dependence of the typical output variables of the sensors with the permittivity of the surrounding medium, microwave sensors are good permittivity sensors, useful for the measurement of any variable related to it, or influenced by it (e.g. material composition and defects in samples). Variables such as temperature, humidity, and pressure do also alter the permittivity of dielectric materials. Therefore, microwave sensors can also be applied to the measurement of such (and many other) variables. Nevertheless, for that purpose, specific sensitive materials,[8] with strong dependence of their permittivity, or conductivity, with the variable under measurement, are typically used in order to boost up the sensitivity.

There are many types of non-remote microwave sensors, and such sensors can be classified according to various criteria. The next section discusses such categorization schemes. Then additional sections are dedicated to briefly discuss the most relevant types and methods of non-remote microwave sensing, i.e. cavity sensors, the Nicolson–Ross–Weir method, coaxial probe sensors, and planar sensors (the subject of this book).

1.2.2.1 Classification of Non-remote Microwave Sensors

A first, and probably the most remarkable, binary categorization of non-remote microwave sensors distinguishes between planar and nonplanar sensors. Planar microwave sensors are devices implemented in planar technology, using additive or subtractive techniques, typically, although not exclusively, based on transmission line elements, e.g. microstrip or coplanar waveguide (CPW), usually combined with other planar elements, such as resonators.[9] Nonplanar sensors, by contrast, are based on waveguides, resonant cavities, or coaxial probes, typically, i.e. nonplanar structures.

Both planar and nonplanar microwave sensors can be either resonant or nonresonant. In general, resonant sensors are more sensitive and accurate than nonresonant sensors [170]. Concerning size and cost, a relevant aspect from a practical viewpoint, resonant sensors are usually smaller and cheaper.[10] However, resonant sensors cannot be used to perform broadband characterization of materials (e.g. dielectric spectroscopy [171]), due to their narrowband nature. In resonant sensors, at least one resonant sensing element is used, and the output variable is, typically, although not exclusively, or not always, the resonance frequency, influenced by the dielectric properties of the material or medium surrounding the resonator. In nonresonant microwave sensors, the amplitude and/or phase of the reflection and/or transmission coefficient are the typical output variables, also influenced by the material or medium surrounding the sensing element (in contact or contactless), for example a transmission line section or a waveguide section.

Another binary classification scheme considers the operation mode of the sensor, either transmission or reflection. Transmission-mode sensors are two-port devices where the output variable is a certain characteristic of the transmission coefficient (magnitude, phase, resonance frequency, quality factor, etc.). Reflective-mode sensors are, in general, one-port structures, where the information of interest is contained in the reflection coefficient. Reflective-mode sensors are especially suited for applications requiring "sensing probes." A coaxial probe is an example of reflective-mode sensor that can be in contact with the sample under test, or introduced in it (e.g. for liquid sensing). Submersible reflective-mode sensors based on transmission lines have also been reported [172]. It should be mentioned that there are microwave material characterization techniques, such as the so-called Nicolson-Ross-Weir method (to be discussed later) [173, 174], that use two-port structures,

---

8 Such materials, sometimes designated as "smart" materials, are discussed in Chapter 7.

9 In most planar sensors, a transmission line (or a transmission-line-based structure) is used for exciting the sensing element, usually a planar resonator. Nevertheless, there are planar sensors where the excitation of the sensing element is carried out by irradiating it with an electromagnetic signal (using an antenna). In such sensors, a transmission line is, in general, not required, and the sensor typically consists of a planar resonant element, an array of resonant elements, a planar antenna, or any other pattern (e.g. a frequency selective surface, or a metasurface) whose characteristics can be altered by the target or by the stimulus. Frequency-domain chipless-RFID backscattering sensors, to be analyzed in Chapter 7, constitute an example of this type of planar sensors. There are also planar microwave sensors based exclusively on transmission line elements.

10 Resonant cavities are an exception.

typically a coaxial line or a waveguide, and retrieve the information (permittivity and permeability of the sample under test) from both the measured reflection and transmission coefficients.

Finally, non-remote microwave sensors can be contact or contactless. It is not always necessary for the sample under test, or target, to be in contact with the sensing element in order to obtain, or guarantee, the functionality of the sensor. Proximity might be enough to alter the electromagnetic field generated by the sensor and thus obtain the measurement of the parameter of interest. Most microwave sensors are guide-wave or transmission-line-based sensors operating under contact or contactless, i.e. by proximity. Nevertheless, there is another class of contactless sensors, i.e. the so-called free-space sensors [175–179], where the sample under test is located in the path between a transmitter and a receiver antenna [Fig. 1.7(a)]. The amplitude and phase of the signal collected in the receiver antenna depend on the dielectric characteristics, as well as dimensions and shape, of the sample under test. Consequently, by measuring the transmission coefficient, information relative to the sample of interest can be inferred. Such free-space sensors can also be implemented as reflective-mode structures, similar to a monostatic or a bistatic RADAR configuration. Free-space sensors are indeed remote sensors, as far as the sample is "distant" from the transmitter and receiver antennas (i.e. the sample does not interact with the near-field). However, these permittivity sensors are not true remote sensors, such as RADARs, LIDARs, or SONARs, with targets typically located "very far" from the sensor.

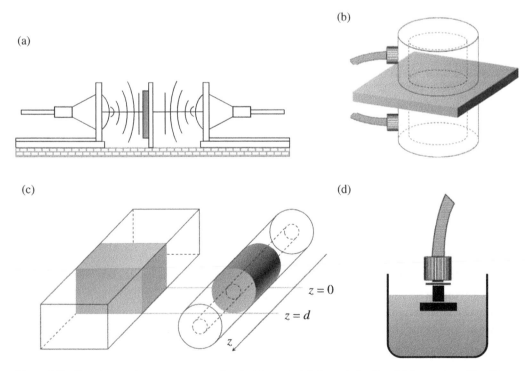

Figure 1.7 Nonplanar microwave sensors devoted to material characterization. (a) Free-space; (b) split cylinder resonator; (c) T/R NRW sensor; (d) coaxial probe.

1.2.2.2 Resonant Cavity Sensors

Resonant cavities are closed metallic structures, generally implemented by means of a waveguide section (rectangular or circular) short-circuited at both ends [180, 181]. Resonant cavities are able to confine electromagnetic energy corresponding to different resonant modes,[11] with frequencies that depend on the geometry of the cavity and dielectric constant of the material filling the cavity. For a rectangular waveguide cavity resonator with width a, height b, and length d, the frequencies of the different modes, TE_{mnl} or TM_{mnl}, are given by [181]

$$f_{mnl} = \frac{c}{2\sqrt{\varepsilon_r \mu_r}} \cdot \sqrt{\left(\frac{m}{a}\right)^2 + \left(\frac{n}{b}\right)^2 + \left(\frac{l}{d}\right)^2} \tag{1.1}$$

where c is the speed of light in vacuum, ε_r and μ_r are the relative permittivity and permeability, respectively, of the material filling the cavity, and m, n, and l are the mode numbers. For a circular waveguide cavity resonator (or cylindrical cavity), the resonance frequencies of the different modes are given by expressions similar, but different, to (1.1) (such expressions, different for the TE_{mnl} and TM_{mnl} modes, can be found in [181]). In view of (1.1), it is obvious that from the resonance frequency of a certain mode, it is possible to obtain the dielectric constant of the material filling the waveguide (note that for a dielectric material, $\mu_r = 1$). Thus, the resonant cavity, either rectangular or cylindrical, is a dielectric constant sensor.

Some power dissipation is inevitable due to the finite conductivity of the metallic walls of the cavity and to the dielectric losses of the material filling the cavity, if it is present. It is well known that air-filled metallic waveguide resonators, or cavities, exhibit very high unloaded quality factors by virtue of the small losses associated with the metallic walls. However, such quality factor is obviously degraded if the cavity is filled, or partially filled, with a (lossy) dielectric material. Let us assume that a rectangular cavity resonator is completely filled with a dielectric material with loss factor $\tan\delta$. The unloaded quality factor is given by

$$Q_0 = \left(\frac{1}{Q_c} + \frac{1}{Q_d}\right)^{-1} \tag{1.2}$$

where Q_c is the unloaded quality factor due to the lossy conductor, but lossless dielectric, and Q_d is the unloaded quality factor associated with the lossy dielectric with perfect conducting metallic walls. For the TE_{10l} mode,[12] Q_c is [181]

$$Q_c = \frac{(kad)^3 b\eta}{2\pi^2 R_s} \frac{1}{(2l^2 a^3 b + 2bd^3 + l^2 a^3 d + ad^3)} \tag{1.3}$$

whereas Q_d is given by

$$Q_d = \frac{1}{\tan\delta} \tag{1.4}$$

regardless of the resonant cavity mode [181]. In (1.3), $k = \omega\sqrt{\varepsilon_r}/c$, where ω is the angular resonance frequency of the mode, $R_s = \sqrt{\omega\mu_0/2\sigma}$ is the surface resistance of the metallic walls, μ_0 and σ being the permeability of vacuum and the conductivity of the metallic walls, respectively, and $\eta = \eta_0/\sqrt{\varepsilon_r}$ is the wave impedance of the dielectric material, where η_0 is the wave impedance of free space. By

---

11 The electromagnetic fields in the cavity are excited through external coupling, by means of a small aperture, a small wire probe (e.g. a coaxial probe), or a loop.

12 For a rectangular cavity resonator with dimensions satisfying $b < a < d$, the dominant TE mode (i.e. the one exhibiting the lowest resonance frequency) is the TE_{101} mode [181].

measuring the unloaded quality factor of the cavity resonator, the loss tangent of the material filling the waveguide can be inferred.[13] The unloaded quality factor Q_0 is dominated by Q_d, since $Q_d << Q_c$, even for low-loss dielectrics. For that reason, cavity resonators are appropriate devices to accurately determining the loss tangent of low-loss materials, where other methods find some difficulty.[14] Moreover, due to their inherent high quality factor, cavity resonators provide a precise measurement of the resonance frequency of the mode of interest. Consequently, the dielectric constant of the MUT can also be determined with a high level of precision.

The main limitation of the resonant cavity method, as it has been presented, concerns the fact that the MUT must occupy the whole volume of the cavity, and this is not practical in most situations. However, resonant cavity sensors can also be applied to the dielectric characterization of material samples occupying a small portion of the cavity volume. In this case, the technique, usually referred to as perturbation method [181, 183, 184], is based on the variation experienced by both the resonance frequency and quality factor of the cavity, when it is loaded (or perturbed) with the sample under test. In this technique, it is assumed that the fields in the cavity loaded with the sample are slightly perturbed as compared with those of the unloaded (unperturbed) cavity, a reasonable approximation in many practical situations (small samples, or samples with small dielectric constants). An approximate expression providing the resonance frequency of the perturbed cavity was derived in [181]. Let us call such (angular) frequency ω_p, and the resonance frequency of the unperturbed cavity ω_0. The fractional change experienced by the resonance frequency by perturbing the cavity is given by

$$\frac{\omega_p - \omega_0}{\omega_0} = - \frac{\int_{V_0} \left(\Delta\varepsilon \left| \vec{E}_0 \right|^2 + \Delta\mu \left| \vec{H}_0 \right|^2 \right) dv}{\int_{V_0} \left(\varepsilon \left| \vec{E}_0 \right|^2 + \mu \left| \vec{H}_0 \right|^2 \right) dv} \tag{1.5}$$

where V_0 is the volume of the cavity, $\vec{E}_0$ and $\vec{H}_0$ are the fields of the unperturbed cavity, ε and μ are the permittivity and permeability, respectively, of the unperturbed cavity ($\varepsilon = \varepsilon_0$ and $\mu = \mu_0$ for air-filled cavities), and $\Delta\varepsilon$ and $\Delta\mu$ are the changes in the permittivity and permeability, respectively, caused by the perturbation. This is the general expression obtained in [181], which considers possible variations in both the permittivity and permeability of the cavity. Nevertheless, for dielectric materials, $\Delta\mu = 0$, and the expression can be further simplified. The integrals in (1.5) extend over the whole volume of the cavity, but it is obvious that the decrease in the resonance frequency of the perturbed cavity is more significant if the permittivity perturbation occurs in regions where the electric field intensity is large.

A variant of the resonant cavity sensor, that exists commercially [185] and is very appropriate for the nondestructive and accurate measurement of the complex permittivity (or dielectric constant and loss tangent) of low-loss and narrow dielectric slabs, is the so-called split cylinder resonator [186–191].[15] This permittivity sensor uses a cylindrical cavity separated into two halves [Fig. 1.7 (b)], with the sample placed in the gap between both parts. Typically, one of the halves is fixed, whereas the other one is movable, in order to accommodate samples of different thickness.

---

13 Techniques to measure the unloaded quality factor can be found in several textbooks, e.g. [181, 182], and are out of the scope of this book.
14 Planar resonant methods can also be applied to the determination of the dielectric constant and loss tangent of dielectric materials, as it will be shown later in this book. However, these planar sensors are implemented on lossy (although typically low-loss) dielectric substrates. Therefore, the determination of the loss tangent with such planar techniques is not as accurate as with the resonant cavity technique, especially for low-loss materials.
15 The split cylinder resonator can be found as integral part of many microwave laboratories for the accurate measurement of the dielectric constant and loss tangent of (uncladded) microwave substrates.

Concerning the sample geometry, the only requirement is that the sample must be flat and it must extend beyond the diameter of the two cavity sections. Thus, it is not necessary to cut a small piece of the sample in order to introduce it in any cavity or to machine it in order to adapt it to the volume of the cavity. In this method, the TE_{011} resonance is excited,[16] and from the measurement of the resonance frequency and quality factor, the complex permittivity of the sample under test is obtained, as detailed in several sources [189–191] corresponding to research activities mainly carried out at the National Institute of Standards and Technology (NIST), Boulder, Colorado. For resonance excitation and measuring purposes, small coupling loops (introduced through small holes in the side of each cylinder half) are used in the commercial version of [185].

Another technique for the nondestructive measurement of the complex permittivity of narrow dielectric slabs is the split (or single) post-dielectric resonator [192–195]. Other dielectric resonator sensing techniques can be found in the literature [196–199]. In these techniques, the sample under test creates a "dielectric resonator," and hence the samples need to be machined in the form of discs or cylinders in order to perform the measurements. Dielectric resonator techniques are very accurate, but determination of the complex permittivity of the samples from the measurement of the resonance frequency and quality factor requires, in general, complex numerical electrodynamic methods [200].

1.2.2.3 The Nicolson–Ross–Weir (NRW) Method

The Nicolson-Ross-Weir (NRW) method is a nonresonant reflection/transmission (R/T) technique for the measurement of the complex permittivity and permeability of the sample under test [178, 179, 201–204]. It utilizes closed-form expressions providing the material properties as a function of the measured S-parameters in the considered two-port test structure, typically a coaxial line or a closed waveguide, containing the sample under test. Such sample should be made of a homogeneous and isotropic material occupying completely the region between the planes $z = 0$ and $z = d$ of the guided wave structure, while the regions $z < 0$ and $z > d$ are assumed to be air-filled [Fig. 1.7(c)]. Let us clarify that the method applies to any guided wave structure operated in the TEM mode (e.g. coaxial lines, or striplines) and to guided wave structures with either a single TE or TM mode present in all regions (e.g. a rectangular waveguide). Nevertheless, the NRW method also contemplates that the specimen under test can be illuminated in free space by either a parallel- or a perpendicular-polarized plane wave that is incident at a certain angle from the normal to the material. In this case, the material should be semi-infinite in the transverse (x and y) directions.

Let us next derive the relevant expressions of the NRW method by considering a TEM guided wave structure (the expressions for TE/TM waveguides or for free-space can be found, e.g., in [203, 204]). The S-parameters, S_{11} and S_{21}, of the guided wave filled with the sample (of finite length d) can be expressed as [178, 203, 204]

$$S_{11} = \frac{\Gamma\left(1 - T^2\right)}{1 - T^2\Gamma^2} \tag{1.6a}$$

$$S_{21} = \frac{T(1 - \Gamma^2)}{1 - T^2\Gamma^2} \tag{1.6b}$$

where Γ is the interfacial reflection coefficient, defined as

$$\Gamma = \frac{Z_1 - Z_0}{Z_1 + Z_0} \tag{1.7}$$

---

16 This mode provides maximum field intensity in the position of the sample under test.

and T is the propagation factor, i.e.

$$T = e^{-\gamma d} = e^{-jkd} \tag{1.8}$$

In (1.7), Z_1 and Z_0 are the transmission line characteristic impedances in the region with and without sample, respectively. In (1.8), γ is the propagation constant, and k is the wavenumber. From (1.6), Γ and T can be isolated, i.e.

$$\Gamma = K \pm \sqrt{K^2 - 1} \tag{1.9}$$

$$T = \frac{S_{11} + S_{21} - \Gamma}{1 - (S_{11} + S_{21})\Gamma} \tag{1.10}$$

where K is a factor that depends on the S-parameters

$$K = \frac{S_{11}^2 - S_{21}^2 + 1}{2S_{11}} \tag{1.11}$$

Thus, by measuring the scattering parameters of the sample-loaded guided wave structure, K, and hence Γ and T, can be obtained. Note that the sign in (1.9) is determined by forcing $|\Gamma| < 1$, as corresponds to a passive sample.

Once Γ and T are known, the propagation factor can be obtained, i.e.[17]

$$k = \frac{1}{d}[j\ln|T| - \phi] \tag{1.12}$$

where T has been expressed in phasor notation as $T = |T|e^{j\phi}$. Finally, the complex relative permittivity and permeability of the sample are calculated according to [203, 204]

$$\varepsilon_r = \frac{k}{k_0}\frac{1 - \Gamma}{1 + \Gamma} \tag{1.13a}$$

$$\mu_r = \frac{k}{k_0}\frac{1 + \Gamma}{1 - \Gamma} \tag{1.13b}$$

where $k_0 = \omega\sqrt{\mu_0\varepsilon_0}$ is the free space wavenumber.

Additional literature related to the NRW technique and related T/R methods for retrieving material properties can be found in [205–216]. In particular, a simplified version of the NRW method that employs only a single-port network analyzer for data collection is reported in [216]. The NRW method has also been applied to retrieve the constitutive parameters of engineered materials and metamaterials [206, 213, 214].

1.2.2.4 Coaxial Probe Sensors

The open-ended coaxial probe [Fig. 1.7(d)] is a reflective-mode nonresonant sensor very useful for the *in vivo*[18] (and *ex vivo*) dielectric characterization of solid, semisolid, and liquid samples over broad frequency bands. It consists of a coaxial line truncated with an open-end termination, which

---

17 Actually, depending on the sample dimension d, an additional phase term $2\pi n$ (n being an integer) should be added in (1.12) [203, 204]. This means that, if the electrical length kd of the sample is unknown, there is an intrinsic indeterminacy (designated as branching) in the method. This indeterminacy is solved by considering thin samples, with d smaller than a half-wavelength in the analyzed frequency range, so that $n = 0$, and thereby (1.12) can be used as it is written (i.e. without such additional phase term). Nevertheless, techniques to solve this branching issue that do not need to consider thin samples have been reported [205, 206].

18 Many biological samples (e.g. tissues) demand in vivo characterization techniques, and coaxial probes are very appropriate for this type of materials.

should be in contact with the sample under test (solid or semisolid samples), or submersed in it (liquid samples). Typically, a flat metallic flange transversally extends the ground plane at the open end, as depicted in Fig. 1.7(d). As compared to cavity sensors or to T/R guided wave sensors exploiting the NRW method, a relevant advantage of coaxial probes is the fact that there is no need to machine the sample under test to fit it to the sensing element, i.e. the coaxial probe method is non-destructive. Nevertheless, for solid or semisolid samples, good contact with the probe is required (uneven sample surfaces in contact with the probe can result in measurement inaccuracies), and for liquid samples, air bubbles should be avoided. The coaxial probe technique is accurate for moderate and high-loss materials, but it cannot compete against resonant cavities for the dielectric characterization of low-loss materials.

Coaxial probes are sensors able to provide the complex dielectric constant ($\varepsilon^* = \varepsilon' - j\varepsilon''$) of the sample material from a simple measurement of the reflection coefficient determined by the impedance seen from the aperture (looking at the sample). Such impedance depends on the dielectric properties of the sample. Thus, the complex dielectric constant can be retrieved from the collected data of the reflection coefficient. Due to its inherent measurement simplicity, the coaxial probe method has been extensively used for the dielectric characterization of different types of materials, with special emphasis on biological samples (it is impossible to enumerate the huge number of references, but a representative list is [217–227]). There are available commercial versions that automatically provide the complex permittivity of the sample from the reflection coefficient data acquired with a vector network analyzer (VNA) [228]. Obviously, such commercial coaxial probe sensor kits include algorithms (based on numerical methods) that convert the measured reflection coefficient to permittivity. The relation between the complex permittivity of the sample and the impedance seen from the open-end termination of the coaxial line is not simple, and there are different methods/models that have been developed to infer the complex permittivity from that impedance (inverse problem methods). The analysis of such methods is out of the scope of this book, but is available in different sources [228–241]. The main drawback of the coaxial probe method concerns the relative complexity of the numerical techniques necessary to retrieve the permittivity of the material sample. Such complexity is inherent to inverse problems, but it is magnified by the rich phenomenology of the physical system, where radiation from the aperture, or the generation of higher order modes, among other effects, is present and adds difficulty to the electromagnetic modeling of the structure and its solution.

Other sensors similar to the coaxial probes are the open-ended rectangular or circular waveguide probes, proposed many years ago [242–246], but still of interest today for different applications (crack and corrosion detection, microwave imaging, etc.), as revealed by the recent activity on these sensors [247–252].

1.2.2.5 Planar Sensors

Planar microwave sensors is the subject of this book. Therefore, as it has been previously mentioned, this section merely intends to provide a general and succinct overview of this type of sensors. An exhaustive analysis and study is left for the subsequent chapters (and sensor categorization, attending to different criteria, is left for a dedicated unit, Section 1.3). Nevertheless, the research activity in planar microwave sensors has experienced such a significant growth in recent years that it is impossible to cover all the reported sensor types, working principles, performance characteristics, and their main applications in a chapter section, or even in a complete book. Thus, the present subsection and Section 1.3 (in brief), and the entire book (exhaustively), focus on the main sensing strategies, working principles, and applications of microwave sensors implemented in planar technology, according to the authors' criteria and perspective. Naturally, many ideas, sensor prototypes and applications, reported by researchers working in this field, will not be included (nor cited) in this book. Indeed, the literature related to planar microwave sensors is innumerable and widespread among very different disciplines, including not only RF/microwave engineering but also electrical and electronics engineering, telecommunications, biology, chemistry, mechanics, medicine, physics, civil engineering, space, automotive engineering, agriculture, robotics, etc. This further hinders the inclusion, and discussion, of the whole research activity on the topic of this book in a single document. Nevertheless, the objective of the authors in writing this book has been to provide a representative, and significant, subset of the research activity and achievements on planar microwave sensors (by the authors and other groups) to date, with special emphasis on discussing those aspects related to sensor performance.

As noted at the beginning of this chapter, one of the main reasons explaining the increasing research activity in planar microwave sensors in recent years is their potential in the framework of the IoT and the "Smart World." Key aspects to the envisioned leadership of planar microwave sensors in next future are their low cost and profile, their compatibility with many different types of substrates (including flexible, organic, and textile/fabric substrates) and fabrication technologies (subtractive and additive), as well as the unique properties of microwaves, e.g. propagation across (and penetration in) many substances, and interaction with matter at different scales. Wireless connectivity, inherent to microwave technology, and the easy integration with the associated electronics (for sensor feeding, post-processing and, eventually, for communication purposes) are additional (and even more important) aspects that explain the increasing interest for planar microwave sensors in Academia and Industry. As an enabling (although not exclusive) technology, planar microwave sensors may contribute to the deployment of the IoT and ease the path toward the "Smart World" in many different fields (e.g. health, agriculture, civil engineering, automotive industry, and space). An additional attractive aspect of planar microwave sensing technology concerns the possibility of implementing eco-friendly sensors using biodegradable/recyclable substrates and organic conductive inks. In a future interconnected and intelligent world, with thousands of sensing elements (present everywhere, in everything, and in everyone), the development of "green sensors" is a demanding need. It is also a big challenge, due to the inevitable performance degradation related to the use of "green" materials.[19]

A planar microwave sensor can be defined as a device where the sensing element/s is/are implemented in planar technology and operates at microwave frequencies. Nevertheless, planar microwave sensors may include additional elements, necessary for sensing, that are not necessarily planar. For example, as it will be later shown, there are liquid sensors implemented in planar form (e.g. by means of resonator-loaded transmission line) that include fluidic channels in order to force the liquid under test to be in contact, or in proximity, to the sensitive element [18, 19]. The associated sensor electronics, needed for signal generation, post-processing, and communication purposes in a real scenario, represents also the addition of nonplanar elements, since it is typically based on commercial integrated circuits (voltage controlled oscillators – VCOs – microcontrollers, detectors, etc.), that must be soldered to the sensor substrate (or implemented in an independent electronic module). Indeed, in any microwave non-remote sensor (planar or nonplanar), three main modules, or blocks, can be distinguished (see Fig. 1.8): (i) the electromagnetic block, constituted by the sensing element and, eventually, by other components necessary to perform the sensing function through microwaves (fluidics channels, tuning elements, etc.); (ii) the electronic

---

19 Recyclable substrate materials, such as paper or compostable materials, exhibit high loss factors, as compared with other materials typically used for the implementation of microwave components (e.g. RT/duroid) or general purpose circuits (e.g. *FR4*). Organic conductive inks exhibit smaller conductivities than silver-based inks, in turn smaller than those of metallic laminates, such as copper, gold, or aluminum. These high-loss factors and low conductivities of green "materials" limit the achievable performance of microwave components based on their use.

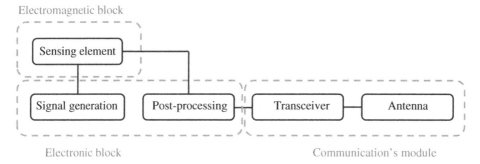

Figure 1.8 General block diagram of a microwave non-remote sensor.

block, in charge of signal generation and post-processing; and (iii) the communication's module (including the transceiver and the antenna). Eventually, depending on the type of sensor and its complexity, mechanical elements/accessories should also be required (for instance, capillaries, Teflon screws, mechanical connectors, and holders, among other elements, are part of microwave fluidic sensors [18, 19]). In this book, the emphasis is on the electromagnetic block and its optimization, despite the fact that in some of the reported prototypes, part of the electronics is also included (e.g. in certain sensors, particularly, microwave encoders, amplitude modulator detectors are essential to infer the envelope function needed to retrieve the information relative to the input variable, see [41, 72] and Chapter 4).

There are many different types of planar microwave sensors, exploiting different working principles, and useful for many diverse scenarios and applications. However, planar microwave sensors essentially detect, or measure, the input variable through the changes that such variable generate in the immediate environment of the sensing element. Such sensing element can be a planar resonator, a planar transmission line, a planar antenna, or a combination of various planar elements. Planar microwave sensors are essentially (at least in most cases) permittivity sensors, where the presence of a certain material in proximity, or in contact, with the sensitive element modifies its (measurable) characteristics (phase, resonance frequency, quality factor, magnitude, etc.). For example, an open transmission line, e.g. a microstrip line or a CPW, covered with a certain material (sample) exhibits transmission and reflection coefficients with phase and magnitude (measurable quantities by means of a VNA) that depend on the complex dielectric constant of the sample. This is an example of a nonresonant planar microwave permittivity sensor [Fig. 1.9(a)]. The complex permittivity can also be retrieved by means of resonant structures. For example, in an open transmission line loaded with a planar resonator (e.g. a split-ring resonator), the resonance frequency and the quality factor depend on the dielectric constant and loss tangent of the material in contact with the resonator [Fig 1.9(b)]. These two illustrative examples, and many other permittivity sensors, will be studied in more detail later in this book, but are pointed out in this short overview of planar microwave sensors for clarification purposes.

Since the permittivity of a material correlates with many other variables, e.g. temperature, pressure, humidity, material composition, and presence of certain analytes or defects, it follows that permittivity sensors are useful in multiple applications. As indicated before, to enhance the sensor sensitivity, it is sometimes necessary to coat the sensitive elements with "smart" films, with highly sensitive permittivity to the variables of interest (see Section 7.3 in Chapter 7, where an exhaustive list of materials of interest for sensing is reported). Motion sensors (i.e. displacement and velocity sensitive devices) can also be implemented by means of planar microwave sensors. In this case, the relative position between the movable part of the sensor and the sensitive (fixed) element

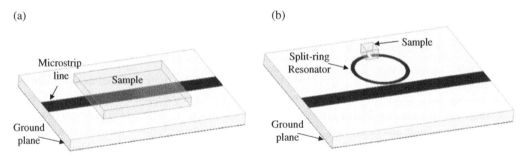

Figure 1.9 Sketch of a planar nonresonant (a) and resonant (b) sensor.

determines the "effective" permittivity in the surrounding environment of the sensitive element, and the position and velocity can be determined from the measurement of a certain characteristic of it.

Most planar microwave sensors are open structures where the sample under test does not need to be machined, contrary to cavity- or waveguide-based sensors. Nevertheless, similar to coaxial probe sensors, in case of solid samples, it is in general necessary that the samples exhibit at least a flat surface, the one that should be in contact, or close, to the sensing element. In contact sensors, in order to minimize the effects of the air gap, or space between the sample and the sensing element, it is necessary to press the sample against the sensor's surface. The thickness and transverse dimensions of the sample can be, in principle, arbitrary. However, by considering semi-infinite samples, i.e. in practice with dimensions exceeding the regions where the electromagnetic fields extend, it is possible to infer analytical expressions linking the output and input variables (as it will be shown in future chapters). For liquid samples, or gases, the air gap is not an issue. However, in the case of liquids, the use of dry films coating the sensing element might be necessary in order to prevent from substrate absorption.[20] Such films are typically very thin, but tend to degrade somehow the sensitivity. The problem of substrate absorption is solved by considering contactless liquid sensors, where, rather than fluidic channels, or liquid holders (containers), the liquid is syringed to a small capillary, or pipe, located close to the sensitive element. However, the sensitivity of these contactless liquid sensors is, in general, worse than those sensitivities achievable with contact-based sensors. The reason is that the electromagnetic field intensity decays with the distance from the sensing element. Thus, in general, for sensitivity optimization, contact sensors are preferred. There is, however, a class of planar microwave sensors where contact is not possible, or is not convenient, i.e. those sensors devoted to the measurement of displacements, velocities, and proximity. In this case, in order to avoid friction, the movable element should be separated from the fixed (sensitive) element. Linear or angular microwave rotary encoders, studied in Chapter 4, are examples of contactless displacement and velocity sensors, where the motion of the movable element is in a plane parallel to the one of the sensitive element. In other motion sensors, e.g. certain proximity sensors, the motion of the movable element is in the orthogonal direction to the plane of the sensing element [253]. Thus, in this case, the device is intrinsically contactless (except for the extreme position where the movable element is in contact with the sensor surface, usually the reference position).

---

20 For aqueous liquid samples, hydrophobic substrates also prevent from substrate absorption.

Another important characteristic of planar microwave sensors is the possibility of device implementation in flexible substrates. This opens the path toward the implementation of conformal and wearable sensors, of interest, e.g. in applications related to health monitoring, and food control [14, 15, 254–259].

For all the reasons explained in the preceding paragraphs, and for their potential as key enablers of the IoT and the "Smart World," planar microwave sensors are the subject of an intensive, and increasing, research activity. Planar sensors can be useful for a wide variety of appliances in many different sectors. Their low cost, small dimensions, compatibility with other technologies, robustness against harsh and hostile environments, potentially wireless connectivity, and the possibility to implement "green" sensing nodes are key advantages as compared with nonplanar sensors. Nevertheless, nonplanar sensors, such as resonant cavities, may exhibit superior performance, which may be needed in certain applications (at the penalty of higher cost and other limitative aspects). For example, the accurate measurement of the complex permittivity of low-loss samples should be preferably carried out by means of resonant cavities, with very high-quality factors, as compared with those of planar resonators [260]. Nevertheless, planar sensors similar to resonant cavities, implemented in substrate integrated waveguide (SIW) technology, have been reported [261, 262] and have demonstrated to exhibit reasonable accuracy for the determination of the dielectric constant and loss tangent of medium loss dielectrics.

The present section should be simply understood as a brief overview of planar microwave sensors at the same depth level of previous sections devoted to other microwave sensor types (nonplanar sensors and remote sensors). In Section 1.3, various classification schemes for planar microwave sensors, obeying different criteria, are reported. One of such classification schemes, the one that considers the working principle for sensor categorization, is adopted to present the different planar sensors throughout the different chapters of this book.

1.3 Classification of Planar Microwave Sensors

Planar microwave sensors can be classified according to various criteria. Some of them coincide with those of Section 1.1.2, relative to the general categorization of sensors and sensing technologies, or with those of Section 1.2.2.1, devoted to the classification of non-remote microwave sensors, whereas other classification schemes are specific of planar microwave sensors. Let us mention, however, that some of the classification schemes of Section 1.1.2 do not apply to planar microwave sensors, or are not representative (i.e. do not provide a useful separation among different categories). For example, according to the definition of active and passive sensors of Section 1.1.2, most (if not all) planar microwave sensors are active, i.e. they require a microwave source for the generation of the electromagnetic fields needed for sensing. Even battery-free RFID sensors, cataloged as passive (in the terminology of RFID), are indeed active, since the sensing tag must be fed by the interrogation signal, generated by the reader (see further details in Chapter 7). Nevertheless, such RFID sensing tags, powered by the interrogation signal, do not include a source of power, and for that reason, they are considered as passive in the usual RFID nomenclature (the one adopted in Chapter 7). Concerning the binary analog/digital categorization of sensors, also discussed in Section 1.1.2, most planar microwave sensors are analog, with the exception of certain types of RFID sensors where the retrieved information is digitally encoded, and microwave encoders, based on pulses. Other binary classification schemes, more interesting for planar microwave sensors, are contact/contactless, wired/wireless, single-ended/differential, resonant/nonresonant,

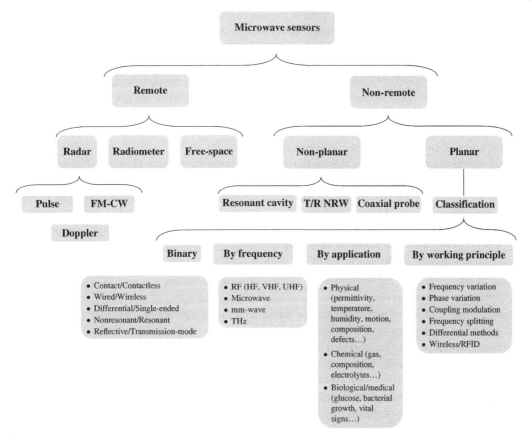

Figure 1.10 General classification of microwave sensors. As indicated in the text (see end of Section 1.2.2.1), there is some controversy in regard to the categorization of free-space permittivity sensors, but it has been opted to consider such sensors as remote in this scheme.

and reflective-mode/transmission-mode.[21] For any of these binary divisions, many different types of sensors can be identified that represent each binary group. Planar microwave sensors can also be cataloged by frequency of operation, by application, or by their working principle (the most convenient scheme, according to the authors' criteria, for the analysis and study of these sensors, and the one considered in this book). Let us next briefly discuss such binary and nonbinary classification schemes (the classification of planar microwave sensors is depicted in Fig. 1.10, where it is shown that such sensors are a subclass of non-remote sensors, in turn a subset of microwave sensors, i.e. all microwave sensor types discussed in this chapter, non-remote and remote, are included in Fig. 1.10).

1.3.1 Contact and Contactless Sensors

Planar microwave sensors are non-remote sensors that need contact or proximity with the sample under test, or analyte. The reason is that sensing in these devices is based on the interaction of the electromagnetic fields generated in a planar structure with the sample, and such fields decay with

---

21 Some of these binary classification schemes do also apply to non-remote (not necessarily planar) microwave sensors, as discussed in Subsection 1.2.2.1, but such schemes are further discussed in this section, where the specificities related to planar technology are considered.

distance. As anticipated in Section 1.2.2.5, for the measurement of material parameters (permittivity), determination of material properties (composition, defect detection, etc.), or for retrieving/detecting other physical, chemical, or biological variables with influence on the properties of a certain ("smart") material (e.g. temperature, humidity, and gas composition), contact is, in general, preferred. The reason is that higher levels of sensitivity result when the sample, or "smart" material, contacts the sensitive part of the sensor. Nevertheless, contact sensors are not exempt of certain difficulties, as also discussed in Section 1.2.2.5, particularly, the air gap effect for solid samples and substrate absorption for liquids (solutions to these issues were pointed out in that section and will be further detailed in subsequent chapters).

Most planar microwave sensors devoted to dielectric characterization and material composition (of solids, liquids, and gases) reported in the literature are contact sensors. Many different examples of such sensors can be found throughout this book. Nevertheless, there are also examples of planar microwave sensors devoted to material characterization, where direct contact between the sample and the sensitive element is inexistent (most of these sensors are used for the characterization and monitoring of liquids). For example, there are liquid sensors where, rather than through fluidic channels, the liquids are forced to be in proximity, but contactless, to the sensitive element by means of small pipes or capillaries [263]. In these sensors, substrate absorption is avoided, and dry films are not required (i.e. higher levels of tolerance against aging effects are expected in such sensors). Other contactless planar microwave liquid sensors are reported in [264–271]. Let us clarify that in certain contactless fluidic sensors, the element used to drive the liquid sample (channel, container, tube, pipe, etc.) is embedded in the planar structure (although, certainly, there is no contact between the sample and the sensitive part of the sensor). Such sensors, and also those where the driving element of the liquid is not embedded in the substrate, but separated from it, can be designated as noninvasive sensors, because the liquid under test is protected by the driving element (it is not "invaded," and hence perturbed, by the sensor). By contrast, contact liquid sensors are invasive, even if a dry film prevents from a direct contact between the liquid under test and the sensitive element. Another subtle terminology distinguishes between intrusive and nonintrusive sensors. In the former, the sensitive part of the sensor is placed, or mounted, within the fluidic element driving the liquid (but not necessarily in contact with the liquid sample). If, alternatively, the sensitive part of the sensor is not located inside the fluidic element, the sensor is classified as nonintrusive. Most planar microwave liquid sensors are nonintrusive (intrusive sensors are uncommon). Figure 1.11 shows schematically the differences between these types of liquid sensors. The microfluidic sensors and other liquid sensors reported later in different chapters of this book are contact sensors, and consequently invasive, but nonintrusive. Being invasive, the sensitivity is optimized, as far as there are no layers (with the exception of the dry film, if it is present) between the sensing element and the fluidic sample.

For the characterization of solids, direct contact with the sample is in general preferred (despite the mentioned air gap effect) because it enhances the sensitivity. However, there is a class of planar microwave sensors, motion sensors, which is intrinsically contactless. In these sensors, two parts can be distinguished: the stator, which includes the sensitive element, and the movable unit, usually (although not exclusively) made of a dielectric material in planar microwave motion sensors (such movable part is usually designated as rotor in angular displacement and velocity sensors). The relative motion between the stator and the movable element can be of different types, i.e. three dimensional, two dimensional, or one dimensional (linear). For two-dimensional and one-dimensional motion sensors, where the plane of movement of the mobile unit is parallel to the substrate where the sensing element is present, contact is potentially possible. However, friction should be avoided in order to prevent from wearing in both the stator and the movable element. Microwave encoders, either linear or angular, constitute a good example of these contactless sensors [41, 72–75],

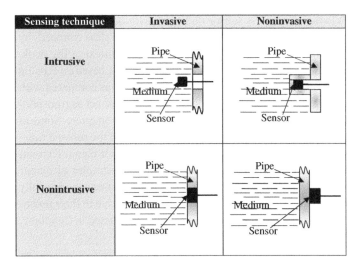

Sensing technique	Invasive	Noninvasive
Intrusive	Pipe / Medium / Sensor	Pipe / Medium / Sensor
Nonintrusive	Pipe / Medium / Sensor	Pipe / Medium / Sensor

Figure 1.11 Distinction between invasive/noninvasive and intrusive/nonintrusive sensors.

but there are many other displacement and velocity sensors (linear, angular, and two dimensional) where the relative displacement between the stator and the movable part proceeds in a plane parallel to the sensor substrate (see, for example [272, 273]). In other sensors, the dominant relative motion between both sensing units is vertical [253] (i.e. orthogonal to the plane of the substrate), and, naturally, these sensors are contactless. A possible sensing mechanism in these contactless motion sensors is the variation in the "effective" permittivity of the immediate environment surrounding the sensitive element caused by the motion of the movable part (e.g. made of a dielectric material) [74]. It is also possible, and generally more sensitive, to infer the relative position and velocity by etching planar resonators, or other metallic inclusions, in the movable part, that perturb the sensing element, usually a planar resonator, through their motion [41, 72]. For example, in many sensors reported later in this book, inter-resonator coupling, modulated by the relative position/orientation between the fixed (sensitive) and the movable resonator/s, modifies the resonance characteristics, and such position/orientation can be inferred, e.g., from the magnitude variation of the sensor's response at a fixed frequency (such sensors will be studied in detail later in this book, see Chapter 4). In other motion sensors, the resonance frequency is the output variable [274].

1.3.2 Wired and Wireless Sensors

As indicated in Section 1.1.2, wireless sensors are characterized by the fact that the measured sensing data, collected by the sensor node (adopting the terminology of WSNs), are sent to a central unit, located at certain distance, through a wireless link. However, the sensing element, or node, should operate by either contact or contactless. Thus, wireless sensors should not be confused with contactless sensors, despite the fact that in wireless sensors, data transfer proceeds contactless. The "wireless" or "wired" attribute in sensors refers to the type of communication (data transfer) between the sensing element and the central unit, not to the transduction procedure or mechanism (contact or contactless) to retrieve the sensing data from the sample under test or stimulus. The typical architecture of classical wireless sensor nodes in WSN, which is explained in detail in Chapter 7, includes the sensor unit, a battery (for sensor feeding), a microcontroller, and a transceiver (for communication purposes). The sensor unit can be of any type, not necessarily planar, and not necessarily a microwave sensor (despite the fact that RF/microwaves are involved in the communication link). Thus, these "classical" wireless sensor nodes are within the interest of this book, as far as the sensing unit is a planar microwave sensor. However, the wireless connectivity does not represent a distinctive aspect that deserves a special analysis in the context of the book. There is, however, another type of wireless sensors, i.e. the so-called RFID sensors, based on sensing tags, either battery-free or with battery, chipped or chipless. The main difference as compared with the classical sensor nodes of WSNs is that RFID sensor tags, at least in most cases, act as transponders, sending the data to the central unit (the reader) in response to an interrogation query. Some of these RFID sensor tags can be categorized as planar microwave sensors, provided the sensing function is performed by means of a planar component, e.g. the tag antenna, or another planar element (typically a resonator). Chapter 7, devoted to RFID sensors, includes many examples of planar microwave wireless sensors, with chipped and chipless tags. Chipless RFID sensor tags are fully planar low-cost devices that have attracted an increasing interest in recent years due to their potential as key enablers for the implementation of "green" IoT systems, as will be discussed in Chapter 7, although there are still many challenging issues to solve.

1.3.3 Single-Ended and Differential-Mode Sensors

In the general classification of sensors of Section 1.1.2, a binary distinction was made between relative (or incremental-type) and absolute sensors. In the framework of microwave sensors, the terms "differential-mode" (equivalent to relative) and "single-ended" (in reference to absolute) are preferred, or, at least, widely accepted in the literature. As indicated in Section 1.1.2, differential-mode sensors are interesting in applications where cross-sensitivities to undesired (usually external) stimuli (e.g. ambient factors, such as temperature or humidity) are significant. Typically, such environmental variables do not vary at the typical scales of the sensors. Therefore, despite the fact that a hypothetical sensing element might be sensitive to the effects of such variables (e.g. permittivity sensors can potentially exhibit a non-negligible cross-sensitivity to humidity and temperature, depending on the considered material), by using two sensing units, rather than one, the undesired variables will affect equally both elements. Thus, the differential output variable, or difference between the output variables of each sensing unit, will be minimally perturbed by the undesired external stimulus.

True differential-mode sensors consist of two independent (in principle identical) sensor elements (sensor pair). The input variable is defined as the difference between the input variables of both sensors, whereas the output variable is the difference between the respective output variables. Namely, in a differential sensor, both the input and the output variables are incremental. Usually, one of the sensors of the pair, and the corresponding input variable (well known), is considered as reference, and therefore the absolute value of the magnitude of interest can be inferred from the measurement of the output (differential) variable, correlated with the input (differential) variable by the transfer function. One interesting utility of differential-mode sensors is as comparator, a device able to detect relative differences, or variations, between a reference stimulus (a sample, or a variable) and the stimulus of interest. For example, defect detectors based on differential-mode sensors, able to discriminate the presence of tiny defects in samples, are later reported in this book (see, e.g., Chapter 3). Differential-mode sensors are also interesting for the measurement of solute concentration in diluted liquid solutions, where the reference is the solvent (Chapter 6 reports diverse examples of such sensors).

Let us clarify that there are certain planar microwave sensors that have been designated as differential-mode devices [275, 276], despite the fact that such sensors do not use two independent and noninteracting sensor units. Some frequency-splitting sensors, to be discussed in Chapter 5, are representative of such sensors. There are also frequency-splitting sensors consisting in two-port planar devices that divide the input signal between two parallel branches (by means of a splitter) that converge at the output port (combiner), where the output signal is collected (such devices have been designated as splitter/combiner frequency-splitting sensor in [18, 277], see also Chapter 5). Each branch is equipped with an identical sensing element (a resonator), sensitive to the stimulus (permittivity in most reported frequency-splitting sensors). Thus, these sensors consist of two sensing elements, but not of two independent and noninteracting sensor units (as a true differential-mode sensor requires).[22] Naturally, a differential input signal can be defined in frequency-splitting sensors, i.e. the difference in the stimulus of both sensing elements. The output variable is the frequency difference between the two notches (or peaks in certain implementations) that are generated by an asymmetry in the stimulus of both sensing elements (i.e. when the input variable in not null). In this sense, it is reasonable to consider that these frequency-splitting sensors are differential-mode devices. However, such sensors do not use a differential sensor pair, with two identical, independent, and noninteracting sensor units. It is also remarkable that, in frequency-splitting sensors, coupling between the sensing elements, if it is present, may degrade sensor performance (specifically the sensitivity) at small perturbations (or for small values of the differential input variable). Chapter 5 discusses this aspect and its solution (see also [277]).

Obviously, the main drawback of differential-mode sensors, as compared with their single-ended counterparts, is their larger dimensions and cost. In this regard, frequency-splitting sensors may represent an interesting trade-off, since similar characteristics to those of differential-mode sensors can be achieved with such sensors, yet keeping sensor dimensions and cost within reasonable limits (similar to those of single-ended sensors).

1.3.4 Resonant and Nonresonant Sensors

In Section 1.2.2.1, it was already pointed out that a relevant partition of non-remote microwave sensors differentiates between resonant and nonresonant sensors. The former tend to be, in general, more accurate and especially useful for the dielectric characterization of low-loss materials at a certain frequency, or discrete set of frequencies. The resonant cavity sensor (see Section 1.2.2.2) is a good example of a highly accurate resonant sensor. By contrast, nonresonant methods, though less accurate, offer the possibility to characterize the materials over broad frequency bands. The waveguide-based T/R sensor (see Section 1.2.2.3) and the coaxial probe (presented in Section 1.2.2.4) are canonical nonresonant sensors, the latter exhaustively used due to their facility of use (especially for liquid samples, where submersing the probe suffices to perform the measurements[23]).

The cited sensors in the previous paragraph are microwave (non-remote) sensors, either resonant or nonresonant, but nonplanar. Concerning resonant and nonresonant planar sensors, many strategies (or sensing approaches), topologies, and technologies have been reported in the literature (it is impossible to cite all of them, but many of these sensors are studied in the next chapters of this book and reported in references therein).

Nonresonant planar microwave sensors essentially consist in transmission line sections, either open (microstrip or CPW, typically) or closed (e.g. strip lines). Sensing is based on the changes in the characteristic impedance and complex propagation constant of the line, caused by the material or sample under test, which in turn affect the magnitude and phase of the reflection and transmission coefficients. The advantage of open lines is that sample preparation and placement is, in

---

22 Indeed, unless the device is adequately designed, the two resonance frequencies in splitter/combiner frequency-splitting sensors are consequence of the interference (i.e. interaction) between both branches. For this main reason, such sensors cannot be considered true differential sensors (despite the fact that each branch, independently, has the functionality of a single-ended frequency-variation sensor).

23 Although calibration and specific algorithms to retrieve the material parameters are needed.

general, simple. Namely, the material to be tested (either solid or liquid) is typically placed on top of the line, as a superstrate.[24] The thickness of the sample can be, in principle, arbitrary, although considering semi-infinite samples tends to facilitate the determination of the variable of interest from the measurement of the output variable/s (e.g. the phase and/or the magnitude of the reflection or transmission coefficient) and the use of simple analytical expressions [278]. Most nonresonant planar microwave sensors are contact devices, but it is possible to implement noncontact sensors, as well. For example, noninvasive and nonintrusive liquid sensors can be implemented by using capillaries or micro pipes to drive the fluid toward the sensitive region of the sensor. For closed transmission lines, micromachining techniques can be applied in order to generate micro-cavities in the substrate, either for direct sample placement (solids) or for the insertion of capillaries (liquids). Clearly, nonresonant sensors based on closed transmission lines are more complex. However, because the sample is typically buried in the substrate, the sensitivity of these sensors can be superior to those sensitivities achievable in the open-line counterparts.

In other nonresonant planar sensors, rather than ordinary lines, artificial transmission lines [279] are used for sensing. As compared to ordinary lines, the dispersion and characteristic impedance of these artificial lines can be further engineered. As a result, highly sensitive sensors based on such artificial lines can be implemented. Chapter 3 reports examples of phase-variation sensors implemented by means of artificial lines, specifically, slow-wave, electro-inductive-wave (EIW), and composite right/left-handed (CRLH) transmission-line-based sensors [280–283], with good sensitivities. However, the design of these sensors is, in general, more complex, since artificial lines consist, typically, in a host (ordinary) line periodically loaded with reactive elements (usually semi-lumped components, such as capacitors, inductors, and/or resonators).[25] It should also be mentioned that artificial lines are intrinsically narrow (or moderate) band structures. Therefore, the intrinsic broadband functionality of nonresonant planar microwave sensors does not apply if artificial lines are considered (indeed, the phase-variation sensors based on the indicated artificial lines, and reported in Chapter 3, operate at a single frequency, the one that optimizes the sensitivity).

With regard to resonant planar microwave sensors, most approaches are based on transmission lines loaded with, or coupled to, electrically small resonators, either metallic or slot (also called defect-ground-structure – DGS) resonators, the sensing elements. Most sensors are based on two-port structures and operate in transmission, but reflective-mode sensors have also been reported (such distinction between transmission-mode and reflective-mode sensors is the subject of the next subsection). The resonant sensing elements can also be distributed, typically quarter- or half-wavelength, resonators. Naturally, these distributed components tend to increase sensor dimensions, or, more specifically, the size of the sensing area, but their modeling (needed to predict the sensor behavior, i.e. sensitivity) is simple. Moreover, sensors based on fully distributed components (resonators) are, in general, robust against fabrication-related tolerances. Typically, the output variables in resonant sensors are the resonance frequency and the quality factor, or magnitude of the resonance peak or notch (Chapter 2 describes various sensors of this type). Nevertheless, there are sensors that exploit the phase (e.g. the phase of the reflection coefficient in reflective-mode

---

24 In case of CPWs, or other planar transmission lines open on both substrate sides, sample placement at either the top or the bottom sides is possible. However, for sensitivity optimization, it is convenient to locate the material sample in contact with the substrate interface where the electromagnetic fields are more intense.

25 EIW transmission lines are an exception. Such artificial lines are propagative structures consisting in a chain of coupled slot resonators, but are also subjected to certain design difficulty. Moreover, such lines exhibit very narrow propagation bands. It should also be mentioned that, despite the fact that EIW transmission lines are based on chains of coupled resonators, EIW-based sensors are not considered to be resonant sensors, since, rather than a resonance phenomenon, such sensors utilize the enhanced dispersion, intrinsic to these lines, for sensing.

resonant sensors), rather than the frequency. As an example, Chapter 3 reports highly sensitive phase-variation reflective-mode resonant sensors, based on open-ended quarter-wavelength and half-wavelength resonators,[26] and based on planar semi-lumped resonators (specifically, open-complementary split-ring resonators – OCSRRs). Frequency-splitting sensors, referred to before and exhaustively discussed in Chapter 5, are also resonant devices, where the canonical output variable is the difference in the split resonance frequencies generated by asymmetric dielectric loads in the sensing resonators. Coupling- modulation sensors, the subject of Chapter 4, are also resonant sensors, but in this case, the canonical output variable is the magnitude of the transmission, or reflection coefficient, modulated by the coupling between the sensing resonator and a transmission line, or another resonator. Typically, such coupling correlates with a spatial variable, such as a linear or angular displacement, and hence, these coupling-modulation sensors are mostly applied to displacement and velocity sensing. Chapters 2–6, and references therein, include many resonant planar microwave sensors.

Let us also mention that there are resonant planar sensors, similar to cavity resonators, implemented in SIW technology [261, 262, 268, 284–287]. By using low-loss substrates, such sensors can operate at relatively high frequencies and exhibit good quality factors, thereby being useful for the dielectric characterization of moderate and low-loss dielectric materials. However, SIW technology is more costly and complex than other planar technologies, either open or closed, for transmission line implementation (note that in SIW cavities, metallic vias and transitions are typically required). Concerning sample placement in SIW cavities, the procedure is similar to the one in closed transmission lines. For liquid measurements, capillaries parallel or transversally oriented to the plane of the substrate, and penetrating it in the region of the cavity where the field intensity is maximum for the considered mode, are used [285, 287]. These SIW-based resonant sensors are not considered further in this book, but the interested readers can find detailed information in various sources [261, 262, 268, 284–287].

Most planar microwave sensors discussed throughout this book are resonant and implemented by means of electrically small resonators. Such sensors are attractive because they combine good performance and size, and can be applied to many different scenarios and applications (e.g. dielectric characterization of solids and liquids, determination of analytes in bio-samples, and measurement of velocities and displacements, to cite some of them).

1.3.5 Reflective-Mode and Transmission-Mode Sensors

The last binary categorization of planar microwave sensors divides them between reflective-mode and transmission-mode sensors. The latter are two-port planar structures, resonant or nonresonant, closed or open. In transmission-mode nonresonant sensors, the output variable/s is/are typically the magnitude or/and the phase of the transmission coefficient, whereas the output variable/s in resonant sensors is/are the frequency, or/and the magnitude or/and the quality factor of the resonance notch, or peak. Nevertheless, sensing data can also be retrieved from the reflection coefficient (as redundant, or complementary, information). In transmission-mode sensors, a VNA is generally used to retrieve the sensing data (i.e. the transmission coefficient) at prototype level (or in proof-of-concept demonstrators), but this represents a high-cost solution in a real scenario. Magnitude and/or phase detectors can be cascaded to the output port of the electromagnetic sensor module, in order to reduce overall costs by avoiding the use of VNAs. For example, in coupling-modulation

---

26 Indeed, such sensors are not categorized as resonant in Chapter 3, since there is not a priori reason that forces to use quarter- or half-wavelength open-ended lines (i.e. resonators). However, as it is demonstrated in Chapter 3, the sensitivity of such sensors is optimized when the length of these lines is either a quarter or a half-wavelength at the operating frequency.

transmission-mode sensors, the canonical output variable is the modulus of the transmission coefficient, but conversion to voltage magnitude is possible by connecting an envelope detector to the output port of the sensing module [41, 72].

Reflective-mode sensors (resonant or nonresonant) utilize the same output variables as transmission-mode sensors, but referred to the reflection coefficient (these sensors are one-port structures). However, retrieving the sensing information by avoiding a VNA is more complex in this case, since the same port (indeed the unique port) is used for sensor excitation and for collecting the sensing data simultaneously. Thus, solutions based on the use of circulators or couplers are necessary for that purpose. For example, Section 3.2.2.4 reports a phase-variation differential sensor, operating in reflection mode, where the canonical output variable, i.e. the phase of the reflection coefficient of the sensing one-port structure (an open-ended quarter-wavelength resonator), is converted to magnitude by properly using a rat-race hybrid coupler [288]. The resulting structure is a two-port device, where the magnitude of the transmission coefficient correlates with the phase of the sensing arm. Thus, similar to transmission-mode sensors, it is possible to add further low-cost components to reflective-mode sensors (e.g. an envelope detector) in order to transform the magnitude of the transmission coefficient into a voltage, easily measurable by means of low-cost devices. Reflective-mode planar sensors (resonant or nonresonant) can be canonically used as sensing probes for liquids (similar to coaxial probes), by submersing the sensing element (the probe) in the liquid under test. Various examples of reflective-mode planar sensors are reported in this book, mainly in Chapter 3, devoted to sensing devices based on phase variation. Nevertheless, reflective-mode sensors exploiting coupling modulation between a sensing resonator and a transmission line have also been proposed (see Section 4.3.1 and Fig. 4.8 in Chapter 4).

1.3.6 Sensor Classification by Frequency of Operation

The frequency of operation of planar microwave sensors can be very diverse, ranging from UHF frequencies (or even lower frequencies) up to the millimeter-wave frequency band and beyond. Obviously, microwave sensors (either planar or nonplanar) can be classified according to their frequency of operation. For that purpose, different frequency band designations can be considered. Nevertheless, probably the most commonly accepted adoption, at least in the framework of microwave sensors, is the designation established by the Institute of Electrical and Electronics Engineers (IEEE), depicted in Fig. 1.12, and originally conceived to distinguish the different frequencies used in RADAR. Needless to say that certain broadband (nonresonant) sensors might cover several bands simultaneously, but the operational frequency (or spectrum) of most planar microwave sensors can be circumscribed within a specific frequency band.

When designing a planar microwave sensor, the choice of the operational frequency, or frequency band, depends on many factors. For example, certain applications inexorably dictate the frequency coverage (band) of the sensor, provided the interest is to study the behavior, or characteristics, of a certain material, or variable, within that band. A clear example is dielectric spectroscopy aimed to determine the complex dielectric constant of a certain substance within a predefined band of frequencies. In other cases, frequency regulations determine the operational frequency of the sensor. This is the case, for example, of the chip-based UHF-RFID sensor tags, a class of RFID sensors to be studied in Chapter 7, where the use of commercial chips and readers forces the design of the tag to be operative at specified frequencies within the UHF band (such specific frequencies depend on the different world regions).

However, in most sensing applications, other aspects related to technical issues, size, cost, etc., determine the operational frequency of the sensor. Size and cost are progressively becoming more important, taking into account the continuously increasing need of including sensors and sensor networks everywhere, in everything, and in everyone. Since planar microwave circuits scale with

IEEE Standard Radar Band Nomenclature		
Designation	**Frequency**	**Wavelength**
HF	3 – 30 MHz	100 m –10 m
VHF	30 – 300 MHz	10 m – 1 m
UHF	300 – 1000 MHz	100 cm – 30 cm
L Band	1 – 2 GHz	30 cm – 15 cm
S Band	2 – 4 GHz	15 cm – 7.5 cm
C Band	4 – 8 GHz	7.5 cm – 3.75 cm
X Band	8 – 12 GHz	3.75 cm – 2.5 cm
Ku Band	12 – 18 GHz	2.5 cm – 1.67 m
K Band	18 – 27 GHz	1.67 cm – 1.11 cm
Ka Band	27 – 40 GHz	1.11 cm – 0.75 cm
V Band	40 – 75 GHz	7.5 mm – 4.0 mm
W Band	75 – 110 GHz	4.0 mm – 2.7 mm
mm Band	110 – 300 GHz	2.7 mm – 1 mm

Figure 1.12 Frequency band designations according to the IEEE standard for RADAR (the indicated wavelength corresponds to wave propagation in free space).

the inverse of frequency, high-frequency operation would be, in principle, desirable. However, it is well known that at high frequencies within the microwave band (dozens of GHz and beyond), the performance of planar circuits degrades significantly. Moreover, the cost of the associated sensor electronics (for signal generation, post-processing, and, eventually, communication purposes) increases with frequency. Consequently, a trade-off is necessary, and planar microwave sensors generally operate at frequencies comprised between the UHF and the X bands. As indicated previously, SIW technology exhibits the advantages of planar and waveguide technologies, i.e. reasonable performance at relatively high frequencies (and hence small size) with moderately low cost, but SIW-based sensors are closed structures that require vias and transitions, and their fabrication, as well as sample placement, is complex (at least as compared with other planar sensors).

Despite the trade-off pointed out in the previous paragraph, sensor operation outside the indicated spectral frequency range is necessary in certain applications. For example, within the healthcare domain, subcutaneous sensors and implants are sometimes inductively coupled to the external unit and typically operate at frequencies below (or in the lower region of) the UHF frequency band [289]. There are also examples of near-field wireless sensors for healthcare applications (implants) based on loop inductors, compatible with the near-field-communications (NFC) regulated band at 13.56 MHz (HF band) [290, 291]. Nevertheless, many other implantable sensors utilize planar antennas operating at the Industrial, Scientific, and Medical (ISM) band [292, 293]. At the other extreme of the spectrum (millimeter-wave and THz bands), an innumerable variety of sensors, especially devoted to chemical sensing and biosensing, have also been proposed (see, e.g., [294–297]). At THz frequencies, extremely high sensitivities for the detection of small volumes of substances, i.e. in the nanoscale, can be achieved. Moreover, there are many chemical and biological analytes with spectral signatures at these frequencies. However, these sensors require specific and expensive micro/nanofabrication techniques that generally increase their production cost.

1.3.7 Sensor Classification by Application

Planar microwave sensors can also be designated according to their applications. The list of applications is so diverse that it is impossible to enumerate all of them. Physical variables as diverse as temperature, humidity, force, strain, pressure, velocity, displacements, and many others can be measured by means of planar microwave sensors, mainly exploiting the permittivity changes generated in the immediate surroundings of the sensing elements, as noted before. Nevertheless, there are many other applications in chemistry, biology, medicine, pharmaceutics, etc., based on the dielectric and electromagnetic properties of the materials and analytes under test (dielectric behavior, spectral signatures, absorption, etc.). Determination of material and liquid composition, or the concentration of components in solutions, including bio-samples, as well as the detection of specific specimens (e.g. bacterial growth) are only a few of the multiple possibilities of chemical sensors and biosensors operating at microwaves. The fields of applications are very transversal, including healthcare and medicine, agriculture, food industry and smart packaging, civil engineering and structural health monitoring, motion control, automotive industry, space, and smart cities.

1.3.8 Sensor Classification by Working Principle[27]

Probably the most useful, and convenient, classification of planar microwave sensors is the one that obeys their working principle. This is the one adopted in this book, as seen in the different chapters. Nevertheless, it is difficult in some cases to establish clear frontiers between the different approaches. Specifically, the categorization proposed in this book for the presentation and analysis of the different planar microwave sensors is as follows (by order of appearance in the book chapters, and with some representative references indicated): (i) frequency-variation sensors [19, 298–301], (ii) phase-variation sensors [281, 302–305], (iii) coupling-modulation sensors [41, 72, 272, 306], (iv) frequency-splitting sensors [18, 277, 307, 308], and (v) differential-mode sensors [309–311]. Then, an additional chapter is devoted to RFID sensors [312–316], the relevant feature being their wireless connectivity, rather than the specific working principle (very different among the various considered RFID sensor prototypes). Let us next briefly summarize the cited working principles in order to provide an early bird's-eye overview to the readers (further details are given in the subsequent chapters of the book).

1.3.8.1 Frequency-Variation Sensors

These planar sensors, discussed in Chapter 2, are essentially resonant sensors, where the output variable/s is/are the resonance frequency or/and the resonance notch (or peak) magnitude. The resonance frequency is mainly correlated with the dielectric constant of the material in contact, or in the immediate environment, with the sensing element, a planar resonator. The resonance magnitude depends principally on the loss factor of the sample (loss tangent or imaginary part of the complex dielectric constant). Alternatively to the resonance magnitude, the quality factor can be used as output variable, as it also correlates with losses. Probably frequency-variation sensors dominate the planar microwave sensor arena, and the main reasons are their small size, low cost, and high achievable accuracy and sensitivity. However, frequency-variation sensors require a frequency-varying source for measuring purposes. Depending on the input dynamic range, the required bandwidth coverage might be relatively narrow, but such sensors are not single-frequency sensors by nature. For input variables experiencing wide spans, and depending on the sensitivity, the output dynamic frequency range might be so broad, that wideband VCOs might be needed for sensing, with a penalty in overall sensor costs (at prototype level, sensor characterization or validation is usually performed by means of VNAs). Most planar frequency-variation microwave sensors are transmission-mode structures, but there are also many examples of reflective-mode sensors in the literature (see Chapter 2, and references therein, for further details).

---

27 This subsection presents the classification used in this book to introduce the different considered planar microwave sensors, grouped by working principle. There are many references of each subclass, which are cited in the subsequent chapters of the book. Therefore, this subsection only quotes a few (but representative) number of works of each sensor type, in order to avoid an excessive number of references in this introductory chapter.

1.3.8.2 Phase-Variation Sensors

In phase-variation sensors, the main output variable is the phase of either the transmission or the reflection coefficient of the sensing device, which depends on the material surrounding the sensitive element, either a planar resonator or a transmission line section. Similar to frequency-variation sensors, the magnitude of the reflection or transmission coefficient can also be useful to gain further insight on material properties (loss tangent). Potentially, very high sensitivities can be obtained with phase-variation sensors implemented by means of transmission line sections, but at the expense of large sensing areas (note that a strong phase variation requires long lines). Meandering is a possibility, as discussed in Chapter 3, in order to avoid oversized sensors [302]. Nevertheless, a technique is proposed in Chapter 3 that boosts up the sensitivity without the need to extend the area of the sensing region [305]. These highly sensitive devices are reflective-mode structures consisting in step impedance quarter-wavelength line sections terminated with a resonant element (the sensitive part), either semi-lumped or distributed (e.g. an open-ended quarter-wavelength or half-wavelength line section). As compared with frequency-variation sensors, phase-variation sensors operate at a single frequency. There is no need to perform a frequency scan, as it is needed to identify the resonance frequency position in frequency-variation sensors. Nevertheless, phase-variation sensors can perform their functionality at different (selectable) frequencies (potentially, with a loss of performance if the frequency is different from the nominal one). Sensor sensitivity in phase-variation sensors can also be enhanced by means of artificial lines exhibiting high dispersion, as noted before [280–283]. Small variations in the permittivity of the sample under test, placed on top of such artificial lines, generate significant phase changes. Nevertheless, in some of these sensors, reported in Chapter 3, the output variable is not a phase, but a magnitude (inferred by means of phase-to-magnitude conversion techniques). Moreover, artificial lines exhibit pass bands and stop bands, and in some cases (e.g. EIW transmission lines [283]), the fundamental (lower frequency) pass band is very narrow. Thus, in sensors based on such artificial lines, the magnitude or/and frequency of the main pass band vary with the stimulus and can be considered to be the output variable/s.

1.3.8.3 Coupling-Modulation Sensors

In most coupling-modulation sensors, the variation in the coupling level between, at least, a planar resonator and a transmission line (host line), caused by the measurand, is the physical sensing mechanism. Such varying electromagnetic coupling modifies the magnitude of the frequency response of the sensing structure, usually, but not necessarily, a transmission-mode device. Thus, the canonical output variable in coupling-modulation sensors is the magnitude of the transmission (or reflection) coefficient (though magnitude to voltage conversion is possible, as indicated before). The strongest variation, or excursion, experienced by the magnitude of the transmission (or reflection) coefficient occurs at resonance or nearby frequencies. Thus, coupling-modulation sensors are typically designed to operate at a single frequency, tuned to the vicinity of resonance. Relative displacements (linear or angular) between the sensing resonator and the host line modulate line-to-resonator's coupling. Thus, most coupling-modulation sensors are devoted to displacement and velocity sensing. It should be mentioned that the resonance frequency in these structures might experience variations due to coupling modulation, but such changes are usually small and hence not useful for sensing. The main aim in coupling-modulation sensors is to generate the strongest possible excursion in the magnitude of the transmission (or reflection) coefficient at the design frequency, caused by the input variable (usually a relative movement). Although not unique, a possible coupling modulation strategy exploits the electromagnetic symmetry properties of transmission lines loaded with a single symmetric resonator [279, 317–319]. A section is dedicated to this interesting topic in Chapter 4. In brief, in a transmission line symmetrically loaded with a planar symmetric resonator, the line is not necessarily able to excite the resonator (even at the fundamental resonance or certain harmonics). This occurs if the symmetry plane of the line and one of the resonator are of different electromagnetic sort, one a magnetic wall and the other one an electric wall. Under these circumstances, line-to-resonator coupling is inexistent, and the line is transparent. However, any relative motion (linear, angular, or both) between the line and the resonator that truncates the symmetry activates their coupling, modulated by the level of asymmetry. The consequence is a variation in the response of the line, usually manifested by a notch with a depth dependent on the magnitude of the perturbation (i.e. dependent on the variation of the resonator position with regard to the reference, or symmetric, position). Obviously, in such coupling-modulation motion sensors, the host line and the symmetric resonator must be etched in independent substrates in relative motion. Such symmetry perturbation can also be achieved by loading asymmetrically the (fixed and symmetrically placed) resonator with a dielectric material, but this strategy has been rarely used, since for the dielectric characterization of materials, there are approaches that are more accurate.

Chapter 4 reports various examples of coupling-modulation sensors based on symmetry perturbation. In other coupling-modulation sensors, the level of proximity between the line (the static part) and the resonator (the movable part) is the coupling mechanism.[28] This is the case of the so-called electromagnetic, or microwave, encoders, either angular or linear, exhaustively discussed in Chapter 4, and potential candidates to replace the well-known optical encoders in certain applications or environments subjected to extreme ambient conditions or pollution. Such encoders essentially consist in a chain (linear or circular) of metallic inclusions (usually resonators) in relative motion over the sensitive part of the sensor (a transmission line or a resonator-loaded transmission line) in proximity but contactless with it. By chain (or encoder) motion, the response (magnitude) of the line at the operating frequency is modulated, and the result is an amplitude modulated (AM) signal at the output port of the line, with peaks or dips in the envelope function correlated with the presence of inclusions in the encoder chain. Thus, the information relative to the velocity and displacement can be easily inferred, through simple post-processing algorithms.

An example of a coupling-modulation sensor devoted to permittivity sensing is also reported in Chapter 4 [320]. As mentioned, coupling-modulation sensors are single-frequency sensors, requiring simple electronics for signal generation. However, amplitude (or magnitude) measurement are prone to the harmful effects of noise or electromagnetic interferences (by contrast, frequency or phase measurements exhibit major tolerance against such effects). Despite that fact, let us mention that electromagnetic encoders have demonstrated very good performance and robustness, even by implementing the encoders in conveyor or elevator belts made of rubber, by means of additive (printing) processes.

1.3.8.4 Frequency-Splitting Sensors

Frequency-splitting sensors, the topic of Chapter 5, are symmetric planar structures loaded with pairs of identical resonators. If such resonators are identically loaded, or perturbed, a single resonance notch, or peak, in the frequency response arises. However, such resonance splits into two notches, or peaks, if the resonators are asymmetrically loaded, or perturbed. Thus, the difference in the resonance frequencies or/and resonance magnitudes can be used as output variable/s for sensing. Such sensors are mostly utilized as quasi-differential permittivity sensors, or for the

---

28 It is possible to implement motion (or proximity) sensors similar to coupling-modulation sensors by replacing the resonant elements with other metallic or dielectric inclusions, as far as such inclusions are able to alter the frequency response of the line (sometimes loaded with a resonant element) and generate a significant excursion in the transmission or reflection coefficient at a certain frequency.

measurement of related variables, but frequency-splitting sensors can also be applied to the measurement of displacements (see, for example, Section 5.6.2). Although frequency-splitting sensors are not true differential sensors, as discussed before, they exhibit some of the advantageous aspects of them, particularly, good tolerance against cross-sensitivities of common-mode stimulus (ambient factors, typically). Similar to frequency-variation sensors, frequency-splitting sensors need excitation signals covering, at least, the frequency range corresponding to the output (differential) dynamic range, but are quite robust against the effects of noise and electromagnetic interference. As compared with differential-mode sensors, the dimensions of frequency-splitting sensors tend to be smaller, since they use a unique element for resonator's excitation (a transmission-line-based structure) plus two sensing resonators, rather than two independent sensor units.

1.3.8.5 Differential-Mode Sensors

Differential-mode sensors, the subject of Chapter 6, were defined in Section 1.3.3. Actually, differential sensing is not exactly a working principle, but rather a strategy to cancel (or minimize) the effects of undesired common-mode stimulus (that may perturb or even obscure the accuracy of measurements), and/or to retrieve the value of a variable in comparison to a reference (incremental measurements or comparison functionality). Thus, differential sensors can exploit any of the working principles indicated in the previous sections, provided the independent sensor units of the differential pair are (individually) based on them.[29] Exceptionally, there are examples of differential sensors in Chapter 3, where the exploited principle is phase variation.[30] However, most differential-mode sensors discussed in this book are jointly presented in a dedicated Chapter 6. Such differential sensors utilize two independent sensor units (a requisite of a true differential sensor). However, in some cases, the sensor includes additional elements, or components, devoted to collect the sensing data in a single port, and/or to use a unique port for sensor feeding.

In the devices reported in Chapter 6, the sensor units (pair) are essentially transmission line sections loaded with planar resonators. In some cases, they operate in transmission; in other cases, such units are reflective-mode structures (loaded or terminated with a resonant element). The essential phenomenon involved in sensing is the relative variation of the resonance frequency and magnitude (or quality factor) in the sensor units. Obviously, the differential output variable/s of the sensor can be the differential resonance frequency or/and the differential resonance magnitude (or quality factor). However, in most sensors, the considered output variable is the cross-mode transmission coefficient (in four-port transmission-mode differential sensors) [309, 310] or the cross-mode reflection coefficient (in two-port reflective-mode differential sensors) [321]. Such cross-mode coefficients (see details in Chapter 6) are differential variables by nature. The magnitude and the frequency position of the maximum in the modulus of the cross-mode coefficients are the usual output variables, which correlate with the considered measurands, typically the real and imaginary parts of the complex dielectric constant of the sample under test (a liquid in most of the sensors of Chapter 6). In certain cases, however, a single output variable suffices for sensing. This is the case of solute concentration measurements in binary mixtures of liquids, where the usual output variable is the magnitude of the cross-mode transmission, or reflection, coefficient [309, 310].

---

29 The exception is frequency splitting, which is indeed a quasi-differential sensing approach by itself, as discussed in the text.

30 The reason for including some differential sensors in Chapter 3 is that such sensors are based on phase variation, the subject of that chapter, and are indeed very similar to other single-ended phase-variation sensors reported in Chapter 3. Thus, for thematic coherence, the authors have (exceptionally) preferred to report and discuss such differential-mode, and phase-variation, sensors in Chapter 3, rather than in Chapter 6.

Note that in differential transmission-mode four-port sensing structures, without further sensor components, sensor validation by measuring the cross-mode transmission coefficient requires four-port measurements. Sensor's excitation proceeds at the input differential port, and data is collected at the output port pair. Alternatively, two-port measurements of the transmission coefficient of each sensor unit, and subsequent subtraction is possible, but this is not a real-time measurement. By contrast, in differential reflective-mode two-port sensors, two-port measurements are needed in order to directly infer the cross-mode reflection coefficient (sensor feeding and data collection are realized at the same differential port). Obviously, a two-step single-port measurement (at either port), followed by a subtraction, is an alternative. It is clear, according to these words, that measurements using differential sensors are not exempt of certain complexity, at least as compared with the simplicity that represents sensor feeding and data retrieving in a single-ended device. Thus, to simplify the measurement in microwave differential sensors, they are sometimes equipped with additional planar components, such as dividers, couplers, and isolators [322]. With such additional elements, the differential sensors (either transmission-mode or reflective) are transformed to two-port structures, where the natural output variable, usually the transmission coefficient (an easily measurable quantity), is related to the cross-mode transmission, or reflection, coefficient of the differential sensor pair. As previously indicated, the transmission coefficient can be further processed and converted into other variables, such as a voltage, which can even be more easily measured. It is noteworthy that the addition of the cited components increases the whole sensor dimensions, but not the size of the sensing area. Chapter 6 reports an example of a differential reflective-mode sensor, where it is shown that a rat-race coupler suffices to transform the device to a two-port structure, providing the cross-mode reflection coefficient of the sensor pair (see Fig. 6.29). Other examples of differential sensors (in reflection or transmission) that utilize rat-race hybrid couplers for similar purposes can be found in Chapter 3. Chapter 6 also includes a differential sensor where a power splitter and a slotline-to-microstrip transition are conveniently combined with the pair of sensing elements in order to implement a two-port structure (see Fig. 6.31).

1.3.8.6 RFID Sensors

RFID sensors, as a subclass of wireless sensors, were already (succinctly) discussed in the binary categorization of planar microwave sensors of Section 1.3.2, devoted to wired and wireless sensors. Thus, like differential sensors, RFID sensors do not exploit a specific working principle (though in many cases, it is frequency variation). Nevertheless, in coherence with the previous sections, with a correspondence with the subsequent chapters, the authors have opted to consider RFID sensors, the subject of Chapter 7, in the present section list. As mentioned, the main relevant characteristic of RFID sensors is the wireless connectivity between the sensing element (the tag) and the central unit (the reader, or interrogator). Such feature, combined with the low tag cost and size, the low energy consumption of the tags (or null, in battery-free tags), and the potential to implement full planar (chipless), and "green," sensor tags, makes RFID sensors very attractive for IoT applications. As discussed in Chapter 7, there are many types of RFID sensors (see Fig. 7.9), and not all of them (the tags) can be considered to be planar, according to the previously indicated description of planar microwave sensors. Nevertheless, almost all RFID sensor types are, at least, briefly discussed in Chapter 7, though the main emphasis is on the so-called chipless-RFID sensors. Chipless-RFID sensor tags are full planar devices that probably constitute the main subject of research in the field of wireless sensors nowadays, and where more efforts are required in order to address the many challenges that the massive deployment of IoT needs.

1.4 Comparison of Planar Microwave Sensors with Other Sensing Technologies

Comparing different technologies, approaches, or devices is not straightforward, as there are many factors that can influence the assessment (application, working principle, etc.). In general, useful quantitative comparisons require that the subjects (sensors in our case) subjected to evaluation are judged using similar standards, and this is not always possible. To be more specific, comparing the performance of two different sensors (which necessarily implies a quantitative analysis) is meaningless, unless a common indicator of such performance can be identified. For example, a figure of merit in optical and microwave rotary encoders, which determines the angular resolution, is the number of pulses per revolution. Thus, optical and microwave encoders can be meaningfully compared, at least, with the light of such shared figure of merit (indeed, a performance parameter), despite the fact that both types of encoders use very different technologies and working principles. Obviously, comparing other quantitative characteristics, such as size and cost, is always possible, since these are aspects transversal to any technology, approach, or device. Rotary encoders constitute a good example where dissimilar sensors in terms of the exploited technology (but devoted to the same application) can be quantitatively compared. But, conversely, there are examples of sensors based on the same technology and focused on the same application, where performance comparison (e.g. considering the sensitivity as the main indicator) might not be representative to determine the preferred option. For example, there are many different types of microwave sensors (planar and nonplanar) useful for dielectric characterization of materials. However, the output variables in the specific sensor implementations under analysis might be diverse (e.g. a phase, a magnitude, and a frequency), and this hinders the comparison, as far as the sensitivities are given by different variables. Nevertheless, even in such cases, it is a common practice in the literature to include comparative tables with the sensor performance. The main reason is that such tables provide a fast quantitative overview of the potential of the considered realizations, but not necessarily give an insightful information (representing a true and faithful quantitative comparison) between them, at least not in all cases.

A qualitative comparison is more feasible, even when very dissimilar sensors, or sensing approaches, or technologies, are considered. Nevertheless, such qualitative comparison is useful to gain insight on the relative advantages and limitations, as well as potential, of the different technologies, or approaches, rather than specific implementations. The main aim of this section is to provide a non-exhaustive qualitative comparison of planar microwave sensors, the subject of this book, with other sensor types, including nonplanar microwave sensors, and other sensing technologies. To make the comparison insightful, a specific set of sensor applications, where microwave sensors are relevant, is considered, in particular, dielectric characterization of materials (and related variables), displacements and velocities (i.e. measurements related to motion), and chemical/bio sensing.

In regard to dielectric characterization of materials, the most extended measuring techniques/devices are resonant cavities, T/R waveguide methods (i.e. mainly the NRW approach), coaxial probes, the free-space method, and planar microwave sensors. All these techniques are non-remote, with the exception of the free-space method, which can be considered to be remote to some extent. The reason is that in free-space dielectric characterization, the sample can be located at distances from the sensor where interaction with the microwave radiation (generated typically by means of a horn antenna) is through the far field. Such dielectric characterization (microwave) techniques have been briefly discussed in Sections 1.2 and 1.3, where some of their advantages and limitations have been pointed out. Nevertheless, let us summarize them in tabular form (see Table 1.1, inspired by the comparison presented in [323], but extended).

Table 1.1 Qualitative comparison of microwave techniques for dielectric characterization of materials.

Method	Materials	Frequency range	Advantages	Drawbacks
Resonant cavity	Mainly solids	Single frequency	• Easy sample preparation • Suitable for low-loss materials • Very accurate method • High-temperature capability • No repetitive calibration • Very useful for the characterization of low-loss microwave substrates	• Measurements only at single frequencies • Suitable for small-sized samples • Bulky and expensive • Complex design and fabrication
T/R (NRW)	Solids and liquids	Broadband	• Useful for both permittivity and permeability measurements • High-frequency operation • Useful for engineered materials	• Require sample machining • Branching issue • Bulky and expensive (if waveguides are used) • Complex design and fabrication
Coaxial probe	Solids and liquids	Broadband	• Easy to use • Nondestructive • Potential for *in vivo* measurements • Simple sample preparation (only a flat surface for solids) • Submersible in liquids • High accuracy for high-loss materials • Very useful for bio-samples • Moderate cost	• Need calibration • Air gap issue (for solids) • Complex algorithms for retrieving the permittivity
Free space	Solids	Broadband	• Wide frequency ranges • Nondestructive • Easy sample preparation • Noncontacting method • High-temperature capability	• Require large flat and solid materials • Diffraction problem • Expensive
Planar resonant	Solids and liquids	Single-frequency or narrow band	• Compatible with subtractive and additive process and other technologies (e.g. microfluidics) • Small size and low profile • Potential for conformal, wearable sensors and implants • Potential for "green" sensors • Potential for integrating the associated electronics in the same substrate • Easy sample preparation and low-sized samples • Low cost • Good accuracy • High versatility (measurement of many physical variables by using smart materials)	• Air gap issue (for solids) • Absorption (in contact liquid sensors) • Single frequency operation or narrow band operation • Subjected to detuning effects

Table 1.1 (Continued)

Method	Materials	Frequency range	Advantages	Drawbacks
			• Potential for wireless and RFID sensors • Very useful for the determination of material composition and characterization of liquid mixtures	
Planar nonresonant	Solids and liquids	Broadband	• Compatible with subtractive and additive process and other technologies (e.g. microfluidics) • Moderate size and low profile • Potential for conformal, wearable sensors and implants • Potential for "green" sensors • Potential for integrating the associated electronics in the same substrate • Easy sample preparation • Low cost • Very simple sensor design[a] • High versatility (measurement of many physical variables by using smart materials) • Potential of wireless and RFID sensors • Useful for dielectric spectroscopy	• Air gap issue (for solids) • Absorption (in contact liquid sensors) • Limited accuracy[b] • Broadband, but up to dozens of GHz[c]

[a] Except for those sensors based on artificial transmission lines.
[b] In general, it is considered that for the determination of the complex permittivity of materials (including the real and the imaginary parts), resonant sensors are more accurate. Nevertheless, sensor accuracy, especially in regard to the imaginary part of the complex permittivity, or loss tangent, is mainly determined by the losses in the sensing element. Thus, for the accurate determination of the loss factor in low-loss materials, resonant cavities are the best solution, but, at the expense of high cost.
[c] Nevertheless, it does not mean that all nonresonant planar microwave sensors exploit the potentially broadband operability. Indeed, many nonresonant planar sensors, e.g. the phase-variation sensors studied in Chapter 3, operate at a single frequency, optimized to boost up the sensitivity.
Source: Based on [323].

In view of Table 1.1, it is clear that planar microwave sensors (either resonant or nonresonant) exhibit many advantageous aspects, as compared with the other (nonplanar) techniques. Let us mention, additionally, that planar microwave sensors exhibit a high versatility (intrinsic to planar implementation) relative to their operational mode, reflection or transmission, single-ended or differential, contact or contactless, wired or wireless. These attributes are indicative of the importance of planar microwave sensors as key components for the measurement of material properties and related variables, and justify the significant research activity on the topic in recent years.

Another field where microwave sensors, especially planar sensors, have found diverse applications is in short-range motion measurements. Proximity sensors, angular displacement and velocity sensors, vibration sensors, etc., can be implemented by means of planar microwave sensors made of a static part, containing the sensing element, and a movable part, in relative motion. Examples of such sensors are electromagnetic encoders, presented in Chapter 4, but there are other planar microwave sensors, most of them reported in that chapter, useful for the accurate determination of variables related to motion. The natural competitors of these short-range motion sensors are optical sensors (e.g. optical encoders), magnetic sensors (e.g. tachometers), and resistive sensors (e.g. potentiometers). Nevertheless, in order to provide a wide overview of the sensing technologies devoted to the measurement of velocities and displacements (and related variables), let us extend the (qualitative) comparative analysis to long-range (remote) motion sensors, i.e. RADARs, LIDARs, and SONARs. Such analysis is presented in Table 1.2 and reveals that microwave technologies (mainly microwave encoders for short-range measurements, and RADAR sensors for long-ranging) constitute a good option that combines high performance, with moderate or low cost, and robustness against harsh conditions, intrinsic to microwaves.

Table 1.2 Qualitative comparison of sensor technologies for displacement, velocity, and ranging measurements.

Range	Technology	Advantages	Drawbacks
Short range (non-remote)	Planar microwave (electromagnetic encoders)	• Small size and low cost • Operation in harsh environments and subjected to pollution • Medium resolution • Potentially unlimited dynamic range (for both displacement and velocity)[a] • Contactless • Operation at a single frequency • Potential to implement the encoder (movable part) in very different dielectric materials (including plastics and rubber)	• Difficult to implement absolute encoders • Issues related to misalignments and vibrations • Potential wearing related to (undesired) friction • Require relatively complex post-processing electronics for retrieving the information relative to motion • No commercially available to date
	Planar microwave (other approaches[b])	• Small size and low cost • Operation in harsh environments and subjected to pollution • Contactless • Simple post-processing electronics • Easy design and fabrication	• Limited resolution and sensitivity • Limited dynamic range • Sometimes a frequency scan is needed for sensing (specially in frequency-variation or frequency-splitting displacement sensors)
	Optical encoders	• Small size • Extremely high resolution • Both incremental type and absolute encoders are possible • Potentially unlimited dynamic range (for both displacement and velocity)[a] • Contactless • Commercially available	• Relatively high cost • Low robustness in hostile environments (with pollution, grease, dirtiness) or in ambient subjected to radiation and extreme temperatures
	Magnetic (Hall effect sensors)	• Very robust • Operation in harsh environments and subjected to pollution • Versatile (linear or angular position and velocity, proximity, etc.). • Operate under static fields • Contactless • Commercially available	• Expensive • Require magnets • Limited accuracy and resolution

Table 1.2 (Continued)

Range	Technology	Advantages	Drawbacks
	Electric resistive (potentiometer)	• Small size and low cost • Simple transduction mechanism	• Very limited dynamic range • Require contact
Long range (remote)	RADAR	• Robust against adverse meteorological conditions • Good directivity (especially in millimeter wave RADAR) • RF and microwaves penetrate media that are not transparent for light (use for buried objects) • Versatility (useful for both medium-range and very long-range applications, depending on the operational frequency) • Very extended technology for military applications, navigation, automotive industry, space, civil engineering, geology and archeology (e.g. ground penetrating RADAR), and study of the atmosphere and Earth	• Susceptible to signal interference • Limited accuracy • Difficult to resolve targets obscured by conducting materials
	LIDAR	• Very accurate and fast • Very selective • Very good directivity	• Need a region transparent to light between the source and the object • Of no use for buried objects or under adverse meteorological conditions
	SONAR	• Sound waves propagate very well in water (specially devoted to underwater communication purposes and ranging)	• Limited accuracy • Limited selectivity and directivity

[a] In linear encoders, the displacement that can be measured depends only on the length of the encoder. In rotary encoders, the number of revolutions can be as high as desired, but methods are needed to infer the cumulative number of revolutions. Nevertheless, encoders (either electromagnetic or optical) are considered to be short-range measuring devices since proximity between the movable part and the static part is needed. By unlimited velocity, the authors mean within reasonable limits considering the typical applications of linear or angular encoders (servomechanisms, conveyor belts, elevators, inertial wheels, etc.).

[b] By other approaches, the authors refer to short-range planar microwave displacement and velocity sensors implemented by means of a transmission line, or a resonator-loaded transmission line element (static part), whose characteristics (resonance frequency, magnitude or phase of the transmission or reflection coefficient, etc.) are determined by the relative displacement and/or orientation between the static part and the movable part, in most cases a dielectric slab with an etched, or printed, inclusion (or a set of inclusions), usually metallic resonators. Examples of such short-range displacement and velocity sensors are reported in Chapter 4, since most of them are based on coupling modulation.

Concerning chemical and biological sensors, planar microwave technology can be useful depending on the specific application. Nevertheless, the leading technologies for sensing chemical and biological species (analytes) are those based on optical and electrochemical methods [324]. Indeed, electrochemical biosensors dominate the commercial market, but their electromagnetic counterparts (including RF, microwave, millimeter-wave, THz, and optical biosensors) are very attractive due to several beneficial aspects, such as low cost, minimal invasiveness, real-time operability, and (potentially) label-free functionality. Mechanical sensors, e.g. based on piezoelectric materials, such as quartz, or micro-cantilevers, are also interesting for the monitoring of mass changes related, e.g., to certain bio-processes such as analyte binding to bio-recognition molecules.

The main advantages of optical and electrochemical sensors over microwave sensors for the detection of specific analytes, as most applications in chemical sensing and biosensing require, are their high selectivity and sensitivity. For example, in many biosensors, bio-recognition elements, also known as bio-receptors, specific for the target analyte, are immobilized onto the transducer (e.g. optical or electrochemical) forming a functionalized surface (detection region), and the sample containing the analyte is directly contacted with that surface, or allowed to flow across it. The transducer then generates a signal (e.g. optical or electrical) that depends on the changes in that surface, caused by the affinity of interaction between the analyte and its bio-receptor, and thus the concentration of the specific analyte in the sample can be determined. Sensor sensitivity is sometimes optimized by means of labels, i.e. components sensitive to the considered transduction mechanism that can bind to the analyte, but this increases complexity, measuring time, and cost (thus, label-free biosensing is in general preferred).

To detect biological changes at the micro/nanometer scale, as many applications require, optical sensors are, in general, more sensitive than other electromagnetic sensors (THz, millimeter-wave, and microwave sensors). Optical biosensors essentially sense optical properties of biomatter, whereas the THz, millimeter-wave, and microwave counterparts employ permittivity sensing, mostly using planar interdigital capacitors, resonators, antennas, and transmission line structures. The size of these planar (sensitive) components determines the discrimination capability of the sensor. Therefore, it follows that the detection of micro/nanoscale analytes requires high-frequency sensors, with dimensions of the sensitive structures close to those of the analytes.

For the detection of chemical species, e.g. gases, functionalized materials that strengthen certain properties (e.g. electromagnetic or electrical) when such substances are in contact with them can be used. Selectivity in chemical sensing and biosensing can (potentially) be achieved also by means of specific reagents, that is, substances or compounds added to the sample (e.g. a liquid solution containing the analyte/s) in order to cause a chemical reaction. Nevertheless, such reagents tend to be expensive, and the presence of multiple analytes in a certain sample may degrade measurement accuracy/selectivity (due to cross-sensitivities to multiple analytes). This is a limitation of microwave biosensing and chemical sensing. For example, the measurement of the concentration of electrolytes in blood or urine, or other components such as glucose, is typically carried out by means of electrochemical sensors, commercially available but expensive, though very selective and sensitive. However, the determination of the complete composition of such (blood or urine) samples by means of (cheaper) microwave methods is by far more complicated, since such techniques are based on permittivity measurements, as mentioned, and the real and the imaginary parts of the dielectric constant of the sample are influenced by the various components present in the sample in a complex and *a priori* unknown way.

Through microwaves, it is possible, however, to detect the collective changes of permittivity, even in small samples, provided the frequency of operation of the sensor is high enough (as indicated before). In this regard, millimeter-wave and THz sensors can be useful in certain applications involving small-sized analytes or samples [325–329]. Nevertheless, let us mention that planar microwave sensors operating at moderate (microwave) frequencies have been demonstrated to be useful for biosensing and chemical sensing (a subset of recent works is given in [269, 298, 330–335]). For example, in [298], two-dimensional arrays of split-ring resonators have been applied to the analysis of organic tissues. Such sensor can be useful for the detection of malignant tissues, similar to the work presented in [330], where a CPW-based sensor is devoted to the detection of several cancer cells. In [332], a SIW-based cavity biosensor useful for the analysis of small-sized liquid samples is proposed. In [333, 334], it is demonstrated the potential of split-ring resonator fluidic sensors for the determination of electrolyte concentration in liquid (aqueous) solutions, and it is shown the potential of the proposed devices to determine the total concentration of electrolytes in animal (horse) urine (but not the individual concentration of electrolytes, as noted in the preceding paragraph). Finally, papers [269, 335] are devoted to testing the concentration and growth of pathogenic bacteria (particularly, *Escherichia coli*) in different medium solutions by using also split-ring resonator fluidic sensors. These works point out the high potential of planar microwave sensors based on electrically small resonators, such as split-ring resonators and other planar resonators, as it will be shown throughout this book. Indeed, there are various review papers on the topic of planar microwave resonant sensors (including not only biosensors) [336–341].

References

1 L. Yan, Y. Zhang, L. T. Yang, and H. Ning, *The Internet of Things: From RFID to the Next-Generation Pervasive Networked Systems: Wireless Networks and Mobile Communications*, Auerbach Publications, Taylor & Francis Group, Boca Raton, FL, USA, 2008.

2 W. E. Zhang, Q. Z. Sheng, A. Mahmood, D. H. Tran, M. Zaib, S. A. Hamad, A. Aljubairy, A. A. F. Alhazmi, S. Sagar, and C. Ma, "The 10 research topics in the internet of things," *2020 IEEE 6th International Conference on Collaboration and Internet Computing (CIC)*, Atlanta, GA, USA, 2020, pp. 34–43.

3 S. Vashi, J. Ram, J. Modi, S. Verma, and C. Prakash, "Internet of Things (IoT): A vision, architectural elements, and security issues," *2017 International Conference on I-SMAC (IoT in Social, Mobile, Analytics and Cloud) (I-SMAC)*, Palladam, India, 2017, pp. 492–496.

4 V. Tsiatsis, S. Karnouskos, J. Holler, D. Boyle, and C. Mulligan, *Internet of Things: Technologies and Applications for a New Age of Intelligence*, 2nd ed., Academic Press, 2018.

5 J. Davies and C. Fortuna, Ed, *The Internet of Things: From Data to Insight*, John Wiley, Hoboken, NJ, USA, 2020.

6 Q. F. Hassan, *Internet of Things A to Z: Technologies and Applications*, John Wiley, Hoboken, NJ, USA, 2018.

7 L. T. Yang, B. Di Martino, and Q. Zhang, "Internet of everything," *Mob. Inf. Syst.*, vol. 2017, Article ID 8035421, 3 pages, 2017.

8 Y. Liao, F. Deschamps, E. de Freitas Rocha Loures, and L. F. Pierin-Ramos, "Past, present and future of Industry 4.0 – a systematic literature review and research agenda proposal," *Int. J. Prod. Res.*, vol. 55, no. 12, pp. 3609–3629. Mar. 2017.

9 S. Vaidyaa, P. Ambadb, and S. Bhosle, "Industry 4.0 – a glimpse", *Procedia Manuf.*, vol. 20, pp. 233–238, 2018.

10 E. G. Popkova, Y. V. Ragulina, and A. V. Bogoviz, *Industry 4.0: Industrial Revolution of the 21st Century*, Springer, 2019.

11 D. Galar-Pascual, P. Daponte, and U. Kumar, *Handbook of Industry 4.0 and SMART Systems*, CRC Press, Taylor and Francis Group, Boca Raton, FL, USA, 2019.

12 M. Wollschlaeger, T. Sauter, and J. Jasperneite, "The future of industrial communication: automation networks in the era of the internet of things and industry 4.0," *IEEE Ind. Electron. Mag.*, vol., 11, pp. 17–27, 2017.

13 L. Sharma, Ed., *Towards Smart World: Homes to Cities Using Internet of Things*, Chapman and Hall/ CRC Press, Taylor and Francis Group, Boca Raton, FL, USA, 2021.

14 P. Wei, B. Morey, T. Dyson, N. Mcmahon, Y. Hsu, S. Gazman, L. Klinker, B. Ives, K. Dowling, and C. Rafferty, "A conformal sensor for wireless sweat level monitoring," *2013 IEEE Sensors*, Baltimore, MD, USA, 3–6 Nov. 2013.

15 L. Su, X. Huang, W. Guo, and H. Wu, "A flexible microwave sensor based on complementary spiral resonator for material dielectric characterization," *IEEE Sensors J.*, vol. 20, no. 4, pp. 1893–1903, Feb. 2020.

16 L. Roselli, Ed., *Green RFID Systems*, Cambridge University Press, 2014.

17 S.F. Kamarudin, M. Mustapha, and J.K. Kim, "Green strategies to printed sensors for healthcare applications," *Polym. Rev.*, vol. 61, pp. 116–156, 2021.

18 P. Vélez, L. Su, K. Grenier, J. Mata-Contreras, D. Dubuc, and F. Martín, "Microwave microfluidic sensor based on a microstrip splitter/combiner configuration and split ring resonators (SRRs) for dielectric characterization of liquids," *IEEE Sensors J.*, vol. 17, no. 20, pp. 6589–6598, Oct. 2017.

19 A. Ebrahimi, J. Scott, and K. Ghorbani, "Ultrahigh-sensitivity microwave sensor for microfluidic complex permittivity measurement," *IEEE Trans. Microw. Theory Techn.*, vol. 67, no. 10, pp. 4269–4277, Oct. 2019.

20 D. B. Wang, X. P. Liao, and T. Liu, "Optimization of indirectly-heated type microwave power sensors based on GaAs micromachining," *IEEE Sensors J.*, vol. 12, no. 5, pp. 1349–1355, May 2012.

21 S. Guha, F. I. Jamal, K. Schmalz, C. Wenger, and C. Meliani, "CMOS lab on a chip device for dielectric characterization of cell suspensions based on a 6 GHz oscillator," *2013 European Microwave Conference*, Nuremberg, Germany, Oct. 2013, pp. 471–474.

22 K. Grenier. D. Dubuc, P. Poleni, M. Kumemura, H. Toshiyoshi, T. Fujii, and H. Fujita, "Resonant based microwave biosensor for biological cells discrimination," *2010 IEEE Radio and Wireless Symposium (RWS)*, New Orleans, LA, USA, Jan. 2010, pp. 523–526.

23 Q. Tang, M. Liang, Y. Lu, P. K. Wong, G. J. Wilmink, D. D. Zhang, and H. Xin, "Microfluidic devices for terahertz spectroscopy of live cells toward lab-on-a-chip applications," *Sensors*, vol. 16, no. 4, p. 476, 2016.

24 M. E. Gharbi, M. Martinez-Estrada, R. Fernández-García, and I. Gil, "Determination of salinity and sugar concentration by means of a circular-ring monopole textile antenna-based sensor," *IEEE Sensors J.*, vol. 21, no. 21, pp. 23751–23760, Nov. 2021.

25 D. Elsheikh and A. R. Eldamak, "Microwave textile sensors for breast cancer detection," *2021 38th National Radio Science Conference (NRSC)*, Mansoura, Egypt, July 2021, pp. 288–294.

26 A. Mason, S. Wylie, O. Korostynska, L. Cordova-Lopez, and A. Al-Shamma, "Flexible e-textile sensors for real-time health monitoring at microwave frequencies," *Int. J. Smart Sens. Intell. Syst.*, vol. 7, pp. 31–47, 2014.

27 N. C. Karmakar, E. M. Amin, and J. K. Saha, *Chipless RFID Sensors*, John Wiley, Hoboken, NJ, USA, 2016.

28 E. M. Amin, J. K. Saha, and N. C. Karmakar, "Smart sensing materials for low-cost chipless RFID sensor," *IEEE Sensors J.*, vol. 14, pp. 2198–2207, 2014.

29 K. Mc Gee, P. Anandarajah, and D. Collins, "A review of chipless remote sensing solutions based on RFID technology," *Sensors*, vol. 19, paper 4829, 2019.

30 V. Mulloni and M. Donelli, "Chipless RFID sensors for the Internet of Things: challenges and opportunities," *Sensors*, vol. 20, paper 2135, 2020.

31 F. Martín, C. Herrojo, J. Mata-Contreras, and F. Paredes, *Time-Domain Signature Barcodes for Chipless-RFID and Sensing Applications*, Springer, 2020.

32 S. Dey, J. K. Saha, and N. C. Karmakar, "Smart sensing: chipless RFID solutions for the internet of everything," *IEEE Microw. Mag.*, vol. 16, no. 10, pp. 26–39, Nov. 2015.

33 M. Forouzandeh and N. C. Karmakar, "Chipless RFID tags and sensors: a review on time-domain techniques," *Wirel. Power Transf.*, vol. 2, no. 2, pp. 62–77, Oct. 2015.

34 A. Ramos, A. Lazaro, D. Girbau, and R. Villarino, *RFID and Wireless Sensors Using Ultra-Wideband Technology*, ISTE Press/Elsevier, London, UK, 2016.

35 L. Cui, Z. Zhang, N. Gao, Z. Meng, and Z. Li, "Radio frequency identification and sensing techniques and their applications - a review of the state-of-the-art," *Sensors*, vol. 19, paper 4012, 2019.

36 F. Costa, S. Genovesi, M. Borgese, A. Michel, F.A. Dicandia, and G. Manara, "A review of RFID sensors, the new frontier of Internet of Things," *Sensors*, vol. 21, paper 3138, 2021.

37 N. Khalid, R. Mirzavand, and A. K. Iyer, "A survey on battery-less RFID-based wireless sensors," *Micromachines*, vol. 12, paper 819, 2021.

38 J. Wilson, Ed., *Sensor Technology Handbook*, Elsevier, Oxford, UK, 2005.

39 R. M. White, "A sensor classification scheme," *IEEE Trans. Ultrason., Ferroelectr. Freq. Control*, vol. 34, no. 2, pp. 124–126, Mar. 1987.

40 E. Eitel, "Basics of rotary encoders: overview and new technologies," *Machine Design Magazine*, 7, May 2014.

41 J. Mata-Contreras, C. Herrojo, and F. Martín, "Application of split ring resonator (SRR) loaded transmission lines to the design of angular displacement and velocity sensors for space applications," *IEEE Trans. Microw. Theory Techn.*, vol. 65, no. 11, pp. 4450–4460, Nov. 2017.

42 W. Wang, and X. Wang, Ed., *Contactless Vital Signs Monitoring*, Academic Press, 2021.

43 P. Kauch, M. Dovica, S. Slosarčk, and J. Kováč, "Comparision of contact and contactless measuring methods for form evaluation," *Procedia Eng.*, vol. 48, pp. 273–279, 2012.

44 X. Liang, H. Li, W. Wang, Y. Liu, R. Ghannam, F. Fioranelli, and H. Heidari, "Fusion of wearable and contactless sensors for intelligent gesture recognition," *Adv. Intell. Syst.*, vol.1, paper 1900088, 2019.

45 M. Nouman, S. Y. Khoo, M. P. Mahmud, and A. Z. Kouzani, "Recent advances in contactless sensing technologies for mental health monitoring," *IEEE Internet Things J.*, doi: https://doi.org/10.1109/JIOT.2021.3097801.

46 W. Dargie and C. Poellabauer, *Fundamentals of Wireless Sensor Networks: Theory and Practice*, John Wiley, Hoboken, NJ, USA, 2010.

47 A. Rida, L. Yang, and M. Tentzeris, *RFID-enabled Sensor Design and Applications*, Artech House, Norwood, MA, USA, 2010.

48 G. Marrocco, "Pervasive electromagnetics: sensing paradigms by passive RFID technology," *IEEE Wirel. Commun.*, vol. 17, no. 6, pp. 10–17, Dec. 2010.

49 S. Karuppuswami, A. Kaur, H. Arangali, and P. P. Chahal, "A hybrid magnetoelastic wireless sensor for detection of food adulteration," *IEEE Sensors J.*, vol. 17, no. 6, pp. 1706–1714, Mar. 2017.

50 V. Palazzi, F. Gelati, U. Vaglioni, F. Alimenti, P. Mezzanotte, and L. Roselli, "Leaf-compatible autonomous RFID-based wireless temperature sensors for precision agriculture," *Proc. 2019 IEEE Topical Conference on Wireless Sensors and Sensor Networks (WiSNet)*, Orlando, FL, USA, Jan. 2019.

51 S. N. Daskalakis, G. Goussetis, S. D. Assimonis, M. M. Tentzeris, and A. Georgiadis, "A UW backscatter-morse-leaf sensor for low-power agricultural wireless sensor networks," *IEEE Sensors J.*, vol. 18, pp. 7889–7898, 2018.

52 A. Ramos, D. Girbau, A. Lazaro, and R. Villarino, "Wireless concrete mixture composition sensor based on time-coded UWB RFID," *IEEE Microw. Wireless Compon. Lett.*, vol. 25, no. 10, pp. 681–683, Oct. 2015.

53 J. Zhang, "Development of a non-contact infrared thermometer," *Proc. 2017 International Conference Advanced Engineering and Technology Research (AETR 2017)*, Shenyang, China, Atlantis Press, 2018.

54 G. Marques and R. Pitarma, "Non-contact infrared temperature acquisition system based on internet of things for laboratory activities monitoring," *Procedia Computer Sci.*, vol. 155, pp. 487–494, 2019.

55 F. J. Hyde, *Thermistors*, Iliffe, 1971.

56 D. D. Pollock, *Thermocouples; Theory and Properties*, CRC Press, Taylor and Francis Group, Boca Raton, FL, USA, 1991.

57 J. Muñoz-Enano, P. Vélez, M. Gil, J. Mata-Contreras, and F. Martín, "Microwave comparator based on defect ground structures," *2019 European Microwave Conference in Central Europe* (EuMCE), Prague, Czech Republic, May 2019, pp. 244–247.

58 J. Fraden, *Handbook of Modern Sensors. Physics, Designs, and Applications*, 4th ed., Springer, New York, USA, 2010.

59 R. Pallás-Areny, J. G. Webster, *Sensors and Signal Conditioning*, 2nd ed., John Wiley, 2012.

60 S. Yurish, *Sensors and Signals*, IFSA Publishing, 2016.

61 K. Yallup and L. Basiricò, *Sensors for Diagnostics and Monitoring*, CRC Press, Taylor and Francis, Boca Raton, FL, USA, 2018.

62 R. Narayanaswamy and O. S. Wolfbeis, *Optical Sensors*, Springer, Heidelberg, Germany, 2004.

63 J. Haus, *Optical Sensors: Basics and Applications*, John Wiley, Hoboken, NJ, USA, 2010.

64 Z. Fang, K. Chin, R. Qu, and H. Cai, *Fundamentals of Optical Fiber Sensors*, John Wiley, Hoboken, NJ, USA, 2012.

65 E. Udd and W. B. Spillman Jr., *Field Guide to Fiber Optic Sensors*, SPIE Digital Library, 2014.

66 J. L. Santos and F. Farahi, Ed., *Handbook of Optical Sensors*, CRC Press, Taylor and Francis, Boca Raton, FL, USA, 2015.

67 B. Dhar Gupta, A. Mohan Shrivastav, and S. Prasood Usha, *Optical Sensors for Biomedical Diagnostics and Environmental Monitoring*, CRC Press, Taylor and Francis, Boca Raton, FL, USA, 2017.

68 I. R. Matias, S. Ikezawa, and J. Corres, *Fiber Optic Sensors: Current Status and Future Possibilities*, Springer, 2017.

69 R. De La Rue, H. P. Herzig, and M. Gerken, Ed., *Biomedical Optical Sensors: Differentiators for Winning Technologies*, Springer, 2020.

70 E. M. Petriu, "Reconsidering natural binary encoding for absolute position measurement application," *IEEE Trans. Instrum. Meas.*, vol. 38, pp. 1014–1016, 1989.

71 T. Ueda, F. Kohsaka, T. Iino, K. Kazami, and H. Nakayama, "Optical absolute encoder using spatial filter," *Proc. SPIE Photomechanics and Speckle Metrology*, San Diego, USA, August 1987; pp. 217–221.

72 J. Mata-Contreras, C. Herrojo, and F. Martín, "Detecting the rotation direction in contactless angular velocity sensors implemented with rotors loaded with multiple chains of split ring resonators (SRRs)," *IEEE Sensors J.*, vol. 18, pp. 7055–7065, 2018.

73 C. Herrojo, F. Paredes, and F. Martín, "Double-stub loaded microstrip line reader for very high data density microwave encoders," *IEEE Trans. Microw. Theory Techn.* vol. 67, pp. 3527–3536, 2019.

74 C. Herrojo, F. Paredes, and F. Martín, "3D-printed all-dielectric electromagnetic encoders with synchronous reading for measuring displacements and velocities," *Sensors*, vol. 20, paper 4837, 2020.

75 C. Herrojo, F. Paredes, and F. Martín, "Synchronism and direction detection in high-resolution/high-density electromagnetic encoders," *IEEE Sensors J.*, vol. 21, pp. 2873–2882, 2021.

76 P. Dong and Q. Chen, *LiDAR Remote Sensing and Applications*, CRC Press, Taylor and Francis, Boca Raton, FL, USA, 2018.

77 P. F. McManamon, *LiDAR Technologies and Systems*, SPIE Digital Library, 2019.

78 F. Ligler and C. Taitt, Ed., *Optical Biosensors*, 2nd ed., Elsevier, 2008.

79 D. Tosi, *Optical Fiber Biosensors: Device Platforms, Biorecognition, Applications*, Academic Press, 2021.

80 V. N. Konopsky and E. V. Alieva, "Photonic crystal surface waves for optical biosensors," *Anal. Chem.*, vol. 79, pp. 4729–4735, 2007.

81 B. Špačková, P. Wrobel, M. Bocková, and J. Homola, "Optical biosensors based on plasmonic nanostructures: a review," *Proc. IEEE*, vol. 104, no. 12, pp. 2380–2408, Dec. 2016.

82 N. Khansili, G. Rattu, and P. M. Krishna, "Label-free optical biosensors for food and biological sensor applications," *Sensors Actuators B Chem.*, vol. 265, pp. 35–49, 2018.

83 R. Peltomaa, B. Glahn-Martínez, E. Benito-Peña, and M. C. Moreno-Bondi. "Optical biosensors for label-free detection of small molecules," *Sensors*, vol. 18, paper 4126, 2018.

84 V. Garzón, D. G. Pinacho, R. H. Bustos, G. Garzón, and S. Bustamante, "Optical biosensors for therapeutic drug monitoring," *Biosensors*, vol. 9, paper 132, 2019.

85 Z. Liao, Y. Zhang, Y. Li, Y. Miao, S. Gao, F. Lin, Y. Deng, and L. Geng, "Microfluidic chip coupled with optical biosensors for simultaneous detection of multiple analytes: a review," *Biosens. Bioelectron.*, vol.126, pp. 697–706, 2019.

86 Y. Chen, J. Liu, Z. Yang, J. S. Wilkinson, and X. Zhou, "Optical biosensors based on refractometric sensing schemes: a review," *Biosens. Bioelectron.*, vol.144, paper 111693, 2019.

87 C. Chen and J. Wanga, "Optical biosensors: an exhaustive and comprehensive review," *Analyst*, vol. 145, pp. 1605–1628, 2020.

88 Y. T. Chen, Y. C. Lee, Y. H. Lai, J. C. Lim, N. T. Huang, C. T. Lin, and J. J. Huang, "Review of integrated optical biosensors for point-of-care applications," *Biosensors*, vol. 10, paper 209, 2020.

89 R. Boll, K. J Overshott, W. Göpel, J. Hesse, J. N. Zemel, Ed., *Sensors: A Comprehensive Survey, Vol. 5, Magnetic Sensors*, VCH, Weinheim, Germany, 1989.

90 J. E. Lenz, "A review of magnetic sensors," *Proc. IEEE*, vol. 78, no. 6, pp. 973–989, Jun. 1990.

91 J. Lenz and S. Edelstein, "Magnetic sensors and their applications," *IEEE Sensors J.*, vol. 6, no. 3, pp. 631–649, Jun. 2006.

92 H. G. Ramos and A. Lopes-Ribeiro, "Present and future impact of magnetic sensors in NDE," *Procedia Eng.*, vol. 86, pp. 406–419, 2014.

93 L. A. Francis and K. Poletkin, *Magnetic Sensors and Devices: Technologies and Applications*, CRC Press, Taylor & Francis Group, Boca Raton, FL, USA, 2018.

94 H. Heidari and V. Nabaei, *Magnetic Sensors for Biomedical Applications*, John Wiley/IEEE Press, Hoboken, NJ, USA, 2020.

95 P. Ripka, Ed., *Magnetic Sensors and Magnetometers*, Artech House, Norwood, MA, USA, 2021.

96 S. Tumanski, "Induction coil sensors—a review," *Meas. Sci. Technol.*, vol. 18, p. R31, 2007.

97 P. G. Slade, Ed., *Electrical Contacts: Principles and Applications*, 2nd ed., CRC Press, Taylor & Francis Group, Boca Raton, FL, USA, 2014.

98 E. Ramsden, *Hall-Effect Sensors: Theory and Application*, Newnes/Elsevier, Burlington, MA, USA, 2006.

99 G. Pepka, "Position and level sensing using Hall-effect sensing technology," *Sens. Rev.*, vol. 27, pp. 29–34, 2007.

100 A. Elzwawy, H. Piskin, N. Akdogan, M. Volmer, G. Reiss, L. Marnitz, A. Moskalsova, O. Gurel, and J. M. Schmalhorst, "Current trends in planar Hall effect sensors: evolution, optimization, and applications," *J. Phys. D. Appl. Phys.*, vol. 54, paper 353002, 2021.

101 P. P. Freitas, R. Ferreira, S. Cardoso, and F. Cardoso, "Magnetoresistive sensors," *J. Phys. Condens. Matter*, vol. 19, no. 16, paper 165221, 2007.

102 L. Jogschies, D. Klaas, R. Kruppe, J. Rittinger, P. Taptimthong, A. Wienecke, L. Rissing, and M. C. Wurz, "Recent developments of magnetoresistive sensors for industrial applications," *Sensors*, vol. 15, pp. 28665–28689, 2015.

103 C. Zheng, K. Zhu, S. C. De Freitas, J. Y. Chang, J. E. Davies, P. Eames, P. P. Freitas, O. Kazakova, C. Kim, C. W. Leung, S. H. Liou, A. Ognev, S. N. Piramanayagam, P. Ripka, A. Samardak, K. H. Shin, S. Y. Tong, M. J. Tung, S. X. Wang, S. Xue, X. Yin, and P. W. T. Pong, "Magnetoresistive sensor

development roadmap (non-recording applications)," *IEEE Trans. Magn.*, vol. 55, no. 4, pp. 1–30, Apr. 2019.

104 M. Julliere, "Tunneling between ferromagnetic films," *Phys. Lett.*, vol. 54A, pp. 225–226, 1975.

105 S. Ikeda, J. Hayakawa, Y. Ashizawa, Y.M. Lee, K. Miura, H. Hasegawa, M. Tsunoda, F. Matsukura, and H. Ohno, "Tunnel magnetoresistance of 604% at 300 K by suppression of Ta diffusion in CoFeB/MgO/CoFeB pseudo-spin-valves annealed at high temperature," *Appl. Phys. Lett.*, vol. 93, paper 082508, 2008.

106 D. S. Ballantine, Jr., R. M. White, S. J. Martin, A. J. Ricco, E. T. Zellers, G. C. Frye, and H. Wohltjen, *Acoustic Wave Sensors: Theory, Design and Physico-Chemical Applications*, Academic Press, 1996.

107 H. Lee, *Acoustical Sensing and Imaging*, CRC Press, Taylor & Francis Group, Boca Raton, FL, USA, 2016.

108 N. Dey, A. S. Ashour, W. S. Mohamed, N. G. Nguyen, *Acoustic Sensors for Biomedical Applications*, Springer, 2019.

109 A. D'Amico and E. Verona, "SAW sensors," *Sensors Actuators*, vol. 17, pp. 55–66, 1989.

110 M. Thompson and D. C. Stone, *Surface-Launched Acoustic Wave Sensors: Chemical Sensing and Thin-Film Characterization*, John Wiley, Hoboken, NJ, USA, 1997.

111 C. C. W. Ruppel and T. A. Fjeldly, Ed., *Advances in Surface Acoustic Wave Technology, Systems and Applications*, World Scientific, 2001.

112 R. B. Priya, T. Venkatesan, G. Pandiyarajan, and H. M. Pandya, "A short review of SAW sensors," *J. Environ. Nanotechnol.*, vol. 4, no. 4, pp. 15–22, 2015.

113 B. Liu, X. Chen, H. Cai, M. M. Ali, X. Tian, L. Tao, Y. Yang, and T. Ren, "Surface acoustic wave devices for sensor applications," *J. Semicond.*, vol. 37, no. 2, p. 021001, 2016.

114 J. Devkota, P. R. Ohodnicki, and D. W. Greve, "SAW sensors for chemical vapors and gases," *Sensors*, vol. 17, no. 4, p. 801, 2017.

115 G. Blokdyk, *Surface Acoustic Wave SAW -Based Sensors A Complete Guide*, 5STARCooks, 2018.

116 P. A. Lieberzeit, C. Palfinger, F. L. Dickert, and G. Fischerauer, "SAW RFID-tags for mass-sensitive detection of humidity and vapors," *Sensors*, vol. 9, pp. 9805–9815, 2009.

117 K. Along, Z. Chenrui, Z. Luo, L. Xiaozheng, and H. Tao, "SAW RFID enabled multi-functional sensors for food safety applications," *2010 IEEE International Conference on RFID-Technology and Applications*, Guangzhou, China, June 2010, pp. 200–204.

118 A. Kang, C. Zhang, X. Ji, T. Han, R. Li, and X. Li, "SAW-RFID enabled temperature sensor," *Sensors Actuators A Phys.*, vol. 201, pp. 105–113, 2013.

119 H. Chambon, P. Nicolay, C. Floer, and A. Benjeddou, "A package-less SAW RFID sensor concept for structural health monitoring," *Mech. Adv. Mater. Struct.*, vol. 28, pp. 648–655, 2021.

120 I. J. Busch-Vishniac, *Electromechanical Sensors and Actuators*, Springer, 1999.

121 S. P. Beeby, G. Ensel, M. Kraft, and N. White, *MEMS Mechanical Sensors*, Artech House, 2004.

122 T. Thundat, P. I. Oden, and R. J. Warmack, "Microcantilever sensors," *Microscale Thermophys. Eng.*, vol. 1, pp. 185–199, 1997.

123 H. P. Lang and C. Gerber, Microcantilever sensors, in: P. Samorì (eds) *STM and AFM Studies on (Bio) molecular Systems: Unravelling the Nanoworld, Topics in Current Chemistry*, vol. 285. Springer, Berlin, Heidelberg, 2008.

124 M. Spletzer, A. Raman, H. Sumali, and J. P. Sullivan, "Highly sensitive mass detection and identification using vibration localization in coupled microcantilever arrays," *Appl. Phys. Lett.*, vol. 92, paper 114102, 2008.

125 J. Mouro, R. Pinto, P. Paoletti, and B. Tiribilli, "Microcantilever: dynamical response for mass sensing and fluid characterization," *Sensors*, vol. 21, paper 115, 2021.

126 A. S. Fiorillo, C. D. Critello, and S. A. Pullano, "Theory, technology and applications of piezoresistive sensors: A review," *Sensors Actuators A Phys.*, vol. 281, pp. 156–175, 2018.

127 B. F. Gonçalves, J. Oliveira, P. Costa, V. Correia, P. Martins, G. Botelho, and S. Lanceros-Mendez, "Development of water-based printable piezoresistive sensors for large strain applications," *Compos. Part B*, vol. 112, pp. 344–352, 2017.

128 C. Steinem and A. Janshoff, *Piezoelectric Sensors*, Springer, 2007.

129 S. J. Rupitsch, *Piezoelectric Sensors and Actuators: Fundamentals and Applications*, Springer, Berlin, Germany, 2019.

130 W. Y. Du, *Resistive, Capacitive, Inductive, and Magnetic Sensor Technologies*, CRC Press, Taylor & Francis Group, Boca Raton, FL, USA, 2015.

131 L. K. Baxter, *Capacitive Sensors: Design and Applications*, Wiley-IEEE Press, 1997.

132 T. Gray, *Projected Capacitive Touch*, Springer, 2019.

133 A. Othman, N. Hamzah, Z. Hussain, R. Baharudin, A. D. Rosli, and A. I. C. Ani, "Design and development of an adjustable angle sensor based on rotary potentiometer for measuring finger flexion," *2016 6th IEEE International Conference on Control System, Computing and Engineering (ICCSCE)*, Penang, Malaysia, 2016, pp. 569–574.

134 I. H. Woodhouse, *Introduction to Microwave Remote Sensing*, CRC Press, Taylor & Francis Group, Boca Raton, FL, USA, 2006.

135 T. Lillesand, R. W. Kiefer, and J. Chipman, *Remote Sensing and Image Interpretation*, 7th ed., John Wiley, Hoboken, NJ, USA, 2015.

136 P. S. Thenkabail, *Remote Sensing Handbook*, CRC Press, Taylor & Francis Group, Boca Raton, FL, USA, 2015.

137 J. B. Campbell and R. H. Wynne, *Introduction to Remote Sensing*, 5th ed., The Guilford Press, 2011.

138 F. M. Henderson and A. J. Lewis, Ed., Principles and Applications of Imaging Radar, *in* Manual of Remote Sensing *(vol. 2)*, 3rd ed., Wiley, 1998.

139 M. A. Richards, J. A. Scheer, and W. A. Holm, *Principles of Modern Radar: Vol. 1: Basic Principles*, SciTech Publishing, New York, USA, 2016.

140 M. Skolnik, *Radar Handbook*, 3rd ed., McGraw-Hill, New York, USA, 2008.

141 C. Nguyen and J. Park, *Stepped-Frequency Radar Sensors*, Springer, 2016.

142 P. V. Jain and P. Heydari, *Automotive Radar Sensors in Silicon Technologies*, Springer, 2013.

143 W.L. Wolf, *Introduction to Radiometry*, SPIE Digital Library, 1998.

144 W. McCluney, *Introduction to Radiometry and Photometry*, 2nd ed., Artech House, 2015.

145 C. T. Swift, L. S. Fedor, and R. O. Ramseier, "An algorithm to measure sea ice concentration with microwave radiometers," *J. Geophysical Research*, vol. 90, pp. 1087–1099, 1985.

146 D. D. Turner, S. A. Clough, J. C. Liljegren, E. E. Clothiaux, K. E. Cady-Pereira, and K. L. Gaustad, 'Retrieving liquid water path and precipitable water vapor from the atmospheric radiation measurement (ARM) microwave radiometers," *IEEE Trans. Geosci. Remote Sens.*, vol. 45, no. 11, pp. 3680–3690, Nov. 2007.

147 T. Meissner and F. J. Wentz, "Wind-vector retrievals under rain with passive satellite microwave radiometers," *IEEE Trans. Geosci. Remote Sens.*, vol. 47, no. 9, pp. 3065–3083, Sep. 2009.

148 D. Cimini, T. J. Hewison, L. Martin, J. Güldner, C. Gaffard, and F. S. Marzano, "Temperature and humidity profile retrievals from ground-based microwave radiometers during TUC," *Meteorol. Z.*, vol. 15, pp. 45–56, 2006.

149 H. M. Jol, Ed., *Ground Penetrating Radar Theory and Applications*, Elsevier, Amsterdam, The Netherlands, 2009.

150 W. Zhao, E. Forte, M. Pipan, and G. Tian, "Ground Penetrating Radar (GPR) attribute analysis for archaeological prospection," *J. Appl. Geophys.*, vol. 97, pp. 107–117, 2013.

151 A. Benedetto and L. Pajewski, *Civil Engineering Applications of Ground Penetrating Radar*, Springer, 2015.

152 J. Hasch, E. Topak, R. Schnabel, T. Zwick, R. Weigel, and C. Waldschmidt, "Millimeter-wave technology for automotive radar sensors in the 77 GHz frequency band," *IEEE Trans. Microw. Theory Techn.*, vol. 60, no. 3, pp. 845–860, Mar. 2012.

153 A. Bartsch, F. Fitzek, and R. H. Rasshofer, "Pedestrian recognition using automotive radar sensors," *Adv. Radio Sci.*, vol. 10, pp. 45–55, 2012.

154 W. Menzel and A. Moebius, "Antenna concepts for millimeter-wave automotive radar sensors," *Proc. IEEE*, vol. 100, no. 7, pp. 2372–2379, Jul. 2012.

155 J. Steinbaeck, C. Steger, G. Holweg, and N. Druml, "Next generation radar sensors in automotive sensor fusion systems," *2017 Sensor Data Fusion: Trends, Solutions, Applications (SDF)*, Bonn, Germany, Oct. 2017.

156 F. Roos, J. Bechter, C. Knill, B. Schweizer, and C. Waldschmidt, "Radar sensors for autonomous driving: modulation schemes and interference mitigation," *IEEE Microw. Mag.*, vol. 20, no. 9, pp. 58–72, Sep. 2019.

157 M. Cheney and B. Borden, *Fundamentals of Radar Imaging*, The Society for Industrial and Applied Mathematics, Philadelphia, PA, USA, 2009.

158 L. C. Potter, E. Ertin, J. T. Parker, and M. Cetin, "Sparsity and compressed sensing in radar imaging," *Proc. IEEE*, vol. 98, no. 6, pp. 1006–1020, Jun. 2010.

159 P. K. M. Nkwari, S. Sinha, and H. C. Ferreira, "Through-the-wall radar imaging: a review," *IETE Tech. Rev.*, vol. 35, pp. 631–639, 2018.

160 K. S. Chen, *Principles of Synthetic Aperture Radar Imaging: A System Simulation Approach*, CRC Press, Taylor & Francis Group, Boca Raton, FL, USA, 2016.

161 M. Jankiraman, *FMCW Radar Design*, Artech House, Norwood, MA, USA, 2018.

162 R. J. Doviak and D. S. Zrnic, *Doppler Radar and Weather Observations*, 2nd ed., Dover Publications Inc, 2006.

163 M. N. Cohen, Pulse compression in radar systems, in J. L. Eaves and E. K. Reedy (eds) *Principles of Modern Radar*, Springer, Boston, MA, USA, 1987.

164 T. T. Mar and S. S. Mon, "Pulse compression method for radar signal processing," *Int. J. Sci. Eng. Appl.*, vol. 3, pp. 31–35, 2014.

165 D. Deiana, E. M. Suijker, R. J. Bolt, A. P. M. Maas, W. J. Vlothuizen, and A. S. Kossen, "Real time indoor presence detection with a novel radar on a chip," *2014 International Radar Conference*, Lille, France, Oct. 2014.

166 J. A. Nanzer, "A review of microwave wireless techniques for human presence detection and classification," *IEEE Trans. Microw. Theory Techn.*, vol. 65, no. 5, pp. 1780–1794, May 2017.

167 P. Nuti, E. Yavari, and O. Boric-Lubecke, "Doppler radar occupancy sensor for small-range motion detection," *2017 IEEE Asia Pacific Microwave Conference (APMC)*, Kuala Lumpur, Malaysia, Nov. 2017, pp. 192–195.

168 C. Gouveia, J. Vieira, and P. Pinho, "A review on methods for random motion detection and compensation in bio-radar systems," *Sensors*, vol. 19, paper 604, 2019.

169 S. Dias Da Cruz, H. Beise, U. Schröder, and U. Karahasanovic, "A theoretical investigation of the detection of vital signs in presence of car vibrations and RADAR-based passenger classification," *IEEE Trans. Veh. Technol.*, vol. 68, no. 4, pp. 3374–3385, Apr. 2019.

170 F. Martín, P. Vélez, and M. Gil, "Microwave sensors based on resonant elements," *Sensors*, vol. 20, p. 3375, 2020.

171 F. Kremer and A. Schönhals, Ed., *Broadband Dielectric Spectroscopy*, Springer, 2003.

172 X. Zhang, C. Ruan, W. Wang, and Y. Cao, "Submersible high sensitivity microwave sensor for edible oil detection and quality analysis," *IEEE Sensors J.*, vol. 21, no. 12, pp. 13230–13238, Jun. 2021.

173 A. M. Nicolson and G. F. Ross, "Measurement of the intrinsic properties of materials by time-domain techniques," *IEEE Trans. Instrum. Meas.*, vol. 19, no. 4, pp. 377–382, Nov. 1970.

174 W. B. Weir, "Automatic measurement of complex dielectric constant and permeability at microwave frequencies," *Proc. IEEE*, vol. 62, no. 1, pp. 33–36, Jan. 1974.

175 D. K. Ghodgaonkar, V. V. Varadan, and V. K. Varadan, "A free-space method for measurement of dielectric constants and loss tangents at microwave frequencies," *IEEE Trans. Instrum. Meas.*, vol. 37, pp. 789–793, 1989.

176 V. V. Varadan, R. Hollinger, D. Ghodgaonkar, and V. K. Varadan, "Free-space, broadband measurements of high-temperature, complex dielectric properties at microwave frequencies," *IEEE Trans. Instrum. Meas.*, vol. 40, pp. 842–846, Oct. 1991.

177 K. A. Jose, V. K. Varadan, and V. V. Varadan, "Wideband and noncontact characterization of the complex permittivity of liquids," *Microw. Opt. Technol. Lett.*, vol. 30, pp. 75–79, 2001.

178 I. S. Seo, W. S. Chin, and D. G. Lee, "Characterization of electromagnetic properties of polymeric composite materials with free space method," *Compos. Struct.*, vol. 66, pp. 533–542, 2004.

179 F. J. F. Gonçalves, A. G. M. Pinto, R. C. Mesquita, E. J. Silva, and A. Brancaccio, "Free-space materials characterization by reflection and transmission measurements using frequency-by-frequency and multi-frequency algorithms," *Electronics*, vol. 7, no. 10, paper 260, 2018.

180 R. E. Collin, *Foundations for Microwave Engineering*, 2nd ed., Wiley/IEEE Press, New York, NY, 2001.

181 D. M. Pozar, *Microwave Engineering*, 4th ed., John Wiley, Hoboken, NJ, USA, 2012.

182 J. Hong and M. J. Lancaster, *Microstrip Filters for RF/Microwave Applications*, John Wiley, Hoboken, NJ, USA, 2001.

183 R. A. Waldron, "Perturbation theory of resonant cavities," *Proc. Inst. Electr. Eng.*, vol. 107, pp. 272–274, 1960.

184 A. D. Vyas, V. A. Rana, D. H. Gadani, and A. N. Prajapati, "Cavity perturbation technique for complex permittivity measurement of dielectric materials at X-band microwave frequency," *2008 International Conference on Recent Advances in Microwave Theory and Applications*, Jaipur, India, Nov. 2008, pp. 836–838.

185 Keysight *85072A 10-GHz Split Cylinder Resonator, Technical Overview*, Application Note, Keysight, June 2018.

186 Association Connecting Electronic Industries PC-TM-650, 2.5.5.13, *Relative Permittivity and Loss Tangent Using a Split-Cylinder Resonator*, Jan. 2007, https://doczz.net/doc/6544488/ipc-tm-650. Accessed November 10, 2021.

187 G. Kent, "An evanescent-mode tester for ceramic dielectric substrates," *IEEE Trans. Microw. Theory Techn.*, vol. 36, no. 10, pp. 1451–1454, Oct. 1988.

188 G. Kent, "Nondestructive permittivity measurement of substrates," *IEEE Trans. Instrum. Meas.*, vol. 45, no. 1, pp. 102–106, Feb. 1996.

189 M. D. Janezic and J. Baker-Jarvis, "Full-wave analysis of a split-cylinder resonator for nondestructive permittivity measurements," *IEEE Trans. Microw. Theory Techn.*, vol. 47, no. 10, pp. 2014–2020, 1999.

190 M. D. Janezic, *Nondestructive Relative Permittivity and Loss Tangent Measurements using a Split-Cylinder Resonator*, Ph.D. Thesis, University of Colorado at Boulder, 2003.

191 M. D. Janezic, E. F. Kuester, and J. Baker-Jarvis, "Broadband complex permittivity measurements of dielectric substrates using a split-cylinder resonator," *IEEE MTT-S International Microwave Symposium Digest*, Fort Worth, TX, USA, Jun. 2004, pp. 1817–1820.

192 J. Krupka, R. G. Geyer, J. Baker-Jarvis, and J. Ceremuga, "Measurements of the complex permittivity of microwave circuit board substrates using split dielectric resonator and reentrant cavity techniques," *Proc. DMMA'96 Conference*, Bath, UK, Sept. 1996, pp. 21–24.

193 J. Krupka, A. P. Gregory, O. C. Rochard, R. N. Clarke, B. Riddle, and J. Baker-Jarvis, "Uncertainty of complex permittivity measurements by split-post dielectric resonator technique," *J. Eur. Ceram. Soc.*, vol. 21, pp. 2673–2676, 2001.

194 J. Krupka and J. Mazierska, "Contactless measurements of resistivity of semiconductor wafers employing single-post and split-post dielectric-resonator techniques," *IEEE Trans. Instrum. Meas.*, vol. 56, no. 5, pp. 1839–1844, Oct. 2007.

195 J. Krupka, "Contactless methods of conductivity and sheet resistance measurement for semiconductors, conductors and superconductors", *Meas. Sci. Technol.*, vol. 24, paper 062001, 2013.

196 W. E. Courtney, "Analysis and evaluation of a method of measuring the complex permittivity and permeability of microwave insulators," *IEEE Trans. Microw. Theory Techn.*, vol. 18, pp. 476–485, 1970.

197 J. Krupka, K. Derzakowski, B. Riddle, and J. Baker-Jarvis, "A dielectric resonator for measurements of complex permittivity of low loss dielectric materials as a function of temperature," *Meas. Sci. Technol.*, vol. 9, pp. 1751–1756, 1998.

198 J. Krupka, K. Derzakowski, A. Abramowicz, M. E. Tobar, and R. G. Geyer, "Whispering gallery modes for complex permittivity measurements of ultra-low loss dielectric materials," *IEEE Trans. Microw. Theory Techn.*, vol. 47, pp. 752–759, 1999.

199 J. Krupka, K. Derzakowski, M. E. Tobar, J. Hartnett, and R. G. Geyer, "Complex permittivity of some ultralow loss dielectric crystals at cryogenic temperatures," *Meas. Sci. Technol.*, vol. 10, pp. 387–392, 1999.

200 J. Krupka, "Microwave measurements of electromagnetic properties of materials," *Materials*, vol. 14, no. 17, paper 5097, 2021.

201 J. Baker-Jarvis, M. D. Janezic, J. H. Grosvenor, and R. G. Geyer, "Transmission/reflection and short-circuit line methods for measuring permittivity and permeability," *National Institute of Standards and Technology (NIST) Technical Note* 1355, pp. 104–105, 1993.

202 L. F. Chen, C. K. Ong, C. P. Neo, V. V. Varadan, and V. K. Varadan, *Microwave Electronics: Measurement and Materials Characterization*, John Wiley & Sons Ltd., West Sussex, England, 2004.

203 E. J. Rothwell, J. L. Frasch, S. M. Ellison, P. Chahal, and R. O. Ouedraogo, "Analysis of the Nicolson-Ross-Weir method for characterizing the electromagnetic properties of engineered materials," *Prog. Electromagn. Res.*, vol. 157, pp. 31–47, 2016.

204 F. Costa, M. Borgese, M. Degiorgi, and A. Monorchio, "Electromagnetic characterisation of materials by using transmission/reflection (T/R) devices," *Electronics*, vol. 6, no. 4, paper 95, 2017.

205 O. Luukkonen, S. I. Maslovski, and S. A. Tretyakov, "A stepwise Nicolson–Ross–Weir-based material parameter extraction method," *IEEE Antennas Wirel. Propag. Lett.*, vol. 10, pp. 1295–1298, 2011.

206 G. Angiulli and M. Versaci, "Retrieving the effective parameters of an electromagnetic metamaterial using the Nicolson-Ross-Weir method: an analytic continuation problem along the path determined by scattering parameters," *IEEE Access*, vol. 9, pp. 77511–77525, 2021.

207 J. Baker-Jarvis, E. J. Vanzura, and W. A. Kissick, "Improved technique for determining complex permittivity with the transmission/reflection method," *IEEE Trans. Microw. Theory Techn.*, vol. 38, pp. 1096–1103, 1990.

208 ASTM D 5568.01. Standard Test Method for Measuring Relative Complex Permittivity and Relative Magnetic Permeability of Solid Materials at Microwave Frequencies; Annual Book of ASTM Standards; American Society for Testing and Materials (ASTM), West Conshohocken, PA, USA, 2003.

209 A. H. Boughriet, C. Legrand, and A. Chapoton, "Noniterative stable transmission/reflection method for low-loss material complex permittivity determination," *Trans. IEEE Microw. Theory Techn.*, vol. 45, pp. 52–57, 1997.

210 S. Jenkins, T. E. Hodgetts, R. N. Clarke, and A. W. Preece, "Dielectric measurements on reference liquids using automatic network analyzers and calculable geometries," *Meas. Sci. Technol.*, vol. 1, pp. 691–702, 1990.

211 A. L. de Paula, M. C. Rezende, and J. J. Barroso, "Modified Nicolson-Ross-Weir (NRW) method to retrieve the constitutive parameters of low-loss materials," *2011 SBMO/IEEE MTT-S International Microwave and Optoelectronics Conference (IMOC 2011)*, Natal, Brazil, 2011, pp. 488–492.

212 A. N. Vicente, G. M. Dip, and C. Junqueira, "The step by step development of NRW method," *2011 SBMO/IEEE MTT-S International Microwave and Optoelectronics Conference (IMOC 2011)*, Natal, Brazil, 2011, pp. 738–742.

213 V. H. Nguyen, M. H. Hoang, H. P. Phan, T. Q. V. Hoang and T. P. Vuong, "Measurement of complex permittivity by rectangular waveguide method with simple specimen preparation," *2014 International Conference on Advanced Technologies for Communications (ATC 2014)*, Hanoi, Vietnam, Oct. 2014, pp. 397–400.

214 Y. Shi, T. Hao, L. Li, and C.H. Liang "An improved NRW method to extract electromagnetic parameters of metamaterials", *Microw. Opt. Technol. Lett.*, vol. 58, pp. 647–652, Mar. 2016.

215 J. Y. Wang, J. Tao, L. Severac, D. Mesguich, and C. Laurent, "Microwave characterization of nanostructured material by modified Nicolson-Ross-Weir method," *Ampere Newsletter*, vol. 94 pp. 35–41, Dec. 2017.

216 S. Sahin, N. K. Nahar, and K. Sertel, "A simplified Nicolson–Ross–Weir method for material characterization using single-port measurements," *IEEE Trans. Terahertz Sci. Techn.*, vol. 10, no. 4, pp. 404–410, Jul. 2020.

217 M. A. Stuchly and S. S. Stuchly, "Coaxial line reflection methods for measuring dielectric properties of biological substances at radio and microwave frequencies—A review," *IEEE Trans. Instrum. Meas.*, vol. 29, pp. 176–183, 1980.

218 E. Burdette, F. Cain, and J. Seals, "In vivo probe measurement technique for determining dielectric properties at VHF through microwave frequencies," *IEEE Trans. Microw. Theory Techn.*, vol. 28, pp. 414–427, 1980.

219 T. W. Athey, M. A. Stuchly, and S. S. Stuchly, "Measurement of radio frequency permittivity of biological tissues with an open-ended coaxial line: Part 1," *IEEE Trans. Microw Theory Techn.*, vol. 30, no. 1, pp. 82–86, Jan. 1982.

220 S. Gabriel, R. W. Lau, and C. Gabriel, "The dielectric properties of biological tissues: II. measurements in the frequency range 10 Hz to 20 GHz," *Phys. Med. Biol.*, vol. 41, pp. 2251–2269, 1996.

221 D. Popovic, M. Okoniewski, D. Hagl, J.H. Booske, and S. C. Hagness, "Volume sensing properties of open ended coaxial probes for dielectric spectroscopy of breast tissue," *Proc. IEEE Antennas and Propagation Society*, Boston, MA, USA, July 2001, pp. 254–257.

222 A. Peyman, S. Holden, and C. Gabriel, *Mobile Telecommunications and Health Research Programme: Dielectric Properties of Tissues at Microwave Frequencies*, Microwave Consultants Limited, London, UK, 2005.

223 B. Filali, F. Boone, J. Rhazi, and G. Ballivy, "Design and calibration of a large open-ended coaxial probe for the measurement of the dielectric properties of concrete," *IEEE Trans. Microw. Theory Techn.*, vol. 56, no. 10, pp. 2322–2328, Oct. 2008.

224 S. A. Komarov, A. S. Komarov, D. G. Barber, M. J. L. Lemes, and S. Rysgaard, "Open-ended coaxial probe technique for dielectric spectroscopy of artificially grown sea ice," *IEEE Trans. Geosci. Remote Sens.*, vol. 54, no. 8, pp. 4941–4951, Aug. 2016.

225 A. La Gioia, E. Porter, I. Merunka, A. Shahzad, S. Salahuddin, M. Jones, and M. O'Halloran, "Open-ended coaxial probe technique for dielectric measurement of biological tissues: challenges and common practices", *Diagnostics*, vol. 8, paper 40, 2018.

226 V. Guihard, F. Taillade, J. P. Balayssac, B. Steck, J. Sanahuja, and F. Deby, "Permittivity measurement of cementitious materials with an open-ended coaxial probe," *Constr. Build. Mater.*, vol. 230, paper 116946, 2020.

227 C. Aydinalp, S. Joof, and T. Yilmaz, "Towards non-invasive diagnosis of skin cancer: sensing depth investigation of open-ended coaxial probes," *Sensors*, vol. 21, paper 1319, 2021.

228 Keysight Technologies, N1501A Dielectric Probe Kit:10 MHz to 50 GHz, Technical Overview, 2020.

229 J. R. Mosig, J. E. Besson, M. Gex-Fabry, and F. E. Gardiol, "Reflection of an open-ended coaxial line and application to nondestructive measurement of materials," *IEEE Trans. Instrum. Meas.*, vol. IM-30, no. 1, pp. 46–51, Mar. 1981.

230 L. S. Anderson, G. B. Gajda, and S. S. Stuchly, "Analysis of open-ended coaxial line sensor in layered dielectric," *IEEE Trans. Instrum. Meas.*, vol. 35, pp. 13–18, 1986.

231 D. K. Misra, "A quasi-static analysis of open-ended coaxial lines," *IEEE Trans. Microw. Theory Techn.*, vol. 35, no. 10, pp. 925–928, Oct. 1987.

232 T. E. Hodgetts, *The Open-Ended Coaxial Line: A Rigorous Variational Treatment*, Royal Signals and Radar Establishment Memorandum, nr.4331, Royal Signals and Radar Establishment, Malvern, UK, 1989.

233 J. P. Grant, R. N. Clarke, G. T. Symm, and N. M. Spyron, "A critical study of the open-ended coaxial-line sensor technique for RF and microwave complex permittivity measurements," *J. Phys. E Sci. Instrum.*, vol. 22, pp. 757–770, 1989.

234 S. Jenkins, A. W. Preece, T. E. Hodgetts, G. T. Symm, A. G. P. Warham, and R. N. Clarke, "Comparison of three numerical treatments for the open-ended coaxial line sensor," *Electron. Lett.*, vol. 26, pp. 234–236, 1990.

235 J. Baker-Jarvis, M. D. Janezic, P. D. Domich, and R. G. Geyer, "Analysis of an open-ended coaxial probe with lift-off for nondestructive testing," *IEEE Trans. Instrum. Meas.*, vol. 45, pp. 711–718, 1994.

236 C. Gabriel, T. Y. Chan, and E. H. Grant, "Admittance models for open ended coaxial probes and their place in dielectric spectroscopy," *Phys. Med. Biol.*, vol. 39, pp. 2183–2200, 1994.

237 P. De Langhe, L. Martens, and D. De Zutter, "Design rules for an experimental setup using an open-ended coaxial probe based on theoretical modelling," *IEEE Trans. Instrum. Meas.*, vol. 43, no. 6, pp. 810–817, Dec. 1994.

238 D. Berube, F. M. Ghannouchi, and P. A. Savard, "Comparative study of four open-ended coaxial probe models for permittivity measurements of lossy dielectric/biological materials at microwave frequencies," *IEEE Trans. Microw. Theory Techn.*, vol. 44, pp. 1928–1934, 1996.

239 D. Misra, "On the measurement of the complex permittivity of materials by an open-ended coaxial probe," *IEEE Microw. Guid. Wave Lett.*, vol. 5, pp. 161–163, 1995.

240 R. Zajícek, L. Oppl, and J. Vrba, "Broadband measurement of complex permitivity using reflection method and coaxial probes," *Radioengineering*, vol. 17, pp. 14–19, 2008.

241 M. D. Perez Cesaretti, *General Effective Medium Model for the Complex Permittivity Extraction with an Open-Ended Coaxial Probe in Presence of a Multilayer Material under Test*. Ph.D. Thesis, University of Bologna, Bologna, Italy, 2012.

242 M. C. Decreton and F. E. Gardiol, "Simple nondestructive method for the measurement of complex permittivity," *IEEE Trans. Instrum. Meas.*, vol. IM-23, pp. 434–438, Dec. 1974.

243 M. C. Decreton and M. S. Ramachandraiah, "Nondestructive measurement of complex permittivity for dielectric slabs," *IEEE Trans. Microwave Theory Tech.*, vol. MTT-23, pp. 1077–1080, Dec. 1975.

244 M. S. Ramachandraiah and M. C. Decreton, "A resonant cavity approach for the nondestructive determination of complex permittivity at microwave frequencies," *IEEE Trans. Instrum. Meas.*, vol. IM-24, pp. 287–291, Dec. 1975.

245 M. Gex-Fabry, J. R. Mosig, and F. E. Gardiol, "Reflection and radiation of an open-ended circular waveguide; application to nondestructive measurement of materials," *Arch. E lek. Übertragung*, vol. 33, pp. 473–478, 1979.

246 S. Bakhtiari, S. Ganchev, and R. Zoughi, "Open-ended rectangular waveguide for nondestructive thickness measurement and variation detection of lossy dielectric slab backed by a conducting plate," *IEEE Trans. Instrum. Meas.*, vol. 42, pp. 19–24, 1993.

247 F. Mazlumi, S. H. H. Sadeghi, and R. Moini, "Interaction of an open-ended rectangular waveguide probe with an arbitrary-shape surface crack in a lossy conductor," *IEEE Trans. Microw. Theory Techn.*, vol. 54, no. 10, pp. 3706–3711, Oct. 2006.

248 J. W. Stewart, *Simultaneous Extraction of the Permittivity and Permeability of Conductor-Backed Lossy Materials Using Open-Ended Waveguide Probes*, Ph. D. Thesis, Air Force Institute of Technology, 2006.

249 K. M. Donnell, A. McClanahan and R. Zoughi, "On the crack characteristic signal from an open-ended coaxial probe," *IEEE Trans. Instrum. Meas.*, vol. 63, no. 7, pp. 1877–1879, Jul. 2014.

250 K. T. M. Shafi, M. S. ur Rahman, M. Abou-Khousa, and M. R. Ramzi, "Applied microwave imaging of composite structures using open-ended circular waveguide," *2017 IEEE International Conference on Imaging Systems and Techniques (IST)*, Beijing, China, Oct. 2017.

251 M. R. Ramzi, M. Abou-Khousa, and I. Prayudi, "Near-field microwave imaging using open-ended circular waveguide probes," *IEEE Sensors J.*, vol. 17, no. 8, pp. 2359–2366, Apr. 2017.

252 R. Sutthaweekul and G. Y. Tian, "Steel corrosion stages characterization using open-ended rectangular waveguide probe," *IEEE Sensors J.*, vol. 18, no. 3, pp. 1054–1062, Feb. 2018.

253 F. Paredes, C. Herrojo, J. Mata-Contreras, and F. Martín, "Near-field chipless-RFID sensing and identification system with switching reading," *Sensors*, vol. 18, p. 1148, 2018.

254 H. Tao, M. A. Brenckle, M. Yang, J. Zhang, M. Liu, S. M. Siebert, R. D. Averitt, M. S. Mannoor, M. C. McAlpine, J. A. Rogers, D. L. Kaplan, and F. G. Omenetto, "Silk-based conformal, adhesive, edible food sensors," *Adv. Matter.*, vol. 24, pp. 1067–1072, Feb. 2012.

255 G. Ekinci, A. Calikoglu, S. N. Solak, A. D. Yalcinkaya, G. Dundar, and H. Torun, "Split-ring resonator-based sensors on flexible substrates for glaucoma monitoring," *Sens. Actuator A: Phys.*, vol. 268, pp. 32–37, Dec. 2017.

256 C. Lee, C. Wu, and Y. Kuo, "Wearable bracelet belt resonators for noncontact wrist location and pulse setection," *IEEE Trans. Microw. Theory and Techn.*, vol. 65, no. 11, pp. 4475–4482, Nov. 2017.

257 A. R. Eldamak and E. C. Fear, "Conformal and disposable antenna-based sensor for non-invasive sweat monitoring," *Sensors*, vol. 18, no. 12, paper 4088, 2018.

258 C. H. Tseng, T. J. Tseng, and C. Z. Wu, "Cuffless blood pressure measurement using a microwave near-field self-injection-locked wrist pulse sensor," *IEEE Trans. Microw. Theory Techn.*, vol. 68, no. 11, pp. 4865–4874, Nov. 2020.

259 C. Tseng and C. Wu, "A novel microwave phased- and perturbation-injection-locked sensor with self-oscillating complementary split-ring resonator for finger and wrist pulse detection," *IEEE Trans. Microw. Theory Techn.*, vol. 68, no. 5, pp. 1933–1942, May 2020.

260 A. K. Jha, N. K. Tiwari, and M. J. Akhtar, "Accurate microwave cavity sensing technique for dielectric testing of arbitrary length samples," *IEEE Trans. Instrum. Measur.*, vol. 70, pp. 1–10, 2021.

261 N. K. Tiwari, A. K. Jha, S. P. Singh, Z. Akhter, P. K. Varshney, and M. J. Akhtar, "Generalized multimode SIW cavity-based sensor for retrieval of complex permittivity of materials," *IEEE Trans. Microw. Theory Techn.*, vol. 66, no. 6, pp. 3063–3072, Jun. 2018.

262 E. Massoni, G. Siciliano, M. Bozzi, and L. Perregrini, "Enhanced cavity sensor in SIW technology for material characterization," *IEEE Microw. Wireless Compon. Lett.*, vol. 28, no. 10, pp. 948–950, Oct. 2018.

263 E. L. Chuma, Y. Iano, G. Fontgalland, and L. L. Bravo Roger, "Microwave sensor for liquid dielectric characterization based on metamaterial complementary split ring resonator," *IEEE Sensors J.*, vol. 18, no. 24, pp. 9978–9983, Dec. 2018.

264 A. Javed, A. Arif, M. Zubair, M. Q. Mehmood, and K. Riaz, "A low-cost multiple complementary split-ring resonator-based microwave sensor for contactless dielectric characterization of liquids," *IEEE Sensors J.*, vol. 20, no. 19, pp. 11326–11334, Oct. 2020.

265 K. Grenier, D. Dubuc, P. Poleni, M. Kumemura, H. Toshiyoshi, T. Fujii, and H. Fujita, "New broadband and contact less RF/microfluidic sensor dedicated to bioengineering," *2009 IEEE MTT-S International Microwave Symposium Digest*, Boston, MA, USA, 2009, pp. 1329–1332.

266 T. Nacke, A. Barthel, B. P. Cahill, M. Meister, and Y. Zaikou, "High frequency fluidic and microfluidic sensors for contactless dielectric and in vitro cell culture measurement applications," *J. Phys. Conf. Ser.*, vol. 434, paper 012034, Heilbad Heiligenstadt, Germany, Apr. 2013.

267 M. H. Zarifi and M. Daneshmand, "Liquid sensing in aquatic environment using high quality planar microwave resonator," *Sensors Actuators B Chem.*, vol. 225, pp. 517–521, 2016.

268 Z. Wei, J. Huang, J. Li, G. Xu, Z. Ju, X. Liu, and X. Ni, "A high-sensitivity microfluidic sensor based on a substrate integrated waveguide re-entrant cavity for complex permittivity measurement of liquids," *Sensors*, vol. 18, no. 11, paper 4005, 2018.

269 R. Narang, S. Mohammadi, M. M. Ashani, H. Sadabadi, H. Hejazi, M. H. Zarifi, and A. Sanati-Nezhad, "Sensitive, real-time and non-intrusive detection of concentration and growth of pathogenic bacteria using microfluidic-microwave ring resonator biosensor," *Sci. Rep.*, vol. 8, paper 15807, 2018.

270 S. Mohammadi, A. V. Nadaraja, D. J. Roberts, and M. H. Zarifi, "Real-time and hazard-free water quality monitoring based on microwave planar resonator sensor," *Sensors Actuators A Phys.*, vol. 303, paper 111663, 2020.

271 M. C. Jain, A. V. Nadaraja, R. Narang, and M. H. Zarifi, "Rapid and real-time monitoring of bacterial growth against antibiotics in solid growth medium using a contactless planar microwave resonator sensor," *Sci. Rep.*, vol. 11, paper 14775, 2021.

272 J. Naqui and F. Martín, "Transmission lines loaded with bisymmetric resonators and their application to angular displacement and velocity sensors," *IEEE Trans. Microw. Theory Techn.*, vol. 61, no. 12, pp. 4700–4713, Dec. 2013.

273 A. K. Horestani, J. Naqui, D. Abbott, C. Fumeaux, and F. Martín, "Two-dimensional displacement and alignment sensor based on reflection coefficients of open microstrip lines loaded with split ring resonators," *Electron. Lett.*, vol. 50, no. 8, pp. 620–622, Apr. 2014.

274 C. Mandel, B. Kubina, M. Schüßler, and R. Jakoby, "Passive chipless wireless sensor for two-dimensional displacement measurement," *Proc. 41st European Microwave Conference*, Manchester, UK, Oct. 2011, pp. 79–82.

275 A. Ebrahimi, G. Beziuk, J. Scott, and K. Ghorbani, "Microwave differential frequency splitting sensor using magnetic-LC resonators," *Sensors*, vol. 20, p. 1066, 2020.

276 A. Ebrahimi, J. Scott, and K. Ghorbani, "Differential sensors using microstrip lines loaded with two split-ring resonators," *IEEE Sensors J.*, vol. 18, pp. 5786–5793, 2018.

277 L. Su, J. Mata-Contreras, J. Naqui, and F. Martín, "Splitter/combiner microstrip sections loaded with pairs of complementary split ring resonators (CSRRs): modeling and optimization for differential sensing applications," *IEEE Trans. Microw. Theory and Techn.*, vol. 64, no. 12, pp. 4362–4370, Dec. 2016.

278 J. Muñoz-Enano, J. Martel, P. Vélez, F. Medina, L. Su, and F. Martín, "Parametric analysis of the edge capacitance of uniform slots and application to frequency-variation permittivity sensors," *Appl. Sci.*, vol. 11, paper 7000, 2021.

279 F. Martín, *Artificial Transmission Lines for RF and Microwave Applications*, John Wiley, Hoboken, NJ, USA, 2015.

280 J. Coromina, J. Muñoz-Enano, P. Vélez, A. Ebrahimi, J. Scott, K. Ghorbani, and F. Martín, "Capacitively-loaded slow-wave transmission lines for sensitivity improvement in phase-variation permittivity sensors," *50th European Microwave Conference*, Utrecht, The Netherlands, Sep. 2020 (held as virtual conference in Jan. 2021).

281 A. Ebrahimi, J. Coromina, J. Muñoz-Enano, P. Vélez, J. Scott, K. Ghorbani, and F. Martín, "Highly sensitive phase-variation dielectric constant sensor based on a capacitively-loaded slow-wave transmission line," *IEEE Trans. Circ. Syst. I: Regular Papers*, vol. 68, no.7, pp. 2787–2799, July 2021.

282 C. Damm, M. Schüßler, M. Puentes, H. Maune, M. Maasch, and R. Jakoby, "Artificial transmission lines for high sensitive microwave sensors," *2009 IEEE Sensors*, Christchurch, New Zeland, 2009, pp. 755–758.

283 M. Gil, P. Vélez, F. Aznar, J. Muñoz-Enano, and F. Martín, "Differential sensor based on electro-inductive wave (EIW) transmission lines for dielectric constant measurements and defect detection," *IEEE Trans. Ant. Propag.*, vol. 68, pp. 1876–1886, Mar. 2020.

284 K. Saeed, R. D. Pollard, and I. C. Hunter, "Substrate integrated waveguide cavity resonators for the complex permittivity characterisation of material," *IEEE Trans. Microw. Theory Techn.*, vol. 56, pp. 2340–2347, Oct. 2008.

285 K. Saeed, M. F. Shafique, M. B. Byrne, and I. C. Hunter, "Planar microwave sensors for complex permittivity characterization of materials and their applications," *Appl. Meas. Sys.*, 2012, pp. 319–350.

286 C. Liu and F. Tong, "An SIW resonator sensor for liquid permittivity measurements at C band," *IEEE Microw. Wirel. Compon. Lett.*, vol. 25, no. 11, pp. 751–753, Nov. 2015.

287 E. Silavwe, N. Somjit, and I. D. Robertson, "A microfluidic-integrated SIW lab-on-substrate sensor for microliter liquid characterization," *IEEE Sensors J.*, vol. 16, no. 21, pp. 7628–7635, Nov. 2016.

288 C. Herrojo, P. Vélez, J. Muñoz-Enano, L. Su, P. Casacuberta, M. Gil, and F. Martín, "Highly sensitive defect detectors and comparators exploiting port imbalance in rat-race couplers loaded with step-impedance open-ended transmission lines," *IEEE Sensors J.*, vol. 21, no. 23, pp. 26731–26745, 2021.

289 R. S. Hassan, J. Lee, and S. Kim, "A minimally invasive implantable sensor for continuous wireless glucose monitoring based on a passive resonator," *IEEE Antennas Wirel. Propag. Lett.*, vol. 19, no. 1, pp. 124–128, Jan. 2020.

290 Z. Xiao, X. Tan, X. Chen, S. Chen, Z. Zhang, H. Zhang, J. Wang, Y. Huang, P. Zhang, L. Zheng, and H. Min, "An implantable RFID sensor tag toward continuous glucose monitoring," *IEEE J. Biomed. Health Inform.*, vol. 19, no. 3, pp. 910–919, May 2015.

291 A. Lazaro, M. Boada, R. Villarino, and D. Girbau, "Study on the reading of energy-harvested implanted NFC tags using mobile phones," *IEEE Access*, vol. 8, pp. 2200–2221, 2020.

292 J. Malik, S. K. Oruganti, D. Paul, and F. Bien, "Fully implantable miniature RF sensor for continuous glucose monitoring system," *2019 IEEE International Symposium on Radio-Frequency Integration Technology* (RFIT), Nanjing, China, Aug. 2019.

293 N. A. Malik, P. Sant, T. Ajmal, and M. Ur-Rehman, "Implantable antennas for bio-medical applications," *IEEE J. Electromagn., RF and Microw. Med. Biol.*, vol. 5, no. 1, pp. 84–96, Mar. 2021.

294 J. F. O'Hara, R. Singh, I. Brener, E. Smirnova, J. Han, A. J. Taylor, and W. Zhang, "Thin-film sensing with planar terahertz metamaterials: sensitivity and limitations," *Opt. Express*, vol. 16, pp. 1786–1795, 2008.

295 C. Sabah and H.G. Roskos, "Terahertz sensing application by using planar split-ring-resonator structures," *Microsyst. Technol.*, vol. 18, pp. 2071–2076, 2012.

296 X. Chen and W. Fan, "Ultrasensitive terahertz metamaterial sensor based on spoof surface plasmon," *Sci. Rep.*, vol. 7, paper 2092, 2017.

297 M. Beruete and I. Jáuregui-López, "Terahertz sensing based on metasurfaces," *Adv. Opt. Mater.*, vol. 8, paper 1900721, Feb. 2020.

298 M. Puentes, M. Maasch, M. Schubler, and R. Jakoby, "Frequency multiplexed 2-dimensional sensor array based on split-ring resonators for organic tissue analysis," *IEEE Trans. Microw. Theory Techn.*, vol. 60, no. 6, pp. 1720–1727, Jun. 2012.

299 C. S. Lee and C. L. Yang, "Complementary split-ring resonators for measuring dielectric constants and loss tangents," *IEEE Microw. Wireless Compon. Lett.*, vol. 24, no. 8, pp. 563–565, Aug. 2014.

300 M. H. Zarifi, S. Deif, M. Abdolrazzaghi, B. Chen, D. Ramsawak, M. Amyotte, N. Vahabisani, Z. Hashisho, W. Chen, and M. Daneshmand, "A microwave ring resonator sensor for early detection of breaches in pipeline coatings," *IEEE Trans. Ind. Electron.*, vol. 65, pp. 1626–1635, 2017.

301 L. Su, J. Mata-Contreras, P. Vélez, A. Fernández-Prieto, and F. Martín, "Analytical method to estimate the complex permittivity of oil samples," *Sensors*, vol. 18, no. 4, p. 984, Mar. 2018.

302 J. Muñoz-Enano, P. Vélez, M. Gil Barba, J. Mata-Contreras, and F. Martín, "Differential-mode to common-mode conversion detector based on rat-race hybrid couplers: analysis and application to differential sensors and comparators," *IEEE Trans. Microw. Theory Techn.*, vol. 68, no. 4, pp. 1312–1325, Apr. 2020.

303 J. Muñoz-Enano, P. Vélez, L. Su, M. Gil, and F. Martín, "A reflective-mode phase-variation displacement sensor," *IEEE Access*, vol. 8, pp. 189565–189575, Oct. 2020.

304 A. K. Jha, A. Lamecki, M. Mrozowski, and M. Bozzi, "A highly sensitive planar microwave sensor for detecting direction and angle of rotation," *IEEE Trans. Microw. Theory Techn.*, vol. 68, no. 4, pp. 1598–1609, Apr. 2020.

305 J. Muñoz-Enano, P. Vélez, L. Su, M. Gil, P. Casacuberta, and F. Martín, "On the sensitivity of reflective-mode phase variation sensors based on open-ended stepped-impedance transmission lines: theoretical analysis and experimental validation," *IEEE Trans. Microw. Theory Techn.*, vol. 69, no. 1, pp. 308–324, Jan. 2021.

306 J. Naqui and F. Martín, "Angular displacement and velocity sensors based on electric-LC (ELC) loaded microstrip lines," *IEEE Sensors J.*, vol. 14, no. 4, pp. 939–940, Apr. 2014.

307 J. Naqui, C. Damm, A. Wiens, R. Jakoby, L. Su, J. Mata-Contreras, and F. Martín, "Transmission lines loaded with pairs of stepped impedance resonators: modeling and application to differential permittivity measurements," *IEEE Trans. Microw. Theory Techn.*, vol. 64, no. 11, pp. 3864–3877, Nov. 2016.

308 W. Liu, J. Zhang and K. Huang, "Dual-band microwave sensor based on planar rectangular cavity loaded with pairs of improved resonator for differential sensing applications," *IEEE Trans. Instrum. Meas.*, vol. 70, pp. 1–8, 2021.

309 P. Vélez, K. Grenier, J. Mata-Contreras, D. Dubuc, and F. Martín, "Highly-sensitive microwave sensors based on open complementary split ring resonators (OCSRRs) for dielectric characterization and solute concentration measurements in liquids," *IEEE Access*, vol. 6, pp. 48324–48338, Dec. 2018.

310 P. Vélez, J. Muñoz-Enano and F. Martín, "Electrolyte Concentration Measurements in DI Water with 0.125 g/L Resolution by means of CSRR-Based Structures," *49th European Microwave Conference (EuMC)*, Paris, France, Oct. 2019, pp. 340–343.

311 J. Muñoz-Enano, P. Vélez, M. Gil, and F. Martín, "An analytical method to implement high sensitivity transmission line differential sensors for dielectric constant measurements," *IEEE Sensors J.*, vol. 20, pp. 178–184, Jan. 2020.

312 D. Girbau, Á. Ramos, A. Lazaro, S. Rima, and R. Villarino, "Passive wireless temperature sensor based on time-coded UWB chipless RFID tags," *IEEE Trans. Microw. Theory Techn.*, vol. 60, no. 11, pp. 3623–3632, Nov. 2012.

313 E. M. Amin, M. S. Bhuiyan, N. C. Karmakar, and B. Winther-Jensen, "Development of a low cost printable chipless RFID humidity sensor," *IEEE Sensors J.*, vol. 14, no. 1, pp. 140–149, Jan. 2014.

314 A. Lázaro, R. Villarino, F. Costa, S. Genovesi, A. Gentile, L. Buoncristiani, and D. Girbau, "Chipless dielectric constant sensor for structural health testing," *IEEE Sensors J.*, vol. 18, no. 13, pp. 5576–5585, Jul. 2018.

315 W. M. Abdulkawi and A. F. A. Sheta, "Chipless RFID sensors based on multistate coupled line resonators," *Sens. Act. A: Phys.*, vol. 309, paper 112025, Jul. 2020.

316 J. Yeo, J. I. Lee, and Y. Kwon, "Humidity-sensing chipless RFID tag with enhanced sensitivity using an interdigital capacitor structure," *Sensors*, vol. 21, paper 6550, 2021.

317 J. Naqui, M. Durán-Sindreu, and F. Martín, "Novel sensors based on the symmetry properties of split ring resonators (SRRs)," *Sensors*, vol. 11, pp. 7545–7553, 2011.

318 J. Naqui and F. Martín, "Microwave sensors based on symmetry properties of resonator-loaded transmission lines: a review," *J. Sens.*, vol. 2015, Article ID 741853, 10 pages, 2015.

319 J. Naqui, *Symmetry Properties in Transmission Lines Loaded with Electrically Small Resonators: Circuit Modeling and Applications*, Springer, 2016.

320 P. Vélez, J. Muñoz-Enano, A. Ebrahimi, C. Herrojo, F. Paredes, J. Scott, K. Ghorbani, and F. Martín, "Single-frequency amplitude-modulation sensor for dielectric characterization of solids and microfluidics," *IEEE Sensors J.*, vol. 21, no. 10, pp. 12189–12201, May 2021.

321 A. Ebrahimi, F. J. Tovar-López, J. Scott, and K. Ghorbani, "Differential microwave sensor for characterization of glycerol-water solutions," *Sensors Actuators B Chem.*, vol. 321, paper 128561, Oct. 2020.

322 J. Muñoz-Enano, P. Vélez, M. Gil, and F. Martín, "Microfluidic reflective-mode differential sensor based on open split ring resonators (OSRRs)," *Int. J. Microw. Wireless Technol.*, vol. 12, pp. 588–597, Sep. 2020.

323 M. T. Khan and S. M. Ali, "A brief review of measuring techniques for characterization of dielectric materials," *Int. J. Inf. Technol. Electr. Eng.*, vol. 1, no. 1, 2012.

324 P. Mehrotra, B. Chatterjee, and S. Sen, "EM-wave biosensors: a review of RF, microwave, mm-wave and optical sensing," *Sensors*, vol. 19, no. 5, paper 1013, 2019.

325 A. Menikh, "Terahertz-biosensing technology: progress, limitations, and future outlook," in M. Zourob and A. Lakhtakia (eds) *Optical Guided-Wave Chemical and Biosensors II*, Springer Series on Chemical Sensors and Biosensors (Methods and Applications), vol 8, Springer, Berlin, Heidelberg, 2010.

326 S. Park, J. Hong, S. Choi, H. Kim, W. Park, S. Han, J. Park, S. Lee, D. Kim, and Y. H. Ahn, "Detection of microorganisms using terahertz metamaterials," *Sci. Rep.*, vol. 4, paper 4988, 2014.

327 W. Xu, L. Xie, J. Zhu, L. Tang, R. Singh, C. Wang, Y. Ma, H. T. Chen, and Y. Ying, "Terahertz biosensing with a graphene-metamaterial heterostructure platform," *Carbon*, vol. 141, pp. 247–252, 2019.

328 H. Zhou, C. Yang, D. Hu, D. Li, X. Hui, F. Zhang, M. Chen, and X. Mu, "Terahertz biosensing based on bi-layer metamaterial absorbers toward ultra-high sensitivity and simple fabrication," *Appl. Phys. Lett.*, vol. 115, paper 143507, 2019.

329 H. Ou, F. Lu, Z. Xu, and Y. S. Lin, "Terahertz metamaterial with multiple resonances for biosensing application," *Nanomaterials*, vol. 10, no. 6, paper 1038, 2020.

330 H. W. Wu, "Label-free and antibody-free wideband microwave biosensor for identifying the cancer cells," *IEEE Trans. Microw. Theory Techn.*, vol. 64, pp. 982–990, 2016.

331 A. Salim, and S. Lim, "Recent advances in the metamaterial-inspired biosensors," *Biosens. Bioelectron.*, vol. 117, pp. 398–402, 2018.

332 A. Salim, S.H. Kim, J. Y. Park, and S. Lim, "Microfluidic biosensor based on microwave substrate-integrated waveguide cavity resonator," *J. Sens.*, vol. 2018, pp. 1–13, 2018.

333 P. Vélez, J. Muñoz-Enano, K. Grenier, J. Mata-Contreras, D. Dubuc, and F. Martín, "Split ring resonator (SRR) based microwave fluidic sensor for electrolyte concentration measurements," *IEEE Sensors J.*, vol. 19, no. 7, pp. 2562–2569, April 2019.

334 J. Muñoz-Enano, P. Vélez, M. Gil, E. Jose-Cunilleras, A. Bassols, and F. Martín, "Characterization of electrolyte content in urine samples through a differential microfluidic sensor based on dumbbell-shaped defect ground structures," *Int. J. Microw. Wirel. Technol.*, vol. 12, no. 9, pp. 817–824, 2020.

335 M. C. Jain, A. V. Nadaraja, B. M. Vizcaino, D. J. Roberts and M. H. Zarifi, "Differential microwave resonator sensor reveals glucose-dependent growth profile of E. coli on solid agar," *IEEE Microw. Wireless Compon. Lett.*, vol. 30, no. 5, pp. 531–534, May 2020.

336 A. A. Bahar, Z. Zakaria, A. A. Isa, E. Ruslan, and R. A. Alahnomi, "A review of characterization techniques for material's properties measurement using microwave resonant sensor," *J. Telecommun., Electron. Comput. Eng.*, vol. 7, no. 2, pp. 1–6, Dec. 2015.

337 N. A. Rahman, Z. Zakaria, R. A. Rahim, Y. Dasril, A. Azuan, and M. Bahar, "Planar microwave sensors for accurate measurement of material characterization: a review," *TELKOMNIKA*, vol.15, pp. 1108–1118, Sep. 2017.

338 J. Muñoz-Enano, P. Vélez, M. Gil, and F. Martín, "Planar microwave resonant sensors: a review and recent developments," *Appl. Sci.*, vol. 10, p. 2615 (29 pages), 2020.

339 M. G. Mayani, F. J. Herraiz-Martínez, J. M. Domingo, and R. Giannetti, "Resonator-based microwave metamaterial sensors for instrumentation: survey, classification, and performance comparison," *IEEE Trans. Instrum. Meas.*, vol. 70, pp. 1–14, Art no. 9503414, 2021.

340 C. G. Juan, B. Potelon, C. Quendo, and E. Bronchalo, "Microwave planar resonant solutions for glucose concentration sensing: a systematic review," *Appl. Sci.*, vol. 11, no. 15, paper 7018, 2021.

341 R. A. Alahnomi, Z. Zakaria, Z. M. Yussof, A. A. Althuwayb, A. Alhegazi, H. Alsariera, and N. A. Rahman, "Review of recent microwave planar resonator-based sensors: techniques of complex permittivity extraction, applications, open challenges and future research directions," *Sensors*, vol. 21, no. 7, paper 2267, 2021.

2

Frequency-Variation Sensors

Frequency variation is probably the most extended working principle in planar microwave sensors, at least in applications devoted to dielectric characterization and material composition determination. Frequency-variation sensors can be implemented by means of planar resonators (the sensitive element), with resonance frequency and quality factor determined by the dielectric properties of the medium surrounding the resonator. For this main reason, such sensors are suitable for the measurement of variables related to permittivity, such as the dielectric constant and loss tangent of materials (solids or liquids), or the concentration of solute in liquid solutions, among others. In these sensors, resonator's excitation is mostly achieved by means of a planar transmission line. Frequency-variation sensors have also been applied to the measurement of spatial variables, i.e. linear or angular displacements. In such sensors, the movable part, usually made of a dielectric material, contains the resonant element, whereas the static part consists of a transmission line. This chapter reports several implementations of frequency-variation sensors devoted to different applications, including dielectric characterization of solids and liquids, material composition measurements, analysis of biosamples, and measurement of spatial variables. However, prior to the review of all these implementations and applications, the working principle of frequency-variation sensors is presented in detail. Moreover, various planar resonant elements useful for sensing are presented and discussed, and a sensitivity analysis is carried out. The chapter includes also a method to selectively determining the composition of samples, by exploiting the different dependence of the permittivity on frequency of the different materials forming the composite. In these latter sensors, operation at different frequency bands is required, as it will be shown.

2.1 General Working Principle of Frequency-Variation Sensors

The working principle of frequency-variation sensors is the frequency shift and notch/peak magnitude variation in the resonance of a resonant (sensing) element, generated by the variable to be measured. In microwave resonators (either implemented in planar form, or as cavity, waveguide or dielectric resonators [1]), the fundamental resonance frequency and peak/notch magnitude depend on the dielectric properties of the material surrounding (or constituting) the resonator or in proximity to it. Consequently, frequency-variation resonant sensors are typically applied to the dielectric characterization of materials, i.e. the determination of their permittivity, or to the measurement of variables related to it, e.g. material composition. The aim of this chapter is the study of frequency-variation sensors implemented in planar technology, in consonance with the orientation of the

Planar Microwave Sensors, First Edition. Ferran Martín, Paris Vélez, Jonathan Muñoz-Enano, and Lijuan Su.
© 2023 John Wiley & Sons, Inc. Published 2023 by John Wiley & Sons, Inc.

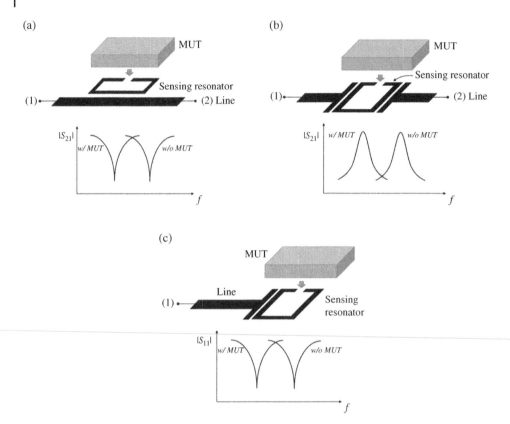

Figure 2.1 Sketch showing the working principle of frequency-variation sensors corresponding to different configurations. (a) Transmission mode and stopband configuration; (b) transmission mode and passband configuration; (c) reflective mode.

book. Although cavity and waveguide resonators for the measurement of the dielectric constant and loss tangent of samples are well known (see, e.g., [2–5] and Chapter 1), and are commercially available [6], the interest in planar resonant sensors is justified for their low cost and low profile. Moreover, planar sensors can be fabricated on flexible substrates, opening the possibility to the implementation of conformal sensors.

In planar resonant sensors, resonator's excitation is generally achieved by means of a transmission line, typically a microstrip or a coplanar waveguide (CPW) transmission line, operating either in transmission or in reflection. Figure 2.1 depicts a sketch of the typical configuration of a frequency-variation transmission-mode and reflective-mode sensor. Actually, for operation in transmission mode, the resonance frequency can be visualized as a notch (stopband configuration) or as a peak (passband configuration). In reflection mode, a notch in the reflection coefficient typically arises as consequence of losses, including radiation losses (if they are present), ohmic and dielectric losses. The frequency position of this notch or peak depends on the dielectric constant of the material surrounding the resonator (material under test, MUT). Therefore, it is obvious that with these configurations, sensing the real part of the permittivity is possible. Nevertheless, the magnitude of the notch, peak, or the quality factor depends mainly on the loss factor of the MUT. Thus, it follows that the imaginary part of the permittivity, or the loss tangent, of the MUT can be inferred from the measurement of the notch or peak magnitude, or from the quality factor.

Depending on the specific configuration, the resonance frequency may depend significantly on the relative position and orientation between the transmission line and the resonant element. Under these conditions, frequency-variation sensors can be applied to the measurement of spatial variables. For that purpose, the resonant element must be etched in an independent substrate, in relative motion to the substrate of the transmission line. Nevertheless, it should be mentioned that these spatial sensors are useful for the measurement of low-range displacements. For linear displacements, the relative position between the line and the resonator should be restricted to distances ensuring line-to-resonator coupling. For angular displacements, varying the relative orientation of the resonator with regard to the line axis may generate a variation in the resonance frequency, but the dependence of the frequency shift with the angular displacement is, in general, soft.

2.2 Transmission-Line Resonant Sensors

In this section, frequency-variation sensors based on resonator-loaded transmission lines are reviewed. The choice of the planar resonant element is important, since it determines the size of the sensor, or, more specifically, the size of the sensing region, as well as the sensitivity, and other important aspects. Thus, the next section is devoted to the analysis of some planar resonators useful for sensing. Then a sensitivity analysis is carried out, and, finally, several implementations are reported.

2.2.1 Planar Resonant Elements for Sensing

There are many types of planar resonators useful for sensing. Planar resonators can be classified according to various criteria. Among them, the electrical size can be used to distinguish between distributed and semi-lumped resonators. The canonical distributed resonators are the quarter-wavelength and the half-wavelength transmission line resonators. These resonators are extensively used for the implementation of microwave circuits and filters [1, 7], their design is very simple, and resonator detuning caused, e.g. by fabrication-related tolerances, is limited. However, these resonators scale with frequency, and their size may be unacceptably large in certain sensing applications involving moderate and low frequencies (within the microwave range) and requiring small sensing regions. Thus, in general, for microwave sensing based on frequency variation, planar semi-lumped resonators are preferred [8, 9]. Semi-lumped resonators are electrically small planar resonators. Their size, e.g. the diameter in case of circularly shaped resonators or the longer side length in case of rectangular-shaped resonators, is significantly smaller than the guided wavelength at their fundamental resonance frequency. Both distributed and semi-lumped resonators can be described by an LC resonant tank in the vicinity of resonance. However, in semi-lumped resonators, the validity of the lumped element model extends significantly beyond the fundamental resonance frequency, as compared with their distributed counterparts.

There are many types of planar semi-lumped resonators, but such resonators can be categorized as either metallic resonators (Fig. 2.2) or slot (or defect ground structure – DGS) resonators (Fig. 2.3). The main advantage of DGS resonators over metallic resonators concerns the fact that the former are etched, in general, in the ground plane of the transmission lines, usually microstrip lines. Consequently, planar sensors based on DGS resonators are less sensitive to the effects of the MUT, or additional required mechanical parts, on the transmission line.[1] The reason is that the

---

1 For example, in microfluidic sensors, devoted to the characterization of liquids, fluidic channels plus mechanical accessories are required. By placing them on the ground plane of the considered transmission line, on top of the DGS sensing resonator, such mechanical parts do not alter the characteristics of the line.

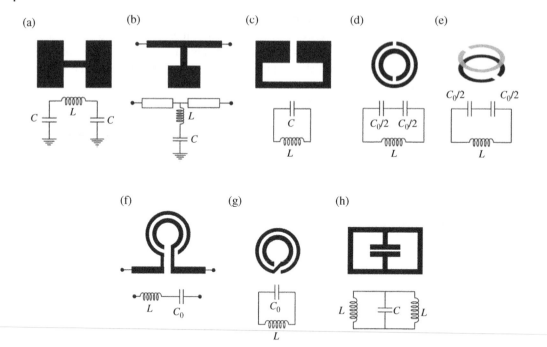

Figure 2.2 Typical topologies and the corresponding circuit models (with losses excluded) of some representative electrically small planar metallic resonators useful for sensing. (a) Step-impedance resonator (SIR); (b) step-impedance shunt stub (SISS) loading a microstrip line; (c) folded SIR; (d) split-ring resonator (SRR); (e) broadside-coupled split-ring resonator (BC-SRR); (f) open split-ring resonator (OSRR); (g) two-turn spiral resonator (2-SR); (h) electric-LC (ELC) resonator.

ground plane provides backside isolation, thereby preventing from any influence of the elements located in the back substrate side on the electrical characteristics of the line.

2.2.1.1 Semi-Lumped Metallic Resonators

Among the metallic resonators, the step-impedance resonator (SIR) [10], see Fig. 2.2(a), and the shunt-connected SIR, usually designated as step-impedance shunt stub (SISS) [11], see Fig. 2.2 (b), exhibit a small electrical size by virtue of the impedance contrast between the narrow and wide strips. That is, the larger the impedance contrast, the smaller the electrical size. In the limit of small impedance contrasts, the SIR resonator resembles a half-wavelength resonator. Both the SIR and the half-wavelength resonator exhibit an electric wall at their symmetry plane for the fundamental resonance, and both resonators can be excited by means of a time-varying electric field orthogonally oriented to the electric wall. For planar sensing, a potential configuration of SIR-based (or half-wavelength resonator based) sensors consists of a gap-coupled structure, e.g. by means of a microstrip line, providing a bandpass response. Transversally etching these resonant elements in the back substrate side of a slot line constitutes an alternative for sensing, but, in this case, the response exhibits a notch. Analogously, SISS resonators exhibit a behavior similar to the one of quarter-wavelength open-ended shunt stub resonators, but with an electrically smaller size by virtue of the impedance contrast. If the impedance contrast is high, the inductive and capacitive part of the SISS can be physically separated, as discussed in [11]. The narrow strip provides the inductive behavior, whereas the wide strip accounts for the capacitance. This differentiation between the

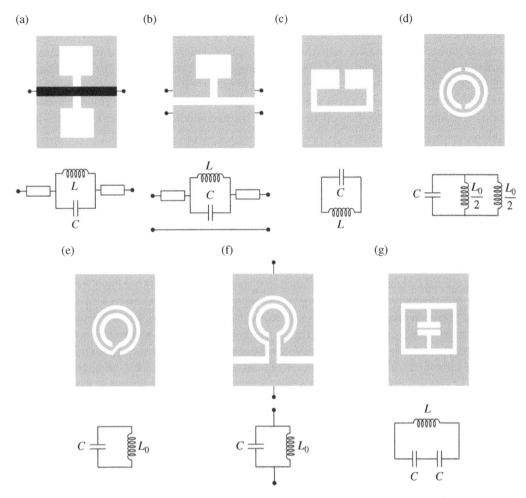

Figure 2.3 Typical topologies and the corresponding circuit models (with losses excluded) of some representative electrically small planar defect ground structure (DGS), or slotted, resonators useful for sensing. (a) Dumbbell-shaped defect ground structure (DB-DGS) resonator; (b) step-impedance slotted stub loading a slot line; (c) folded DB-DGS resonator; (d) complementary split ring resonator (CSRR); (e) two-turn complementary spiral resonator (2-CSR); (f) open complementary split-ring resonator (OCSRR); (g) magnetic-LC (MLC) resonator.

capacitance and the inductance applies also to SIR resonators with high-impedance contrast. The circuit models of these resonant elements are also depicted in Fig. 2.2(a) and (b). In the circuit model of Fig. 2.2(a), it has been assumed that the SIR is on top of a ground plane. In the circuit model of Fig. 2.2(b), the access lines to the SISS resonator are included.

Another interesting planar resonator for sensing purposes is the so-called folded SIR (Fig. 2.2c) [12]. The canonical topology is identical to the one of the SIR resonator, but the capacitive patches are oriented face to face. The folded SIR exhibits also an electric wall in the symmetry plane at the fundamental resonance, and therefore this resonant particle can be driven by means of an electric field perpendicular to the symmetry plane. However, it can also be driven by means of a time-varying magnetic field applied orthogonally to the plane of the resonator. Thus, if a folded SIR is etched in proximity of a microstrip line, where a significant component of the magnetic field orthogonal to the particle (and line) plane is expected, it can be magnetically driven, thereby generating a notch at the fundamental resonance. The folded SIR can also be combined with slot lines or with CPW transmission lines for sensing purposes. In the former case, the resonator should be etched in the back substrate side, with the symmetry plane preferably aligned with the symmetry plane of the line, where the electric field lines are orthogonal. In combination with CPW transmission lines, a pair of folded SIRs are convenient in order to preserve the symmetry of the line (thereby preventing from the appearance of the parasitic slot mode). By symmetrically etching the resonators below the slots of the CPW, they can be excited by the magnetic field generated by the line, and, depending on the orientation, also by the electric field (in this later case, mixed coupling arises [13]). In the folded SIR, the sensitive part of the particle to dielectric loads is the slot region between the patches. This region exhibits an edge-coupled capacitance, with a significant portion of the electric field lines above the slot. Thus, it is expected that the sensitivity of the folded SIR can be improved as compared with the one of SIR or SISS resonators. The circuit model of the folded SIR is simply a closed resonator, where the dimensions of the narrow strip determine the inductance, whereas the dimensions of the patches and their separation provide the capacitance.

Another set of electrically small planar metallic resonators is constituted by those resonant elements originally devoted to the implementation of metamaterials [14, 15]. Among them, the most genuine representative resonator is the so-called split-ring resonator (SRR), see Fig. 2.2(d), formerly used for the implementation of artificial media exhibiting negative permeability [14, 16–20]. Similar to the folded SIR, such resonant particle can be excited by means of an axial time-varying magnetic field or by means of an electric field applied in the plane of the particle in the direction orthogonal to the symmetry plane, an electric wall at the fundamental resonance. Thus, for sensing purposes, the SRR can be combined with microstrip lines, slot lines, or CPW transmission lines, among other lines, and the relative orientation/position between the lines and the resonator should follow the guidelines indicated in the precedent paragraph in reference to folded SIRs. It is important to mention that the electrical size of the SRRs depends on the distance between the concentric rings. Reducing such distance enhances the electric coupling between both rings, thereby driving the fundamental resonance toward lower frequencies (and hence decreasing the electrical size). In practice, the distance between the rings, or inter-rings space, has an inferior limit, dictated by the minimum slot dimensions that can be resolved by the available technology. For most available technologies, the inter-rings coupling associated with such minimum slot width is limited, and the electrical size of the SRRs is not significantly smaller than the one corresponding to the larger loop (a circular distributed resonator). In [21, 22], the broadside-coupled SRR (BC-SRR) was proposed as a means to enhance the electric coupling between the pair of (identical) rings. In this particle, see Fig. 2.2(e), the rings are etched at both sides of the substrate. The behavior is very similar to the one of the SRR,[2] but if the substrate is very narrow, the capacitance between the rings can be made very large by virtue of the broadside coupling. Thus, very electrically small resonators can be implemented with this strategy. However, for sensing purposes, the broadside capacitance limits the sensitivity of the resonance frequency with regard to changes in the dielectric load, since the field lines are mainly concentrated in the substrate region. Nevertheless, with advanced technologies, such as micromachining, sandwiching the MUT between the rings in a BC-SRR-based sensor is possible. In this case, the sensitivity can be significant, but at the expense of higher fabrication costs and complexity. Both the SRR and the BC-SRR are described by the (identical) closed

---

2 Actually, the BC-SRR presents a fundamental difference as compared with the SRR. Whereas the SRR can be driven by means of a uniform electric field applied parallel to the plane of the particle, in the direction orthogonal to the symmetry plane, the BC-SRR cannot be excited by such an electric field, since it exhibits inversion symmetry. The BC-SRR is indeed a non-bianisotropic resonator, with a behavior roughly identical to the one of the so-called non-bianisotropic split-ring resonator (NB-SRR) [23, 24], which can only be excited by means of a time-varying magnetic field with a non-negligible component in the axial direction.

resonant tanks depicted in Fig. 2.2(d) and (e). Note that, in these models, as described in detail in [14, 15, 25], the inductance is the one corresponding to a single ring with average radius and width identical to the one of the rings forming the SRR, or BC-SRR. The capacitance is expressed as the series connection of the edge capacitances of the two halves of the particle (separated by the symmetry plane). Thus, the total capacitance of the SRR (or BC-SRR) is four times smaller than the edge capacitance of the whole circumference, C_0.

Figure 2.2(f) depicts the so-called open SRR (OSRR) [26]. This particle was inspired by the SRR, but the OSRR is an open resonator that can be driven by means of a current or a voltage source. It consists of a pair of concentric hooks, as shown in Fig. 2.2(f). As it is demonstrated in [14, 15], the inductance of the OSRR is identical to the one of the SRR (provided ring dimensions and inter-rings space are identical). However, the capacitance of the OSRR is four times larger than the capacitance of the SRR (considering that the particles are also implemented on identical substrates). The reason is that the displacement current between both concentric rings of the OSRR is distributed throughout the whole circumference, i.e. the capacitance of the OSRR is C_0. This means that the OSRR is electrically smaller than the SRR by a factor of two. Because an open-series LC resonator can describe the OSRR, such resonant element can be applied to the implementation of transmission-mode frequency-variation sensors operating in passband configuration (indeed, it has been demonstrated that OSRRs are of interest for the implementation of compact bandpass filters [27]).

If the connecting terminals of the OSRR are short-circuited, the resulting particle is the so-called two-turn spiral resonator (2-SR), see Fig. 2.2(g), a closed resonator that can be driven by a time-varying magnetic field [28, 29]. Obviously, the inductance and capacitance of the 2-SR are identical to those of the OSRR. Therefore, the 2-SR is smaller than the SRR by a factor of two as well, and, consequently, this particle can be of interest for the implementation of transmission-mode sensors based on closed resonators (coupled with transmission lines and exhibiting notched responses) with very small sensing regions.

Let us finally consider the so-called electric-LC (ELC) resonator [30], see Fig. 2.2(h). The ELC is a bisymmetric resonator consisting of a pair of contacting capacitively loaded metallic loops. Note that the capacitance in Fig. 2.2(h) can be replaced with a patch capacitance. The resulting particle is also an ELC, but, in this case, based on a pair of oppositely oriented and contacting folded SIRs. This particle is, obviously, electrically larger than the folded SIR. However, the fact that the ELC exhibits two orthogonal symmetry planes is interesting for some applications. At the fundamental resonance, the vertical symmetry plane is a magnetic wall, whereas the horizontal symmetry plane is an electric wall. At that frequency, the electric currents in both loops flow in opposite directions, one clockwise and the other one counter-clockwise, and the resonator can be driven either by an electric field applied in the plane of the particle, in the direction orthogonal to the electric wall, or by counter magnetic fields applied at both sides of the magnetic wall. The ELC resonator can be applied to the implementation of frequency-variation sensors, e.g. by loading a CPW transmission line with such particle, with the magnetic wall of the ELC aligned with the axial symmetry plane of the line (also a magnetic wall). By this means, with a single resonant element, the ELC-loaded line exhibits axial symmetry, thereby preventing from the appearance of the parasitic slot mode. Nevertheless, the ELC resonator is not the preferred option for the implementation of frequency-variation sensors due to the relatively large electrical size of this particle. However, the ELC resonator is interesting for the implementation of sensors based on symmetry properties, as will be later shown in this book (see Chapter 4, devoted to coupling-modulation sensors). A planar resonant element exhibiting similar properties to the ELC resonator but with substantially smaller electrical size is the S-shaped SRR (S-SRR) [31–33] (like the ELC resonator, the S-SSR is of interest for the implementation of microwave sensors based on coupling modulation).

2.2.1.2 Semi-Lumped Slotted Resonators

The previous metallic resonators have their complementary, or dual, counterparts (the exception is the BC-SRR, with rings not being coplanar). Figure 2.3 depicts the corresponding topologies, where it can be appreciated that the metals and apertures (slots) are interchanged with regard to the metallic particles. That is, the complementary resonators are the negative images of the metallic resonators, and their electromagnetic behavior is roughly dual [34–37].[3] This means that if the field $\vec{F} = \left(\vec{E}, \vec{H}\right)$ is a solution for the original structure, its dual field, $\vec{F}'$, defined by

$$\vec{F}' = \left(\vec{E}', \vec{H}'\right) = \left(-\sqrt{\frac{\mu}{\varepsilon}} \cdot \vec{H}, \sqrt{\frac{\varepsilon}{\mu}} \cdot \vec{E}\right) \tag{2.1}$$

is the solution for the complementary structure to a good approximation. The general consequence of duality is that driving the complementary resonators can be achieved following the same procedure as in the metallic resonators, but with the roles of the electric and magnetic fields interchanged. Nevertheless, this aspect requires further clarification, at least in some cases, to be discussed next.

Let us start with the complementary counterpart of the SIR resonator. This particle (Fig. 2.3a), designated as dumbbell-shaped defect ground structure (DB-DGS) resonator [38, 39], has been applied to the design of circuits and filters [40, 41], where it is typically etched transversally in the ground plane of a microstrip line. The DB-DGS resonator loading a microstrip line is also of interest for the implementation of frequency-variation permittivity sensors, due to the highly achievable sensitivity (an aspect that will be discussed later). In the limit where the width of the apertures at the extremes coincides with the width of the central slot, the resulting structure is the slot resonator, which is indeed the dual counterpart of the half-wavelength resonator [42]. The negative image of the SISS resonator of Fig. 2.2(b) is depicted in Fig. 2.3(b). Actually, Fig. 2.2 (b) depicts a microstrip line loaded with a SISS, since a stub needs a host line for connection. Similarly, the structure of Fig. 2.3(b) is a stub-loaded slot line, where the stub is a slot made of two sections of unequal width. In [43], a CPW symmetrically loaded with a pair of such slotted stubs is reported. This complementary structure is rarely used for sensing, since the sensing element, the stub, is coplanar to the line (either a slot line or a CPW). For sensing purposes using DGS resonators, microstrip lines, with the resonant elements etched in the backside of the line (the ground plane), are preferred. The circuit models of these resonators, also included in Fig. 2.3(a) and (b), are parallel resonant tanks.

Figure 2.3(c) depicts the complementary version of the folded SIR (which can be canonically designated as folded DB-DGS resonator). From duality considerations, it follows that this resonator exhibits a magnetic wall in the symmetry plane at the fundamental resonance. It can be driven either by a time-varying electric field applied in the direction orthogonal to the plane of the particle or by a magnetic field applied parallel to the plane of the particle in the direction orthogonal to the magnetic wall. The folded DB-DGS resonator is very appropriate for ground plane etching in a microstrip line, beneath the strip, where there is a significant component of the electric field in the vertical direction able to excite the particle. The circuit model of the particle is a closed resonant tank, where the metallic path between the inner and the outer metallic regions of the particle determines the inductance, whereas the capacitance is given by the width and length of the narrow (folded) slot.

---

3 The finite thickness and conductivity of metal layers, and the presence of a dielectric substrate with finite thickness, are the causes of departure from perfect duality.

The metallic resonators inspired by metamaterials have also their complementary counterparts. Thus, the complementary SRR (CSRR), see Fig. 2.3(d), was first presented in [44] for the implementation of transmission line metamaterials exhibiting negative effective permittivity. The CSRR exhibits a similar behavior to the folded DB-DGS resonator, with a magnetic wall in the symmetry plane at the fundamental resonance. Thus, CSRR excitation by means of external fields proceeds exactly according to the same mechanisms indicated in the previous paragraph. It should be mentioned that there are many reported implementations of microwave sensors based on CSRR-loaded microstrip lines. Such sensors are mainly devoted to permittivity sensing and material characterization based on the working principle subject of this chapter, i.e. frequency variation. Nevertheless, a smaller electrical size is achieved by replacing the CSRR with a two-turn complementary spiral resonator (2-CSR), see Fig. 2.3(e) [45], the dual counterpart of the 2-SR. The open complementary SRR (OCSRR), see Fig. 2.3(f) [46], is the dual particle of the OSRR. The OCSRR is also an open resonator that can be modeled by means of a parallel LC resonant tank between the two terminals. The capacitance is identical to the one of the CSRR, whereas the inductance is four times larger. Indeed, the electrical size of the OCSRR is identical to the electrical size of the 2-CSR. This can be concluded from duality arguments, as far as the CSRR and the 2-CSR are the complementary counterparts of the SRR and 2-SR, respectively. If the electrical size of the SRR is twice the one of the 2-SR, or OSRR, it follows that such relation between the electrical sizes should apply to the dual particles. However, whereas the inductance of the SRR is identical to the one of the 2-SR, or OSRR, the capacitance being four times smaller, for the complementary resonators, the reactive elements of the model interchange the roles, i.e. the capacitances are identical, whereas the inductance is four times smaller in the CSRR. This is also a consequence of duality, namely if the structures are dual, their lumped element models are also circuit duals [36]. In the case of the dual particles, if the inductance of the whole circumference is designated by L_0, it follows that the total inductance of the CSRR is the parallel connection of the two inductances corresponding to the two halves (each of value $L_0/2$), and hence the total inductance of the CSRR is $L_0/4$. By contrast, for the OCSRR, or 2-CSR, the total inductance is L_0, since the metallic path between the inner and the outer metallic regions of these particles is the one corresponding to the whole circumference.

Let us finally consider the magnetic-LC (MLC) resonator (Fig. 2.3g) [47, 48], the dual counterpart of the ELC resonator. In the MLC resonator, the orthogonal electric and magnetic walls are interchanged with regard to those in the ELC resonator, as expected from duality. The MLC resonator can thus be driven by counter electric fields applied at both sides of the electric wall, in the direction perpendicular to the plane of the particle, or by means of a uniform magnetic field parallel to the plane of the particle and orthogonal to the magnetic wall. The MLC resonator is able to generate a notch at the fundamental resonance in differential microstrip transmission lines, provided the resonator is conveniently oriented, i.e. with its electric wall aligned with the axial symmetry plane of the differential line, also an electric wall for the differential mode. Nevertheless, the MLC resonator does not seem to offer competitive advantages over other DGS resonators for microwave sensing exploiting frequency variation. Similar to ELC resonators, the relative orientation between a host transmission line, not necessarily a differential line, and the MLC resonator determines the coupling level. Therefore, these resonators are useful for the implementation of displacement sensors based on symmetry disruption. Note that the MLC is described by a closed resonant circuit, with two series-connected capacitors, series connected to an inductor. The inductor describes the metal path between the metallic regions within the slotted loops, whereas the capacitances are the edge capacitances of both loops.

Although additional metallic and DGS resonators useful for sensing based on frequency variation can be envisaged, the resonant elements discussed before, depicted in Figs. 2.2 and 2.3, constitute a representative set of electrically small sensing particles. The specific lumped element circuit models of such resonators when they load a certain transmission line will be discussed when needed, not only in this chapter but also in the subsequent chapters of this book. Such circuits are useful to gain insight on sensor performance, particularly the sensitivity, which is the subject of the next section.

2.2.2 Sensitivity Analysis

In frequency-variation sensors, the canonical output variable is the resonance frequency of the sensing resonator. Other output variables, such as the quality factor, or the notch/peak magnitude, can also be used for sensing to the extent that it is necessary to measure more than one variable. Indeed, most frequency-variation sensors implemented by means of resonator-loaded lines (operating either in transmission or in reflection) are devoted to permittivity measurements and to the determination of related variables. In particular, for the measurement of the complex permittivity of materials (or the dielectric constant and loss tangent), with two different parameters involved, both the resonance frequency and the notch/peak magnitude, or quality factor, have been considered for sensing. Nevertheless, in these sensors, the key sensitivity is the one corresponding to the derivative of the resonance frequency, f_0, with the dielectric constant of the MUT, ε_{MUT}. Obviously, the dielectric constant of the MUT mainly influences the resonance frequency. The loss tangent, intimately related to material losses, has a major influence on the quality factor and notch/peak magnitude, and the sensitivity of any of these output variables with the loss factor of the MUT is also a sensor performance parameter. Moreover, depending on the specific sensing structure and considered MUTs, the cross sensitivities cannot be neglected. Nevertheless, being designated the sensors under study as frequency-variation sensors, it follows that the canonical sensitivity is the one defined above, i.e. $S = df_0/d\varepsilon_{MUT}$. It is obvious that this sensitivity depends on the resonance frequency of the resonant element loaded with the reference material, or surrounded by air. That is, the variation experienced by the resonance frequency with the dielectric constant of the MUT is expected to be larger in sensors based on high-frequency resonators. Therefore, for the sake of comparison, a relative sensitivity can be defined, i.e.

$$\overline{S} = \frac{1}{f_0} \frac{df_0}{d\varepsilon_{MUT}} \tag{2.2}$$

A variation in the dielectric constant of the MUT modifies the capacitance of the considered resonator, provided the MUT is in the region of influence of the resonant element, and this in turn produces a shift in the resonance frequency. Thus, the above relative sensitivity can be expressed as

$$\overline{S} = \frac{1}{f_0} \frac{df_0}{dC'} \frac{dC'}{d\varepsilon_{MUT}} \tag{2.3}$$

where C' is the capacitance of the resonant element. The first derivative in (2.3) depends on the specific configuration of the line and sensing resonator, and it will be discussed later. Let us next focus on the calculation of the second derivative, $dC'/d\varepsilon_{MUT}$, by considering a set of simplifying, but reasonable, approximations and conditions:

- The analysis is restricted to fully planar resonant sensing elements etched in a single metallic layer and exhibiting an edge capacitance. Namely, resonant elements such as the BC-SRR and the microstrip SISS resonator, both exhibiting a broadside capacitance, are excluded in this analysis. Moreover, the metal is considered an ideal (perfect) conductor, with negligible resistivity and thickness.

- The MUT is considered to be semi-infinite in the vertical direction (i.e. the direction orthogonal to the plane of the resonator) and uniform, with transverse dimensions extending beyond the region occupied by the resonator.
- The substrate (also uniform) is thick enough, so that the influence of any potential metallic pattern in the face opposite to the one where the resonator is etched (e.g. the line strip) can be neglected.[4] In practice, this means that the substrate can also be considered to be semi-infinite in the vertical direction, and the electric field generated by the resonator in the opposite side can be neglected. For metallic resonators, it is assumed that any coplanar metallic pattern (including the line) is distant enough, thereby not affecting the resonator capacitance.
- The resonator capacitance and the line capacitance are uncoupled, that is the capacitance of the resonator can be calculated as if the line was not present.

Figure 2.4 illustrates a sketch of a slot resonator (a folded DB-DGS resonator) etched in the ground plane of a microstrip line, as well as the electric field lines generated by the line and those present in the slot of the particle. With the considered dimensions and the corresponding field lines, the previous conditions are satisfied to a good approximation. Since the material above (substrate) and below (MUT) the resonant element is uniform and semi-infinite, and the electric field generated by the line does not reach the slot region (except at the indicated circles), it is reasonable to assume that the plane of the particle is a magnetic wall. Thus, the electric field lines in the slots are approximately tangential to the plane of the slots (plane of the particle), whereas the magnetic field lines are roughly orthogonal to that plane. Under these conditions, the capacitance of the resonant element can be separated in two parts, i.e. the capacitances associated with the lower and upper half-spaces (representing the contributions of the MUT and substrate, respectively), both connected in parallel.

The calculation of the total capacitance of the resonator as a function of the transverse geometry and dielectric constant of the substrate material, ε_r, and MUT, ε_{MUT}, can be carried out, e.g., by means of conformal mapping [49]. Indeed, by using such an approach, it is possible to obtain the capacitance by considering a finite thickness for the substrate and MUT to a good approximation. Nevertheless, obtaining the resonator's capacitance as a function of the transverse geometry, ε_r, and ε_{MUT} is not necessary for our purposes, i.e. the calculation of the relative sensitivity. By considering the aforementioned approximations, the capacitance of the resonator loaded with a certain MUT, C', can be easily expressed as a function of the capacitance of the bare resonator, C, and the involved dielectric constants. The capacitances C' and C are simply given by (see Fig. 2.5)

$$C' = C_{subs} + C_{MUT} \tag{2.4a}$$

$$C = C_{subs} + C_{air} \tag{2.4b}$$

where C_{subs} is the capacitance corresponding to the substrate half-space, and C_{MUT} and C_{air} are the capacitances corresponding to the opposite half-space when the MUT is present and absent (i.e. resonator surrounded by air, or bare resonator), respectively. It is obvious that if the substrate and MUT are semi-infinite, and uniform, and the metal thickness is null, the electric field

---

4 If the substrate is thick enough, the presence of a metal in the opposite side to the resonator does not alter the capacitance of the particle (this capacitance is neither altered by the presence of a coplanar metal pattern, provided it is distant enough). However, if such metal is the line strip or the ground plane of the transmission line to which the resonator is coupled, or connected, the electric and magnetic fields generated by the line are expected to reach the region occupied by the resonator (a necessary condition for resonator's excitation in resonator-coupled lines). Nevertheless, the electric field generated by the line does not necessarily reach (or significantly reach) the capacitive region of the resonator (see the last approximation). Under these conditions, the mutual coupling between the line capacitance and the capacitance of the resonator can be neglected.

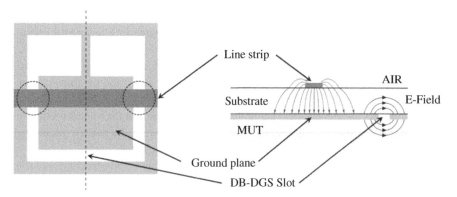

Figure 2.4 Sketch of the top view and cross-sectional view (corresponding to the indicated dashed plane) of a microstrip line loaded with a folded DB-DGS resonator etched in the ground plane. The electric field lines are indicated. The dashed circles indicate the slot regions under the influence of the electric field generated by the line. These regions represent a small portion of the whole slot. The upper level metallization (strip of the line) is depicted in dark gray, whereas the light gray color corresponds to the lower level metallization (ground plane).

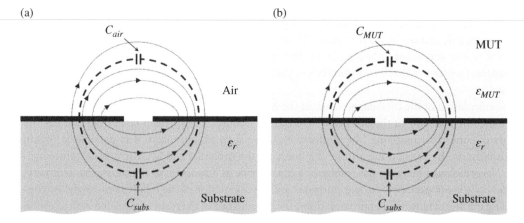

Figure 2.5 Sketch of the cross-sectional view of the slot region between two coplanar infinitesimally thin metals on top of a semi-infinite substrate (exhibiting an edge capacitance) and electric field lines. (a) Bare structure; (b) structure loaded with a semi-infinite MUT. The contributions to the total capacitance, as given by (2.4), are also indicated.

distribution in any of the two half-spaces is a mirror image of the field distribution in the opposite half-space. Therefore, the capacitance of the substrate and the capacitance of the MUT are related to the capacitance of air by

$$C_{subs} = C_{air} \cdot \varepsilon_r \tag{2.5a}$$

$$C_{MUT} = C_{air} \cdot \varepsilon_{MUT} \tag{2.5b}$$

and combining (2.4) and (2.5), the capacitance of the resonator loaded with the MUT can be written as a function of the capacitance of the bare resonator as

$$C' = C \frac{\varepsilon_r + \varepsilon_{MUT}}{\varepsilon_r + 1} \tag{2.6}$$

Thus, a variation in the dielectric constant of the MUT modifies the total capacitance of the resonant element according to

$$\frac{dC'}{d\varepsilon_{MUT}} = \frac{C}{\varepsilon_r + 1} \tag{2.7}$$

since C and ε_r are constant. Obviously, C can be determined numerically from the transverse geometry and the dielectric constant of the substrate, but it can be alternatively inferred from the response of the line loaded with the bare resonator (either simulated or measured), through a parameter extraction procedure [15, 50–52], an aspect to be discussed later.

Concerning the first derivative in (2.3), df_0/dC', it depends on the specific sensor configuration. Nevertheless, let us compare three different configurations of sensors based on DGS resonators and implemented in microstrip technology. Figure 2.6 depicts these sensors, as well as the corresponding lumped element circuit models. The considered sensing resonators are the CSRR, the DB-DGS resonator, and the OCSRR. It should be mentioned that the circuit model of the CSRR-loaded line is also valid for a microstrip line loaded with a folded DB-DGS resonator. All these sensing structures operate in transmission and exhibit a notched response for the CSRR- and the DB-DGS-loaded lines, whereas the OCSRR-loaded line exhibits a bandpass-type behavior. The output variable, f_0, is the frequency position of either the notch, or transmission zero, in the CSRR- and DB-DGS-based structures, or the peak (revealed as a reflection zero) in the OCSRR-loaded line. For the CSRR-loaded line, the notch frequency is given by

$$f_0 = \frac{1}{2\pi\sqrt{L(C_{line} + C')}} \tag{2.8a}$$

whereas for the OCSRR- and the DB-DGS-loaded microstrip line, the respective peak and notch frequencies are simply

$$f_0 = \frac{1}{2\pi\sqrt{LC'}} \tag{2.8b}$$

In (2.8a) and (2.8b), L is the inductance of the resonant element (not dependent on ε_{MUT}), and C_{line} in (2.8a) is the capacitance of the line section lying within the CSRR area.

Using (2.8a), it follows that

$$\frac{df_0}{dC'} = -\frac{1}{2} \cdot \frac{f_0}{C_{line} + C'} \tag{2.9}$$

and introducing (2.9) and (2.7) in (2.3), the relative sensitivity for the CSRR-loaded line is found to be

$$\overline{S} = -\frac{1}{2} \cdot \frac{C}{C_{line}(\varepsilon_r + 1) + C(\varepsilon_r + \varepsilon_{MUT})} \tag{2.10}$$

It is obvious that the relative sensitivity for the OCSRR- and the DB-DGS-loaded microstrip lines can be inferred from (2.10) by simply forcing $C_{line} = 0$, i.e.

$$\overline{S} = -\frac{1}{2} \cdot \frac{1}{\varepsilon_r + \varepsilon_{MUT}} \tag{2.11}$$

This result indicates that in the OCSRR- and DB-DGS-loaded lines, the relative sensitivity does not depend on the transverse geometry of the resonator (provided the semi-infinite approximation for both the MUT and the substrate prevails). Moreover, for sensitivity optimization, low dielectric constant substrates should be used, an expected result. Expression (2.11) also indicates that the

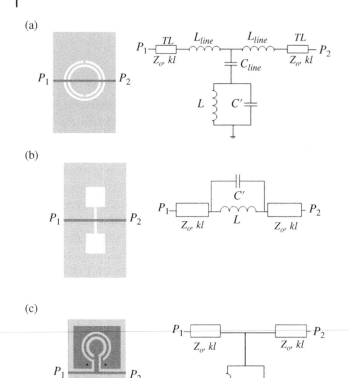

Figure 2.6 Typical topology and circuit model of a microstrip line loaded with a CSRR (a), DB-DGS resonator (b), and OCSRR (c).

magnitude of the relative sensitivity decreases as the dielectric constant of the MUT increases. It follows by comparing (2.10) and (2.11) that the relative sensitivity is superior in the OCSRR- and DB-DGS-based structures. This is because the capacitive dependence of the resonance frequency in such structures is entirely given by the sensing capacitance, C' (see expression 2.8b). By contrast, in the CSRR-loaded line, the line capacitance, C_{line}, in part obscures the variation of the resonance frequency caused by the changes in the sensing capacitance (see expression 2.8a), thereby degrading the sensitivity. Thus, from this analysis, it follows that for sensitivity optimization, resonator-loaded transmission lines exhibiting a resonance frequency with a capacitive dependence exclusively on the sensing capacitance are preferred.

The previous analysis was validated through electromagnetic simulation in [9, 53] by considering the above-cited DGS resonators. The specific topologies are depicted in Fig. 2.7, where the dimensions and substrate parameters are indicated. In [53], the objective was also to demonstrate that the magnitude of the relative sensitivity in a DB-DGS-loaded microstrip line is higher than in a CSRR-loaded line (provided the substrates are identical), in coherence with the analysis. The frequency responses (transmission coefficients) inferred from the electromagnetic simulations of the structures of Figs. 2.7(a) and (b) are depicted in Fig. 2.8. Such responses are compared with those obtained by means of circuit simulation with the extracted parameters (inferred through the method reported in [15], and depicted in Table 2.1). It should be mentioned that the models of Fig. 2.6 are lossless. However, the electromagnetic simulations of Fig. 2.8 do include the effects of losses.

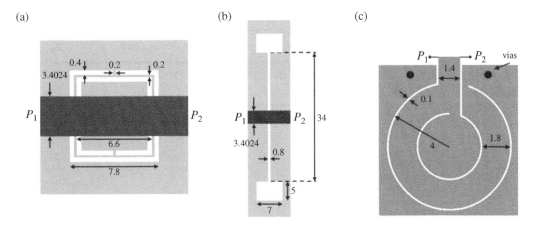

Figure 2.7 Specific sensor topologies (layouts) used for validation of the sensitivity analysis. (a) CSRR-based sensor; (b) DB-DGS-based sensor; (c) OCSRR-based sensor. Dimensions are given in mm. The considered substrate is the *Rogers RO4003C* with dielectric constant ε_r = 3.55, thickness h = 1.524 mm, and loss tangent tanδ = 0.0023.

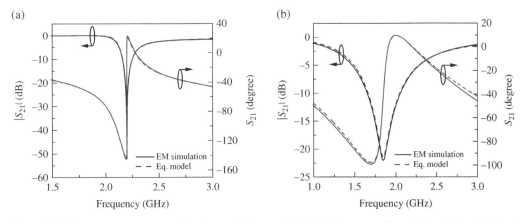

Figure 2.8 Frequency responses of the transmission coefficient (electromagnetic and circuit simulation) of the structures of Fig. 2.7 for unloaded (a) CSRR and (b) DB-DGS. *Source:* Reprinted with permission from [53]; copyright 2019 IEEE.

Thus, coherently, the circuit simulations were carried out by including a resistance R in parallel with the reactive elements of the resonator. Such resistance, determined by curve fitting and also shown in Table 2.1, accounts for resonator's losses (line losses are neglected). The excellent agreement between the circuit and electromagnetic simulation reveals that the circuit models provide an accurate description of the structures in the region of interest.

The simulated response of the sensors that result by considering the resonant elements covered with semi-infinite MUTs of different dielectric constant are depicted in Fig. 2.9. As expected, the resonance frequency decreases as the dielectric constant of the MUT increases. From the dependence of the resonance frequency with the dielectric constant of the MUT, the relative sensitivity, as given by (2.2), was inferred in [53]. The results, depicted in Fig. 2.10, reveal that the magnitude of the relative sensitivity is superior in the sensor based on the DB-DGS, in accordance with the theory. Figure 2.10 includes the analytical solution given by (2.10) or (2.11), and the agreement with

Table 2.1 Extracted circuit parameters of the structures of Fig. 2.7(a) and (b).

	L(nH)	C(pF)	R(kΩ)	L_{line} (nH)	C_{line} (pF)
Fig. 2.7(a)	2.312	1.51	106.1	2.214	0.759
Fig. 2.7(b)	5.07	1.463	1.08	N/A	N/A

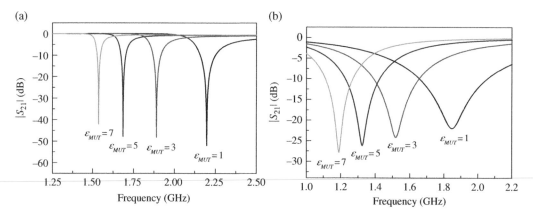

Figure 2.9 Simulated frequency responses of the transmission coefficient of the CSRR-based microstrip sensor (a) and DB-DGS-based microstrip sensor (b) for MUTs with different dielectric constants. *Source:* Reprinted with permission from [53]; copyright 2019 IEEE.

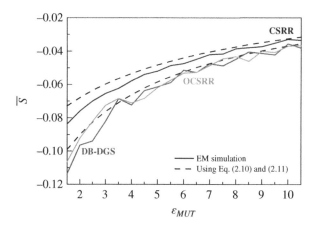

Figure 2.10 Dependence of the relative sensitivity on the dielectric constant of the (semi-infinite) MUT for the sensors depicted in Fig. 2.7. *Source:* Reprinted with permission from [9]; copyright 2020 MDPI.

the results inferred from full-wave electromagnetic simulation is reasonable (a perfect agreement is not expected as far as the approximations needed for the validity of the analysis are not strictly fulfilled).

According to the previous analysis, since the resonance frequency of the OCSRR- and DB-DGS-loaded line is given by (2.8b), the relative sensitivity of both structures should be identical, provided the considered substrates are also identical and the previous approximations are reasonably satisfied. In [9], the OCSRR structure of Fig. 2.7(c) was simulated by considering various MUTs with different dielectric constants, and the relative sensitivity was inferred by means of (2.2). The results, also included in Fig. 2.10, are in reasonable agreement with the analytical expression (2.11) and with the relative sensitivity of the DB-DGS.

From the analysis reported in this section, valid under the approximations and conditions listed before, it can be concluded that for certain resonant sensing elements, such as the DB-DGS, the OCSRR, and other planar resonators with resonance frequency given by (2.8b), the relative sensitivity does not depend on resonator's geometry. It only depends on the dielectric constant of the substrate (the magnitude of $\overline{S}$ increases by decreasing ε_r), and, obviously, it varies with the dielectric constant of the MUT, as inferred from (2.11). One of the considered conditions is the validity of the semi-infinite MUT and substrate approximation. Concerning the MUT, its thickness might not be under control in certain scenarios. However, there are situations where it is potentially possible to guarantee an MUT thickness enough to satisfy the semi-infinite approximation requirement (e.g. in microfluidic sensors, the MUT thickness is determined by the height of the channel, and the MUT is semi-infinite in submersible sensors). By contrast, the thickness of the substrate is typically limited to values not exceeding 2 mm, and the semi-infinite substrate approximation cannot always be guaranteed. Rigorously speaking, if either the MUT or the substrate (or both) have a limited thickness, and the electric field generated in the resonant element reaches the interface opposite to the plane of the resonator (in either the MUT or substrate), the plane of the resonator is no longer a magnetic wall. Consequently, the separation of the resonator's capacitance in two parts is not strictly valid. However, as it was demonstrated in [54], for any reasonable sensing resonator geometry and substrate thickness, the plane of the resonant element is a quasi-magnetic wall, and the resonator's capacitance can be considered to be given by the addition of the capacitances of the two halves to a very good approximation regardless of the substrate and MUT thickness. However, if the semi-infinite MUT and substrate approximation is not satisfied, expressions (2.10) and (2.11) are not valid.

In [54], the previous sensitivity analysis was extended to the case of sensors with finite substrate (but with semi-infinite MUT). The main relevant contribution of that work concerns the fact that by introducing the concept of equivalent dielectric constant of the substrate, $\varepsilon_{r,eq}$, the relative sensitivity as given by (2.10) and (2.11) can be easily reformulated by simply replacing ε_r with $\varepsilon_{r,eq}$ (this applies also to the capacitance given by 2.6). The equivalent dielectric constant of the substrate is defined as the dielectric constant that should exhibit a semi-infinite substrate in order to provide the same contribution to the capacitance of the sensing resonator (obviously, in the limit of thick substrates, $\varepsilon_{r,eq} \rightarrow \varepsilon_r$). In particular, for DB-DGS- and OCSRR-based sensors, the relative sensitivity should be calculated according to

$$\overline{S} = -\frac{1}{2} \cdot \frac{1}{\varepsilon_{r,eq} + \varepsilon_{MUT}} \tag{2.12}$$

For the determination of $\varepsilon_{r,eq}$, numerical methods can be used [54]. However, $\varepsilon_{r,eq}$ can be calculated by considering a MUT with a well-known dielectric constant. Using (2.6) with ε_r replaced with $\varepsilon_{r,eq}$, and isolating the latter, one obtains

$$\varepsilon_{r,eq} = \frac{\frac{C'}{C} - \varepsilon_{MUT}}{1 - \frac{C'}{C}} = \frac{f_{0,air}^2 - \varepsilon_{MUT}f_{0,MUT}^2}{f_{0,MUT}^2 - f_{0,air}^2} \tag{2.13}$$

where $f_{0,MUT}$ and $f_{0,air}$ are the resonance frequencies corresponding to the sensor loaded with the MUT and to the bare sensor, respectively. Once the equivalent dielectric constant is found, the dielectric constant of an unknown MUT is obtained by isolating it from (2.6), i.e.

Table 2.2 Geometry and substrate parameters for the considered DB-DGS-based sensors.

Sensor	$f_{0,air}$ (GHz)	W_s (mm)	ε_r	h (mm)	S (mm)	l (mm)	W_a (mm)	l_a (mm)
A	3.204	3.91	2.20	1.270	0.300	21	3.7	3.7
B	3.226	0.562	3.55	0.254	0.300	20	3.5	3.5
C	3.22	1.13	3.55	0.508	0.600	21.6	3.5	3.5
D	3.24	0.368	6.15	0.254	0.200	20	2.3	2.3
E	3.28	1.184	10.2	1.270	0.200	13.5	1.7	1.7
F	3.29	1.184	10.2	1.270	0.300	15	1.6	1.6

$$\varepsilon_{MUT} = \left(\varepsilon_{r,eq} + 1\right) \frac{f_{0,air}^2}{f_{0,MUT}^2} - \varepsilon_{r,eq} \tag{2.14}$$

Note that (2.13) and (2.14) are valid provided the resonance frequency of the considered resonator is given by (2.8b).

A parametric analysis carried out in [54] reveals that the introduction of the equivalent dielectric constant of the substrate predicts very accurately the relative sensitivity. However, it should be taken into account that the equivalent dielectric constant depends on the ratio between the capacitance slot width and the thickness of the substrate. Thus, it depends on the geometry, and, consequently, the relative sensitivity depends on the geometry as well. Nevertheless, sensors with different geometries but with the same value of the equivalent dielectric constant should provide the same relative sensitivity, as demonstrated in [54]. Moreover, as anticipated before, if the substrate is semi-infinite, the equivalent dielectric constant is the one of the substrate ($\varepsilon_{r,eq} = \varepsilon_r$), and the relative sensitivity does not depend on the geometry.

Table 2.2 shows the parameters of several sensing structures based on a microstrip line loaded with a DB-DGS resonator (W_s is the strip width, whereas the DB-DGS is characterized by the width S and length l of the slot, and the width W_a and length l_a of the apertures – square-shaped in all cases). The electromagnetically simulated responses for MUTs with different dielectric constants (all semi-infinite in the vertical direction) are depicted in Fig. 2.11 [54], where the equivalent dielectric constant is indicated. Figure 2.12(a) depicts the dependence of the resonance frequency with the dielectric constant of the MUT for the different sensors, whereas Fig. 2.12(b) depicts the relative sensitivity. The agreement between the relative sensitivity inferred from the derivative of the simulated data points and the analytical expression (2.12) is excellent. Moreover, it can be appreciated from Fig. 2.12(b) that those sensors exhibiting the same (or roughly the same) equivalent dielectric constant exhibit an undistinguishable relative sensitivity, thereby pointing out the validity of the analysis.

2.2.3 Sensors for Dielectric Characterization

There are many planar frequency-variation sensors (implemented with different resonant elements) reported in the literature [55–80]. In this section, two of such sensors, both devoted to dielectric characterization, are presented in detail. In the first sensor, the sensitive resonant element is a CSRR, whereas in the second implementation, a DB-DGS resonator is considered. As compared with other frequency-variation sensors, in the sensors reported in this section, the complex permittivity of the considered MUT is determined from an analytical method.

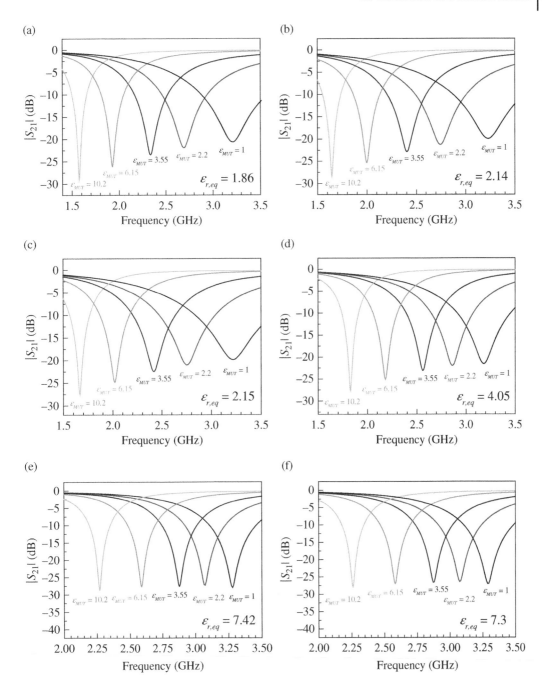

Figure 2.11 Responses of the sensors for different dielectric constants of the MUT (indicated). (a) Sensor A; (b) sensor B; (c) sensor C; (d) sensor D; (e) sensor E; (f) sensor F. *Source:* Reprinted with permission from [54]; copyright 2021 MDPI.

2.2.3.1 CSRR-Based Microstrip Sensor

The considered sensor, reported in [65], is based on a square-shaped CSRR, similar to the one of the structure of Fig. 2.7(a). However, the microstrip line is embedded in the substrate for the reasons that will be justified later. The sensing method is based on the measurement of the transmission coefficient, particularly on the determination of the notch frequency and magnitude, from which

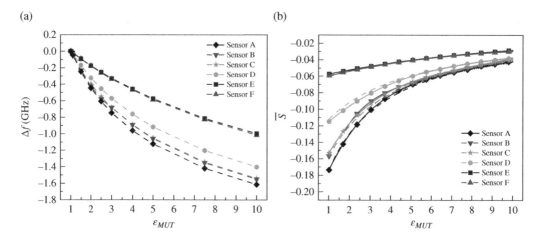

Figure 2.12 Variation of the resonance frequency with the dielectric constant of the MUT (a) and relative sensitivity (b) for sensors A, B, C, D, E, and F. The theoretical relative sensitivity inferred from expression (2.12) is also included in (b), in dashed line. *Source:* Reprinted with permission from [54]; copyright 2021 MDPI.

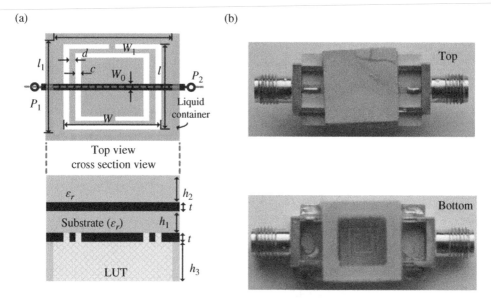

Figure 2.13 Top and cross-sectional views (a) and photograph (b) of the CSRR-loaded embedded microstrip line. Dimensions are: $W = l = 5.92$ mm, $c = d = 0.5$ mm, $W_0 = 1.15$ mm, $W_1 = l_1 = 9.92$ mm, $h_1 = 1.27$ mm, $h_2 = 3.81$ mm, $h_3 = 7.2$ mm and $t = 35$ μm. The substrate is *Rogers RO3010* with dielectric constant $\varepsilon_r = 10.2$, and the container is *FR4*.

the dielectric constant and the loss tangent of the MUT can be extracted. The specific sensing structure is depicted in Fig. 2.13, where the presence of a liquid container in the ground plane, surrounding the CSRR, can be appreciated (i.e. the structure is devoted to the dielectric characterization of liquid samples). Moreover, there is a dielectric slab on top of the line, identical to the substrate of the structure. By this means, the dielectric surrounding the metallic strip of the line can be considered to be homogeneous, and the line is embedded in the substrate. This homogeneity of the dielectric material surrounding the line is necessary for the application of the analytical method, but such homogeneity is necessary only in the regions where the electromagnetic field generated by the line is present. The thickness of the dielectric layer on top of the line, h_2, indicated in the caption of Fig. 2.13, is enough to ensure that the field lines do not reach the air/dielectric interface. On the other hand, the height of the liquid container, h_3, also indicated in the caption of Fig. 2.13, was chosen in order to guarantee the uniformity of the liquid under test (LUT) on top of the CSRR (a minimum LUT depth so as to ensure that the electric field lines generated in the slots of the CSRR do not cross the air/LUT interface is necessary).

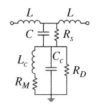

Figure 2.14 Lumped element equivalent circuit of the CSRR-loaded embedded microstrip line, without liquid in the container.

The presence of the liquid modifies (increases) the capacitance of the CSRR (through the effects of its dielectric constant, ε_{LUT}), and consequently the resonance frequency of this element shifts down. On the other hand, the notch depth is intimately related to losses, and such losses are influenced by the loss tangent of the LUT ($\tan\delta_{LUT}$). Therefore, the determination of such loss parameter ($\tan\delta_{LUT}$) is possible. Obviously, by obtaining the variation of the notch frequency and depth of liquid samples with well-known complex permittivity, calibration curves can be generated, and the measurement of the dielectric constant and loss tangent of the LUT can be achieved. Nevertheless, a technique that does not require calibration (as long as the measurement of the dielectric constant and loss tangent of the LUT is done from an analytical method) is reported in this section [65]. The method requires that the complex permittivity of the substrate material is accurately known, the case of the typical low-loss commercial microwave substrates. It should be mentioned that the use of low-loss substrates (i.e. with small loss tangent) is necessary, especially if the materials under test exhibit moderate or low losses. Otherwise, substrate losses may degrade the sensitivity of the method for the measurement of material losses (LUT in the sensor under consideration).

The sensing method, to be discussed next, is based on the lumped element equivalent circuit model of the structure of Fig. 2.13, depicted in Fig. 2.14 (the nomenclature and formulation – equations – adopted in [65] are reproduced here). This model is a generalization of the model of a CSRR-loaded line of Fig. 2.6, by including the effects of losses. The reactive elements of the model are L_c and C_c, the inductance and capacitance, respectively, of the CSRR, L, the inductance of the conductive strip (line), and C, the capacitance between such strip and the inner region of the CSRR.[5] Losses include substrate losses, through R_s, plus CSRR losses, modeled by R_M (ohmic losses) and R_D (dielectric losses of the substrate). Radiation losses are excluded since electromagnetic simulations of the structure by excluding dielectric and ohmic losses (not shown) indicate that unitarity is preserved in the region of interest. The sensing method is based on the following equations providing the real (2.15) and the imaginary part (2.16) of the impedance of the shunt branch [81]:

$$R_{eq} = \frac{1}{R_s C^2 \omega^2} + \frac{R_M R_D^2 + R_D L_c^2 \omega^2}{R_D^2 (1 - L_c C_c \omega^2)^2 + (L_c \omega + R_M R_D C_c \omega)^2} \tag{2.15}$$

$$\chi_{eq} = -\frac{1}{C\omega} + \frac{R_D^2 L_c \omega (1 - L_c C_c \omega^2) - R_M R_D (L_c \omega + R_M R_D C_c \omega)}{R_D^2 (1 - L_c C_c \omega^2)^2 + (L_c \omega + R_M R_D C_c \omega)^2} \tag{2.16}$$

---

5 Note that C in the circuit model of Fig. 2.14 is the line capacitance, not the capacitance of the bare sensing resonator (CSRR), designated as C_c in this model. This clarification is pertinent since in the sensitivity analysis carried out in the Subsection 2.2.2, C was considered to be the capacitance of the bare resonator.

where ω is the angular frequency. Another necessary equation in this analytical method is the expression providing the magnitude of the transmission coefficient at the frequency, ω_0, where $\chi_{eq} = 0$, that is [81],

$$|S_{21}|_{\omega_0} = \frac{2Z_0 R_{eq}}{\sqrt{\left(2Z_0 R_{eq} + Z_0^2 - L^2\omega_0^2\right)^2 + \left[2L\omega_0\left(Z_0 + R_{eq}\right)\right]^2}} \tag{2.17}$$

Z_0 being the reference impedance of the ports, typically $50\,\Omega$.

Let us consider the structure of Fig. 2.13 without LUT in the container (unloaded CSRR). From the measured frequency response, the reactive parameters (L_c, C_c, L, and C) can be inferred using the parameter extraction method reported in [50]. On the other hand, due to the uniformity of the dielectric material surrounding the strip line (with well-known loss parameter, $\tan\delta$), R_S and R_D can be obtained by means of [81]

$$\tan\delta = \frac{1}{Q_s} = \frac{1}{R_s C\omega} \tag{2.18}$$

$$\tan\delta = \frac{1 + \varepsilon_r}{R_D C_c \omega \varepsilon_r} \tag{2.19}$$

where Q_s is the substrate quality factor and ε_r is the dielectric constant of the substrate. Expression (2.18) is strictly valid in an embedded microstrip line assuming that slots are not present in the ground plane. This last condition is not satisfied due to the presence of the CSRR. However, as long as the side dimension of the CSRR, l, is high as compared with c and d, expression (2.18) provides a good approximation of the loss factor of the substrate. Expression (2.19), on the other hand, is derived as follows. The capacitance of the unloaded CSRR (i.e. without LUT) is given by the contribution of the substrate, C_{c_subs}, plus the contribution of the air region, C_{c_air} (see Fig. 2.15a), i.e.

$$C_c = C_{c\_subs} + C_{c\_air} \tag{2.20}$$

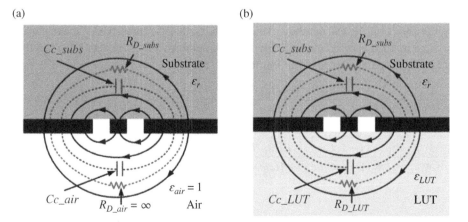

Figure 2.15 Cross-sectional view of the slot region of the CSRR, with electric field lines and contributions to the total CSRR capacitance and dielectric resistance. (a) Empty container; (b) full container.

Since the capacitance of the air region is related to the capacitance of the substrate by $C_{c\_subs} = \varepsilon_r \cdot C_{c\_air}$, it follows that

$$\varepsilon_r C_c = C_{c\_subs}(1 + \varepsilon_r) \tag{2.21}$$

The loss tangent of the substrate can also be expressed as

$$\tan \delta = \frac{1}{R_D C_{c\_subs} \omega} \tag{2.22}$$

where it is assumed that the field lines do not cross the air/substrate interface of the CSRR slots, and the metal thickness is neglected. The actual metal thickness is $t = 35$ μm, as indicated in the caption of Fig. 2.13, but this thickness is small enough and hence the above assumption is justified. Note also that $R_D = R_{D\_subs}$ (provided air can be considered as a perfect isolator, with $R_{D\_air} = \infty$), and for this reason, R_{D_subs} in the denominator of (2.22) can be replaced by R_D, as indicated. By isolating C_{c_subs} in (2.21) and introducing the resulting expression in (2.22), equation (2.19) is finally obtained. Thus, from (2.18) and (2.19), the dielectric loss parameters Rs and R_D can be obtained. Finally, the ohmic resistance R_M is obtained from the measurement of the transmission coefficient at ω_0, also given by (2.17). From (2.17), the real part of the shunt impedance, R_{eq}, can be obtained, and from it, R_M can be inferred using (2.15).

Once R_M is known, by loading the container with the LUT, all the model parameters remain invariable except C_c and R_D, influenced by the properties of LUT. Let us consider that C_c is the CSRR capacitance with the container empty (as defined before) and C_c' is the CSRR capacitance with the presence of the LUT. The notch (angular) frequencies in both cases are given by

$$\omega_0 = \frac{1}{\sqrt{L_c(C + C_c)}} \tag{2.23a}$$

$$\omega_0' = \frac{1}{\sqrt{L_c(C + C_c')}} \tag{2.23b}$$

The capacitance C_c' is given by

$$C_c' = C_{c\_subs} + C_{c\_LUT} = C_{c\_subs} + \frac{\varepsilon_{LUT}}{\varepsilon_r} C_{c\_subs} \tag{2.24}$$

where ε_{LUT} is the dielectric constant of the LUT. With (2.21) and (2.24), C_c' can be expressed in terms of C_c as follows,

$$C_c' = C_c \frac{\varepsilon_r + \varepsilon_{LUT}}{\varepsilon_r + 1} \tag{2.25}$$

which is indeed equivalent to (2.6), i.e. the semi-infinite substrate and MUT approximation is adopted. By introducing (2.25) in (2.23b), the dielectric constant of the LUT can be isolated, resulting in the following expression[6]

$$\varepsilon_{LUT} = 1 + \frac{\left(\omega_0'^{-2} - \omega_0^{-2}\right)}{L_c C_c}(1 + \varepsilon_r) \tag{2.26}$$

Therefore, from the measurement of the notch frequencies for the empty and full container, the dielectric constant of the LUT can be inferred using expression (2.26).

---

6 Note that expression (2.26) is different from (2.14). The reason is that (2.14) is valid when the resonance frequency is given by (2.8b), different from (2.8a) and (2.23).

The dielectric loss of the CSRR with the presence of the LUT, R_D', can be obtained from (2.17) and (2.15), with R_D replaced with R_D' in (2.15). Since R_D is known, the loss associated with the LUT, R_{D_LUT}, can be inferred, and, from it, the loss tangent of the LUT can be obtained. Namely,

$$R_D' = \frac{R_D R_{D\_LUT}}{R_D + R_{D\_LUT}} \tag{2.27}$$

since $R_D = R_{D\_subs}$, as indicated before. From (2.27), R_{D_LUT} can be isolated, and once R_{D_LUT} is known, the loss tangent of the LUT can be inferred according to

$$\tan \delta_{LUT} = \frac{1}{R_{D\_LUT} C_{c\_LUT} \omega} \tag{2.28}$$

with C_{c_LUT} given by

$$C_{c\_LUT} = C_c' \frac{\varepsilon_{LUT}}{\varepsilon_r + \varepsilon_{LUT}} \tag{2.29}$$

The frequency response of the empty sensing structure of Fig. 2.13 is depicted in Fig. 2.16. This includes the measured response (inferred from the *Keysight N5221A* vector network analyzer), the electromagnetically simulated response (inferred from *CST Microwave Studio suite 2010*), and the circuit response (inferred from the element values of the circuit model, shown in Table 2.3). The considered substrate, with dimensions indicated in the caption of Fig. 2.13, has a dielectric constant of $\varepsilon_r = 10.2$ and a loss tangent of $\tan\delta = 0.0023$. The measured response, with notch frequency at $f_0 = \omega_0/2\pi = 2.54$ GHz and notch depth of -24.4 dB, is in very good agreement with the circuit simulation (note that parameter extraction was carried out from the experimental data).

By adding olive oil in the liquid container, the circuit response changes to the one indicated in Fig. 2.17, where the notch frequency moves to $f_0' = \omega_0'/2\pi = 2.37$ GHz, and the measured notch level is found to be -16.26 dB. From these values, using (2.26), the dielectric constant of the substrate is found to be $\varepsilon_{LUT} = 2.93$. From (2.15) and (2.17), the dielectric loss of the CSRR with the

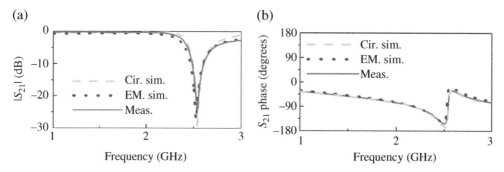

Figure 2.16 Transmission coefficient of the empty sensing structure of Fig. 2.13. (a) Magnitude; (b) phase. The phase has been obtained after reference plane shift. *Source:* Reprinted with permission from [65]; copyright 2018 MDPI.

Table 2.3 Circuit parameters corresponding to the empty sensing structure of Fig. 2.13.

L (nH)	C (pF)	C_c (pF)	L_c (nH)	C_{c_subs} (pF)	R_s (Ω)	R_D (Ω)	R_M (Ω)
3.70	1.09	8.07	0.43	7.35	25015	3709	0.0147

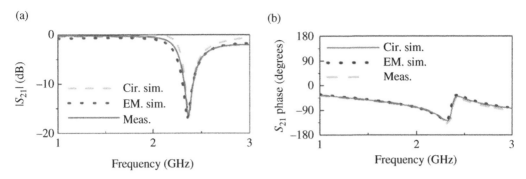

Figure 2.17 Transmission coefficient of the sensing structure with olive oil in the container. (a) Magnitude; (b) phase. The phase has been obtained after reference plane shift. *Source:* Reprinted with permission from [65]; copyright 2018 MDPI.

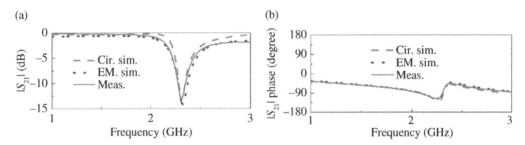

Figure 2.18 Transmission coefficient of the sensing structure with castor oil in the container. (a) Magnitude; (b) phase. The phase has been obtained after reference plane shift. *Source:* Reprinted with permission from [65]; copyright 2018 MDPI.

presence of the LUT is found to be $R_D' = 530.84\ \Omega$, and, using (2.27), the resulting loss associated to the oil is $R_{D\_LUT} = 619.5\ \Omega$. From (2.28), using (2.29), the loss tangent is found to be $\tan\delta_{LUT} = 0.103$. The dielectric constant and loss tangent of olive oil obtained by the proposed method are in good agreement with independent results, measured with the dielectric probe kit *Keysight 85070E* at the resonance frequency of the loaded CSRR [67], i.e. $\varepsilon_r = 2.89$, and $\tan\delta = 0.116$.

The complex permittivity of castor oil was also obtained in [65]. The corresponding response is depicted in Fig. 2.18, where it can be appreciated that the notch frequency shifts down to $f_0' = \omega_0'/2\pi = 2.31$ GHz, and the measured notch level is found to be -13.05 dB. With these values and the analytical expressions of the method, the resulting values of the dielectric constant and loss tangent are found to be $\varepsilon_{LUT} = 3.64$ and $\tan\delta_{LUT} = 0.139$. The values measured with the dielectric probe kit *85070E* [67] are in this case $\varepsilon_r = 3.32$, and $\tan\delta = 0.105$. Thus, the prediction of the method is not so good in this case, but it provides a reasonable estimation. It should be mentioned that there are several sources of errors, as discussed in [65], including certain tolerance in the dielectric constant of the substrate material, the limited homogeneity of the substrate surrounding the line strip (due to the fact that certain air gap effect may appear), and the fact that the semi-infinite approximation of the LUT is not strictly valid.

2.2.3.2 DB-DGS-Based Microstrip Sensor
The DB-DGS-based sensor, reported and experimentally validated in [80], is similar to the structure depicted in Fig. 2.7(b). The photograph and the specific dimensions can be seen in Fig. 2.19. The

(a)   (b)

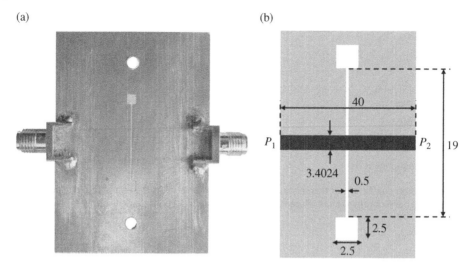

Figure 2.19 Backside photograph (a) and layout (b) of the DB-DGS-based frequency-variation permittivity sensor. The dimensions are indicated in mm. The sensor was implemented in the *Rogers 4003C* substrate with dielectric constant $\varepsilon_r = 3.55$, thickness $h = 1.524$ mm, and loss factor $\tan\delta = 0.0022$. The ground plane in (b) is depicted in gray.

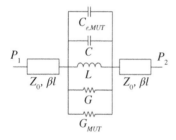

Figure 2.20 Circuit model of the sensor of Fig. 2.19.

dielectric constant of the MUT (provided it is semi-infinite) is determined by means of expression (2.14), from the measurement of the resonance (notch) frequency of the bare sensor and sensor loaded with the MUT (the equivalent dielectric constant can be inferred using (2.13) by measuring the response of a sample with well-known dielectric constant). Nevertheless, let us report a method to determine the loss tangent of the MUT, related to the magnitude of the measured notch. For that purpose, the circuit model of the DB-DGS, including the presence of the MUT, is necessary (see Fig. 2.20). The capacitance C and the inductance L are the reactive elements of the bare DB-DGS resonator, whereas the conductance G accounts for substrate losses (it is assumed that metallic losses have negligible effect). The dielectric material (MUT) in contact with the DB-DGS is taken into account by including two additional elements in the model, i.e. the capacitance $C_{e,MUT}$ [7] and the conductance G_{MUT}. The former accounts for the enhancement of the DB-DGS capacitance due to the presence of the MUT, and it is related to its dielectric constant, ε_{MUT}. The conductance G_{MUT} describes the effects of losses in the MUT, and it is related to its loss tangent, $\tan\delta_{MUT}$. Note that $C_{e,MUT}$ and G_{MUT} are null in the bare

---

7 The capacitance $C_{e,MUT}$ should not be confused with the capacitance C_{MUT}, as defined in Fig. 2.5, i.e. the contribution to the total capacitance corresponding to the half-space where the MUT is present. The total capacitance with the presence of the MUT can be expressed in terms of $C_{e,MUT}$ as $C'_{MUT} = C + C_{e,MUT}$, or, alternatively, in terms of C_{MUT}, it can be written as $C'_{MUT} = C_{subs} + C_{MUT}$, where C_{subs} is the substrate contribution to the total capacitance (also defined in Fig. 2.5). Note that if the MUT is absent (bare sensor), $C_{MUT} = C_{air}$ (also defined in Fig. 2.5), but $C_{e,MUT} = 0$ (as indicated in the text), and $C'_{MUT} = C$. It should be taken into account that in the original source [80] where the sensor reported in this subsection was presented, the capacitance $C_{e,MUT}$ was designated as C_{MUT} (for this main reason, the present clarifying comment is very pertinent).

sensor. Finally, βl and Z_0 are the electrical length and characteristic impedance, respectively, of the microstrip line, β and l being the phase constant and the physical length, respectively, of the line.

For the determination of the loss tangent of the MUT from the analytical method to be described next, it is necessary that the characteristic impedance of the line is set to the port impedance. For this reason, such impedance has been designated as Z_0 in the model. With such matching between the line and the ports, the line sections at both sides of the resonant element introduce a phase shift in the transmission coefficient, but do not have any influence on its magnitude, given by

$$|S_{21}| = \left| \frac{1}{1 + \dfrac{1}{2Z_0 Y_{MUT}}} \right| \tag{2.30}$$

where Y_{MUT} is the admittance of the DB-DGS loaded with the MUT sample, i.e.

$$Y_{MUT} = G'_{MUT} + j\omega C'_{MUT}\left(1 - \frac{\omega_{0,MUT}^2}{\omega^2}\right) \tag{2.31}$$

where $G'_{MUT} = G + G_{MUT}$, $C'_{MUT} = C + C_{e,MUT}$, $\omega = 2\pi f$ is the angular frequency, and $\omega_{0,MUT} = 2\pi f_{0,MUT}$ is the resonance (angular) frequency of the DB-DGS loaded with the MUT. Expression (2.31) is also valid for the bare sensor, with $G_{MUT} = 0$ (i.e. $G'_{MUT} = G$), $C_{e,MUT} = 0$ (that is $C'_{MUT} = C$), and $\omega_{0,MUT}$ replaced with $\omega_{0,air}$, the resonance (angular) frequency of the bare structure. For the sensor loaded with the MUT sample, the notch magnitude, given by the transmission coefficient evaluated at $\omega = \omega_{0,MUT}$, is

$$|S_{21}|_{\omega_{0,MUT}} = \frac{2Z_0 G'_{MUT}}{2Z_0 G'_{MUT} + 1} \tag{2.32a}$$

Similarly, the notch magnitude for the bare sensor is

$$|S_{21}|_{\omega_{0,air}} = \frac{2Z_0 G}{2Z_0 G + 1} \tag{2.32b}$$

and combining (2.32a) and (2.32b), the conductance contribution of the MUT is found to be

$$G_{MUT} = G'_{MUT} - G = \frac{1}{2Z_0} \cdot \left\{ \frac{|S_{21}|_{\omega_{0,MUT}}}{1 - |S_{21}|_{\omega_{0,MUT}}} - \frac{|S_{21}|_{\omega_{0,air}}}{1 - |S_{21}|_{\omega_{0,air}}} \right\} \tag{2.33}$$

The loss tangent of the MUT is given by

$$\tan \delta_{MUT} = \frac{G_{MUT}}{\left(C'_{MUT} - C_{subs}\right)\omega_{0,MUT}} \tag{2.34}$$

where $C'_{MUT} - C_{subs}$ is the contribution of the MUT subspace to the capacitance of the DB-DGS. Note that C_{subs}, given by[8]

$$C_{subs} = C \frac{\varepsilon_{r,eq}}{\varepsilon_{r,eq} + 1} \tag{2.35}$$

is the substrate contribution to the capacitance of the DB-DGS resonator. Introducing (2.35) and C'_{MUT}, given by[9]

---

8 This expression is inferred by inverting (2.21) and replacing ε_r with $\varepsilon_{r,eq}$. Note also that in (2.21), the capacitance of the bare resonator (a CSRR) is designated as C_c.

9 This expression is indeed (2.6), with C' and ε_r replaced with C'_{MUT} and $\varepsilon_{r,eq}$, respectively.

$$C'_{MUT} = C \frac{\varepsilon_{r,eq} + \varepsilon_{MUT}}{\varepsilon_{r,eq} + 1} \tag{2.36}$$

in (2.34), the following expression is obtained

$$\tan \delta_{MUT} = \frac{G_{MUT}(\varepsilon_{r,eq} + 1)}{C \varepsilon_{MUT} \omega_{0,MUT}} \tag{2.37}$$

Finally, introducing (2.33) in (2.37), the loss tangent is found to be

$$\tan \delta_{MUT} = \frac{\varepsilon_{r,eq} + 1}{2 Z_o C \varepsilon_{MUT} \omega_{0,MUT}} \cdot \left\{ \frac{|S_{21}|_{\omega_{0,MUT}}}{1 - |S_{21}|_{\omega_{0,MUT}}} - \frac{|S_{21}|_{\omega_{0,air}}}{1 - |S_{21}|_{air}} \right\} \tag{2.38}$$

Thus, the loss tangent of the MUT can be inferred from the measured notch magnitudes of the bare sensor, $|S_{21}|_{\omega_{0,air}}$, and sensor loaded with the MUT, $|S_{21}|_{\omega_{0,MUT}}$, since all the parameters preceding the brackets in (2.38) are either known or previously inferred, e.g. ε_{MUT}.

The experimental validation of the sensor of Fig. 2.19 and the reported analytical method for the determination of the dielectric constant and loss tangent of the MUT was carried out in [80] by considering several samples with well-known parameters (ε_{MUT} and tan δ_{MUT}). In particular, uncladded microwave substrates were used, with several pieces stacked in order to achieve a sample thickness enough to guarantee the validity of the semi-infinite MUT approximation. DI water was also considered as MUT (or LUT), and in this case the sensor was equipped with a liquid contained similar to that of Fig. 2.13(b) (see details in [80]). Table 2.4 indicates the considered MUT materials and the nominal values of their dielectric constants and loss tangents. Nevertheless, the method reported before was also applied by considering the electromagnetically simulated responses (using *ANSYS HFSS*) of the sensor loaded with the considered samples (note that in simulation, inaccuracies related to connectors, air gap effects, etc., are precluded).

The measured and electromagnetically simulated responses of the sensor loaded with the different MUT samples, as well as the bare sensor, are depicted in Fig. 2.21, where the good agreement between the two sets of results (simulated and measured) can be appreciated (the exception is *FR4*, where there is some shift attributed to the tolerance in the nominal value of the dielectric constant). The equivalent dielectric constant of the substrate, necessary for the calculation of ε_{MUT} by means of (2.14) and for the determination of tan δ_{MUT} through (2.38), was estimated to be $\varepsilon_{r,eq} = 2.15$. Moreover, in the denominator of the first term in the right-hand-side member of (2.38), the capacitance of the bare resonator, C, should be introduced. This parameter, inferred by parameter extraction, was found to be $C = 0.73$ pF. Application of (2.14) and (2.38) by considering the simulated and measured notch frequencies and magnitudes provides the dielectric constant and loss tangents

Table 2.4 Dielectric constant and loss tangent of the MUTs used for validation of the sensor of Fig. 2.19.

	RO4003C	FR4	RO3010	DI Water
ε_{MUT} (nom.)	3.55	4.40	10.2	80.9
$\tan\delta_{MUT}$ (nom.)	0.0022	0.020	0.0020	0.042
ε_{MUT} (sim.)	3.57	4.44	10.6	79.5
$\tan\delta_{MUT}$ (sim.)	0.0030	0.035	0.0022	0.050
ε_{MUT} (exp.)	3.53	4.05	10.0	78.2
$\tan\delta_{MUT}$ (exp.)	0.0017	0.039	0.0022	0.041

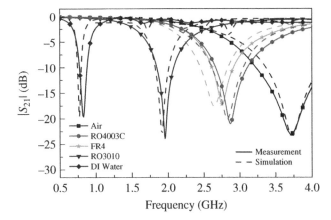

Figure 2.21 Measured and simulated transmission coefficient of the bare DB-DGS-based sensor of Fig. 2.19, and sensor loaded with stacks of *FR4*, *Rogers RO3010*, and *Rogers 4003C*, as well as DI water. *Source:* Reprinted with permission from [80]; copyright 2022 IEEE.

included in Table 2.4. The estimation of the dielectric constant of the MUT is good, except for *FR4* using the measured data (due to the explained reason). Concerning the loss tangent, the inferred results provide a reasonable estimation, but not a very accurate prediction. Note that even for the results inferred from the simulated data, the accuracy is limited, and the main reason is attributed to the high sensitivity of expression (2.38) to small variations in the magnitude of the notch. As mentioned in Chapter 1, resonant cavities provide better accuracy for the determination of the loss tangent. Nevertheless, the reported DB-DGS-based sensor is able to provide the dielectric constant of the MUT with good accuracy and a rough estimation of the loss tangent.

2.2.4 Measuring Material and Liquid Composition

Frequency-variation sensors are very appropriate for the determination of material and liquid composition in solid and liquid homogenous mixtures. The reason is that the permittivity of the composite (MUT) depends on the concentration of the different components. Nevertheless, in sensors exploiting a single resonance, only binary mixtures can be characterized. The determination of the composition in multicomponent samples is more complex. For that purpose, several resonators are needed, or, alternatively, the harmonic frequencies of a single sensing resonator can be used. Using a single resonance frequency in multicomponent mixtures does not allow to distinguish the concentration or volume fraction of each component, since different element combinations may generate the same dielectric constant in the homogenous mixture. However, since the dependence of the permittivity with frequency of the different components of the mixture is expected to be unequal, by measuring the frequency shift of different resonances, it is potentially possible to characterize solid or liquid homogenous mixtures containing more than two different constitutive elements. In this section, an example of a single resonance frequency-variation sensor devoted to the characterization of binary liquid mixtures is presented [78]. Nevertheless, by measuring not only the resonance frequency, but also the notch depth (the sensor exhibits a notched response), it is possible to determine the complex permittivity of the LUT. The characterization of multicomponent mixtures by exploiting several resonances will be considered later in this chapter.

The device subject of this section is a microfluidic sensor based on a SISS resonator loading a microstrip line (the sensor topology is depicted in Fig. 2.22). This sensor exhibits a high relative sensitivity, by virtue of the specific arrangement. That is, the ground plane beneath the SISS is

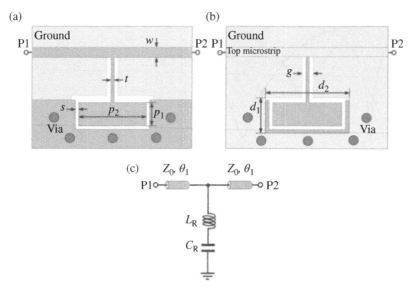

Figure 2.22 Topology of the SISS based frequency-variation sensor reported in [78], including the top (a) and bottom (b) views, and lumped element equivalent circuit model (c). Dimensions (in mm) are $d_1 = 10.8$, $d_2 = 5.2$, $g = 1.2$, $p_1 = 3.6$, $p_2 = 9.3$, $t = 0.4$, and $w = 1.2$. The considered substrate is the *Rogers RT6002* with dielectric constant $\varepsilon_r = 2.94$, thickness $h = 0.508$ mm, and loss tangent $\tan\delta = 0.0012$. *Source:* Reprinted with permission from [78]; copyright 2019 IEEE.

removed, and a metallization is added on the top surface, surrounding the rectangular SISS patch with a gap of width s (see Fig. 2.22). Such metallization is connected to the backside ground plane by means of metallic vias. By these means, the single capacitance of the resonant element is the edge capacitance of the SISS, which can be easily perturbed by means of the MUT (either a solid or a liquid), placed on top of it. With this configuration, the capacitance contributing to the resonance is solely the edge capacitance of the SISS, and therefore, the relative sensitivity is optimized, as discussed before in this chapter. This sensor was equipped with a microfluidic channel in order to perform measurements of the complex permittivity of liquids (Fig. 2.23).

The measured frequency responses of the sensor for different mixtures of DI water and ethanol are depicted in Fig. 2.24(a), whereas Fig. 2.24(b) shows the resonance frequency position and the magnitude of the notch as a function of the volume fraction of water. The curves of Fig. 2.24(b) can be used to determine the volume fraction of water, or ethanol, of unknown mixtures of such components. Nevertheless, the water–ethanol solutions offer a wide range of complex permittivity values, which are appropriate for sensor calibration, and thus use the sensor for determining the complex permittivity of other liquids or liquid solutions. That is, with the set of measurements of Fig. 2.24, a mathematical model for the sensor can be developed. Specifically, a mathematical relation linking the frequency shift and the notch magnitude to the complex relative permittivity of the test water–ethanol solutions was derived in [78]. Since the complex permittivity of water–ethanol as a function of the volume fraction can be independently inferred, a nonlinear least-square curve fitting can be used in order to derive an equation describing the relation between variations of the resonance frequency and notch magnitude as a function of the complex permittivity variations. In [78], the authors used the complex permittivity values of water–ethanol mixtures provided in [82]. The model is as follows:

$$\begin{pmatrix} \Delta\varepsilon'_{MUT} \\ \Delta\varepsilon''_{MUT} \end{pmatrix} = \begin{pmatrix} -93.47 & 2.387 \\ -9.246 & -1.05 \end{pmatrix} \begin{pmatrix} \Delta f_0 \\ \Delta|S_{21}| \end{pmatrix}$$

(2.39)

(a)

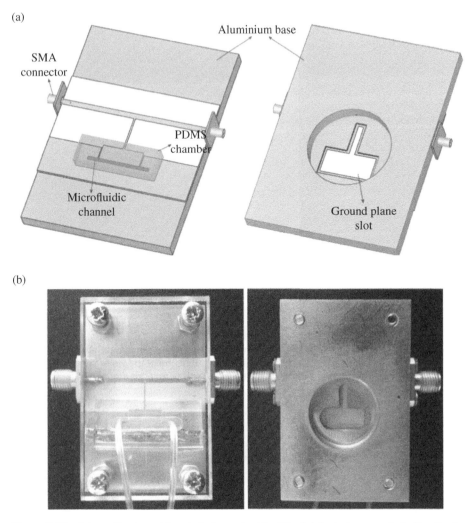

(b)

Figure 2.23 Schematic views (a) and photographs (b) of the fabricated sensor, including the top (left) and bottom (right) views. *Source:* Reprinted with permission from [78]; copyright 2019 IEEE.

where $\Delta\varepsilon'_{MUT} = \varepsilon'_{MUT} - \varepsilon'_{REF}$, $\Delta\varepsilon''_{MUT} = \varepsilon''_{MUT} - \varepsilon''_{REF}$, $\Delta f_0 = f_{0,MUT} - f_{0,REF}$, and $\Delta|S_{21}| = |S_{21,MUT}| - |S_{21,REF}|$. The subscripts MUT and REF denote the MUT and the REF sample, respectively, pure ethanol being the REF sample, and ε' and ε'' are the real and the imaginary parts, respectively, of the complex dielectric constant (of either the MUT or REF sample, depending on the sub-index). With such model, the complex permittivity of other liquid solutions can be inferred from the measured values of the resonance frequency and notch magnitude. For instance, in [78] the authors characterized solutions of water–methanol. The measured resonance frequency and notch magnitude for the different concentrations is shown in Fig. 2.25, whereas Fig. 2.26 depicts the real and the imaginary parts of the permittivity inferred from the mathematical model. In this figure, the values inferred from the literature are also depicted, and the good agreement between both sets of data can be appreciated. It is remarkable that the relative sensitivity of this sensor is very good, as compared with the one of other similar sensors reported in the literature (see [78] for further details).

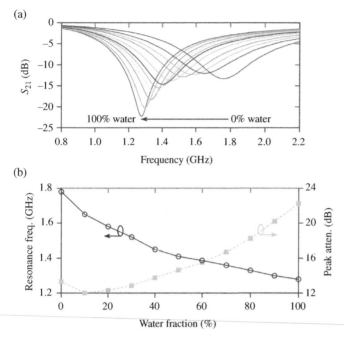

Figure 2.24 Measured frequency response of the sensor for various water–ethanol solutions (a) and representation of the resonance frequency and notch magnitude as a function of the water concentration (b). In (a), the step variation of water content is 10%. *Source:* Reprinted with permission from [78]; copyright 2019 IEEE.

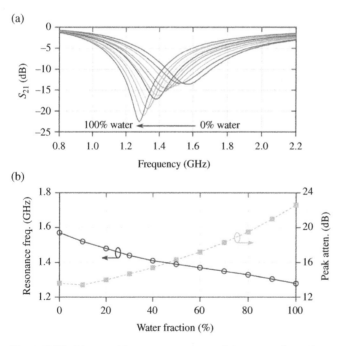

Figure 2.25 Measured frequency response of the sensor for various water–methanol solutions (a) and representation of the resonance frequency and notch magnitude as a function of the water concentration (b). In (a), the step variation of water content is 10%. *Source:* Reprinted with permission from [78]; copyright 2019 IEEE.

(a)

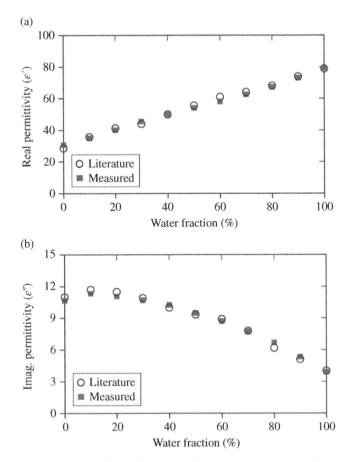

(b)

Figure 2.26 Real (a) and imaginary (b) parts of the complex dielectric constant of water–methanol solutions as a function of the water content, inferred from the data of Fig. 2.25 and the mathematical model developed in [78] (expression 2.39). *Source:* Reprinted with permission from [78]; copyright 2019 IEEE.

2.2.5 Displacement Sensors

Frequency-variation sensors can also be applied to the measurement of short-range displacements. For that purpose, the transmission line and the sensing resonator should be etched in independent substrates. The working principle is based on the resonance frequency shift generated by a relative motion (linear or angular) between the line and the resonator. In [56], a two-dimensional linear displacement frequency-variation sensor based on triangular-shaped CSRRs is reported, whereas [66] presents an angular displacement sensor implemented by means of circularly shaped CSRRs. As an example, let us report in this section the sensor reported in [56].

The displacement sensor of [56] uses triangular CSRRs since a lateral displacement with regard to the line significantly alters the coupling capacitance between the line and the resonant element (C_{line}, see expression 2.8a), and this modifies the resonance frequency, according to expression (2.8a). Figure 2.27 shows in gray color the coupling region between the line and the resonant element, and it is apparent from the figure that a lateral displacement of the CSRR varies the capacitance between the line and the inner region of the CSRR (the model of the structure is the one depicted in Fig. 2.6a). Note that the triangular topology of the CSRR is convenient for sensitivity enhancement (by contrast, square, rectangular, or circular CSRR geometries do not significantly

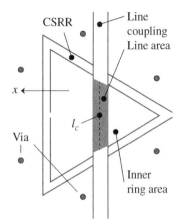

Figure 2.27 Topology of the triangular-shaped CSRR beneath the line with indication of the coupling area. *Source:* Reprinted with permission from [56]; copyright 2011 EuMA.

change the coupling capacitance, C_{line}, under CSRR lateral motion). In order to implement a two-dimensional displacement sensor, a bended line with a pair of orthogonally oriented triangular CSRR was used in [56]. Moreover, two orthogonally polarized receiving and transmitting antennas were added to the input and the output ports of the bended line in order to provide a wireless link with the reader. Figure 2.28 depicts the schematic of the wireless link, the photograph of the fabricated device, as well as the responses inferred for various CSRR positions.

Indeed, by taking into account that the coupling capacitance is roughly proportional to the gray area indicated in Fig. 2.27, it is possible to estimate the variation of the resonance frequency with the lateral displacement. Figure 2.29 depicts such variation of the resonance frequency with the relative lateral shift between the line and the triangular resonator, as inferred analytically. The figure includes also the values that result from electromagnetic simulation and experiment (the agreement achieved by the authors of [56] is reasonably good). Similar to other displacement sensors based on transmission lines loaded with movable resonators, the sensor of Fig. 2.28 exhibits a limited input dynamic range, dictated by the region surrounding the line where the resonant element can be driven. Nevertheless, this sensor is interesting since it utilizes the shape of the resonant element for sensitivity optimization, this being the most relevant aspect justifying its inclusion in this chapter. However, it is considered that a qualitative explanation suffices (further details are given in [56]).

2.2.6 Sensor Arrays for Biomedical Analysis

Sensor arrays able to resolve the dielectric properties of the MUT, and applied to the analysis of organic tissues, have been reported [55, 83]. Such arrays consist of a microstrip line loaded with multiple SRRs, each one exhibiting a different resonance frequency. The SRRs are spatially distributed, forming either one- or two-dimensional arrays, so that a frequency shift of one individual resonant notch indicates the dielectric properties of the MUT and its location within the array. Thus, these devices provide a dielectric image, with a number of pixels given by the number of SRRs within the array.

In this section, the two-dimensional SRR-based sensor array presented in [83] is reported. The photograph of the device, depicted in Fig. 2.30, shows that the SRRs are single-loop rectangular resonators with apertures consisting of parallel strips, the sensitive elements of the array. In [83], an isolation layer was included between the SRR and the tissue in order to prevent from the degradation of the resonant notches (otherwise caused by the high level of losses inherent to the tissues). The interest of the sensor array of Fig. 2.30 in a real scenario is the potential detection of malignant regions (e.g. tumors) in the tissue under study, and this is possible as far as such regions are expected to exhibit different dielectric properties, in particular the real part of the complex dielectric constant, as compared with unaltered tissue. In order to obtain good results and repeatability, the position of the sample of tissue is critical. For this purpose, a specific measurement setup that guarantees the accurate and repeatable positioning of the organic tissue under test was developed in [83], where further details are given.

The validation of the sensor array and experimental setup was carried out with different measurement campaigns. In the first measurements, dielectric loads in the form of perturbers were

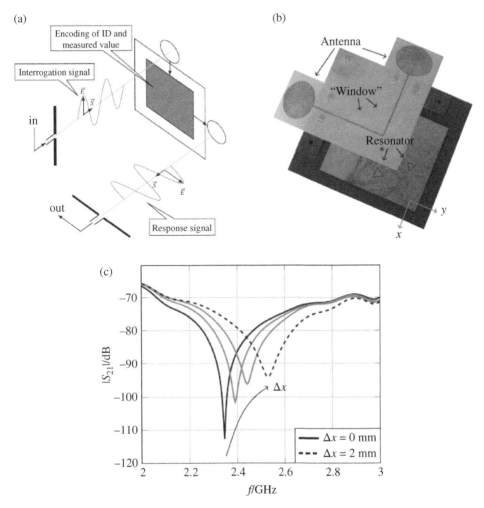

(a)

Encoding of ID and measured value

Interrogation signal

$\vec{E}$
$\vec{S}$

in

out

$\vec{S}$
$\vec{E}$

Response signal

(b)

Antenna

"Window"

Resonator

y
x

(c)

$|S_{21}|$/dB

-70

-80

-90

-100

-110

-120

Δx

—— $\Delta x = 0$ mm
- - - $\Delta x = 2$ mm

2   2.2   2.4   2.6   2.8   3
f/GHz

Figure 2.28 Two-dimensional frequency-variation displacement sensor based on a bended line and a pair of triangular CSRRs. (a) Scheme of the polarization decoupling of interrogation and response signals; (b) photograph of the sensor; (c) frequency response for various lateral displacements in x direction. The perimeters of the used CSRRs are 22 mm for the small ring and 28 mm for the large ring, leading to resonance frequency ranges of 2.4–2.8 GHz and 1.7–2.1 GHz, respectively, dependent on the coupling capacitance. The ultra-wideband (UWB) monopole antennas have been designed to cover at least the concatenation of these two frequency ranges. The considered substrate for the line and antenna layer is the *Rogers 4003C* with dielectric constant ε_r = 3.38, and loss tangent tanδ = 0.0027. The CSRRs were implemented on the *Rogers RT/duroid* 5880 microwave substrate with dielectric constant ε_r = 2.20, and loss tangent tanδ = 0.0009. *Source:* Reprinted with permission from [56]; copyright 2011 EuMA.

selectively placed on top of certain SRRs of the array, whereas the other SRRs were kept uncovered. Figure 2.31 shows that only those SRRs affected by the dielectric perturbers experience a shift in the corresponding resonance frequency. The second measurement was conducted with animal lung tissue. In one case, the tissue was unaltered, whereas in the other one, the same lung tissue with dielectric perturbers placed under the tissue, around the second and third pixel, was considered the MUT. The shift of the notch frequencies can be clearly identified, and therefore, it was proven that monitoring local changes in the tissue samples is possible (see Fig. 2.32). In the third round of measurements, the intention in [83] was to characterize human tissue. However, because obtaining

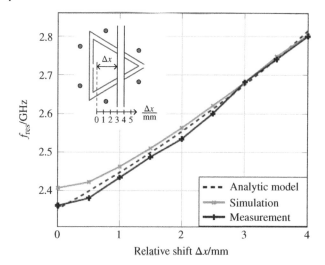

Figure 2.29 Resonance frequency variation with the lateral displacement in *x* direction for the sensor of Fig. 2.28. *Source:* Reprinted with permission from [56]; copyright 2011 EuMA.

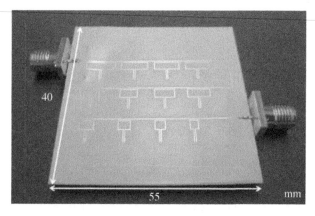

Figure 2.30 Photograph of the two-dimensional sensor array based on 12 SRRs, implemented on the *RT/Duroid 6010* substrate with dielectric constant ε_r = 10.2 and thickness of 0.254 mm. *Source:* Reprinted with permission from [83]; copyright 2012 IEEE.

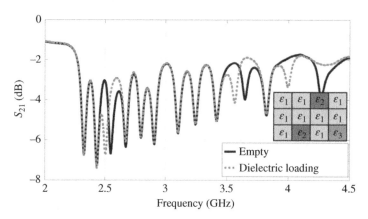

Figure 2.31 Measurements in the sensor array of Fig. 2.30, by selectively adding dielectric perturbers in the third, tenth, and twelfth pixel. The frequencies of the other pixels do no change. *Source:* Reprinted with permission from [83]; copyright 2012 IEEE.

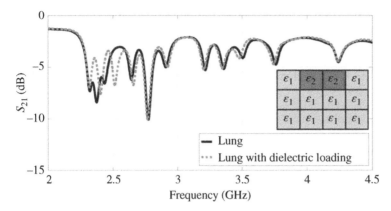

Figure 2.32 Measurements in the sensor array of Fig. 2.30, with animal lung tissue and with animal lung tissue with dielectric loading included around the second and third pixel. *Source:* Reprinted with permission from [83]; copyright 2012 IEEE.

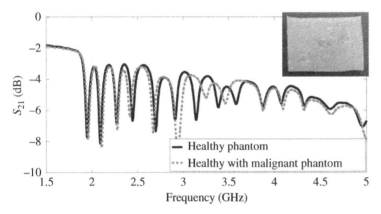

Figure 2.33 Measurements in the sensor array of Fig. 2.30, with a fibro-glandular phantom with and without the malignant tissue. The inset shows a picture of the phantom with parts of malignant tissue (blue). *Source:* Reprinted with permission from [83]; copyright 2012 IEEE.

human tissue is quite difficult and the process is long, phantoms of human tissue were fabricated in [83] using the receipts given in [84, 85]. The phantoms are based on glycerin, and by changing the ratio of water, salt, and polyethylene powder, the dielectric prototypes were controlled to mimic a specific tissue, e.g. skin, muscle, or any organ in the body, and, importantly, including malignant tissue as well. Figure 2.33 shows a measurement result of a phantom that behaves as fibro-glandular tissue, as well as a measurement of the same fibro-glandular phantom with malignant pieces inserted around the seventh, eighth, and ninth unit cell. In both cases, the phantom covers almost the complete sensor and only the resonances of the pixels that interact with the malignant tissue show a significant change. Thus, these results show the potential of these sensor arrays to locally determine potential defects in samples, including malignant regions in tissue samples.

In later works by the authors of [83], sensor operation in two modes was considered, in order not only to characterize the tissue but also to use the array as a therapy method [86, 87]. The idea is first to make a low power measurement, in order to infer the dielectric properties of the tissue, and, once the abnormality is detected (if it is present), then switch the sensor to the second mode of operation.

In this second operation mode, the power should be increased in order to heat the malignant tissue, a technique known as thermal ablation (see further details in [86, 87]).

2.2.7 Multifrequency Sensing for Selective Determination of Material Composition

In the previous section, multifrequency sensing devoted to the spatial determination of the dielectric properties or anomalies in samples (by conveniently distributing the sensing resonators forming a two-dimensional array of pixels), was discussed. In this section, it is shown that by using multiple frequencies, the determination of material composition in multicomponent samples, e.g. liquid solutions with several solutes, is possible. In binary composites, such as a solution of DI water and ethanol, for example, the determination of the volume fraction of ethanol is simple, since the complex dielectric constant of the mixture depends only on the relative concentration of ethanol (and hence a single-frequency measurement suffices). However, if the number of components in the mixture is higher than two, different combinations of such components may provide the same dielectric constant. Thus, the relative concentrations of the different components cannot be merely determined from a single-frequency measurement. For the selective determination of material composition in multicomponent samples, it is necessary to perform multifrequency sensing. The working principle is based on the fact that the dependence of the permittivity with frequency varies from component to component.[10] Consequently, from the frequency shift generated at either frequency, it is possible to obtain the relative concentration of the different components of the sample. Multifrequency sensing was demonstrated in [92, 93], where it was applied to the selective determination of the volumetric fraction of multicomponent liquids. Indeed, multifrequency sensing utilizes the frequency dependence of the dielectric properties of materials, thereby generating a discrete dielectric spectrum (the approach has also been applied to the accurate permittivity characterization of dispersive materials [94–96]). A continuous dielectric spectrum can be generated by tuning the operation frequency of the sensor within a certain range [97], similar to the so-called dielectric spectroscopy [98, 99]. In this section, the specific strategy to selectively determining the relative concentration of the various components of a multicomponent composite is reported, and such strategy is validated by characterizing a ternary liquid mixture [93].

In the present analysis, it is assumed that the multi-component composite under test is uniform, or it can be considered to be uniform at the scale of the sensing region (otherwise, the effect of the composite under test on the different resonance frequencies may depend on the relative position between the sample and the sensing region). On the other hand, the different resonance frequencies required for multivariable sensing can be generated either by multiple resonators, or, alternatively, by a single resonant element, using its harmonics (the approach used in [92, 93]). Let us consider that the composite is formed by n different materials. For the determination of the volume fraction for each component, V_j (with $j = 1...n$), n independent equations are thus needed. However, one of such equations is simply

$$\sum_{j=1}^{n} V_j = 1 \qquad (2.40)$$

as far as the total volume fraction should be the unity. The other $n - 1$ equations required to univocally determine the composition of the sample are generated by measuring the frequency shifts

---

10 The dispersive behavior of the complex permittivity of materials is described by various models, depending on the specific material. Polar liquids, for instance, are typically modeled by the Debye relaxation formula [88, 89]. Depending on the dielectric relaxation and loss factor of the material, variants of the Debye model, such as the so-called Cole-Cole model, provide better accuracy [90, 91].

caused by the composite at the $n - 1$ operational frequencies (i.e. the number of required frequencies for sensing is $n - 1$). Let us designate by $\Delta f_{i,j}$ the frequency shift in the operational frequency f_i caused by the individual material designated as j, and let us call $\Delta f_{i,mix}$ the frequency shift caused by the composite at the frequency f_i. With these designations, the following matrix equation can be written

$$
\begin{pmatrix} \Delta f_{1,mix} \\ \vdots \\ \Delta f_{n-1,mix} \end{pmatrix} = \begin{pmatrix} \Delta f_{1,1} & \cdots & \Delta f_{1,n} \\ \vdots & \ddots & \vdots \\ \Delta f_{n-1,1} & \cdots & \Delta f_{n-1,n} \end{pmatrix} \cdot \begin{pmatrix} V_1 \\ \vdots \\ V_n \end{pmatrix}
\tag{2.41}
$$

corresponding to a set of $n - 1$ linear equations. The first matrix in the right-hand-side member of (2.41) is indeed a calibration matrix that can be easily obtained from the shift that each individual component causes in the considered operational frequencies, as mentioned. With (2.40) and (2.41), the volume fractions, V_j, can be univocally determined, as far as equations (2.41) constitute an independent set of equations. In order to satisfy such requirement, the permittivity of the different components of the composite should vary with frequency (in the considered range) and should be different, the usual situation. To further understand this aspect, let us suppose that the components of the mixture are dispersionless, or at least are dispersionless in the considered frequency range of operational frequencies (i.e. all of them exhibit a permittivity invariable with frequency in that range). Moreover, let us consider that the multifrequency sensor uses the fundamental frequency of a single resonant (sensing) element and its harmonics. In this case, the rows of the calibration matrix are proportional (to a first-order approximation), and hence the volume fractions of the composite cannot be inferred. Thus, this method requires that the components of the mixture exhibit dielectric dispersion within the frequency span covered by the operational frequencies.

As an example of application of the previous multifrequency technique for the selective determination of material composition, let us report the sensor and the ternary composite presented in [93], a mixture of DI water, alcohol, and sugar. The photograph of the fabricated multifrequency sensor is depicted in Fig. 2.34. It consists of a microstrip line coupled with an SRR, with an additional U-shaped strip used to boost-up the sensitivity. The sensor was fabricated on the *Rogers RO3003* substrate with relative permittivity $\varepsilon_r = 3$, loss tangent $\tan\delta = 0.0013$, and thickness $h = 0.76$ mm. With the dimensions of the SRR, given in the caption of Fig. 2.34, the first operational frequency (i.e. the fundamental frequency) is found to be $f_1 = 1.1$ GHz, but the SRR also resonates at $f_2 = 2.2$ GHz, and at higher frequencies. Nevertheless, the third- ($i = 3$) and higher-order harmonics are not used since the sample is a ternary mixture. For sensing purposes, a liquid container was placed on top of the sensing region, the U-shaped strip, as sketched in Fig. 2.34(c).

Figure 2.35 depicts the frequency response (transmission coefficient) of the bare sensor and those responses of the sensor loaded with the individual components constitutive of the sample, i.e. DI water, ethanol, and sugar (indeed, the sugar is a saturated glucose solution). The resonance shifts caused by the different components at each operational frequency are shown in Table 2.5. These frequency shifts correspond to the calibration matrix of expression (2.41). Then, the authors of [93] considered several mixtures of the previous components, and the measured responses (in the vicinity of the harmonics of interest) are depicted in Fig. 2.36. Such shifts are summarized in Table 2.6. The table includes also the shift that should result from the calibration matrix and the nominal volume concentration of each component, and it can be appreciated that the agreement is reasonably good. Finally, Table 2.7 shows the values of the volume fraction of the three components for the different mixtures, as inferred from equations (2.40) and (2.41), using the measured values of the frequency shifts at the two operational frequencies, $\Delta f_{i,mix}$, generated by the mixtures. It can be observed that the values obtained are very similar to the nominal volume fractions.

(a)

(b)

(c)

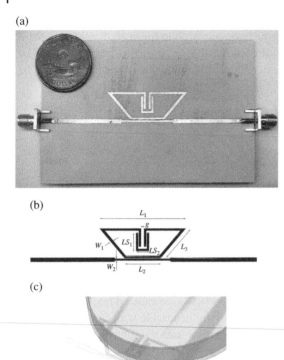

Figure 2.34 Photograph (a), topology (b), and sketch in perspective view including the liquid container (c) of the multifrequency sensor devoted to selectively determining the volume fraction of a liquid mixture. Dimensions are L_1 = 41 mm, L_2 = 17.4 mm, L_3 = 18 mm, LS_1 = 8.4 mm, LS_2 = 6.3 mm, W_1 = 1.2 mm, W_2 = 0.9 mm, and g = 1.5 mm. *Source:* Reprinted with permission from [93]; copyright 2021 Elsevier.

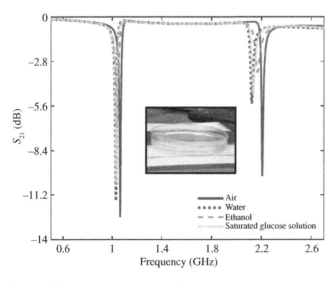

Figure 2.35 Frequency response of the bare sensor and sensor loaded with the different individual components of the mixture. *Source:* Reprinted with permission from [93]; copyright 2021 Elsevier.

Table 2.5 Frequency shift at the operational frequency caused by the different individual components of the composite.

	Water	Ethanol	Saturated glucose solution
Δf_1 (MHz)	34.56	23.62	32.81
Δf_2 (MHz)	84.87	39.81	74.37

Source: From [93]/with Permission of Elsevier.

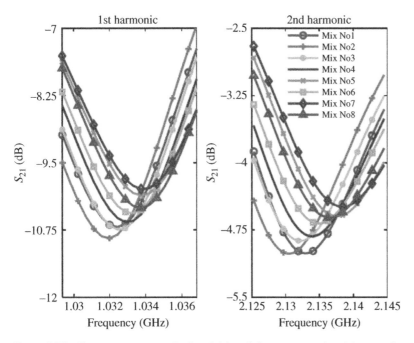

Figure 2.36 Frequency response in the vicinity of the two operational frequencies corresponding to the sensors loaded with mixtures with different combinations of volume fraction of each component. *Source:* Reprinted with permission from [93]; copyright 2021 Elsevier.

Table 2.6 Calculated and measured results of the frequency shift at the operational frequencies for the different samples.

	Harmonic shifts (MHz)							
Sampling harmonic	**Mix 1**	**Mix 2**	**Mix 3**	**Mix 4**	**Mix 5**	**Mix 6**	**Mix 7**	**Mix 8**
Calculated $\Delta f_{1,\ mix}$ (MHz)	32.7	33.05	32.59	32.13	31.32	31.67	30.86	31.21
Calculated $\Delta f_{2,\ mix}$ (MHz)	74.75	76.84	75.11	73.39	69.56	71.66	67.83	69.93
Measured $\Delta f_{1,\ mix}$ (MHz)	32.81	33.25	32.81	32.37	31.5	31.93	31.48	31.51
Measured $\Delta f_{2,\ mix}$ (MHz)	74.37	77	75.25	73.5	69.56	72.62	69.12	70.87

Source: From [93]/with Permission of Elsevier.

Table 2.7 Comparison between the extracted (measured) and nominal volume fraction for the different samples.

Mixture components	Volumetric percentages (%)							
	Mix 1	Mix 2	Mix 3	Mix 4	Mix 5	Mix 6	Mix 7	Mix 8
Water assigned, V. %	20	40	40	40	20	40	20	40
Glucose sol. assigned, V. %	75	55	50	45	60	40	55	35
Ethanol assigned, V. %	5	5	10	15	20	20	25	25
Measured water, V. %	19	44	42.9	43.5	20	39.4	22.3	34.5
Measured glucose sol., V. %	75.3	50.4	46.5	45.3	60.4	43.6	55.7	44
Measured ethanol, V. %	5.7	5.6	10.6	11.2	19.6	17	22	21.5
Average error	0.0647	0.1010	0.0675	0.1157	0.0088	0.0850	0.0825	0.1781

Source: From [93]/with Permission of Elsevier.

The results of Table 2.7 validate the multifrequency multivariable sensing approach reported in this section. The specific sensor is based on a single sensing resonator (Fig. 2.34) and exploits its harmonics. However, an alternative scheme with several resonant elements tuned to different frequencies can also be envisaged.

2.3 Other Frequency-Variation Resonant Sensors

In Section 2.1, it was indicated that transmission-line-based frequency-variation resonant sensors may operate either in transmission (the most extended operation mode) or in reflection. In the previous section, all the reported examples consist in transmission-mode sensors. Let us report in this section a one-port reflective-mode device, useful as submersible sensor for liquid characterization [100]. Submersible sensors may represent an interesting approach for the analysis of liquid samples, as far as fluidic channels or liquid containers on top of the sensing elements are avoided, thereby simplifying sensor design and fabrication, and reducing costs. Indeed, submersible sensors act as a dielectric probe, able to provide information of the liquid under study. Although the natural approach for the implementation of submersible sensors is one-port reflective-mode devices, transmission-mode submersible sensors have also been reported [67]. In this section, an illustrative example of a one-port reflective-mode submersible sensor, devoted to the characterization of liquids, is reported. Let us also mention that, despite the fact that most planar resonant sensors are implemented by means of transmission lines coupled with the sensing resonator (or resonators), operating either in transmission or in reflection, there are other sensors involving antenna elements rather than transmission lines. Examples of such sensors will also be succinctly reviewed in this section.

2.3.1 One-Port Reflective-Mode Submersible Sensors

In one-port reflective-mode resonant sensors based on frequency variation, the output variables are the frequency position and the magnitude of the notch generated in the reflection coefficient. Such notch is caused by the effects of losses in the sensor structure (including ohmic, dielectric, and radiation losses), as well as in the MUT. The typical configuration of such sensors is of the type depicted in Fig. 2.1(c), i.e. a transmission line terminated with a sensing resonator. The sensor reported in [100],

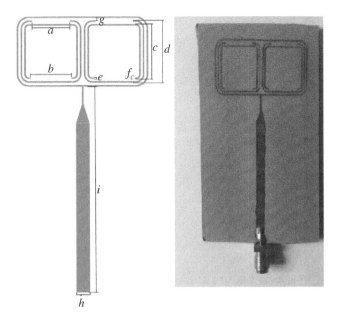

Figure 2.37 Layout and photograph of the reflective-mode submersible frequency-variation sensor based on a microstrip line terminated with a pair of SRRs. The sensor was fabricated on the *Taconic RF-35* substrate, with dielectric constant 3.5, loss tangent 0.0029, substrate thickness 0.5 mm and copper thickness of 35 μm. Dimensions are h = 4.99 mm, the gap of outer ring a = 10.5 mm, the gap of inner ring b = 11.5 mm, the side of the inner ring c = 15.64 mm, the side of the outer ring d = 17.6 mm, the width of the ring e = 0.5 mm, the separation between the rings and the feed line f = 0.56 mm, the separation between the ring and the feed line g = 0.65 mm. *Source:* Reprinted with permission from [100]; copyright 2018 IEEE.

depicted in Fig. 2.37, consists of a transmission line terminated with a pair of coupled SRRs surrounded by a loop structure (to which the resonators are coupled) in contact with the feed line. The dielectric properties of the MUT, solutions of pisco (an alcoholic beverage) in water, modify the resonance frequency and the notch depth, and therefore it is possible to determine the concentration of pisco. Indeed, the paper [100] uses a method that provides the dielectric constant of the solution, based on the measurement of the frequency shift in the SRRs caused by the pisco solution. Such method, reported in [101] in reference to SRR based sensors, is based on a formula equivalent to (2.14). Figure 2.38 depicts the return loss of the sensor when it is submersed in various solutions of pisco and water, and, as expected, the notches[11] shift a quantity that depends on the pisco concentration (further details on this sensor can be found in [100]).

2.3.2 Antenna-Based Frequency-Variation Resonant Sensors

Many planar frequency-variation sensors reported in the literature are based on different configurations of antenna elements coupled with (or loaded with) sensing resonators [102–106] and interdigital capacitors [107–109]. Other implementations exploit backscattering RFID technology,[12] where the antenna and the sensing element are wirelessly connected [110]. In this section, the sensors discussed in [105] and [110] are concisely reported as illustrative examples of antenna-based frequency-variation sensors.

---

11 Note that the considered sensor exhibits two notches within the considered frequency range.
12 RFID sensors, including backscattering chipless-RFID sensors, are the subject of Chapter 7. However, an example of a backscattered sensor is reported in this chapter, since such sensors exploit typically frequency variation.

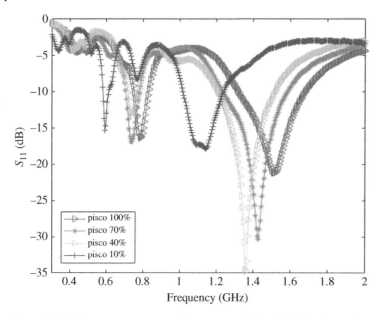

Figure 2.38 Frequency response of the sensor of Fig. 2.37, when it is submersed in solutions of water and pisco with different concentrations (indicated in the inset). *Source:* Reprinted with permission from [100]; copyright 2018 IEEE.

The sensor reported in [105] is a submersible sensor implemented by means of an SRR, the sensing element, excited by two monopole antennas. The notch introduced by the resonator in the transmission coefficient is used for sensing, similar to transmission-line-based resonant sensors. The layout and the photograph of the sensor submersed in a liquid are depicted in Fig. 2.39 (the details of the substrate and dimensions can be found in [105]). Figure 2.40 shows the responses of the sensor when it is submersed in different liquids. These results confirm that the notch frequency decreases with the dielectric constant of the LUT, as expected, and point out the viability of this approach for liquid sensing and characterization. In [102], a similar antenna-based sensor, with the sensing element (also an SRR) and the monopole antennas integrated in the same substrate, was reported, although it was applied to biomolecular sensing. By contrast, in the antenna-based sensor reported in [106], also devoted to the characterization of liquids, two antennas loaded with SRRs are placed at both sides of the liquid container.

The device reported in [110] is a backscattered temperature sensor based on the changes that the temperature generate in the dielectric constant of the sensor substrate. The sensor is a slot radiation patch with an alumina substrate backing, and it is interrogated by a CPW patch antenna, as depicted in Fig. 2.41 (see further details in [110]). Figure 2.42(a) depicts the responses of the sensor for different temperatures, whereas Fig. 2.42(b) shows the dependence of the resonance frequency with the temperature. These results validate the backscattering approach, though the distance between the interrogator antenna and the sensing element is small (5 mm, according to [110]). Other backscattering sensors, devoted to the measurement of other variables, e.g. strain, have also been reported (see for instance [111]). Nevertheless, backscattering sensors are discussed in further detail in Section 7.2.2.2 of Chapter 7, devoted to RFID sensors.

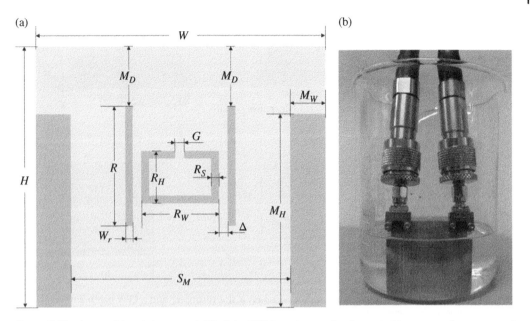

Figure 2.39 Layout (a) and photograph (b) of the SRR-based sensor implemented by means of two monopole antennas. *Source:* Reprinted with permission from [105]; copyright 2019 MDPI.

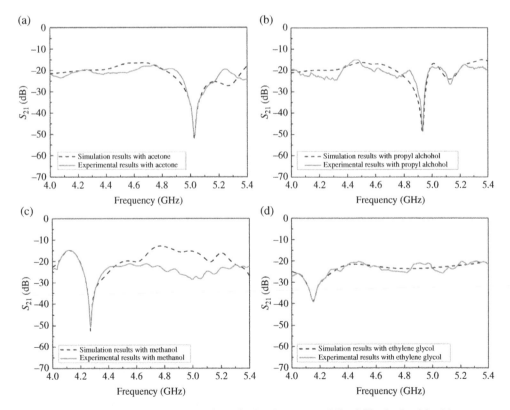

Figure 2.40 Simulated and experimental results for the sensor of Fig. 2.39, obtained for (a) acetone (ε_{MUT} = 20.7), (b) propyl alcohol (ε_{MUT} = 21.8), (c) methanol (ε_{MUT} = 33.1), and (d) ethylene glycol (ε_{MUT} = 37). *Source:* Reprinted with permission from [105]; copyright 2019 MDPI.

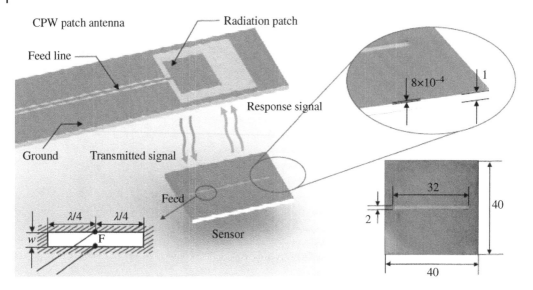

Figure 2.41 Sketch of the backscattering sensor based on a slot resonator and a CPW patch antenna. *Source:* Reprinted with permission from [110]; copyright 2018 MDPI.

2.4 Advantages and Drawbacks of Frequency-Variation Sensors

Frequency variation is probably the most extended working principle in planar microwave sensors. The main reason for such deployment of frequency-variation sensors is their design simplicity combined with their robustness against the effects of electromagnetic interference and noise. As mentioned, the canonical output variable in frequency-variation sensors is the frequency position of the notch, or peak, in the frequency response of the sensor. It is well-known that frequency measurements are less sensitive than amplitude measurements to the effects of noise and electromagnetic interference. Thus, the sensors subject of the present chapter exhibit high robustness against such effects, at least as compared with other sensors to be discussed later in this book (e.g. coupling-modulation sensors). Concerning design simplicity (an additional advantageous aspect of frequency-variation sensors), note that the sensors reported in this chapter are, in general, very simple. In most cases, such sensors are implemented by means of a transmission line loaded with a single sensing resonator. Moreover, there are not restrictions in regard to the shape and symmetry of the resonant element. By contrast, in other planar microwave sensors, such as frequency-splitting or coupling-modulation sensors, the number and shape of the sensing resonant elements should not be arbitrary, i.e. at least two sensing resonators are needed in frequency-splitting sensors, and coupling-modulation sensors, at least those based on electromagnetic symmetry properties, must be implemented by means of symmetric resonators symmetrically loading the host transmission line (these sensor types will be studied later in this book).

Another strong point in regard to frequency-variation sensors is their versatility or capability to measure different kinds of variables. Along this chapter, it has been shown that frequency-variation sensors are especially suitable for the dielectric characterization of solids and liquids, as well as for the determination of material composition, on the basis of complex permittivity measurements. The reason is that the resonance characteristics (frequency, magnitude, and quality factor) of a resonant element are intimately related to the dielectric properties of the surrounding material. However, examples of displacement sensors have also been reported. Moreover, there are other

(a)

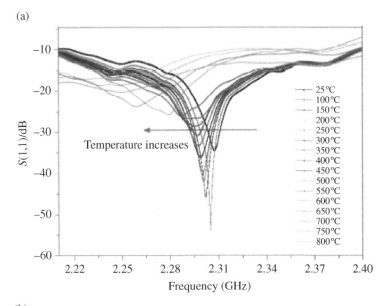

(b)

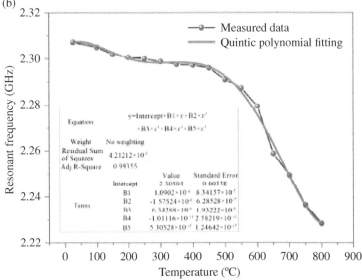

Figure 2.42 Response of the sensor of Fig. 2.41 for various temperatures (a) and dependence of the resonance frequency with temperature (b). *Source:* Reprinted with permission from [110]; copyright 2018 MDPI.

variables, such as strain, pressure, humidity, or temperature, among others, that affect the complex dielectric constant of the materials (or at least of certain functional materials), and, therefore, such variables can also be characterized by means of frequency-variation sensors.

Concerning drawbacks, the main limitative aspect of frequency-variation sensors, as compared with other sensors such as phase-variation or coupling-modulation sensors, concerns the need to perform broadband frequency measurements. That is, frequency-variation sensors are intrinsically broadband devices, since their operational principle relies on the variation of the resonance frequency of the sensing element. In sensors with a large output dynamic range, the frequency span needed for measurements should be accordingly high, and this may represent a cost penalty in a real scenario, where wideband voltage-controlled oscillators (VCOs) should be used for the generation of the feeding signals of the sensors[13]. Since the tuning capability of VCOs is limited, several VCOs, covering different frequency ranges, might be needed in order to cover the entire output dynamic range of the sensor. This solution is feasible, but it is not the most favorable in terms of cost, as compared with the use of a single VCO, as required in the so-called single-frequency sensors (e.g. phase-variation or coupling-modulation sensors), operating at a single frequency. Indeed, electronic frequency tuning is not needed in single-frequency sensors. However, some tuning of the input signal frequency (operating frequency) is necessary in practice, since fabricated sensors are not exempt of certain tolerances (in dimensions and substrate parameters – mainly the dielectric constant) that may affect the operating frequency. On the other hand, it should be mentioned that in multifrequency sensors devoted to multivariable measurements, such as the determination of the composition in multicomponent mixtures, the use of multiple VCOs covering different frequency bands is a due.

References

1 D. M. Pozar, *Microwave Engineering*, 4th ed., John Wiley, Hoboken, NJ, USA, 2012.

2 G. Kent, "An evanescent-mode tester for ceramic dielectric substrates," *IEEE Trans. Microw. Theory Tech.*, vol. 36, pp. 1451–1454, Oct. 1988.

3 G. Kent, "Nondestructive permittivity measurement of substrates," *IEEE Trans. Instrum. Meas.*, vol. 45, pp. 102–106, Feb. 1996.

4 M. D. Janezic and J. Baker-Jarvis, "Full-wave analysis of a split-cylinder resonator for nondestructive permittivity measurements," *IEEE Trans. Microw. Theory Tech.*, vol. 47, pp. 2014–2020, Oct. 1999.

5 M. D. Janezic, *Nondestructive Relative Permittivity and Loss Tangent Measurements using a Split-Cylinder Resonator*, Ph.D. Thesis, University of Colorado at Boulder, 2003.

6 Keysight Technologies, "85072A 10-GHz Split Cylinder Resonator," Technical Overview.

7 J. Hong and M. J. Lancaster, *Microstrip Filters for RF/Microwave Applications*, John Wiley, Hoboken, NJ, USA, 2001.

8 M. Durán-Sindreu, J. Naqui, F. Paredes, J. Bonache, and F. Martín, "Electrically small resonators for planar metamaterial microwave circuit and antenna design: a comparative analysis," *Appl. Sciences*, vol. 2, pp. 375–395, 2012.

9 J. Muñoz-Enano, P. Vélez, M. Gil, and F. Martín, "Planar microwave resonant sensors: a review and recent developments," *Appl. Sciences*, vol. 10, p. 2615 (29 pages), 2020.

10 M. Makimoto and S. Yamashita, "Compact bandpass filters using stepped impedance resonators," *Proc. IEEE*, vol. 67, pp. 16–19, 1979.

11 J. Naqui, M. Durán-Sindreu, J. Bonache, and F. Martín, "Implementation of shunt connected series resonators through stepped-impedance shunt stubs: analysis and limitations," *IET Microw. Ant. Propag.*, vol. 5, pp. 1336–1342, Aug. 2011.

12 J. Naqui and F. Martín, "Mechanically reconfigurable microstrip lines loaded with stepped impedance resonators (SIRs) and potential applications," *Int. J. Ant. Propag.*, vol. 2014, Article ID 346838 (8 pages), 2014.

13 J. Naqui, M. Durán-Sindreu, and F. Martín, "Modeling split ring resonator (SRR) and complementary split ring resonator (CSRR) loaded transmission lines exhibiting cross polarization effects," *IEEE Ant. Wireless Propag. Lett.*, vol. 12, pp. 178–181, 2013.

---

13 For sensor validation purposes, vector network analyzers (VNAs) are typically used. However, these instruments are expensive and cannot be considered in operational environments.

14 R. Marqués, F. Martín, and M. Sorolla, *Metamaterials with Negative Parameters: Theory, Design and Microwave Applications*, John Wiley, Hoboken, NJ, USA, 2007.

15 F. Martín, *Artificial Transmission Lines for RF and Microwave Applications*, John Wiley, Hoboken, NJ, USA, 2015.

16 J. B. Pendry, A. J. Holden, D. J. Robbins, and W. J. Stewart, "Magnetism from conductors and enhanced nonlinear phenomena," *IEEE Trans. Microw. Theory Tech.*, vol. 47, pp. 2075–2084, 1999.

17 D. R. Smith, W. J. Padilla, D. C. Vier, S. C. Nemat-Nasser, and S. Schultz, "Composite medium with simultaneously negative permeability and permittivity," *Phys. Rev. Lett.*, vol. 84, pp. 4184–4187, 2000.

18 R. A. Shelby, D. R. Smith, and S. Schultz, "Experimental verification of a negative index of refraction," *Science*, vol. 292, pp. 77–79, 2001.

19 R. Marqués, J. Martel, F. Mesa, and F. Medina, "Left-handed-media simulation and transmission of EM waves in subwavelength split-ring-resonator-loaded metallic waveguides," *Phys. Rev. Lett.*, vol. 89, paper 183901, Oct. 2002.

20 F. Martín, F. Falcone, J. Bonache, R. Marqués, and M. Sorolla, "Split ring resonator based left handed coplanar waveguide," *Appl. Phys. Lett.*, vol. 83, pp. 4652–4654, Dec. 2003.

21 R. Marques, F. Medina, and R. Rafii-El-Idrissi, "Role of bi-anisotropy in negative permeability and left handed metamaterials," *Phys. Rev. B*, vol. 65, paper 144441, 2002.

22 R. Marques, F. Mesa, J. Martel, and F. Medina, "Comparative analysis of edge- and broadside-coupled split ring resonators for metamaterial design - theory and experiments," *IEEE Trans. Ant. Propag.*, vol. 51, pp. 2572–2581, Oct. 2003.

23 R. Marqués, J. D. Baena, J. Martel, F. Medina, F. Falcone, M. Sorolla, and F. Martin, "Novel small resonant electromagnetic particles for metamaterial and filter design," *Proc. ICEAA'03*, Torino, Italy, 8–12 Sept. 2003, pp. 439–442.

24 J. García-García, F. Martín, J. D. Baena, R. Marqués, and L. Jelinek, "On the resonances and polarizabilities of split ring resonators," *J. Appl. Phys.*, vol. 98, paper 033103, 2005.

25 J. D. Baena, J. Bonache, F. Martín, R. Marqués, F. Falcone, T. Lopetegi, M. A. G. Laso, J. García, I. Gil, M. Flores-Portillo, and M. Sorolla, "Equivalent circuit models for split ring resonators and complementary split rings resonators coupled to planar transmission lines," *IEEE Trans. Microw. Theory Techn.*, vol. 53, pp. 1451–1461, Apr. 2005.

26 J. Martel, R. Marqués, F. Falcone, J. D. Baena, F. Medina, F. Martín, and M. Sorolla, "A new LC series element for compact band pass filter design," *IEEE Microw. Wireless Compon. Lett.*, vol. 14, pp. 210–212, 2004.

27 J. Martel, J. Bonache, R. Marqués, F. Martín, and F. Medina, "Design of wide-band semi-lumped bandpass filters using open split ring resonators," *IEEE Microw. Wireless Compon. Lett.*, vol. 17, pp. 28–30, Jan. 2007.

28 J. D. Baena. R. Marqués, F. Medina, and J. Martel, "Artificial magnetic metamaterial design by using spiral resonators," *Phys. Rev. B*, vol. 69, paper 014402, 2004.

29 F. Falcone, F. Martín, J. Bonache, M. A. G. Laso, J. García-García, J. D. Baena, R. Marqués, and M. Sorolla, "Stop band and band pass characteristics in coplanar waveguides coupled to spiral resonators," *Microw. Opt. Technol. Lett.*, vol. 42, pp. 386–388, Sep. 2004.

30 D. Schurig, J. J. Mock, and D. R. Smith, "Electric-field-coupled resonators for negative permittivity metamaterials," *Appl. Phys. Lett.*, vol. 88, paper 041109, 2006.

31 H. Chen, L. Ran, J. Huangfu, X. Zhang, K. Chen, T. M. Grzegorczyk, and J. A Kong, "Left-handed materials composed of only S-shaped resonators," *Phys. Rev. E*, vol. 70, paper 057605, 2004.

32 H. Chen, L. Ran, J. Huangfu, X. Zhang, K. Chen, T. M. Grzegorczyk, and J. A Kong, "Negative refraction of a combined double S-shaped metamaterial," *Appl. Phys. Lett.*, vol. 86, paper 151909, 2005.

33 J. Naqui, J. Coromina, A. Karami-Horestani, C. Fumeaux, and F. Martín, "Angular displacement and velocity sensors based on coplanar waveguides (CPWs) loaded with S-shaped split ring resonator (S-SRR)," *Sensors*, vol. 15, pp. 9628–9650, 2015.

34 H. G. Booker, "Slot aerials and their relation to complementary wire aerials (Babinet's principle)," *J. IEE*, vol. 93, pt. III-A, no. 4, pp. 620–626, 1946.

35 G. A. Deschamps, "Impedance properties of complementary multiterminal planar structures," *IRE Trans. Ant. Propag.*, vol. AP-7, pp. 371–378, 1959.

36 W. J. Getsinger, "Circuit duals on planar transmission media," *1983 IEEE MTT-S Int. Microwave Symp. Dig. (IMS'1983)*, Boston, MA, USA, May-June 1983, pp. 154–156.

37 F. Falcone, T. Lopetegi, M. A. G. Laso, J. D. Baena, J. Bonache, R. Marqués, F. Martín, and M. Sorolla, "Babinet principle applied to the design of metasurfaces and metamaterials," *Phys. Rev. Lett.*, vol. 93, paper 197401, Nov. 2004.

38 D. Ahn, J. S. Park, C. S. Kim, J. Kim, Y. Qian, and T. Itoh, "A design of the low-pass filter using the novel microstrip defected ground structure," *IEEE Trans. Microw. Theory Techn.*, vol. 49, pp. 86–93, Jan. 2001.

39 C. S. Kim, J. S. Park, D. Ahn, and J. B. Lim, "A novel 1-D periodic defected ground structure for planar circuits," *IEEE Microw. Guided Wave Lett.*, vol. 10, pp. 131–133, Apr. 2000.

40 V. Komarov, O. Barybin, Y. V. Rassokhina, and V. G. Krizhanovski, "Dumbbell-shaped defected ground structure resonator filter for high-efficiency microwave power amplifiers," *2018 Int. Conf. Inf. Telecomm. Technol. Radio Electronics (UkrMiCo)*, Odessa, Ukraine, Sept. 2018.

41 M. K. Khandelwal, B. K. Kanaujia, and S. Kumar "Defected ground structure: fundamentals, analysis, and applications in modern wireless trends," *Int. J. Ant. Propag.*, vol. 2017, paper 2018527 (22 pages), 2017.

42 J. X. Chen, J. Shi, Z. H. Bao, and Q. Xue, "Tunable and switchable bandpass filters using slot-line resonators," *Prog. Electromagn. Res.*, vol. 111, pp. 25–41, 2011.

43 A. M. E. Safwat, F. Podevin, P. Ferrari, and A. Vilcot, "Tunable bandstop defected ground structure resonator using reconfigurable dumbbell-shaped coplanar waveguide," *IEEE Trans. Microw. Theory Techn.*, vol. 54, pp. 3559–3564, Sept. 2006.

44 F. Falcone, T. Lopetegi, J. D. Baena, R. Marqués, F. Martín, and M. Sorolla, "Effective negative-ε stop-band microstrip lines based on complementary split ring resonators," *IEEE Microw. Wireless Compon. Lett.*, vol. 14, pp. 280–282, Jun. 2004.

45 J. Selga, G. Sisó, M. Gil, J. Bonache, and F. Martín, "Microwave circuit miniaturization with complementary spiral resonators (CSRs): application to high-pass filters and dual-band components," *Microw. Opt. Technol. Lett.*, vol. 51, pp. 2741–2745, Nov. 2009.

46 A. Velez, F. Aznar, J. Bonache, M. C. Velázquez-Ahumada, J. Martel, and F. Martín, "Open complementary split ring resonators (OCSRRs) and their application to wideband CPW band pass filters," *IEEE Microw. Wireless Compon. Lett.*, vol. 19, pp. 197–199, Apr. 2009.

47 J. Naqui, M. Durán-Sindreu, and F. Martín, "Differential and single-ended microstrip lines loaded with slotted magnetic-LC (MLC) resonators," *Int. J. Ant. Propag.*, vol. 2013, paper 640514, 8 pages, 2013.

48 A. Ebrahimi, T. C. Baum, K. Wang, J. Scott, and K. Ghorbani, "Differential transmission lines loaded with magnetic LC resonators and application in common mode suppression," *IEEE Trans. Circ. Syst. I: Reg. Papers*, vol. 66, pp. 3811–3821, 2019.

49 E. Chen and S. Y. Chou, "Characteristics of coplanar transmission lines on multilayer substrates: modeling and experiments," *IEEE Trans. Microw. Theory Techn.*, vol. 45, no. 6, pp. 939–945, Jun. 1997.

50 J. Bonache, M. Gil, I. Gil, J. Garcia-García, and F. Martín, "On the electrical characteristics of complementary metamaterial resonators," *IEEE Microw. Wireless Compon. Lett.*, vol. 16, pp. 543–545, Oct. 2006.

51 F. Aznar, M. Gil, J. Bonache, J. D. Baena, L. Jelinek, R. Marqués, and F. Martín, "Characterization of miniaturized metamaterial resonators coupled to planar transmission lines through parameter extraction," *J. Appl. Phys.*, vol. 104, no. 11, p. 114501, Dec. 2008.

52 M. Durán-Sindreu, A. Vélez, F. Aznar, G. Sisó, J. Bonache, and F. Martín, "Application of open split ring resonators and open complementary split ring resonators to the synthesis of artificial transmission lines and microwave passive components," *IEEE Trans. Microw. Theory Techn.*, vol. 57, pp. 3395–3403, Dec. 2009.

53 J. Muñoz-Enano, P. Vélez, C. Herrojo, M. Gil, and F. Martín, "On the sensitivity of microwave sensors based on slot resonators and frequency variation," *2019 Int. Conf. Electromagn. Advan. Applications (ICEAA)*, Granada, Spain, Sept. 2019, pp. 112–115.

54 J. Muñoz-Enano, J. Martel, P. Vélez, F. Medina, L. Su, and F. Martín, "Parametric analysis of the edge capacitance of uniform slots and application to frequency-variation permittivity sensors," *Appl. Science*, vol. 11, no. 15, paper 7000, 2021.

55 M. Puentes, C. Weiss, M. Schussler, and R. Jakoby, "Sensor array based on split ring resonators for analysis of organic tissues," *IEEE MTT-S Int. Microw. Symp. Dig.*, Baltimore, MD, USA, Jun. 2011.

56 C. Mandel, B. Kubina, M. Schüßler, and R. Jakoby, "Passive chipless wireless sensor for two-dimensional displacement measurement," *41st European Microwave Conference*, Manchester, UK, Oct. 2011, pp. 79–82.

57 A. Ebrahimi, W. Withayachumnankul, S. Al-Sarawi, and D. Abbott, "High-sensitivity metamaterial-inspired sensor for microfluidic dielectric characterization," *IEEE Sensors J.*, vol. 14, no. 5, pp. 1345–1351, May 2014.

58 M. Schueler, C. Mandel, M. Puentes, and R. Jakoby, "Metamaterial inspired microwave sensors," *IEEE Microw. Mag.*, vol. 13, no. 2, pp. 57–68, Mar. 2012.

59 M. S. Boybay and O. M. Ramahi, "Material characterization using complementary split-ring resonators," *IEEE Trans. Instrum. Meas.*, vol. 61, no. 11, pp. 3039–3046, Nov. 2012.

60 C. S. Lee and C. L. Yang, "Complementary split-ring resonators for measuring dielectric constants and loss tangents," *IEEE Microw. Wireless Compon. Lett.*, vol. 24, no. 8, pp. 563–565, Aug. 2014.

61 W. Withayachumnankul, K. Jaruwongrungsee, A. Tuantranont, C. Fumeaux, and D. Abbott, "Metamaterial-based microfluidic sensor for dielectric characterization", *Sens. Actuators A*, vol. 189, pp. 233–237, 2013.

62 A. Salim and S. Lim, "Complementary split-ring resonator-loaded microfluidic ethanol chemical sensor", *Sensors*, vol 16, paper 1802, 2016.

63 C. L. Yang, C. S. Lee, K. W. Chen, and K. Z. Chen, "Noncontact measurement of complex permittivity and thickness by using planar resonators," *IEEE Trans. Microw. Theory Techn.*, vol. 64, no. 1, pp. 247–257, Jan. 2016.

64 L. Su, J. Mata-Contreras, P. Velez, and F. Martin, "Estimation of the complex permittivity of liquids by means of complementary Split ring resonator (CSRR) loaded transmission lines," *IEEE MTT-S Int. Microw. Workshop Series on Advanced Materials and Processes (IMWS-AMP 2017)*, Pavia, Italy, Sep. 2017, pp. 20–22.

65 L. Su, J. Mata-Contreras, P. Vélez, A. Fernández-Prieto, and F. Martín, "Analytical method to estimate the complex permittivity of oil samples," *Sensors*, vol. 18, no. 4, paper 984, Mar. 2018.

66 A. K. Jha, N. Delmonte, A. Lamecki, M. Mrozowski, and M. Bozzi, "Design of microwave-based angular displacement sensor," *IEEE Microw. Wireless Compon. Lett.*, vol. 29, no. 4, pp. 306–308, Apr. 2019.

67 G. Galindo-Romera, F. Javier Herraiz-Martinez, M. Gil, J. J. Martinez-Martinez, and D. Segovia-Vargas, "Submersible printed split-ring resonator-based sensor for thin-film detection and permittivity characterization," *IEEE Sensors J.*, vol. 16, no. 10, pp. 3587–3596, May 2016.

68 M. Abdolrazzaghi, M. H. Zarifi, and M. Daneshmand, "Sensitivity enhancement of split ring resonator based liquid sensors," *2016 IEEE Sensors*, Orlando, FL, USA, 30 Oct.–3 Nov. 2016.

69 M. Abdolrazzaghi, M. H. Zarifi, W. Pedrycz, and M. Daneshmand, "Robust ultra-high resolution microwave planar sensor using fuzzy neural network approach," *IEEE Sens. J.*, vol. 17, pp. 323–332, 2016.

70 M. H. Zarifi and M. Daneshmand, "Monitoring solid particle deposition in lossy medium using planar resonator sensor," *IEEE Sens. J.*, vol. 17, pp. 7981–7989, 2017.

71 M. H. Zarifi, S. Deif, M. Abdolrazzaghi, B. Chen, D. Ramsawak, M. Amyotte, N. Vahabisani, Z. Hashisho, W. Chen, and M. Daneshmand, "A microwave ring resonator sensor for early detection of breaches in pipeline coatings," *IEEE Trans. Ind. Electron.*, vol. 65, pp. 1626–1635, 2017.

72 M. Abdolrazzaghi, M. Daneshmand, and A. K. Iyer, "Strongly enhanced sensitivity in planar microwave sensors based on metamaterial coupling," *IEEE Trans. Microw. Theory Tech.*, vol. 66, pp. 1843–1855, 2018.

73 R. A. Alahnomi, Z. Zakaria, E. Ruslan, S. R. Ab Rashid, and A. A. M. Bahar, "High-Q sensor based on symmetrical split ring resonator with spurlines for solids material detection," *IEEE Sensors J.*, vol. 17, no. 9, pp. 2766–2775, May 2017.

74 N. Jankovic and V. Radonic, "A microwave microfluidic sensor based on a dual-mode resonator for dual-sensing applications," *Sensors*, vol. 17, no. 12, paper 2713, Nov. 2017.

75 H. Zhou, D. Hu, C. Yang, C. Chen, J. Ji, M. Chen, Y. Chen, Y. Yang, and X. Mu, "Multi-band sensing for dielectric property of chemicals using metamaterial integrated microfluidic sensor," *Sci. Rep.*, vol. 8, no. 1, paper 14801, Dec. 2018.

76 M. A. H. Ansari, A. K. Jha, Z. Akhter, and M. J. Akhtar, "Multi-band RF planar sensor using complementary split ring resonator for testing of dielectric materials," *IEEE Sensors J.*, vol. 18, no. 16, pp. 6596–6606, Aug. 2018.

77 H. Lobato-Morales, J. H. Choi, H. Lee, and J. L. Medina-Monroy, "Compact dielectric-permittivity sensors of liquid samples based on substrate integrated-waveguide with negative-order-resonance," *IEEE Sensors J.*, vol. 19, no. 19, pp. 8694–8699, Oct. 2019.

78 A. Ebrahimi, J. Scott, and K. Ghorbani, "Ultrahigh-sensitivity microwave sensor for microfluidic complex permittivity measurement," *IEEE Trans. Microw. Theory Tech.*, vol. 67, pp 4269–4277, 2019.

79 Y. Khanna and Y. K. Awasthi, "Dual-band microwave sensor for investigation of liquid impurity concentration using a metamaterial complementary split-ring resonator," *J. Electron. Mater.*, vol. 49, no. 1, pp. 385–394, Jan. 2020.

80 J. Muñoz-Enano, P. Vélez, M. Gil, and F. Martín, "Frequency variation sensors for permittivity measurements based on dumbbell-shaped defect ground structures (DB-DGS): analytical method and sensitivity analysis," *IEEE Sensors J.*, Mar. 2022. DOI: https://doi.org/10.1109/JSEN.2022.3163470.

81 L. Su, J. Mata-Contreras, P. Vélez, and F. Martín, "Estimation of conductive losses in complementary split ring resonator (CSRR) loading an embedded microstrip line and applications", *2017 IEEE MTT-S International Microwave Symposium (IMS'2017)*, Honolulu, HI, USA, Jun. 2017, pp. 476–479.

82 J. Z. Bao, M. L. Swicord, and C. C. Davis, "Microwave dielectric characterization of binary mixtures of water, methanol, and ethanol," *J. Chem. Phys.*, vol. 104, no. 12, pp. 4441–4450, 1996.

83 M. Puentes, M. Maasch, M. Schubler, and R. Jakoby, "Frequency multiplexed 2-dimensional sensor array based on split-ring resonators for organic tissue analysis," *IEEE Trans. Microw. Theory Techn.*, vol. 60, no. 6, pp. 1720–1727, Jun. 2012.

84 A. Trehan and N. Nikolova, *"Summary of Materials and Recipes Available in the Literature to Fabricate Biological Phantoms for RF and Microwave Experiments,"* Ph.D. dissertation, Computational Electromagnetics Lab., Dept. Elect. Comp. Eng., McMaster University, ON, Canada, 2009.

85 A. Trehan and N. Nikolova, *"Numerical and Physical Models for Microwave Breast Imaging,"* Ph.D. dissertation, Department of Electrical and Computer Engineering, McMaster University, 2009.

86 M. Puentes, F. Bashir, M. Maasch, M. Schüßler, and R. Jakoby, "Planar microwave sensor for thermal ablation of organic tissue," *2013 European Microwave Conference*, Nuremberg, Germany, Oct. 2013, pp. 479–482.

87 M. Puentes, M. Maasch, M. Schüssler, C. Damm, and R. Jakoby, "Analysis of resonant particles in a coplanar microwave sensor array for thermal ablation of organic tissue," *2014 IEEE MTT-S International Microwave Symposium (IMS'2014)*, Tampa, FL, USA, June 2014.

88 L. Solymar and D. Walsh, *Electrical Properties of Materials*, 7th ed., Oxford University Press, Oxford, New York, 2004.

89 A. K Jonscher, "Dielectric relaxation in solids," *J. Phys. D: Appl. Phys.*, vol. 32, pp. R57–R70, 1999.

90 K. S. Cole, and R. H. Cole, "Dispersion and absorption in dielectrics I. Alternating current characteristics," *J. Chem. Phys.*, vol. 9, pp. 341–352, 1941.

91 Y. P. Kalmykov, W. T. Coffey, D. S. Crothers, and S. V. Titov, "Microscopic models for dielectric relaxation in disordered systems," *Phys. Rev. E*, vol. 70, no. 4, paper 041103, 2004.

92 N. Hosseini, M. Baghelani, and M. Daneshmand, "Selective volume fraction sensing using resonant-based microwave sensor and its harmonics," *IEEE Trans. Microw. Theory Techn.*, vol. 68, no. 9, pp. 3958–3968, Sep. 2020.

93 N. Hosseini and M. Baghelani, "Selective real-time non-contact multi-variable water-alcohol-sugar concentration analysis during fermentation process using microwave split-ring resonator based sensor," *Sens. Act. A: Phys.*, vol. 325, p. 112695, 2021.

94 A. M. Albishi, S. H. Mirjahanmardi, A. M. Ali, V. Nayyeri, S. M. Wasly, and O. M. Ramahi, "Intelligent sensing using multiple sensors for material characterization," *Sensors*, vol. 19, no. 21, paper 4766, Nov. 2019.

95 T. Mosavirik, M. Soleimani, V. Nayyeri, S. H. Mirjahanmardi, and O. M. Ramahi, "Permittivity characterization of dispersive materials using power measurements," *IEEE Trans. Instrum. Meas.*, vol. 70, pp. 1–8, paper 6005508, 2021.

96 A. A. Abduljabar, H. Hamzah, and A. Porch, "Multi-resonators, microwave microfluidic sensor for liquid characterization," *Microw Opt Technol Lett.*, vol. 63, pp. 1042–1047, 2021.

97 M. A. Rafi, *Tunable Microwave Resonator Using Liquid Metal and 3D Printed Fluidic Channel for the Development of Microwave Spectroscopy*, Master Thesis, University of British Columbia, 2021.

98 K. Grenier, D. Dubuc, T. Chen, F. Artis, T. Chretiennot, M. Poupot, and J. J. Fournie, "Recent advances in microwave-based dielectric spectroscopy at the cellular level for cancer investigations," *IEEE Trans. Microw. Theory Tech.*, vol. 61, no. 5, pp. 2023–2030, 2013.

99 A. P. Saghati, J. S. Batra, J. Kameoka, and K. Entesari, "A metamaterial-inspired wideband microwave interferometry sensor for dielectric spectroscopy of liquid chemicals," *IEEE Trans. Microw. Theory Tech.*, vol. 65, no. 7, pp. 2558–2571, 2017.

100 A. Nuñez-Flores, P. Castillo-Araníbar, A. García-Lampérez, and D. Segovia-Vargas, "Design and implementation of a submersible split ring resonator based sensor for pisco concentration measurements," *2018 IEEE MTT-S Latin America Microwave Conference (LAMC 2018)*, Arequipa, Peru, Dec. 2018.

101 S. P. Chakyar, S. K. Simon, C. Bindu, J. Andrews, and V. P. Joseph, "Complex permittivity measurement using metamaterial split ring resonators," *J. Appl. Phys.*, vol. 121, paper 054101, 2017.

102 H. Torun, F. Cagri Top, G. Dundar, and A. D. Yalcinkaya, "An antenna-coupled split-ring resonator for biosensing", *J. Appl. Phys.*, vol. 116, paper 124701, 2014.

103 A. Verma, S. Bhushan, P. N. Tripathi, M. Goswami, and B. R. Singh, "A defected ground split ring resonator for an ultra-fast, selective sensing of glucose content in blood plasma," *J. Electromagn. Waves Appl.*, vol. 31, no. 10, pp. 1049–1061, Jul. 2017.

104 M. Asad, S. A. Neyadi, O. A. Aidaros, M. Khalil, and M. Hussein, "Single port bio-sensor design using metamaterial split ring resonator," *2016 5th International Conference on Electronic Devices, Systems and Applications* (ICEDSA), Ras AI Khaimah, United Arab Emirates, Dec. 2016.

105 E. Reyes-Vera, G. Acevedo-Osorio, M. Arias-Correa, and D. E. Senior, "A submersible printed sensor based on a monopole-coupled split ring resonator for permittivity characterization," *Sensors*, vol. 19, Apr. 2019.

106 M. Islam, F. Ashraf, T. Alam, N. Misran, and K. Mat, "A compact ultrawideband antenna based on hexagonal split-ring resonator for pH sensor application," *Sensors*, vol. 18, no. 9, paper 2959, Sep. 2018.

107 R. Sujith and R. Augustine, "CPW fed antenna for nearfield sensor applications," *2018 IEEE Conference on Antenna Measurements & Applications* (CAMA), Vasteras, Sweden, Sep. 2018.

108 T. A. Elwi, "Metamaterial based a printed monopole antenna for sensing applications," *Int. J. RF Microw. Comput. Aided Eng.*, vol. 28, no. 7, paper e21470, Sep. 2018.

109 W. H. W. Morshidi, Z. Zaharudin, S. Khan, A. N. Nordin, F. A. Shaikh, I. Adam, and K. A. Kader, "Inter-digital sensor for non-invasive blood glucose monitoring," *2018 IEEE International Conference on Innovative Research and Development* (ICIRD), Bangkok, Thailand, May 2018.

110 F. Lu, H. Wang, Y. Guo, Q. Tan, W. Zhang, and J. Xiong, "Microwave backscatter-based wireless temperature sensor fabricated by an alumina-backed Au slot radiation patch," *Sensors*, vol. 18, paper 242, 2018.

111 R. Melik, E. Unal, N. Kosku Perkgoz, C. Puttlitz, and H. Volkan Demir, "Metamaterial-based wireless strain sensors," *Appl. Phys. Lett.*, vol. 95, paper 011106, 2009.

3

Phase-Variation Sensors

This chapter focuses on phase-variation sensors, mainly, although not exclusively, devoted to the measurement of dielectric properties of solids and liquids and determination of material composition. Most reported phase-variation sensors are implemented by means of transmission lines, working either in transmission or in reflection, on the basis of the phase variation experienced by the line when a certain material (material under test – MUT) lies on the sensing region. As it will be shown throughout this chapter, the sensitivity of transmission line phase-variation sensors increases with the length of the sensing line and/or with the operating frequency. To avoid excessively sized sensors and/or high operational frequencies, several strategies for sensitivity enhancement are discussed in this chapter. Of particular interest are the reflective-mode sensors implemented by means of step-impedance open-ended lines, where unprecedented sensitivities with limited sensor dimensions are achieved, or the sensors based on highly dispersive artificial lines, such as composite right/left handed (CRLH), electro-inductive wave (EIW), or slow-wave transmission lines, exhibiting sensitivities superior to those achievable with ordinary lines. Nevertheless, resonant-type phase-variation sensors are also studied in this chapter. Such resonant sensors are of interest for the dielectric characterization of materials, but they are also useful for sensing angular displacements, provided their working principle is based on the effects that cross-polarization (intimately related to the relative orientation between a host line and the resonant element) causes on the phase of the reflection coefficients. Prototypes for each specific sensor type are reported in the chapter, which concludes with the main relevant advantages and limitations of phase-variation sensors.

3.1 General Working Principle of Phase-Variation Sensors

The general working principle of phase-variation sensors is the change experienced by the phase of the transmission or reflection coefficient of a transmission line (ordinary, meandered, artificial, resonator-loaded, etc.) caused by the variable to be sensed (or measurand). Phase-variation sensors are typically, although not exclusively, used for measuring material properties and composition. The main reason is that the phase of the lines, as well as the characteristic impedance, depends on the permittivity of the MUT, provided it is in close proximity, or in contact, with the line. The typical input variable in phase-variation sensors is thus the dielectric constant, or, more generally, the complex permittivity of the MUT. However, any other parameter related to the dielectric constant can also be measured by means of phase-variation sensors. In particular, these sensors can be applied to detect defects in samples, provided such defects modify the effective dielectric constant of the MUT,

Planar Microwave Sensors, First Edition. Ferran Martín, Paris Vélez, Jonathan Muñoz-Enano, and Lijuan Su.
© 2023 John Wiley & Sons, Inc. Published 2023 by John Wiley & Sons, Inc.

(a)   (b)

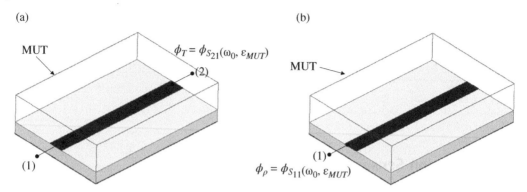

Figure 3.1 Sketch showing the schematic and working principle of transmission-mode (a) and reflective-mode (b) phase-variation sensors based on ordinary lines.

or to determine material composition (e.g. the percentage of solute content in liquid solutions). Nevertheless, the phase of a transmission line (and the impedance) can also be perturbed by means of a movable object closely spaced to it. Consequently, these sensors are also of interest for measuring displacements (either linear or angular), as it will be shown later.

Figure 3.1 shows a sketch showing the working principle of phase-variation sensors, where a transmission-mode (two-port) structure and a reflective-mode (one-port) structure are considered, and the output variables are the phase of the transmission and reflection coefficients, respectively. It should be mentioned, however, that in certain sensor implementations, the phase information is converted into magnitude information. Moreover, phase-variation sensors based on cross-polarized resonant elements, useful for angular displacement measurement, are also included in this chapter. In such sensors, the differential phase of the reflection coefficients at both ports is the output variable.

3.2 Transmission-Line Phase-Variation Sensors

Let us consider a sensing structure implemented by means of an ordinary transmission line with physical length l and phase constant β at a certain frequency f_0 (the operating frequency). The electrical length, or phase, of the line at that frequency is thus $\phi = \beta l$, which in turn can be expressed as [1]

$$\phi = \beta l = \frac{\omega_0}{v_p} l \qquad (3.1)$$

where $\omega_0 = 2\pi f_0$ is the angular frequency and v_p is the phase velocity of the line. According to expression (3.1), any change in the electrical length of the line should be entirely due to a variation in the phase velocity, since the frequency is set to a given value and the physical length is constant (i.e. it does not depend on the input variable, typically the dielectric constant of the MUT, ε_{MUT}). Without loss of generality, for the purpose of the present analysis, let us consider that ε_{MUT} is the input variable, and that the output variable is the phase of the line, ϕ. The sensitivity, the key sensor parameter, defined as the derivative of ϕ with ε_{MUT}, can thus be expressed as

$$S = \frac{d\phi}{d\varepsilon_{MUT}} = \frac{d\phi}{dv_p} \cdot \frac{dv_p}{d\varepsilon_{MUT}} = -\frac{\omega_0}{v_p^2} l \cdot \frac{dv_p}{d\varepsilon_{MUT}} \tag{3.2}$$

Inspection of expression (3.2) reveals that increasing the operating frequency and/or the length of the line improves the sensitivity. For this reason, long-line phase-variation sensors have been reported [2, 3].[1] Typically, such sensors have been implemented by means of meandered line geometries [3], in order to achieve sensing regions with reasonable shape factors.

In view of (3.2), reducing the phase velocity of the line significantly increases the first term of the sensitivity. The phase velocity is given by [1]

$$v_p = \frac{c}{\sqrt{\varepsilon_{eff}}} \tag{3.3}$$

c and ε_{eff} being the speed of light in free space and the effective dielectric constant of the line, respectively. Thus, by increasing the dielectric constant of the substrate, ε_r, intimately related to ε_{eff}, the phase velocity of the line is reduced. However, increasing ε_r tends to degrade the second term of the sensitivity. Therefore, evaluating the effectiveness in reducing the phase velocity of the sensing line for sensitivity improvement requires a detailed analysis of the last term in (3.2). This analysis depends on the type of open line considered (e.g. microstrip and coplanar waveguide).[2] Nevertheless, such term can be written as

$$\frac{dv_p}{d\varepsilon_{MUT}} = \frac{dv_p}{d\varepsilon_{eff}} \cdot \frac{d\varepsilon_{eff}}{d\varepsilon_{MUT}} = -\frac{v_p^3}{2c^2} \cdot \frac{d\varepsilon_{eff}}{d\varepsilon_{MUT}} \tag{3.4}$$

where (3.3) has been used. By introducing (3.4) in (3.2), the sensitivity is found to be

$$S = \frac{\omega_0 v_p l}{2c^2} \cdot \frac{d\varepsilon_{eff}}{d\varepsilon_{MUT}} \tag{3.5}$$

Let us now obtain the second term of the right-hand side member of (3.5) for the most extended types of planar open lines, namely, microstrip lines and coplanar waveguides (CPW). In both cases, a semi-infinite MUT in contact with the line is considered. In practice, this approximation means that the electromagnetic fields generated in the line do not reach the boundaries of the MUT. This approximation will also be applied to the substrate of the CPW transmission line, uncoated in the backside. A cross-sectional view of both line types, covered by the MUT, is depicted in Fig. 3.2.

For microstrip lines, the effective dielectric constant under the considered approximation is given by [1]

$$\varepsilon_{eff} = \frac{\varepsilon_r + \varepsilon_{MUT}}{2} + \frac{\varepsilon_r - \varepsilon_{MUT}}{2} F \tag{3.6}$$

where F is a geometry factor that depends on the width of the line, W, and thickness of the substrate, h, according to

---

1 Increasing the operating frequency of the sensor is another possibility for sensitivity optimization, but this option may represent further costs for signal generation in a real scenario.

2 Note that closed transmission lines, such as striplines, are not useful, in general, for sensing, as far as the presence of the MUT is not able to modify the phase velocity of such lines, given by $c/\sqrt{\varepsilon_r}$. That is, in such lines, the effective dielectric constant is the one of the considered substrate, since the ground plane prevents from any interaction between the electromagnetic fields generated by the line and the MUT. Nevertheless, there are options to locate (embed) the MUT within the substrate, e.g. by means of micromachining, or through micro-tubes (for liquid sensing), and thus alter the effective dielectric constant of the line (and hence the phase velocity). Under these circumstances, sensing by means of closed lines is possible.

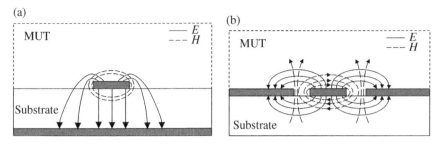

Figure 3.2 Cross-sectional view of the microstrip (a) and CPW (b) transmission lines covered by the MUT. The electromagnetic fields are indicated.

$$F = \left(1 + 12\frac{h}{W}\right)^{-1/2} \tag{3.7a}$$

provided $W > h$. If $W < h$, the geometry factor is

$$F = \left(1 + 12\frac{h}{W}\right)^{-1/2} + 0.04\left(1 - \frac{W}{h}\right)^2 \tag{3.7b}$$

The validity of (3.6) requires that $t \ll h$, as usually occurs, where t is the thickness of the metallic layer. Using (3.6), the second term in (3.5) is found to be

$$\frac{d\varepsilon_{eff}}{d\varepsilon_{MUT}} = \frac{1-F}{2} \tag{3.8}$$

and the sensitivity is thus

$$S = \frac{\omega_0 v_p l}{4c^2}(1 - F) = \frac{\omega_0 l}{2\sqrt{2}c} \cdot \frac{1-F}{\sqrt{\varepsilon_r(1+F) + \varepsilon_{MUT}(1-F)}} \tag{3.9}$$

For CPW transmission lines with $t \ll h$, the effective dielectric constant can be approximated by

$$\varepsilon_{eff} = \frac{\varepsilon_r + \varepsilon_{MUT}}{2} \tag{3.10}$$

and the sensitivity is

$$S = \frac{\omega_0 v_p l}{4c^2} = \frac{\omega_0 l}{2\sqrt{2}c\sqrt{\varepsilon_r + \varepsilon_{MUT}}} \tag{3.11}$$

Despite the fact that by increasing the dielectric constant of the substrate, ε_r, the phase velocity of the line, v_p, is reduced, it can be concluded from (3.9) and (3.11) that the net effect of an increase in ε_r is sensitivity degradation. Therefore, the correct strategy for sensitivity optimization is to choose a low dielectric constant substrate material. Note, however, that the sensitivity depends on ε_{MUT}, and sensitivity enhancement by reducing ε_r is obscured if the considered MUTs exhibit a high dielectric constant value. Additionally, the dielectric constant of the substrate cannot be considered to be a true sensor design parameter (i.e. in certain applications, the substrate material is dictated by external factors). Thus, according to the previous sentence, increasing the line length and/or the operating frequency as much as possible seems to be the single effective approach for sensitivity optimization in sensors based on ordinary sensing lines. In this section, it will be shown that by line meandering, the sensitivity can be significantly optimized, yet keeping sensor dimensions within reasonable values. However, other sensitivity optimization techniques, based on the addition of certain distributed elements to the sensing line (i.e. transmission line-based components), will also be discussed in this section. Section 3.4 shows that tailoring the dispersion diagram in artificial lines is an alternative (and efficient) approach to line meandering in ordinary lines for sensitivity optimization.

The transmission line phase-variation sensors of this section are divided in transmission-mode and reflective-mode devices, and are discussed separately. The previous analysis has been carried out by considering that the output variable is the phase of the line. Although such phase can be inferred from the measured S-parameters, the canonical output variable in transmission-mode two-port sensors is the phase of the transmission coefficient, whereas for reflective-mode one-port sensors (e.g. based on sensing lines terminated by means of an open-circuit), the output variable is the phase of the reflection coefficient (Fig. 3.1). For both sensor types, the general expression for the sensitivity is not as simple as the one given by (3.2), with explicit expressions given by (3.9) and (3.11) for microstrip- and CPW-based sensing structures, respectively. Nevertheless, the conclusions relative to the effects of the operating frequency, line length and substrate dielectric constant on the sensitivity of the phase of the line with the dielectric constant of the MUT, do also apply when the considered output variable is the phase of either the transmission coefficient (transmission-mode sensors) or the reflection coefficient (reflective-mode sensors).

If the sensing lines are matched to the reference impedance of the ports, Z_0 (typically 50 Ω), the electrical length of the line, given by (3.1), coincides (except the sign) with the phase of the transmission coefficient, the usual output variable, as indicated before, in transmission-mode sensors. In reflective-mode sensors based on matched ordinary lines terminated with an open end, the reflection coefficient is

$$\rho = e^{-2j\phi} \tag{3.12}$$

Therefore, the phase of the reflection coefficient, the usual output variable, is merely

$$\phi_\rho = -2\phi \tag{3.13}$$

An inherent advantage of open-ended sensing lines over transmission-mode phase-variation sensors is that the same sensitivity can be roughly achieved by reducing the sensor length by a factor of two. Nevertheless, the fact that the phase of the transmission and reflection coefficient for transmission-mode and reflective-mode sensors, respectively, exhibits such simple dependence with the phase of the line under matching conditions, does not mean, necessarily, that the same dependence applies to the sensitivity of the sensors with regard to the sensitivity of the phase of the line given by (3.2). Moreover, line matching to the ports cannot be maintained by varying the dielectric constant of the MUT (i.e. the characteristic impedance of the sensing line depends on ε_{MUT}). Indeed, in the design of the sensor, a reference (REF) material, with a certain dielectric constant, ε_{REF}, should be considered for setting the electrical parameters of the unperturbed line, that is, the phase and the characteristic impedance of the line loaded with such REF material, to a certain nominal value (called REF phase and characteristic impedance). In particular, the REF material can be air, and, in this case, the REF phase and impedance are those corresponding to the uncovered line.

It is obvious that a variation in the dielectric constant of the MUT modifies the phase and the characteristic impedance of the line, both contributing to the phase of the transmission coefficient (transmission-mode sensors) and to the phase of the reflection coefficient (reflective-mode sensors). Let us next separately calculate the sensitivity for transmission-mode and reflective-mode sensors based on a simple ordinary line with arbitrary electrical length and characteristic impedance.

For the two-port transmission-mode sensor based on an ordinary line, let us call ϕ_T the phase of the transmission coefficient, the output variable. The sensitivity can be expressed as

$$S = \frac{d\phi_T}{d\varepsilon_{MUT}} = \frac{d\phi_T}{d\phi_s} \cdot \frac{d\phi_s}{d\varepsilon_{MUT}} + \frac{d\phi_T}{dZ_s} \cdot \frac{dZ_s}{d\varepsilon_{MUT}} \tag{3.14}$$

where Z_s and ϕ_s are the characteristic impedance and phase, respectively, of the sensing line (the subindex s is used to denote that these electrical parameters correspond to the sensing line). Expression (3.14) is valid for any arbitrary value of the dielectric constant of the MUT and for any arbitrary value of the REF phase and characteristic impedance of the sensing line. The transmission coefficient of a transmission line with arbitrary impedance, Z_s, and phase, ϕ_s, is found to be [1]

$$S_{21} = \frac{1}{\cos \phi_s + j\frac{1}{2} \left(\frac{Z_s}{Z_0} + \frac{Z_0}{Z_s} \right) \sin \phi_s} \tag{3.15}$$

Consequently, the phase of the transmission coefficient is

$$\phi_T \equiv \phi_{S_{21}} = -\arctan \left\{ \frac{1}{2} \left(\frac{Z_s}{Z_0} + \frac{Z_0}{Z_s} \right) \tan \phi_s \right\} \tag{3.16}$$

and the derivatives of ϕ_T with ϕ_s and Z_s that appear in the right-hand side member of (3.14) are found to be

$$\frac{d\phi_T}{d\phi_s} = -\frac{1}{M^{-1} \cdot \cos^2 \phi_s + M \cdot \sin^2 \phi_s} \tag{3.17}$$

and

$$\frac{d\phi_T}{dZ_s} = -\frac{\frac{1}{2}}{1 + M^2 \cdot \tan^2 \phi_s} \cdot \tan \phi_s \cdot \left(\frac{1}{Z_0} - \frac{Z_0}{Z_s^2} \right) \tag{3.18}$$

respectively, where the following dimensionless factor

$$M = \frac{1}{2} \left(\frac{Z_s}{Z_0} + \frac{Z_0}{Z_s} \right) \tag{3.19}$$

is used for simplification purposes.

It is apparent that (3.18) vanishes for $Z_s = Z_0$ and/or for $\phi_s = n \cdot \pi/2$ (n being an integer). Thus, the right-hand side term of the second member in (3.14) is null for these impedance and/or phase conditions, and the sensitivity (3.14) coincides with (3.2), but multiplied by (3.17). Note, however, that (3.17) gives $-M$ and $-M^{-1}$ for $\phi_s = n \cdot \pi$ and $\phi_s = (2n + 1) \cdot \pi/2$, respectively. Thus, unless the sensing line is matched to the ports ($Z_s = Z_0$), a situation that gives $M = 1$, the sensitivity of the (mismatched) sensing line increases for $\phi_s = n \cdot \pi$, and it decreases for $\phi_s = (2n + 1) \cdot \pi/2$ (note that $M \geq 1$). Indeed, choosing the sensing line with a REF phase of $\phi_s = n \cdot \pi$ is interesting because for such phase the injected power is completely transmitted to the output port (except by the effect of losses), and the sensitivity is enhanced by M. However, in practice, M cannot take very high values, because the REF impedance of the sensing line cannot be very extreme, as compared to the impedance of the ports. For sensing lines matched to the ports (i.e. with the REF impedance identical to the impedance of the ports), the usual situation, the sensitivity in the limit of small perturbations ($\varepsilon_{MUT} \rightarrow \varepsilon_{REF}$) coincides with the sensitivity of the phase of the line, given by (3.9) and (3.11). If the matching condition is not preserved, e.g. as consequence of large variations in ε_{MUT}, or because the REF impedance is deliberately larger or smaller than the one of the ports, then the general expression (3.14) should be used. Besides (3.17) and (3.18), the other two terms contributing to the general expression of the sensitivity are

$$\frac{d\phi_s}{d\varepsilon_{MUT}} = \frac{\phi_s}{4\varepsilon_{\textit{eff}}}(1-F) \tag{3.20}$$

and

$$\frac{dZ_s}{d\varepsilon_{MUT}} = -\frac{Z_s}{4\varepsilon_{\textit{eff}}}(1-F) \tag{3.21}$$

for microstrip sensing lines, whereas for CPWs the same expressions apply, but forcing the geometry factor to be null ($F = 0$). In summary, for sensitivity optimization in transmission-mode sensors based on an ordinary line, it is convenient to set the REF phase to $\phi_s = n \cdot \pi$ (thereby avoiding mismatching reflections due to phase matching), and the REF line impedance, Z_s, to the highest possible value[3] [4].

For the reflective-mode one-port sensor based on an open-ended sensing line, the sensitivity can be expressed as

$$S = \frac{d\phi_\rho}{d\varepsilon_{MUT}} = \frac{d\phi_\rho}{d\phi_s} \cdot \frac{d\phi_s}{d\varepsilon_{MUT}} + \frac{d\phi_\rho}{dZ_s} \cdot \frac{dZ_s}{d\varepsilon_{MUT}} \tag{3.22}$$

where ϕ_ρ is the phase of the reflection coefficient, the output variable. The derivatives given by (3.20) and (3.21) are also valid for (3.22). For the calculation of the other two derivatives of (3.22), the phase of the reflection coefficient must be first calculated. The input impedance of an open-ended sensing line with impedance Z_s and phase ϕ_s is [1]

$$Z_{in} = -jZ_s \cot\phi_s \tag{3.23}$$

Therefore, the reflection coefficient can be expressed as

$$\rho = \frac{Z_{in} - Z_0}{Z_{in} + Z_0} = \frac{+Z_0 + jZ_s \cot\phi_s}{-Z_0 + jZ_s \cot\phi_s} \tag{3.24}$$

and the phase of the reflection coefficient is

$$\phi_\rho = 2\arctan\left(\frac{Z_s \cot\phi_s}{Z_0}\right) + \pi = -2\arctan\left(\frac{Z_0 \tan\phi_s}{Z_s}\right) \tag{3.25}$$

From (3.25), the derivatives of ϕ_ρ with ϕ_s and Z_s are found to be

$$\frac{d\phi_\rho}{d\phi_s} = -\frac{2}{\frac{Z_0}{Z_s}\sin^2\phi_s + \frac{Z_s}{Z_0}\cos^2\phi_s} \tag{3.26}$$

and

$$\frac{d\phi_\rho}{dZ_s} = \frac{2}{1 + \left(\frac{Z_s}{Z_0}\right)^2 \cot^2\phi_s} \cdot \frac{\cot\phi_s}{Z_0} \tag{3.27}$$

respectively.

It follows from (3.27) that those phases satisfying $\phi_s = n \cdot \pi/2$ (n being an integer) null the derivative $d\phi_\rho/dZ_s$, and, therefore, the last term of the sensitivity in (3.22) vanishes. However, line

---

3 Indeed, according to (3.19), a low impedance value also boosts up M, and consequently the contribution to the sensitivity given by (3.17), identical to $-M$ for $\phi_s = n \cdot \pi$. However, a decrease in the impedance of the line increases F (or decreases $1-F$), thereby degrading the term (3.20), provided the sensor is implemented in microstrip technology. For sensor implementation in CPW technology, the impedance of the sensing line can be indistinctly high or low for sensitivity optimization.

matching ($Z_s = Z_0$) does not null the derivative of the output variable, ϕ_ρ, with the impedance of the line, Z_s, contrary to transmission-mode sensors, where the equivalent term (i.e. $d\phi_T/dZ_s$) is null for $Z_s = Z_0$. This means that, unless the above-cited phase condition is satisfied, both terms in (3.22) should be taken into account for an accurate prediction of the sensitivity [5].

Let us now determine the optimum conditions for sensitivity optimization in the considered sensors. For that purpose, the explicit dependence of the sensitivity with the involved variables is first calculated by introducing (3.20), (3.21), (3.26), and (3.27) in (3.22). The result can be expressed as

$$S = -\frac{Z_s(1-F)}{2Z_0\varepsilon_{eff}} \cdot \frac{\phi_s\sin^{-2}\phi_s + \cot\phi_s}{1 + \left(\frac{Z_s}{Z_0}\right)^2 \cot^2\phi_s} \tag{3.28}$$

In order to determine the optimum phase, ϕ_s, the derivative of S with that variable is obtained, i.e. [5]

$$\frac{dS}{d\phi_s} = -\frac{Z_s(1-F)\sin^{-2}\phi_s}{Z_0\varepsilon_{eff}} \cdot \frac{\cot\phi_s\left\{\left(\frac{Z_s}{Z_0}\right)^2 \cdot \cot\phi_s + \phi_s\left(\frac{Z_s}{Z_0}\right)^2 - \phi_s\right\}}{\left(1 + \left(\frac{Z_s}{Z_0}\right)^2 \cot^2\phi_s\right)^2} \tag{3.29}$$

The zeros of (3.29) are given by those phases satisfying

$$\cot\phi_s = 0 \tag{3.30a}$$

$$\phi_s\tan\phi_s = \frac{Z_s^2}{Z_0^2 - Z_s^2} \tag{3.30b}$$

and correspond to maxima or minima of the sensitivity. Solution of (3.30a) gives $\phi_s = (2n+1)\cdot\pi/2$, and the sensitivity for such phases is found to be

$$S = -\frac{Z_s(1-F)}{2Z_0\varepsilon_{eff}}\phi_s \tag{3.31a}$$

This value is a maximum, provided the normalized line impedance Z_s/Z_0 is high. Note that the sensitivity for $\phi_s = (2n+1)\cdot\pi/2$ is proportional to the normalized line impedance. By contrast, (3.30b) does not have a canonical solution. Nevertheless, let us consider various cases of potential interest, depending on the value of normalized line impedance. For $Z_s/Z_0 = 1$, the phase $\phi_s = (2n+1)\cdot\pi/2$ is a solution of (3.30b), but the corresponding sensitivity is given by (3.31a) with $Z_s = Z_0$. This case lacks interest, since for such phase condition, the normalized impedance must be high, as indicated before. For $Z_s/Z_0 > 1$, the right-hand side term of (3.30b) tends to -1, and the phase should satisfy $\cot\phi_s = -\phi_s$. Introducing this result in (3.28) gives

$$S = -\frac{Z_0(1-F)}{2Z_s\varepsilon_{eff}}\phi_s \tag{3.31b}$$

and the sensitivity is small, as far as this result has been obtained by considering $Z_s/Z_0 > 1$. Finally, for $Z_s/Z_0 < 1$, the solution of (3.30b) tends to $\phi_s = n\cdot\pi$. For this phase condition, the sensitivity is also given by (3.31b), but in this case the sensitivity is high, since $Z_s/Z_0 < 1$.

To summarize, the sensitivity in reflective-mode sensors based on ordinary open-ended lines in the limit of small perturbations can be optimized by choosing $\phi_s = (2n+1)\cdot\pi/2$ and Z_s as high as possible (as compared to Z_0). The corresponding sensitivity is given by (3.31a). Alternatively, the sensitivity can be optimized by choosing the phase of the sensing line according to (3.30b) and $Z_s < Z_0$. As Z_s decreases, the phase solution approaches $\phi_s = n\cdot\pi$, and for this particular value the sensitivity is given by (3.31b). Since for sensitivity optimization extreme impedances are needed,

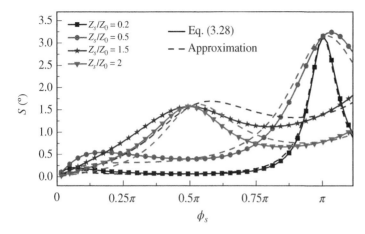

Figure 3.3 Plot of the normalized sensitivity, as given by (3.28), compared to the approximate value that neglects the last term in (3.22), by considering Z_s/Z_0 as a parameter. The normalized sensitivity is the one given by (3.28) divided by the maximum of either (3.31a) or (3.31b), except ϕ_s. *Source:* Reprinted with permission from [6]; copyright 2020 IEEE.

the optimum phase for lines satisfying $Z_s < Z_0$ can be reasonably approximated by the canonical value, $\phi_s = n \cdot \pi$. To gain insight on this aspect, Fig. 3.3 depicts the normalized sensitivity (defined in the caption and inferred from 3.28) as a function of the phase of the sensing line, for different values of the normalized impedance [6]. The figure confirms that as the normalized impedance decreases, the optimum phase approaches $\phi_s = n \cdot \pi$, whereas for $Z_s/Z_0 > 1$, the optimum phases satisfy $\phi_s = (2n + 1) \cdot \pi/2$. The figure also depicts the approximate normalized sensitivity that results by neglecting the second term in the right-hand side member of (3.22), equivalent to null $d\phi_\rho/dZ_s$. As expected, the exact and approximate solutions coincide for phases satisfying $\phi_s = n \cdot \pi/2$. However, it is also found that for extreme line impedances, the curves tend to merge, and the sensitivity can be reasonably estimated by neglecting the effects of the dielectric constant of the MUT on the line impedance.

Let us next review several implementations of transmission-line based phase-variation sensors, including transmission-mode (Section 3.2.1) and reflective-mode (Section 3.2.2) sensors, as well as specific strategies for sensitivity optimization.

3.2.1 Transmission-Mode Sensors

This section is dedicated to transmission-mode phase-variation sensors implemented by means of ordinary lines. In the two reported implementations, the lines are meandered in order to avoid extremely long sensing areas. Moreover, both prototype sensors are based on a differential-mode sensing scheme.[4] The section mainly discusses specific strategies to obtain high sensitivity. The

---

4 Differential-mode sensors are the subject of Chapter 6, mainly focused on resonator-loaded (or terminated) transmission-line differential sensors. In those sensors, the considered output variable is the cross-mode transmission (or reflection) coefficient, whereas in the sensors of this subsection, the output variable is the differential phase between the so-called reference (REF) line and the sensing line. The working principle of the differential sensors reported in this chapter is phase variation, and for this main reason such sensors are included in the present chapter. Indeed, differential sensing is a mode of operation, rather than a sensor working principle. Nevertheless, it has been opted to dedicate a specific Chapter 6 to the topic of differential microwave sensors, since there are aspects common to all differential sensors that justify grouping them in a dedicated thematic unit. In addition, as indicated, in most differential sensors reported in Chapter 6, the output variable is the cross-mode transmission (or reflection) coefficient, an inherent differential variable with specific singularities.

use of these sensors for dielectric characterization of solid samples, as well as their application in microfluidics and defect detection, is also demonstrated.

3.2.1.1 Transmission-Mode Four-Port Differential Sensors

The objective of this section is to report an approach for sensitivity enhancement, and control, in transmission-mode phase-variation sensors. The reported sensor is a differential device able to discriminate tiny differences between the so-called REF sample and the MUT sample. Therefore, this sensor exhibits also very good resolution. The device is a four-port structure based on a pair of meandered lines. An analytical method to determine the length of the sensor lines from the required input dynamic range (the maximum differential dielectric constant), is presented [3]. The utility of the device as a highly sensitive sensor for dielectric characterization of solids, and as a high-resolution comparator, is experimentally validated.

3.2.1.1.1 Sensor Structure and Analysis

Let us consider a four-port structure based on a pair of uncoupled lines aimed to operate as a differential-mode phase-variation sensor able to provide the differential dielectric constant between the REF sample and the MUT (i.e. $\Delta \varepsilon = \varepsilon_{REF} - \varepsilon_{MUT}$, where ε_{REF} and ε_{MUT} are the dielectric constants of the REF and MUT samples, respectively). The REF sample must be placed on top of one of the lines, designated as REF line, whereas the other line (MUT line) must be circumscribed within the sensing region of the MUT. Let us also consider that the interest is to measure small variations of the dielectric constant of the MUT in reference to the one of the REF material, with a maximum input dynamic range designated as $\pm \Delta \varepsilon_{max}$. Under these conditions, the objective is to determine the length, l, of the pair of lines providing an output dynamic range (or maximum differential phase) of $\pm 180°$ at the operating frequency, designated as f_0.

The differential phase, $\Delta \phi = \phi_{REF} - \phi_{MUT}$ (where ϕ_{REF} and ϕ_{MUT} are the electrical lengths of the REF and MUT line, respectively), is determined by the dielectric constants of the REF and MUT samples, and can be expressed as

$$\Delta \phi = \beta_0 l \left(\sqrt{\varepsilon_{eff,REF}} - \sqrt{\varepsilon_{eff,MUT}} \right) \tag{3.32}$$

where $\varepsilon_{eff,REF}$ and $\varepsilon_{eff,MUT}$ are the effective dielectric constants of the REF line and MUT line, respectively, and $\beta_0 = \omega_0/c$ is the phase constant in vacuum. These effective dielectric constants are given by (3.6) or by (3.10), depending on the considered line technology (microstrip or CPW).[5] Obviously, the subindex MUT in (3.6) or (3.10) should be replaced with the subindex REF for the corresponding line.

Introducing the effective dielectric constants of both lines in (3.32), the following result is obtained

$$\Delta \phi = \frac{\beta_0 l}{\sqrt{2}} \left(\sqrt{\varepsilon_r(1+F) + \varepsilon_{REF}(1-F)} - \sqrt{\varepsilon_r(1+F) + \varepsilon_{MUT}(1-F)} \right) \tag{3.33}$$

and assuming that $\Delta \varepsilon \ll \varepsilon_{REF}$, expression (3.33) can be approximated by

$$\Delta \phi = \frac{\beta_0 l}{2\sqrt{2}} \cdot \frac{\Delta \varepsilon (1-F)}{\sqrt{\varepsilon_r(1+F) + \varepsilon_{REF}(1-F)}} \tag{3.34}$$

It can be appreciated that sensor implementation in microstrip technology is considered. As it has been mentioned before, a sensor with very good resolution and sensitivity, able to discriminate and

---

5 The same approximations in reference to the validity of (3.6) and (3.10) are adopted in this analysis.

measure small differential dielectric constants, is pursuit. For this reason, the above-cited approximation ($\Delta\varepsilon \ll \varepsilon_{REF}$) is justified (note that under this approximation, $\Delta\phi$ is proportional to $\Delta\varepsilon$, and, consequently, the sensitivity is constant). Nevertheless, the sensor can also be designed to measure larger input dynamic ranges (differential dielectric constants), at the expense of a degradation in the sensitivity (considering the same output dynamic range, between $+180°$ and $-180°$).

Let us now introduce in (3.34) the maximum input and output dynamic ranges, i.e. $\Delta\varepsilon = \pm\Delta\varepsilon_{max}$ and $\Delta\phi = \pm\pi$, respectively. By isolating the length of the lines, one obtains:

$$l = 2\sqrt{2}\pi \frac{\sqrt{\varepsilon_r(1+F) + \varepsilon_{REF}(1-F)}}{\beta_0 \Delta\varepsilon_{max}(1-F)} \tag{3.35}$$

and the sensitivity is

$$S = \frac{\pi}{\Delta\varepsilon_{max}} = \frac{\beta_0 l}{2\sqrt{2}} \cdot \frac{(1-F)}{\sqrt{\varepsilon_r(1+F) + \varepsilon_{REF}(1-F)}} \tag{3.36}$$

with l given by (3.35). Note that the sensitivity can be made as much high as desired by simply increasing l (provided the operating frequency, and hence β_0, is set to a certain value), but obviously at the expense of sensor size. Inspection of (3.35) reveals that considering substrates with small dielectric constant, ε_r, helps in reducing the length of the lines (this also applies to ε_{REF}, but this parameter is not a design variable).

3.2.1.1.2 Sensor Implementation and Application to Dielectric Characterization and Comparators

Without loss of generality, let us consider sensor implementation in microstrip technology, as mentioned. From the line width, W, the substrate parameters, ε_r and h, the dielectric constant of the REF sample, ε_{REF}, and the considered input dynamic range, $\Delta\varepsilon_{max}$, the line length, l, can inferred [using expression (3.35)]. The reported prototype sensor is implemented in the *Rogers RO4003C* substrate with dielectric constant $\varepsilon_r = 3.5$, thickness $h = 0.8128$ mm, and loss tangent tan $\delta_{SUB} = 0.0027$. The REF sample is considered to be a piece of the *Nelco N4350-13RF* substrate, with dielectric constant $\varepsilon_{REF} = 3.55$, and dissipation factor tan $\delta_{REF} = 0.0065$. The thickness of the REF sample is enough to guarantee a uniform half-space for the electromagnetic field present above the line (in agreement with the aforementioned semi-infinite MUT sample approximation). With these material properties, the width of the REF line necessary to provide a 50-Ω characteristic impedance (i.e. a REF line matched to the ports) is found to be $W = 1.6$ mm. Let us consider that the measurement range of the dielectric constant of the MUT sample corresponds to a differential dielectric constant of $\Delta\varepsilon_{max} = 0.45$, i.e. limited to 12.7% variations of the dielectric constant of the REF sample. Evaluation of (3.35) gives $\beta_0 l = 84.23$ rad. By considering an operating frequency of $f_0 = \omega_0/2\pi = 6$ GHz, the length of the meandered lines is found to be $l = 670.3$ mm. Such long line length is the reason for line meandering, but similar results are achievable by using straight lines (obviously, with the penalty of an excessive sensor dimension in the direction of the line axis).

With the above-indicated length of the REF and MUT lines, line meandering prevents from an excessive elongated sensor. Figure 3.4(a) depicts the photograph of the fabricated bare sensor with relevant dimensions indicated, whereas the sensor loaded with the REF and MUT samples is depicted in Fig. 3.4(b). The differential phase of the structure (obtained from full-wave electromagnetic simulation using *Keysight Momentum*) as a function of the dielectric constant of the MUT, by considering variations in the vicinity of $\varepsilon_{REF} = 3.55$ with $\Delta\varepsilon_{max} = 0.45$, is depicted in Fig. 3.5. Actually, such figure shows the results obtained for different values of the loss tangent of the MUT. The curves are very similar, which means that the loss factor of the MUT does not play an important role on the differential phase. The simulated curves of Fig. 3.5 are in very good agreement with the

(a)

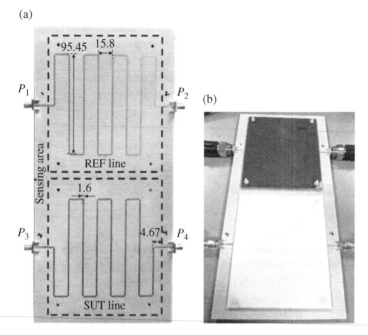

(b)

Figure 3.4 Photograph of the designed differential-mode phase-variation sensor without (a) and with (b) REF and MUT samples on top of it (dimensions are given in mm). In the figure, the MUT is designated as SUT (sample under test), following the nomenclature of the original source. *Source:* Reprinted with permission from [3]; copyright 2020 IEEE.

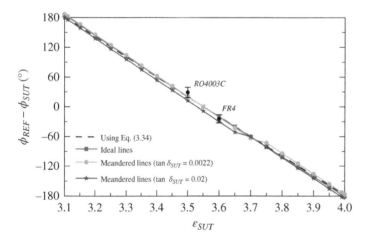

Figure 3.5 Simulated differential phase at $f_0 = 6$ GHz, corresponding to the sensor of Fig. 3.4, as a function of the dielectric constant of the MUT, parameterized by the loss tangent of the MUT. The curve given by expression (3.34) is also included. The mean of the measured differential phases (with $N = 3$ independent measurements) for the indicated materials, as well as the error bars inferred from the standard deviation, are also included. *Source:* Reprinted with permission from [3]; copyright 2020 IEEE.

analytical expression (3.34), also included in the figure. Such agreement validates the previous analysis, and points out that it is not necessary to know in advance the loss tangent of the MUT sample in order to determine its dielectric constant. Note that for the MUT sample with the highest or lowest dielectric constant (corresponding to $\pm\Delta\varepsilon_{max}$), the differential phase is found to be approximately $\Delta\phi = \phi_{REF} - \phi_{MUT} = \pm\pi$. The sensitivity is roughly constant, in agreement with (3.34), and it is found to be $S = d\Delta\phi/d\Delta\varepsilon = -415.6°$ (very close to the theoretical value, i.e. $-400°$). With such high sensitivity, the sensor is able to detect extremely small differential dielectric constants. Let us clarify that, in reality, the differential phase depicted in Fig. 3.5 is the difference in the phases of the transmission coefficients of both lines, but the results are roughly identical since quasi-matching conditions are preserved even for differential dielectric constants close to the extremes of the considered input dynamic range.

With the sensor of Fig. 3.4, it is also possible to determine the loss tangent of the MUT. For that purpose, the output variable should be the magnitude of the transmission coefficient of the MUT line, $|S_{43}|$. Figure 3.6 depicts the simulated $|S_{43}|$ at f_0 as a function of the dielectric constant, by considering the loss tangent of the MUT as a parameter. The dependence of $|S_{43}|$ on the dielectric constant is roughly negligible, that is, $|S_{43}|$ is dictated by the loss factor of the MUT. Consequently, it is confirmed that $|S_{43}|$ at f_0 is the convenient output variable for the determination of the loss tangent of the MUT.

The simulation results of Figs. 3.5 and 3.6 indicate that the prototype sensor of Fig. 3.4 is useful for dielectric characterization of solid samples. However, it has also a high potential as highly sensitive comparator, able to detect small differences between a certain sample (the MUT) and the REF sample. Thus, the sensor is also useful for defect detection. Let us next review in detail the first application (for defect detection, the authors suggest reference [3], where this application is exhaustively detailed).

Experimental validation of the sensor for dielectric constant measurements is demonstrated by first loading the sensing region with a commercial (uncladded) microwave substrate with well-known dielectric constant, i.e. the *FR4* substrate with $\varepsilon_{MUT} = 4.5$. The thickness of the REF sample is 3.3 mm, whereas the one of the MUT is 3.2 mm (achieved by stacking two samples). With these thicknesses, the samples can be roughly considered semi-infinite, as it has been corroborated by means of electromagnetic simulations (not shown), demonstrating that for samples thicker than 3 mm, the phase of the lines does not experience an appreciable variation. It should be indicated that ε_{MUT} (or the corresponding differential dielectric constant, $\Delta\varepsilon$) is out of the considered input dynamic range. However, the measured differential phase, $\Delta\phi$, is a periodic function with $\Delta\varepsilon$

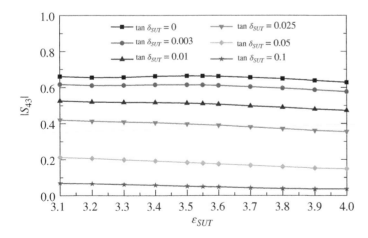

Figure 3.6 Simulated transmission coefficient of the MUT line at f_0 = 6 GHz, as a function of the dielectric constant of the MUT, for different values of the loss tangent of the MUT. *Source:* Reprinted with permission from [3]; copyright 2020 IEEE.

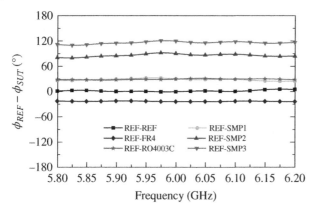

Figure 3.7 Measured differential phase for different MUT samples, including *FR4* and *Rogers RO4003C* slabs, as well as samples identical to the REF sample, but with different densities of holes across the substrate. Measurements have been performed by means of the *Keysight N5221A* four-port vector network analyzer. For *FR4* and *RO4003C*, the mean measured differential phase with $N = 3$ independent measurement is depicted. The density of holes in the "defected" samples is larger for the SMP3 and smaller for the SMP1 samples (see further details in [3]). *Source:* Reprinted with permission from [3]; copyright 2020 IEEE.

(actually, $\Delta\phi$ varies linearly with $\Delta\varepsilon$, but any measured value of $\Delta\phi$ out of the range $[-\pi, +\pi]$ is expressed by its equivalent value within that range). Thus, from the measured differential phase, it is possible to check the validity of the approach (the limitation is dictated by the requirement $\Delta\varepsilon \ll \varepsilon_{REF}$). According to the results of Fig. 3.5, the expected value of the differential phase for $\varepsilon_{MUT} = 4.5$ is $\Delta\phi = -19.18°$ (equivalent to $\varepsilon_{MUT} = 3.6$), whereas the mean of the measured values at $f_0 = 6$ GHz ($N = 3$ independent measurement have been carried out) is $\Delta\phi = -24.23°$ (Fig. 3.7), i.e. in very good agreement, taking into account the relatively large value of ε_{MUT}. The experiment has also been carried out by considering as MUT a piece of the *Rogers RO4003C* substrate with $\varepsilon_{MUT} = 3.5$ (satisfying $\Delta\varepsilon \ll \varepsilon_{REF}$) and thickness 3.05 mm (also achieved by stacking two samples). The resulting mean of the measured differential phases at $f_0 = 6$ GHz (also with $N = 3$ independent measurements) is $\Delta\phi = 28.97°$ (Fig. 3.7), in good agreement with the expected value ($\Delta\phi = 20.34°$). The measured differential phases at $f_0 = 6$ GHz are depicted in Fig. 3.5 to ease the comparison with the theoretical value. Note that from the measured phases, the dielectric constant values inferred from expression (3.34) for *FR4* and *RO4003C* are found to be 4.51 and 3.48, respectively, i.e. very close to the nominal values. In order to minimize the effects of the air gap (between the substrate and the REF and MUT samples), pressure to the MUT and REF samples against the substrate is necessary. This can be achieved by means of Teflon screws [as depicted in Fig. 3.4(b)].

Concerning material losses, Fig. 3.6 can potentially be used to determine the loss tangent from the value of the transmission coefficient (magnitude) of the MUT line. However, discrepancies between the simulated and measured transmission coefficient are inevitable due to the effects of the access lines and connectors (not accounted for in the simulations). Consequently, the magnitude of the transmission coefficient for several samples with different values of the loss tangent has been measured. The obtained $|S_{43}|$ values at $f_0 = 6$ GHz are depicted in Fig. 3.8. From these values, a calibration curve can be generated (see Fig. 3.8, inset), and this curve is useful to determine the loss tangent of the MUT from the measured value of $|S_{43}|$. The considered samples for the generation of the calibration curve are the *FR4* and *Rogers 4003C* substrates indicated in the previous paragraph, as well as the *Nelco N4350-13RF* substrate (the REF sample for the differential phase measurements). Nevertheless, the transmission coefficient of the unloaded MUT line (corresponding to $\tan\delta_{MUT} = 0$) has also been measured, as it provides an additional data point. To validate the

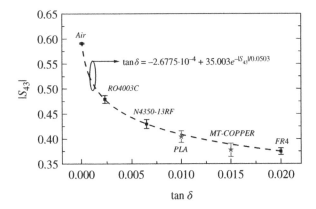

Figure 3.8 Measured transmission coefficient at f_0 = 6 GHz for different (indicated) MUT samples. The mean values, as well as the error bars, obtained from N = 3 independent measurements are depicted. *Source: Reprinted with permission from [3]; copyright 2020 IEEE.*

approach for the determination of the loss tangent, two samples fabricated with a 3D-printer (model *Ultimaker 3 Extended*) have been used as MUT. The loss factor of such samples, measured by means of the *Agilent Keysight 85072A* commercial resonant cavity, are $\tan \delta_{MUT} = 0.010$ (for MUT1, fabricated by considering *PLA* as filament) and $\tan \delta_{MUT} = 0.016$ (for MUT2, fabricated by considering *RS Pro MT-COPPER* as filament). The measured values of the insertion loss at $f_0 = 6$ GHz that are obtained by loading the MUT line with these 3D-printed samples are indicated in Fig. 3.8. It can be seen that the resulting points are in reasonable agreement with the calibration curve. Therefore, it is demonstrated that such curve can be used to estimate the loss tangent of the MUT sample.

3.2.1.2 Two-Port Sensors Based on Differential-Mode to Common-Mode Conversion Detectors and Sensitivity Enhancement

In the second prototype sensor, the four-port differential-mode structure, also implemented by means of meandered lines, is transformed to a two-port device by adding a pair of rat-race hybrid couplers conveniently arranged [7]. The output variable of this sensor is the magnitude of the transmission coefficient at the operating frequency, an easily measurable quantity related to the differential phase.[6] Indeed, this sensor operates as a device able to detect conversion from the differential mode to the common mode (or vice versa) in the pair of balanced lines. Mode conversion is consequence of the phase difference (imbalance) in the lines, e.g. caused by an asymmetric dielectric loading between the REF and MUT line. Let us first present and analyze in detail the proposed mode conversion detector, applicable to detect symmetry perturbations in any four-port balanced structure, in order to infer the transmission coefficient, obviously related to symmetry truncation. After such analysis, which provides the required conditions for sensitivity optimization, a prototype sensor design and several applications are reported.

---

6 The magnitude of the transmission coefficient at a certain frequency can be indirectly inferred from the amplitude of the output voltage wave in response to a harmonic feeding signal tuned to that frequency. For that purpose, a simple harmonic generator and an envelope detector suffice. This aspect is out of the scope of this chapter, but treated in detail in Chapter 4.

3.2.1.2.1 Differential-Mode to Common-Mode Conversion Detector

For mode conversion detection, the balanced circuit under study should be fed by a differential-mode signal, e.g. generated by means of a rat-race balun from a single-ended signal. By connecting the pair of output ports of the balanced circuit under test to one of the two pairs of isolated ports of a second rat-race coupler, conversion to the common-mode (if it exists as a consequence of symmetry perturbation) can be detected, and recorded in the Σ-port. Since the signal level at the output (single-ended) port depends on mode conversion efficiency, the complete two-port structure, including the pair of rat-race couplers plus the balanced circuit in between, can be used as a comparator or differential microwave sensor based on mode conversion [7].

The schematic of the proposed mode conversion detector is depicted in Fig. 3.9, where an arbitrary (not necessarily balanced) four-port network is considered. Such four-port network can be described either by the single-ended S-parameters, or by the mixed-mode S-parameters. According to the port designation (Fig. 3.9), the single-ended S-parameter matrix is

$$\mathbf{S_{se}} = \begin{pmatrix} S_{AA} & S_{AA'} & S_{AB} & S_{AB'} \\ S_{A'A} & S_{A'A'} & S_{A'B} & S_{A'B'} \\ S_{BA} & S_{BA'} & S_{BB} & S_{BB'} \\ S_{B'A} & S_{B'A'} & S_{B'B} & S_{B'B'} \end{pmatrix} \tag{3.37}$$

whereas the mixed-mode S-parameter matrix can be expressed as a combination of four order-2 matrices as follows:

$$\mathbf{S_{mm}} = \begin{pmatrix} \mathbf{S^{dd}} & \mathbf{S^{dc}} \\ \mathbf{S^{cd}} & \mathbf{S^{cc}} \end{pmatrix} = \begin{pmatrix} S_{11}^{dd} & S_{12}^{dd} & S_{11}^{dc} & S_{12}^{dc} \\ S_{21}^{dd} & S_{22}^{dd} & S_{21}^{dc} & S_{22}^{dc} \\ S_{11}^{cd} & S_{12}^{cd} & S_{11}^{cc} & S_{12}^{cc} \\ S_{21}^{cd} & S_{22}^{cd} & S_{21}^{cc} & S_{22}^{cc} \end{pmatrix} \tag{3.38}$$

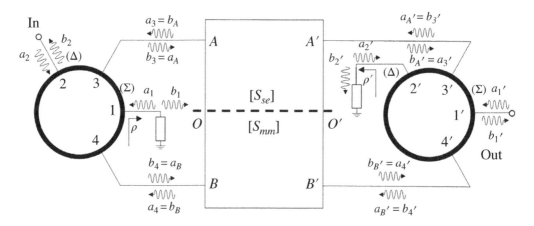

Figure 3.9 Sketch of the proposed mode conversion detector. *Source:* Reprinted with permission from [7]; copyright 2020 IEEE.

where the mixed-mode S-parameters are related to the single-mode S-parameters according to well-known transformations (see, e.g. [8–10]), i.e.

$$\mathbf{S^{dd}} = \frac{1}{2} \begin{pmatrix} S_{AA} - S_{AB} - S_{BA} + S_{BB} & S_{AA'} - S_{AB'} - S_{BA'} + S_{BB'} \\ S_{A'A} - S_{A'B} - S_{B'A} + S_{B'B} & S_{A'A'} - S_{A'B'} - S_{B'A'} + S_{B'B'} \end{pmatrix} \qquad (3.39a)$$

$$\mathbf{S^{cc}} = \frac{1}{2} \begin{pmatrix} S_{AA} + S_{AB} + S_{BA} + S_{BB} & S_{AA'} + S_{AB'} + S_{BA'} + S_{BB'} \\ S_{A'A} + S_{A'B} + S_{B'A} + S_{B'B} & S_{A'A'} + S_{A'B'} + S_{B'A'} + S_{B'B'} \end{pmatrix} \qquad (3.39b)$$

$$\mathbf{S^{dc}} = \frac{1}{2} \begin{pmatrix} S_{AA} + S_{AB} - S_{BA} - S_{BB} & S_{AA'} + S_{AB'} - S_{BA'} - S_{BB'} \\ S_{A'A} + S_{A'B} - S_{B'A} - S_{B'B} & S_{A'A'} + S_{A'B'} - S_{B'A'} - S_{B'B'} \end{pmatrix} \qquad (3.39c)$$

$$\mathbf{S^{cd}} = \frac{1}{2} \begin{pmatrix} S_{AA} - S_{AB} + S_{BA} - S_{BB} & S_{AA'} - S_{AB'} + S_{BA'} - S_{BB'} \\ S_{A'A} - S_{A'B} + S_{B'A} - S_{B'B} & S_{A'A'} - S_{A'B'} + S_{B'A'} - S_{B'B'} \end{pmatrix} \qquad (3.39d)$$

Note that the pair of composite differential- and common-mode ports, necessary to derive the above transformations, are considered to be the ports A–B and A'–B' (driven as a pair). For perfect symmetry conditions, with regard to the indicated bisection plane O–O', the cross-mode (or mode conversion) matrices, $\mathbf{S^{cd}}$ and $\mathbf{S^{dc}}$, are null. By contrast, in the use of the mode conversion detector as differential sensor or comparator (to be discussed later), a symmetry imbalance is expected, provided the sensitive parts of the structure (symmetrically located with regard to the O–O' plane) are loaded with different samples (the REF and the MUT sample). The result of such imbalance is mode conversion, and, consequently, $\mathbf{S^{cd}} \neq \mathbf{0}$ and $\mathbf{S^{dc}} \neq \mathbf{0}$.

The input port of the two-port mode conversion detector is the Δ-port of the first coupler (port 2), whereas the output port is the Σ-port of the second coupler (port 1'). The ports in both couplers are differentiated by means of a super-index prime (') in the second coupler. For a matched isolated port (Σ-port) in coupler 1 (port 1), such coupler acts as a balun, providing out-of-phase signals at ports 3 and 4 (i.e. a pure differential signal in the differential port 3–4, or A–B). Similarly, by terminating the Δ-port in the second coupler (port 2') with a matched load, the Σ-port (port 1') only detects common-mode signals, if they are present, at the composite port 3'–4' (or A'–B'). Therefore, the structure is sensitive to differential-mode to common-mode conversion, potentially caused by symmetry imbalances, either undesired or deliberated. In the former case, the structure can be used to detect fabrication inaccuracies in balanced circuits. The second case is the one of interest in this book, where symmetry truncation is caused by a phase imbalance in a four-port structure consisting of a pair of identical lines.

According to the previous paragraphs, the transmission coefficient of the complete two-port structure (including the pair of couplers and the four-port network in between) should be related to the cross-mode S-parameters of the four-port network, in turn correlated with the level of asymmetry of such network. Thus, let us next calculate such transmission coefficient, by considering the general case of ports 1 and 2' terminated with arbitrary loads.

3.2.1.2.2 *Analysis and Sensitivity Optimization*

Despite the fact that the working principle of the proposed mode conversion detector has been explained before by considering that the isolated ports of the couplers (ports 1 and 2') are terminated with matched loads, it does not mean that these terminations are optimum for sensitivity optimization.[7] Consequently, the following analysis, devoted to obtain the transmission coefficient of the mode conversion detector, is carried out by considering arbitrary loads in ports 1 and 2',

---

7 Indeed, for sensitivity optimization (i.e. obtaining a maximum variation of the modulus of the transmission coefficient as symmetry imbalances increase), the isolated ports of the couplers should not be terminated with matched loads, as it will be later demonstrated.

described by reflection coefficients ρ and ρ', respectively (see Fig. 3.9). Let us call f_0 the operating frequency of the device, at which the length of the ring couplers is exactly 1.5λ (λ being the guided wavelength), and let Z_0 be the reference impedance of the ports. With the port designation of Fig. 3.9, the S-parameter matrix of the couplers (at f_0) is [1]

$$\mathbf{S} = \mathbf{S'} = -\frac{j}{\sqrt{2}} \begin{pmatrix} 0 & 0 & 1 & 1 \\ 0 & 0 & 1 & -1 \\ 1 & 1 & 0 & 0 \\ 1 & -1 & 0 & 0 \end{pmatrix} \tag{3.40}$$

The present analysis considers the normalized amplitudes of the voltage waves incident to (a_i) or reflected from (b_i) the ports, where the subindex i identifies the port. The variable of interest, the transmission coefficient of the whole structure, is

$$S_{1'2} = \frac{b'_1}{a_2}\Big|_{a'_1 = 0} \tag{3.41}$$

where, again, the super index prime is used to distinguish between the normalized amplitudes of the incident and reflected waves in both couplers. According to (3.40), b'_1 is

$$b'_1 = -\frac{j}{\sqrt{2}}(a'_3 + a'_4) \tag{3.42}$$

where the normalized amplitude of the incident voltages at ports $3'$ and $4'$ can be expressed as a function of the elements of the single-ended S-parameter matrix of the four-port network, i.e.

$$a'_3 = S_{A'A}b_3 + S_{A'A'}b'_3 + S_{A'B}b_4 + S_{A'B'}b'_4 \tag{3.43a}$$

$$a'_4 = S_{B'A}b_3 + S_{B'A'}b'_3 + S_{B'B}b_4 + S_{B'B'}b'_4 \tag{3.43b}$$

Note that (3.43) results from the following (trivial) correspondence between normalized amplitudes of incident and reflected voltages waves of the couplers (right-hand side members) and the four-port network (left-hand side members):

$$a_A = b_3 \tag{3.44a}$$

$$b_A = a_3 \tag{3.44b}$$

$$a_{A'} = b'_3 \tag{3.44c}$$

$$b_{A'} = a'_3 \tag{3.44d}$$

$$a_B = b_4 \tag{3.44e}$$

$$b_B = a_4 \tag{3.44f}$$

$$a_{B'} = b'_4 \tag{3.44g}$$

$$b_{B'} = a'_4 \tag{3.44h}$$

The normalized amplitudes of the reflected voltage waves (referred to the couplers) that appear in the right-hand side members in (3.43) are given by:

$$b_3 = -\frac{j}{\sqrt{2}}(a_1 + a_2) = -\frac{j}{\sqrt{2}}(\rho b_1 + a_2) \tag{3.45a}$$

$$b'_3 = -\frac{j}{\sqrt{2}}(a'_1 + a'_2) = -\frac{j}{\sqrt{2}}\rho'b'_2 \tag{3.45b}$$

$$b_4 = -\frac{j}{\sqrt{2}}(a_1 - a_2) = -\frac{j}{\sqrt{2}}(\rho b_1 - a_2) \tag{3.45c}$$

$$b_4' = -\frac{j}{\sqrt{2}}(a_1' - a_2') = -\frac{j}{\sqrt{2}}(-\rho' b_2') \tag{3.45d}$$

with $a_1 = \rho b_1$ and $a_2' = \rho' b_2'$. Note also that $a_1' = 0$ in (3.45), since port 1' should be matched for the calculation of the transmission coefficient [see expression (3.41)]. By introducing the right-hand side terms of (3.45) in (3.43), and the resulting expressions in (3.42), the normalized amplitude of the output voltage wave at port 1' is found to be:

$$b_1' = -\left\{S_{21}^{cc}\rho b_1 + S_{21}^{cd}a_2 + S_{22}^{cd}\rho' b_2'\right\} \tag{3.46}$$

where the correspondence between the mixed-mode and single-ended S-parameters of the four-port network [expressions (3.39)] has been used.

In order to obtain the transmission coefficient of the whole structure using (3.41), the first and third terms of the right-hand side member in (3.46) must be written as a function of a_2. For that purpose, the first step is to express b_1 and b_2', given by

$$b_1 = -\frac{j}{\sqrt{2}}(a_3 + a_4) \tag{3.47a}$$

$$b_2' = -\frac{j}{\sqrt{2}}(a_3' - a_4') \tag{3.47b}$$

in terms of the same variables that appear in the second member of (3.46). The procedure is very similar to the one detailed before to obtain Eq. (3.46). The results are:

$$b_1 = -\left\{S_{11}^{cc}\rho b_1 + S_{11}^{cd}a_2 + S_{12}^{cd}\rho' b_2'\right\} \tag{3.48}$$

$$b_2' = -\left\{S_{21}^{dc}\rho b_1 + S_{21}^{dd}a_2 + S_{22}^{dd}\rho' b_2'\right\} \tag{3.49}$$

where, again, (3.39) has been used. From (3.48), b_1 can be expressed in terms of a_2 and b_2', i.e.

$$b_1 = -\frac{S_{11}^{cd}a_2 + S_{12}^{cd}\rho' b_2'}{1 + S_{11}^{cc}\rho} \tag{3.50}$$

By introducing (3.50) in (3.49), b_2' can be isolated and expressed in terms of a_2 as follows:

$$b_2' = M' \cdot a_2 \tag{3.51}$$

where M', introduced to simplify the notation, depends only on the mixed-mode S parameters of the four-port network, as well as on the reflection coefficients of the loads present at ports 1 and 2', i.e.

$$M' = -\frac{S_{21}^{dd} - \dfrac{\rho S_{21}^{dc} S_{11}^{cd}}{1 + \rho S_{11}^{cc}}}{1 + \rho' S_{22}^{dd} - \dfrac{\rho\rho' S_{21}^{dc} S_{12}^{cd}}{1 + \rho S_{11}^{cc}}} \tag{3.52}$$

From (3.50), b_1 can be expressed in terms of a_2 as

$$b_1 = M \cdot a_2 \tag{3.53}$$

where[8]

$$M = -\frac{S_{11}^{cd} + \rho' M' S_{12}^{cd}}{1 + \rho S_{11}^{cc}} \qquad (3.54)$$

Finally, by introducing (3.51) and (3.53) in (3.46), b_1' can be expressed as proportional to a_2, and the transmission coefficient, given by (3.41), is found to be

$$S_{1'2} = -\left\{S_{21}^{cc}\rho M + S_{21}^{cd} + S_{22}^{cd}\rho' M'\right\} \qquad (3.55)$$

In view of (3.55), it follows that for perfectly balanced four-port networks (i.e. with $\mathbf{S^{cd}} = \mathbf{S^{dc}} = \mathbf{0}$), $S_{1'2} = 0$, as expected (note that $M = 0$ for balanced structures). A detailed analysis of (3.55) reveals that for unbalanced four-port networks, $S_{1'2}$ can also be null. However, this occurs for very specific combinations of port loading (i.e. ρ and ρ') and mixed-mode S-parameters of the four-port network [11]. Thus, when symmetry is truncated, in general $S_{1'2} \neq 0$, and the structure can be used to determine the level of asymmetry of the four-port network under test (mode conversion detector).

For matched terminations at ports 1 and 2' ($\rho = \rho' = 0$), the transmission coefficient of the mode conversion detector is found to be

$$S_{1'2} = -S_{21}^{cd} \qquad (3.56)$$

coinciding with the cross-mode transmission coefficient of the four-port network (except the sign). Thus, the proposed structure provides a straightforward way to obtain the cross-mode transmission coefficient in four-port networks, which is indicative of symmetry imbalances. Note, however, that M and M' depend on various elements of the cross-mode matrices. Therefore, it is not a priori clear that the preferred option for sensitivity optimization is to terminate ports 1 and 2' with matched loads. To determine the convenient terminations at these ports (for sensitivity enhancement), the mixed-mode (or single-ended) S-parameters of the four-port network must be known. Our interest for the present chapter is the four-port network consisting of a pair of matched and uncoupled lines, since such networks can be applied to the implementation of differential sensors and comparators.

For a pair of identical matched and uncoupled lines (Fig. 3.10), the mixed-mode S-parameters that explicitly appear in (3.55) are

$$S_{22}^{cd} = 0 \qquad (3.57a)$$

$$S_{21}^{cd} = \frac{1}{2}\left(e^{-j\phi_A} - e^{-j\phi_B}\right) \qquad (3.57b)$$

$$S_{21}^{cc} = \frac{1}{2}\left(e^{-j\phi_A} + e^{-j\phi_B}\right) \qquad (3.57c)$$

where ϕ_A and ϕ_B are the electrical lengths of the lines between ports A–A' (line A) and B–B' (line B), respectively.[9] The other term needed for the determination of $S_{1'2}$ is M, given by

---

8 Note that the variable M has also been used in (3.19) to designate a dimensionless factor given by an impedance ratio. In (3.54), the same symbol (M) is used, in order to respect the nomenclature of the original source [7] where the present analysis was reported for the first time.

9 The difference in the phase of lines A and B emulates the presence of a different dielectric load in the REF and MUT lines. If the REF line is matched to the port impedance when it is loaded with the REF sample, the presence of a different sample in the MUT line should necessarily alter line matching in such line. Thus, rigorously speaking, (3.57b) and (3.57c) are not valid, but reasonable approximations to the actual values of the considered parameters of the mixed-mode scattering matrix. According to these approximations, it is assumed that the characteristic impedance of the MUT line does not vary significantly when it is loaded with a sample different than the REF sample.

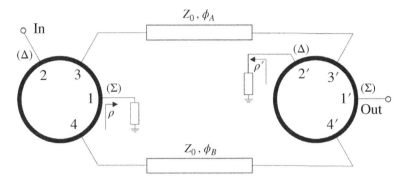

Figure 3.10 Mode conversion detector for a four-port network consisting of a pair of matched and uncoupled lines. *Source:* Reprinted with permission from [7]; copyright 2020 IEEE.

$$M = \frac{\rho'}{4}\left(e^{-j\phi_A} - e^{-j\phi_B}\right) \cdot \left\{ \frac{e^{-j\phi_A} + e^{-j\phi_B}}{1 - \frac{\rho\rho'}{4}\left(e^{-j\phi_A} - e^{-j\phi_B}\right)^2} \right\} \tag{3.58}$$

and the transmission coefficient (3.55) is found to be[10]

$$S_{1'2} = -\frac{1}{2}\left(e^{-j\phi_A} - e^{-j\phi_B}\right) \cdot \left\{ 1 + \frac{\rho\rho'}{4} \cdot \frac{\left(e^{-j\phi_A} + e^{-j\phi_B}\right)^2}{1 - \frac{\rho\rho'}{4}\left(e^{-j\phi_A} - e^{-j\phi_B}\right)^2} \right\} \tag{3.59}$$

Let us now designate the first and second term in the product of the right-hand side member of (3.59) as P and Q, respectively. The modulus of (3.59), the variable of interest for the use of the structure of Fig. 3.10 for sensing purposes, is given by the modulus of P, i.e.

$$|P| = \left| \sin\left(\frac{\phi_A - \phi_B}{2}\right) \right| \tag{3.60}$$

times the modulus of Q. This later quantity cannot be expressed in a straightforward form. However, for small asymmetries (imbalances), corresponding to similar (but not identical) values of the electrical lengths of both lines ($\phi_A \approx \phi_B$), the modulus of the transmission coefficient can be approximated by

$$|S_{1'2}| \approx \left| \frac{\phi_A - \phi_B}{2} \right| \cdot \left| 1 + \rho\rho'e^{-j(\phi_A + \phi_B)} \right| \tag{3.61}$$

The case of small imbalances is justified since the detection of tiny differences between the phases of both lines (e.g. caused by slightly different sample loading in the lines) is of interest for high sensitive detectors and sensors.

It follows from (3.61) that for small asymmetries, termination of any of the isolated ports (1 or 2′) with a matched load (i.e. $\rho \cdot \rho' = 0$) is a sufficient condition to obtain a proportional dependence of the modulus of the transmission coefficient with the phase difference. Thus, by considering the phase difference, $\Delta\phi = \phi_A - \phi_B$, as the input variable, the sensitivity for small perturbations is constant and given by[11]

---

10 Expression (3.59) has been obtained by neglecting losses. This approximation is justified as far as the substrate for device implementation is a low-loss microwave substrate, and provided the imbalances (if they are present) are caused by dielectric loads (in lines A and B) corresponding to low-loss materials. The generalization of (3.59) by considering the lines loaded with lossy materials (e.g. liquids) is carried out in Appendix A of reference [7].

11 Note that the sensitivity as defined in (3.62) is dimensionless since $\Delta\phi$ is given in rad.

$$S = \frac{d|S_{1'2}|}{d(\Delta\phi)} = \frac{1}{2} \tag{3.62}$$

For $\rho \cdot \rho' \neq 0$, the sensitivity (as defined above) for small perturbations is not constant, but it can be either enhanced or degraded, depending on the phase of the lines ($\phi_A \approx \phi_B$).

Let us now consider two canonical cases: (i) $\rho \cdot \rho' = 1$ (corresponding, e.g. to $\rho = \rho' = \pm 1$), and (ii) $\rho \cdot \rho' = -1$ (corresponding, e.g. to $\rho = 1$ and $\rho' = -1$, or to $\rho = -1$ and $\rho' = 1$). In the former case, $|Q| \approx 0$ if $\phi_A \approx \phi_B \approx (2n+1)\cdot\pi/2$ (with $n = 0, 1, 2, 3, ...$), and $|Q| \approx 2$ if $\phi_A \approx \phi_B \approx n\cdot\pi$. Thus, if ports 1 and 2' are left open ($\rho = \rho' = 1$) or grounded ($\rho = \rho' = -1$), sensitivity is optimized when the length of the pair of matched and uncoupled lines is roughly a half wavelength (or a multiple of this length). In this case, the sensitivity in the limit of small asymmetries is $S = 1$. Conversely, for $\rho \cdot \rho' = -1$, $|Q| \approx 2$ if $\phi_A \approx \phi_B \approx (2n+1)\cdot\pi/2$, and $|Q| \approx 0$ if $\phi_A \approx \phi_B \approx n\cdot\pi$. Thus, if one port is left open and the other one is short-circuited to ground, the optimum line length for sensitivity optimization (with $S = 1$, as well) is an odd multiple of a quarter wavelength. It should be mentioned that such canonical cases can also be obtained by means of purely reactive loads. For instance, case (i) results by considering, e.g. $\rho = j$ (pure inductance) and $\rho' = -j$ (pure capacitance). Nevertheless, the most interesting case from a practical viewpoint is case (i) with $\rho = \rho' = 1$, since neither a specific load nor a via is required in the isolated ports (1 and 2') of the couplers.

For the validation of the previous sensitivity analysis, the exact value of the term Q [in brackets in expression (3.59)] is represented in Fig. 3.11. Such term can be rewritten as

$$Q = \frac{1}{1 - \dfrac{\rho\rho'[\cos(\phi_A - \phi_B) + 1]}{2[\rho\rho' + j\sin(\phi_A + \phi_B) + \cos(\phi_A + \phi_B)]}} \tag{3.63}$$

It can be seen in Fig. 3.11 that when $\phi_A \approx \phi_B$, Q approaches 0 or 2 for the phase conditions indicated above, depending on the specific case ($\rho \cdot \rho' = 1$ or $\rho \cdot \rho' = -1$).

Figure 3.12 depicts the exact value of the transmission coefficient [expression (3.59)], as well as the sensitivity, as a function of the phase difference between the lines ($\Delta\phi = \phi_A - \phi_B$) by considering the phase of line A set to a fixed value of $\phi_A = \pi/2$ [Fig. 3.12(a)] and $\phi_A = \pi$ [Fig. 3.12(b)]. For small values of $\Delta\phi$, the sensitivity is optimized for $\rho \cdot \rho' = 1$ and $\phi_A = \pi$ (or $n\pi$), or for $\rho \cdot \rho' = -1$ and $\phi_A = \pi/2$ [or $(2n+1)\pi/2$], in agreement to the previous analysis based on the approximate expression (3.61). Indeed, the transmission coefficient and the sensitivity that results from (3.61), valid for small perturbations, has also been inferred (it is depicted in Fig. 3.12). The agreement is progressively better as $\Delta\phi$ tends to zero, as expected. Figure 3.12 also depicts the transmission coefficient and sensitivity corresponding to $\rho \cdot \rho' = 0$. In this case, the sensitivity varies with $\Delta\phi$, being roughly constant for small perturbations ($\Delta\phi \approx 0$), but it does not depend on the phase of line A.

3.2.1.2.3 Sensor Design

This section reports a sensor design based on the results of the previous sensitivity analysis. The sensor uses a pair of matched and uncoupled lines with the phases set to 2π, and ports 1 and 2' of the couplers are terminated with open-ended loads (i.e. $\rho = \rho' = 1$). The sensor operates at $f_0 = 2$ GHz, and the considered substrate for sensor fabrication is the *Rogers RO4003C* with dielectric constant $\varepsilon_r = 3.55$, thickness $h = 0.8128$ mm and dissipation factor $\tan\delta = 0.0021$. With these substrate parameters and frequency, the dimensions of the rat-race hybrid couplers are those indicated in Fig. 3.13, where the whole fabricated sensor is depicted (the width of the ring lines is the one corresponding to a characteristic impedance of 70.71 Ω). Identical meandered lines of width 1.82 mm (providing a 50-Ω characteristic impedance when they are uncovered) and length of 84.93 mm complete the sensing structure. This line length corresponds to a phase of 2π at f_0,

(a)

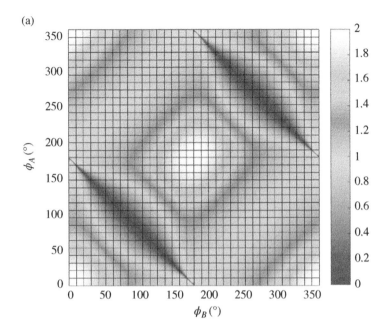

(b)

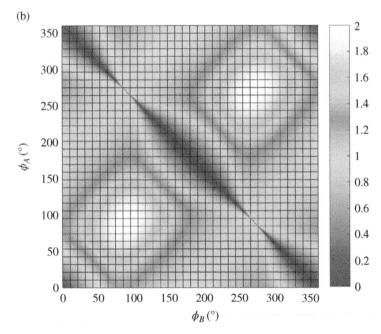

Figure 3.11 Variation of the term Q as a function of the electrical lengths of line A (ϕ_A) and line B (ϕ_B). (a) Q for $\rho \cdot \rho' = 1$; (b) Q for $\rho \cdot \rho' = -1$. *Source:* Reprinted with permission from [7]; copyright 2020 IEEE.

provided the sensing regions (indicated in Fig. 3.13 with dashed rectangles) are covered by a piece of non-metalized *Rogers RO3010* substrate with dielectric constant $\varepsilon_r = 10.2$, thickness $h = 1.27$ mm, and dissipation factor $\tan \delta = 0.0022$, the REF sample in this study. With this electrical length of the lines, the sensitivity is optimized (see the previous section), provided $\rho \cdot \rho' = 1$ (note that the isolated

(a)

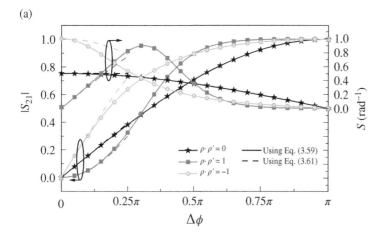

(b)

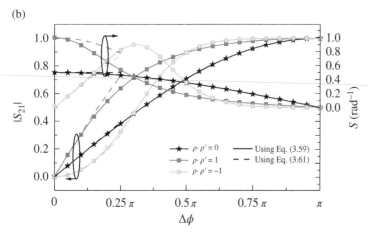

Figure 3.12 Variation of the transmission coefficient and the sensitivity as a function of the phase difference ($\Delta\phi = \phi_A - \phi_B$) by considering the electrical length of line A (ϕ_A) set to a fixed value. (a) $\phi_A = \pi/2$; (b) $\phi_A = \pi$. For comparison purposes, the transmission coefficient and the sensitivity that results when $\rho \cdot \rho' = 0$ is also included in the figures. *Source:* Reprinted with permission from [7]; copyright 2020 IEEE.

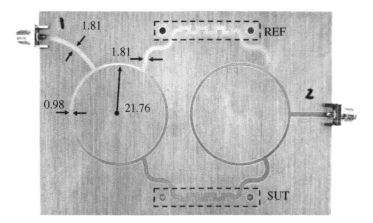

Figure 3.13 Photograph of the fabricated sensor/comparator. Dimensions are given in mm.

ports of both couplers are opened, and thereby $\rho = \rho' = 1$). It should be mentioned that the characteristic impedance of the line varies slightly (decreases) in the sensing region, due to the presence of the REF sample. Namely, in the structure of Fig. 3.13, the characteristic impedance of the pair of lines between the couplers is not uniform since the sensing regions (for the REF and MUT samples) only cover partially these lines. This incomplete line coating is necessary in order to avoid any effect of the REF and MUT samples on the rat-race couplers. Nevertheless, to a first-order approximation, the matching condition can be assumed to be valid, and, thus, the conclusions relative to the previous analysis prevail, as it will be shown later.

Figure 3.14 depicts the simulated (inferred through the *Keysight Momentum* commercial software) and the measured response of the bare device, i.e. without the presence of the REF and MUT samples. The pair of balanced lines exhibits good balance, as revealed by the measured transmission coefficient at f_0, smaller than -60 dB. Such good balance is an essential aspect for sensor/comparator resolution. The measured response that results by loading both lines with the REF sample indicated before is also included in Fig. 3.14. The transmission coefficient at f_0 is -60 dB, pointing out that line balance is preserved by loading the lines with identical samples (to minimize the effects of the air gap, the samples have been attached to the lines by means of Teflon screws).

The sensitivity analysis of the previous section can be validated by simulating the transmission coefficient at f_0 considering the REF line loaded with the REF sample and the MUT line loaded with a hypothetical material of identical dimensions and different dielectric constants. Figure 3.15(a) depicts the modulus of the transmission coefficient at f_0 as a function of the variation of the dielectric constant of the MUT, where such variation is expressed as percentage with regard to the nominal value of the REF sample (with dielectric constant $\varepsilon_r = 10.2$). The magnitude of the transmission coefficient exhibits a roughly linear variation with the differential dielectric constant, and the average sensitivity with this variable is $S = d|S_{21}|/d\%\varepsilon_r = 0.0052$. The phase of line B (the MUT line) for the different MUTs, inferred from independent simulations, has been introduced in expression (3.59) for evaluation of the magnitude of the transmission coefficient (note that the phase of line A does not vary and corresponds to the optimum case for sensitivity optimization, with $\rho \cdot \rho' = 1$, as indicated before). As it can be seen in Fig. 3.15(a), the agreement between the simulated value of the transmission coefficient and the value inferred from (3.59) is good. The magnitude of the transmission coefficient that results by introducing the simulated phases of the MUT line in expression (3.61), the low perturbation approximation, is also depicted in Fig. 3.15(a). This curve is roughly

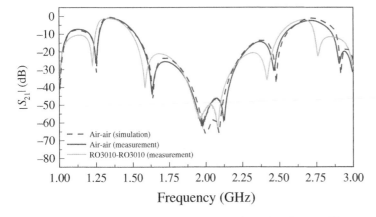

Figure 3.14 Frequency response (measured and simulated) of the fabricated sensor/comparator without line loading, and measured frequency response with the sensing region loaded with the REF sample in both lines. *Source:* Reprinted with permission from [7]; copyright 2020 IEEE.

(a)

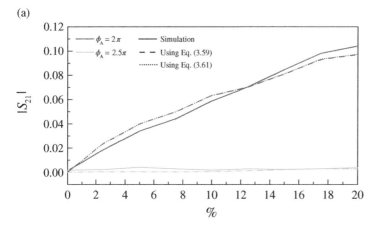

(b)

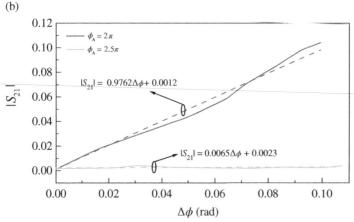

Figure 3.15 (a) Simulated transmission coefficient at f_0, inferred by loading the REF line (*A*) with the REF sample and line *B* with MUT samples of different dielectric constant, expressed as percentage variation with regard to the dielectric constant of the REF sample. (b) Representation of the transmission coefficient at f_0 as a function of the differential phase inferred from the simulations. The transmission coefficient that results by evaluating (3.59) and (3.61) with the phases of line *B* inferred from electromagnetic simulation is also included in (a). *Source:* Reprinted with permission from [7]; copyright 2020 IEEE.

undistinguishable from the one of (3.59) because the phase difference between the lines is small, even for the larger variation of the dielectric constant (20%). These results validate the considered approximation (3.61), at least up to moderate variations in the differential dielectric constant. The simulations that result by considering a structure identical to the one described before, but extending the phase of the lines loaded with the REF sample to 2.5π, i.e. the worst case in terms of sensitivity (with $Q = 0$), are also depicted in Fig. 3.15(a). As it can be seen, the variation of the transmission coefficient for small values of the differential dielectric constant is roughly negligible, in agreement with a negligible value of the sensitivity for small perturbations.

Figure 3.15(b) shows the dependence of the transmission coefficient with the simulated differential phase for the two considered cases in Fig. 3.15(a). For the optimum case (electrical length of the lines of 2π), the average slope is 0.9762, corresponding to an average sensitivity with the differential phase very close to the theoretical value ($S = 1$). This slight discrepancy is in part because the lines are not perfectly matched, as discussed before.

The previous sensitivity analysis and validation refers to the variation of the output variable, the magnitude of the transmission coefficient, with the differential phase [see Fig. 3.15(b)]. However, the actual input variable, that causes a variation in the differential phase of the lines, is the differential dielectric constant between the REF and MUT samples, $\Delta\varepsilon$. Therefore, the sensitivity of interest is actually[12]

$$S = \frac{d|S_{1'2}|}{d(\Delta\varepsilon)} = \frac{d|S_{1'2}|}{d(\Delta\phi)} \cdot \frac{d(\Delta\phi)}{d(\Delta\varepsilon)} \tag{3.64}$$

where the first term in the right-hand side member is comprised between 0 and 1, as discussed before. The second term has been exhaustively studied in Section 3.2.1.1 [particularly, expression (3.36) applies to lines implemented in microstrip technology, under the approximations considered in that section]. Thus, for sensitivity optimization, sensor design should consider long balanced lines, exhibiting an electrical length multiple of π (2π in the fabricated sensor), i.e. the optimum length for open-circuit loads in the isolated ports of the coupler.

3.2.1.2.4 Comparator Functionality

The functionality of the sensor of Fig. 3.13 as comparator is demonstrated by loading line A, the REF line, with the REF sample, and line B subsequently with eight identical samples, but with arrays of drilled holes of different densities (Fig. 3.16). Figure 3.17 depicts the measured transmission coefficients corresponding to the different MUT samples. The transmission coefficient at the operating frequency, f_0, experiences a significant change as the density of holes varies. This comparator is able to detect tiny differences between the REF and MUT samples, as derived from the different transmission coefficient that results when line B is loaded with the REF sample and with the MUT with the smaller density of holes.

3.2.1.2.5 Dielectric Constant Measurements

The sensor of Fig. 3.13 is also useful for the measurement of the dielectric constant of solid samples. The measured responses of the sensor that result by loading line B with different MUT samples with well-known dielectric constant, particularly different types of un-cladded dielectric substrates

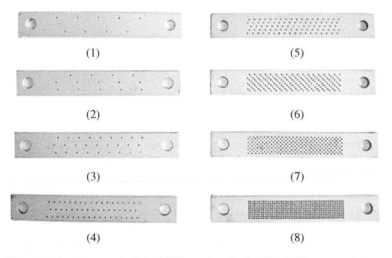

(1)

(5)

(2)

(6)

(3)

(7)

(4)

(8)

Figure 3.16 Photograph of the MUT samples obtained by drilling sparse hole arrays of different densities in samples identical to the REF sample.

---

12 Actually, a differential dielectric constant between the REF sample and the MUT causes also a variation in the differential characteristic impedance of the lines, with influence in the output variable. However, this term can be neglected to a first-order approximation, since quasi-matching conditions are considered in this study.

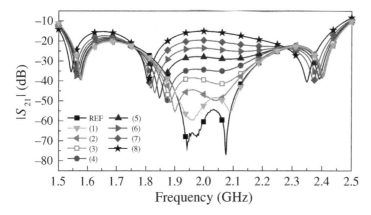

Figure 3.17 Measured transmission coefficient that results by loading line *A* with the REF sample and line *B* with the MUT samples depicted in Fig. 3.16. *Source:* Reprinted with permission from [7]; copyright 2020 IEEE.

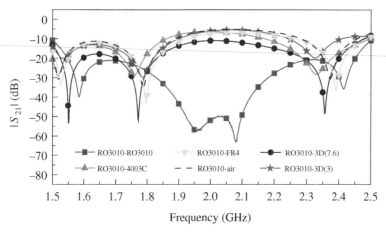

Figure 3.18 Measured transmission coefficient that results by loading line *A* with the REF sample and line *B* with MUT samples corresponding to the indicated materials. *Source:* Reprinted with permission from [7]; copyright 2020 IEEE.

(i.e. *FR4* with $\varepsilon_r = 4.6$ and *RO4003C* with $\varepsilon_r = 3.55$), is depicted in Fig. 3.18. The transmission coefficients at f_0 are depicted in Fig. 3.19. From these values, the calibration curve depicted in such figure is obtained. This curve is useful to determine the dielectric constant of the MUT by measuring the corresponding transmission coefficient magnitude. Note that such curve is useful for the dielectric characterization of materials with dielectric constant smaller than the one of the REF sample, with $\varepsilon_{REF} = 10.2$. Moreover, it is necessary that the thickness of the MUT is comparable to the one of the REF samples and MUTs used for calibration. Nevertheless, if the REF sample and the MUTs are thick enough, such that the electric field lines generated by the lines do not reach the interface between the sample and air, the response of the sensor does not depend on sample thickness (an aspect that has been discussed before in this chapter).

Two MUT samples (designated as *PLA* and *RS Pro MT-COPPER*) fabricated by means of a 3D printer (*Ultimaker 3 Extended*) are used to experimentally validate the designed and fabricated sensor for dielectric constant measurements. These samples exhibit dielectric constants of 3 (for *PLA*) and 7.6 (for *RS Pro MT-COPPER*), as measured by means of a commercial resonant cavity.

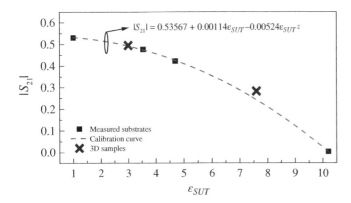

Figure 3.19 Transmission coefficient at f_0 corresponding to the MUT samples of Fig. 3.18. The correlation factor of the calibration curve is $R^2 = 0.9998$. *Source:* Reprinted with permission from [7]; copyright 2020 IEEE.

The thickness of such samples is the same to the one of the REF sample and to those of the samples used for sensor calibration. The measured transmission coefficients corresponding to these samples are also included in Figs. 3.18 and 3.19. In particular, the values at f_0 are in close proximity to the calibration curve (Fig. 3.19), thereby demonstrating that the proposed sensor provides a good estimation of the dielectric constant of the MUT samples. The maximum sensitivity inferred from the calibration curve is $|S| = 0.097$, and the maximum sensitivity, in dB, as derived from the results of Fig. 3.19 at f_0 is 17.62 dB.

As mentioned before, the measurement of the dielectric constant of the MUT from the calibration curve requires that such MUT exhibits a smaller dielectric constant than the one of the REF material. In order to elucidate this aspect, let us consider again that the phase variation of line B does not experience a strong change when it is loaded with the MUT sample. Under these conditions, expression (3.59) can be approximated by

$$S_{1'2} \approx \frac{1}{2} e^{j\left(\frac{\pi}{2} - \phi_A\right)} (\phi_A - \phi_B)\left(1 + \rho\rho' e^{-j(\phi_A + \phi_B)}\right) \tag{3.65}$$

For $\phi_A = n \cdot \pi$ and $\phi_A \approx \phi_B$, the last term in (3.65) is roughly $Q \approx 2$, as discussed before, and the transmission coefficient can be expressed as

$$S_{1'2} \approx j(-1)^n (\phi_A - \phi_B) \tag{3.66}$$

According to (3.66), for n odd, the phase of the transmission coefficient is $+90°$ for $\phi_A < \phi_B$ (corresponding to a smaller dielectric constant for the REF sample), and $-90°$ for $\phi_A > \phi_B$ (with a larger dielectric constant for the REF sample). For n even (the case considered in the reported prototype example of Fig. 3.13, with $n = 2$), the phase of the transmission coefficient is $-90°$ for $\phi_A < \phi_B$, and $+90°$ for $\phi_A > \phi_B$. Hence, the phase of the transmission coefficient at f_0 can be used to distinguish whether the dielectric constant of the MUT is larger or smaller than the one of the REF sample. Despite the fact that the calibration curve of Fig. 3.19 is valid for MUT samples satisfying $\varepsilon_{MUT} < \varepsilon_{REF}$, let us mention that by considering samples with (well-known) dielectric constant larger than the one of the REF sample, a calibration curve for MUT samples satisfying $\varepsilon_{MUT} > \varepsilon_{REF}$ can also be generated. Then, the reported method to determine the relative magnitude of ε_{MUT} with regard to ε_{REF} can be used to determine which of the two calibration curves must be considered. From that curve, and the measurement of the magnitude of the transmission coefficient, the dielectric constant of the MUT can be finally determined.

The prototype sensor discussed in this section is useful to measure the real part of the permittivity of the MUT (or its dielectric constant), provided low-loss MUT samples are considered. It was shown in Section 3.2.1.1 that the reported four-port differential-mode sensor based on a pair of meandered lines was able not only to provide the dielectric constant of the MUT, but also its loss tangent (related to the imaginary part of the permittivity). However, for that purpose, it was necessary to measure the magnitude of the transmission coefficient of the MUT sensing line. Such line is not accessible in the sensor of Fig. 3.13. On the other hand, variations in the loss tangent of the MUT, as compared to the one of the REF sample (a low loss material), are expected to affect the modulus of the transmission coefficient. Consequently, in significantly lossy MUTs, it is not possible to accurately determine the dielectric constant of the MUT because the output variable (the modulus of the transmission coefficient) is affected by it and by the loss tangent. The fact that the output variable depends on both the dielectric constant and the loss tangent of the MUT (for significantly lossy MUTs) prevents from the application of the reported sensor to the determination of the loss tangent of the MUT. In general, for the measurement of both the dielectric constant and the loss tangent of a certain material, resonant methods are preferred (as it was shown in Chapter 2). Nevertheless, it does not mean that differential sensing involving lossy materials, e.g. liquids, with the reported prototype sensor is not possible. On the contrary, the sensor of Fig. 3.13 is very appropriate for measuring the volume fraction of solute in diluted liquid solutions, to be discussed in the next section. Since liquids are lossy materials, line imbalance may be caused not only by variations in the dielectric constant between the REF and MUT sample, mainly affecting the differential phase of the lines, but also by changes in the loss tangent (with effect on the magnitude of the transmission coefficient of the MUT line). The generalization of the formulation of Section 3.2.1.2.2 for the lossy case, detailed in Appendix A of [7], reveals that the same phase conditions for the lines resulting for the lossless case should be applied for sensitivity optimization.

3.2.1.2.6 *Microfluidic Sensor. Solute Concentration Measurements*

For the last campaign of experiments, devoted to the determination of solute content in liquid solutions, the sensor of Fig. 3.13 was redesigned [7]. The considered REF sample is deionized (DI) water, whereas the MUTs are different mixtures of isopropanol in DI water. Thus, for sensitivity optimization, the length of the meandered sensing lines is re-dimensioned, so that the electrical length of the REF line (line A) is $\phi_A = n\cdot\pi$, when such line is loaded with the REF liquid. This gives a total line length of 131.77 mm, corresponding to $n = 3$, i.e. 46.84 mm longer than the one of the sensor of Fig. 3.13. For the determination of the volume fraction of different solutions of isopropanol in DI water, the sensing regions of lines A and B are equipped with microfluidic channels. Such channels and the necessary accessories for liquid injection (through syringes) are described in detail in [12]. It should be mentioned that, in order to avoid liquid absorption by the substrate, a dry film with thickness 50 µm and dielectric constant 3.6 is used. The presence of such film does not substantially modify the electrical characteristics of the lines. The photograph of the sensor, with the channel and other mechanical parts, including the capillaries for liquid injection, is depicted in Fig. 3.20.

For measuring purposes, the REF liquid (pure DI water) is injected in the REF channel of line A, and subsequently different mixtures of isopropanol in DI water are injected in the MUT channel (line B), starting with a null volume fraction (corresponding to the REF liquid), and progressively increasing the volume fraction of isopropanol. The responses for the different mixtures are depicted in Fig. 3.21. The sensor is able to resolve concentrations of isopropanol in DI water as small as 1%. The dependence of the transmission coefficient at f_0 as a function of the isopropanol content is depicted in Fig. 3.22, which includes the calibration curve, i.e.

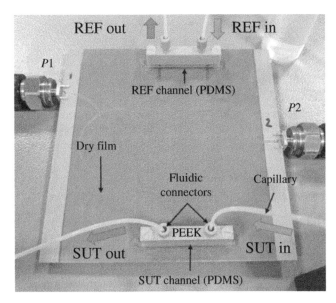

Figure 3.20 Photograph of the sensor of Fig. 3.13 (redesigned as indicated in the text) equipped with fluidic channels and protected with an anti-absorption dry film.

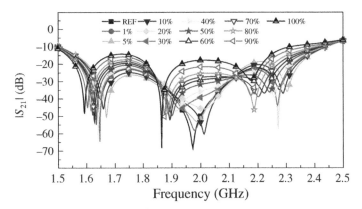

Figure 3.21 Measured transmission coefficient that results by loading line *A* with the REF liquid (DI water) and line *B* with MUT samples corresponding to the indicated volume fractions of isopropanol in DI water. *Source:* Reprinted with permission from [7]; copyright 2020 IEEE.

$$F_v(\%) = 154.1 - 18.0 \cdot 10^{-\left(\frac{S_{21}(\text{dB}) + 0.0803}{1.95}\right)} - 204.8 \cdot 10^{-\left(\frac{S_{21}(\text{dB}) + 0.0803}{580.97}\right)} \tag{3.67}$$

Such curve is useful for the determination of the volume fraction (F_v) of solute in unknown mixtures of isopropanol and DI water.

By properly modifying sensor dimensions, the sensitivity for different REF liquids, determined by the specific intended application, can be optimized. In particular, it is necessary to adjust the electrical length of the REF line loaded with the REF liquid, so that it satisfies $\phi_A \approx n \cdot \pi$. These sensors can be of interest in several industrial processes requiring an accurate determination of solute content in liquid mixtures, for instance, for monitoring variations of alcohol content in wine fermentation processes.

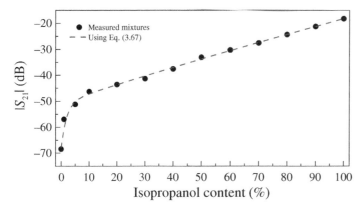

Figure 3.22 Value of the transmission coefficient (dB) at f_0 as a function of the isopropanol content. The correlation factor of the calibration curve is $R^2 = 0.9992$. *Source:* Reprinted with permission from [7]; copyright 2020 IFEE.

3.2.2 Reflective-Mode Sensors

Reflective-mode phase-variation sensors constitute a good alternative to the phase-variation sensors of Section 3.2.1 (implemented by means of transmission lines operating in transmission) for sensitivity enhancement and size reduction. Moreover, reflective-mode sensors are one-port structures, which can result of practical interest in certain applications, e.g. for the implementation of submersible liquid sensors [13]. The simplest configuration of a reflective-mode transmission line-based sensor is an open-ended line. The phase of the reflection coefficient of such line, the usual output variable, is given by expression (3.25), whereas the sensitivity is given by (3.28). As it was demonstrated at the beginning of Section 3.2, the sensitivity can be optimized by using either a high-impedance $(2n + 1) \cdot \pi/2$ open-ended sensing line or a low-impedance $n \cdot \pi$ open-ended sensing line. The sensitivity (for the same dimensions) can be significantly improved as compared to the one of a two-port sensor based on an ordinary transmission line. However, it is possible to further (and significantly) enhancing the sensitivity of reflective-mode transmission line-based sensors by considering a step-impedance configuration [5], to be discussed next.

3.2.2.1 Sensitivity Enhancement by Means of Step-Impedance Open-Ended Lines

The schematic of the step-impedance open-ended line reflective-mode phase-variation sensor is depicted in Fig. 3.23. The structure consists of an open-ended sensing line with characteristic impedance Z_s and electrical length ϕ_s at the operating frequency, cascaded to a set of N line sections with alternating high and low characteristic impedance. Such impedances are designated as Z_i, where $i = 1, 2, \ldots N$ identifies the line section and $i = 1$ corresponds to the line section adjacent to the sensing line. Similarly, ϕ_i denotes the electrical length of line i. The open end of the sensing line and the first step-impedance discontinuity delimit the sensitive region, indicated by the dashed rectangle in Fig. 3.23. It should be emphasized that the impedance Z_s and the phase ϕ_s vary with the dielectric constant of the MUT, contrary to Z_i and ϕ_i, which do not change, provided the sensing region is restricted to the open-ended sensing line exclusively. This means that the design values of Z_s and ϕ_s for sensor implementation are those corresponding to the sensing line loaded with the REF material.

For dielectric constant measurements, the sensitivity is given by (3.22). However, the derivatives $d\phi_\rho/d\phi_s$ and $d\phi_\rho/dZ_s$ are, obviously, more complex than (3.26) and (3.27), due to the presence of the high/low impedance line sections cascaded to the sensing line. It is well known that in any

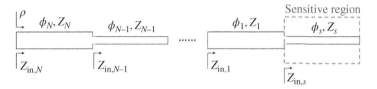

Figure 3.23 Structure of the reflective-mode phase-variation sensor based on a step-impedance open-ended line.

transmission line-based sensor devoted to dielectric constant measurements, elongating the sensing line contributes to sensitivity optimization. However, this strategy increases the size of the sensing region. The relevance of the sensor topology of Fig. 3.23 is the fact that, by engineering the step-impedance structure (through impedance contrast), the first term ($d\phi_\rho/d\phi_s$) can be made extremely large (even by using only few sections of cascaded step-impedance lines). Thus, the implementation of sensors with extraordinary sensitivities and limited size (for both the sensing region and the whole sensor) is possible.

The analysis of the network of Fig. 3.23 with an arbitrary number, N, of step-impedance transmission line sections is complex. Therefore, let us proceed by first studying the circuit with a single step discontinuity ($N = 1$), with the aim of obtaining, and optimizing, the sensitivity.[13] The impedance seen from the input port is

$$Z_{in} = \frac{jZ(Z\tan\phi - Z_s\cot\phi_s)}{Z + Z_s\cot\phi_s\tan\phi} \tag{3.68}$$

In (3.68), the impedance and electrical length of the single line section cascaded to the sensing line have been redefined as $Z_1 = Z$ and $\phi_1 = \phi$ in order to simplify the notation by avoiding excessive subindexes in the formulas. The reflection coefficient is found to be

$$\rho = \frac{-Z_0(Z + Z_s\tan\phi\cot\phi_s) + jZ(Z\tan\phi - Z_s\cot\phi_s)}{+Z_0(Z + Z_s\tan\phi\cot\phi_s) + jZ(Z\tan\phi - Z_s\cot\phi_s)} \tag{3.69}$$

and the phase of the reflection coefficient, the output variable, can be expressed as

$$\phi_\rho = 2\arctan\left(\frac{Z(Z_s\cot\phi_s - Z\tan\phi)}{Z_0(Z + Z_s\tan\phi\cot\phi_s)}\right) + \pi \tag{3.70}$$

Calculation of $d\phi_\rho/d\phi_s$, designated as S_{ϕ_s}, gives

$$S_{\phi_s} \equiv \frac{d\phi_\rho}{d\phi_s} = \frac{-2Z^2Z_sZ_0(1 + \tan^2\phi)}{Z_0^2(Z\sin\phi_s + Z_s\tan\phi\cos\phi_s)^2 + Z^2(Z\tan\phi\sin\phi_s - Z_s\cos\phi_s)^2} \tag{3.71}$$

whereas $d\phi_\rho/dZ_s$ is found to be

$$S_{Z_s} \equiv \frac{d\phi_\rho}{dZ_s} = \frac{2Z_0Z^2\cot\phi_s(1 + \tan^2\phi)}{Z_0^2(Z + Z_s\tan\phi\cot\phi_s)^2 + Z^2(Z_s\cot\phi_s - Z\tan\phi)^2} \tag{3.72}$$

Optimization of the sensitivity that results by introducing (3.71) and (3.72), as well as (3.20) and (3.21), in (3.22), with four design variables (Z, Z_s, ϕ, and ϕ_s), is not easy. Nevertheless, it can be seen from (3.72) that $d\phi_\rho/dZ_s = 0$ for $\phi = (2n + 1)\cdot\pi/2$ and $\phi_s = n\cdot\pi$ or $\phi_s = (2n + 1)\cdot\pi/2$. Such phases of

---

13 The general analysis for a multistep-impedance transmission line structure with an arbitrary number of sections N is carried out later in this subsection.

the sensing line ($\phi_s = n \cdot \pi$ and $\phi_s = (2n + 1) \cdot \pi/2$) are those that maximize the sensitivity for the structure consisting solely of the sensing line, as discussed before, and providing also $d\phi_\rho/dZ_s = 0$. Thus, let us see if the above-cited phase combinations are those that maximize S_{ϕ_s}. Obviously, optimization of S_{ϕ_s} means also to determine the optimum characteristic impedances for the sensing line and for the line cascaded to it (designated as design line in [5]).

It is clear that the structure under study exhibits four degrees of freedom (the impedance and phase of both lines). Nevertheless, let us consider that the impedance and phase of the sensing line are set to certain values. This is justified since the structure of Fig. 3.23 with $N = 1$ is essentially the one corresponding to the reflective mode sensor of Fig. 3.1(b), with the addition of the design line. Thus, it can be interpreted that sensitivity optimization proceeds by tailoring, or engineering, such design line (this justifies the designation given to this line). Nevertheless, this does not mean (at least a priori) that the electrical parameters of the sensing line that maximize S_{ϕ_s} are necessarily those corresponding to the sensor without design line (i.e. with $N = 0$), determined before. Moreover, let us assume that the impedance of the design line is also set to a certain value. According to the previous sentences, for the optimization of S_{ϕ_s}, it is pertinent to calculate the derivative of S_{ϕ_s} with the electrical length of the design line, ϕ. By forcing the resulting expression to be zero, the value of ϕ giving the maximum value of S_{ϕ_s} (provided the other design values, i.e. ϕ_s, Z_s, and Z, are fixed) can be found. However, in order to simplify the calculation, we have inferred the derivative with respect to $\tan\phi$. The following result has been found:

$$\frac{dS_{\phi_s}}{d\tan\phi} = -4\frac{(Z_0^2 Z_A^2 + Z^2 Z_B^2)Z^2 Z_s Z_0 \tan\phi - Z^2 Z_s Z_0(1 + \tan^2\phi)(Z_0^2 Z_A Z_s \cos\phi_s + Z^2 Z_B Z \sin\phi_s)}{(Z_0^2 Z_A^2 + Z^2 Z_B^2)^2}$$

(3.73)

where Z_A and Z_B, defined as

$$Z_A = Z \sin\phi_s + Z_s \tan\phi \cos\phi_s \tag{3.74a}$$

$$Z_B = Z \tan\phi \sin\phi_s - Z_s \cos\phi_s \tag{3.74b}$$

have been used for simplification purposes. The zero (or zeros) in (3.73) are given by those values of $\tan\phi$ that null the numerator. Thus, rearranging the numerator and forcing it to be zero, the following second-order equation ($\tan\phi$ being the unknown) results:

$$ZZ_s \sin\phi_s \cos\phi_s \tan^2\phi + (Z^2\sin^2\phi_s - Z_s^2\cos^2\phi_s)\tan\phi - ZZ_s \sin\phi_s \cos\phi_s = 0 \tag{3.75}$$

and the two solutions ($\phi = \phi_a$ and $\phi = \phi_b$) are

$$\tan\phi_a = \frac{Z_s \cos\phi_s}{Z \sin\phi_s} \tag{3.76a}$$

$$\tan\phi_b = -\frac{Z \sin\phi_s}{Z_s \cos\phi_s} \tag{3.76b}$$

Thus, for ϕ_a and ϕ_b given by (3.76), the sensitivity S_{ϕ_s} (calculated using 3.71), should be a local (or absolute) maximum or minimum. The specific values are:

$$S_{\phi_s}|_{\phi_a} = -\frac{2Z^2 Z_s}{Z_0}\frac{1}{Z^2\sin^2\phi_s + Z_s^2\cos^2\phi_s} \tag{3.77a}$$

$$S_{\phi_s}|_{\phi_b} = -2Z_0 Z_s \frac{1}{Z^2\sin^2\phi_s + Z_s^2\cos^2\phi_s} \tag{3.77b}$$

Nevertheless, in order to be sure that the convenient value of ϕ for the optimization of S_{ϕ_s} is either ϕ_a or ϕ_b, it should be verified that S_{ϕ_s} does not exhibit poles (zeros in the denominator). Inspection of (3.71) reveals that the denominator cannot be zero, unless both squared terms are simultaneously null, and this is not possible.

In view of expressions (3.77), it follows that if $Z < Z_0$, S_{ϕ_s} is larger for ϕ_b, and it is larger for ϕ_a if $Z > Z_0$. Let us now calculate the derivative of expressions (3.77) with ϕ_s, in order to calculate the value of ϕ_s that maximizes, or minimizes, these values. The following values are obtained:

$$\frac{dS_{\phi_s}|_{\phi_a}}{d\phi_s} = \frac{4Z^2 Z_s}{Z_0} \frac{(Z^2 - Z_s^2) \sin \phi_s \cos \phi_s}{(Z^2 \sin^2 \phi_s + Z_s^2 \cos^2 \phi_s)^2} \tag{3.78a}$$

$$\frac{dS_{\phi_s}|_{\phi_b}}{d\phi_s} = 4Z_0 Z_s \frac{(Z^2 - Z_s^2) \sin \phi_s \cos \phi_s}{(Z^2 \sin^2 \phi_s + Z_s^2 \cos^2 \phi_s)^2} \tag{3.78b}$$

and for both expressions the derivative is null if $\phi_s = n \cdot \pi$, or $\phi_s = (2n + 1) \cdot \pi/2$.

Concerning (3.78a), if $\phi_s = (2n + 1) \cdot \pi/2$, then $\phi_a = n \cdot \pi$, according to (3.76). Conversely, if $\phi_s = n \cdot \pi$, then it follows that $\phi_a = (2n + 1) \cdot \pi/2$. The sensitivity S_{ϕ_s} in the former case is

$$S_{\phi_s} = \frac{-2Z_s}{Z_0} = -2\overline{Z_s} \tag{3.79}$$

and the following value results for $\phi_s = n \cdot \pi$, and $\phi_a = (2n + 1) \cdot \pi/2$:

$$S_{\phi_s} = \frac{-2Z^2}{Z_0 Z_s} = -2\frac{\overline{Z}^2}{\overline{Z_s}} \tag{3.80}$$

where the normalized impedances $\overline{Z_s} = Z_s/Z_0$ and $\overline{Z} = Z/Z_0$ have been used. In view of (3.79) and (3.80), it is clear that for $Z > Z_0 > Z_s$, (3.80) is a maximum and (3.79) is a minimum.

Concerning (3.78b), if $\phi_s = (2n + 1) \cdot \pi/2$, then $\phi_b = (2n + 1) \cdot \pi/2$, and $\phi_b = n \cdot \pi$ if $\phi_s = n \cdot \pi$. The sensitivity S_{ϕ_s} in the former case is

$$S_{\phi_s} = \frac{-2Z_0 Z_s}{Z^2} = -2\frac{\overline{Z_s}}{\overline{Z}^2} \tag{3.81}$$

For $\phi_b = n \cdot \pi$ and $\phi_s = n \cdot \pi$, S_{ϕ_s} is found to be:

$$S_{\phi_s} = \frac{-2Z_0}{Z_s} = -2\frac{1}{\overline{Z_s}} \tag{3.82}$$

Inspection of (3.81) and (3.82) reveals that for $Z < Z_0 < Z_s$, (3.81) is a maximum and (3.82) is a minimum.

According to the previous analysis, it can be concluded that to maximize the sensitivity S_{ϕ_s}, there are two optimum combinations regarding the electrical lengths of the sensing and design lines (plus the corresponding multiples):

- Case A: $\phi_s = (2n + 1) \cdot \pi/2$ and $\phi = (2n + 1) \cdot \pi/2$. In this case, the line impedances should satisfy $Z < Z_0 < Z_s$, and the value of S_{ϕ_s} is given by (3.81), corresponding to a maximum. Moreover, the lower the impedance contrast, defined as the ratio between the characteristic impedances of the design and sensing line, the higher the sensitivity S_{ϕ_s}.
- Case B: $\phi_s = n \cdot \pi$ and $\phi = (2n + 1) \cdot \pi/2$. In this case, the line impedances should satisfy $Z > Z_0 > Z_s$, and the value of S_{ϕ_s} is given by (3.80), also corresponding to a maximum. A high impedance contrast enhances S_{ϕ_s} in this case.

Figure 3.24 plots the dependence of S_{ϕ_s} with ϕ_s and ϕ, calculated by means of expression (3.71), for two cases: (i) $Z < Z_0 < Z_s$ and (ii) $Z > Z_0 > Z_s$. The figure validates the previous analysis,

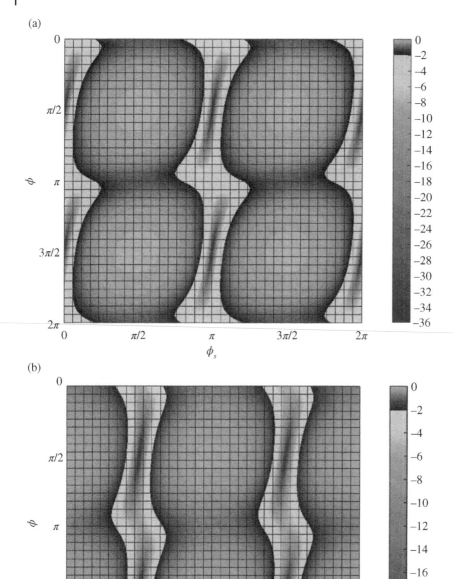

Figure 3.24 Plot (in the form of level-chart) of the sensitivity of the phase of the reflection coefficient with the phase of the sensing line, S_{ϕ_s}, as a function of the electrical lengths of the sensing and design lines, for two cases: (a) $Z = 150\,\Omega$ and $Z_s = 25\,\Omega$; (b) $Z = 25\,\Omega$ and $Z_s = 150\,\Omega$. The reference impedance of the port is $Z_0 = 50\,\Omega$. *Source:* Reprinted with permission from [5]; copyright 2020 IEEE.

confirming that S_{ϕ_s} is optimum (maximum) for the phase combinations predicted by the theory. It is interesting to mention that for any value of ϕ_s, there are two values of ϕ [designated as ϕ_a and ϕ_b in expression (3.76)], where S_{ϕ_s} is either a maximum or a minimum. The positions of these maxima and minima in the ϕ–ϕ_s plane are visible in the top-view plots of Fig. 3.24. Note that these positions converge to the canonical values if the electrical length of the sensing line is either $\phi_s = n\cdot\pi$ or $\phi_s = (2n + 1)\cdot\pi/2$. The maximum value of the sensitivity for the two cases considered in Fig. 3.24, given by expression (3.80) for $Z = 150\,\Omega$ and $Z_s = 25\,\Omega$, and by expression (3.81) for $Z = 25\,\Omega$ and $Z_s = 150\,\Omega$, are $|S_{\phi_s}| = 36$ and $|S_{\phi_s}| = 24$, respectively. These high values of S_{ϕ_s} are achieved by considering moderate line impedances (25 and 150 Ω), corresponding to implementable lines in most microwave substrates.

The phase combinations indicated in cases A and B maximize S_{ϕ_s} and give $d\phi_\rho/dZ_s = 0$. Nevertheless, this does not necessarily mean that these phase combinations are those that optimize the overall sensitivity, S, given by (3.22). To gain insight on this aspect, let as introduce (3.71), (3.72), (3.20), and (3.21) in (3.22). The result can be expressed in terms of S_{ϕ_s} as

$$S = \frac{(1-F)}{4\varepsilon_{\textit{eff}}} S_{\phi_s}\{\phi_s + \sin\phi_s\cos\phi_s\} \tag{3.83}$$

since

$$S_{Z_s} = -\frac{S_{\phi_s}}{Z_s}\sin\phi_s\cos\phi_s \tag{3.84}$$

In (3.83), the dependence of S on ϕ is exclusively within the term S_{ϕ_s}. Therefore, the phases ϕ that provide either a maximum or a minimum in the overall sensitivity S coincide with those phases that maximize or minimize S_{ϕ_s}, given by (3.76). Introducing such phases in (3.83), one obtains

$$S|_{\phi_a} = -\frac{(1-F)}{2\varepsilon_{\textit{eff}}}\frac{Z^2 Z_s}{Z_0}\frac{\phi_s + \sin\phi_s\cos\phi_s}{Z^2\sin^2\phi_s + Z_s^2\cos^2\phi_s} \tag{3.85a}$$

$$S|_{\phi_b} = -\frac{(1-F)}{2\varepsilon_{\textit{eff}}}Z_0 Z_s\frac{\phi_s + \sin\phi_s\cos\phi_s}{Z^2\sin^2\phi_s + Z_s^2\cos^2\phi_s} \tag{3.85b}$$

and these expressions can be rewritten as

$$S|_{\phi_a} = -\frac{Z_s(1-F)}{2Z_0\varepsilon_{\textit{eff}}}\cdot\frac{\phi_s\sin^{-2}\phi_s + \cot\phi_s}{1 + \left(\frac{Z_s}{Z}\right)^2\cot^2\phi_s} \tag{3.86a}$$

$$S|_{\phi_b} = -\frac{Z_s Z_0(1-F)}{2Z^2\varepsilon_{\textit{eff}}}\cdot\frac{\phi_s\sin^{-2}\phi_s + \cot\phi_s}{1 + \left(\frac{Z_s}{Z}\right)^2\cot^2\phi_s} \tag{3.86b}$$

that is, formally identical to (3.28). Apart from the proportionality factor, the unique difference with (3.28) concerns the dimensionless factor of the $\cot^2\phi_s$ in the denominators, i.e. $(Z_s/Z)^2$ [such factor is $(Z_s/Z_0)^2$ in (3.28)]. This means that the values of ϕ_s that maximize, or minimize, expressions (3.86) are also given by (3.30), with Z_0 replaced with Z in (3.30b). Thus, the solutions are $\phi_s = (2n + 1)\cdot\pi/2$ (exact value), and $\phi_s = n\cdot\pi$ (approximate value for $Z_s < Z$).

If $\phi_s = (2n + 1)\cdot\pi/2$, then $\phi_a = n\cdot\pi$, according to (3.76), and the sensitivity in this case, is

$$S = -\frac{Z_s(1-F)}{2Z_0\varepsilon_{\textit{eff}}}\phi_s = -\overline{Z_s}\frac{(1-F)}{2\varepsilon_{\textit{eff}}}\phi_s \tag{3.87}$$

as inferred from (3.85a). If $\phi_s = n\cdot\pi$, then it follows that $\phi_a = (2n + 1)\cdot\pi/2$, and the sensitivity is found to be

$$S = -\frac{Z^2(1-F)}{2Z_0 Z_s \varepsilon_{eff}}\phi_s = -\frac{\overline{Z}^2}{\overline{Z}_s}\frac{(1-F)}{2\varepsilon_{eff}}\phi_s \tag{3.88}$$

Inspection of (3.87) and (3.88) reveals that for $Z > Z_0 > Z_s$, (3.88) is a maximum and (3.87) is a minimum. On the other hand, if $\phi_s = (2n + 1)\cdot\pi/2$, then $\phi_b = (2n + 1)\cdot\pi/2$, and $\phi_b = n\cdot\pi$ for $\phi_s = n\cdot\pi$. The sensitivity in the former case, inferred by means of (3.85b), is

$$S = -\frac{Z_s Z_0(1-F)}{2Z^2 \varepsilon_{eff}}\phi_s = -\frac{\overline{Z}_s}{\overline{Z}^2}\frac{(1-F)}{2\varepsilon_{eff}}\phi_s \tag{3.89}$$

For $\phi_b = n\cdot\pi$ and $\phi_s = n\cdot\pi$, S is found to be

$$S = -\frac{Z_0(1-F)}{2Z_s \varepsilon_{eff}}\phi_s = -\frac{1}{\overline{Z}_s}\frac{(1-F)}{2\varepsilon_{eff}}\phi_s \tag{3.90}$$

and for $Z < Z_0 < Z_s$, (3.89) is a maximum whereas (3.90) is a minimum.

In summary, the phase combinations that maximize S_{ϕ_s}, corresponding to cases A and B, are those that maximize S, as well. Namely, $\phi_s = (2n + 1)\cdot\pi/2$ and $\phi = (2n + 1)\cdot\pi/2$, with the line impedances satisfying $Z < Z_0 < Z_s$ [and sensitivity given by (3.89)], or $\phi_s = n\cdot\pi$ and $\phi = (2n + 1)\cdot\pi/2$, with the line impedances set according to $Z > Z_0 > Z_s$ [and sensitivity given by (3.88)]. For such phases, $d\phi_p/dZ_s = 0$, as indicated before, and the optimum overall sensitivities can be simply calculated from the product of (3.80), or (3.81), and (3.20), giving (3.88), or (3.89), respectively. The sensitivities (3.88) and (3.89) can be substantially improved as compared to those of the open-ended sensing lines without the presence of the design line [given by (3.31a) and (3.31b)]. The reason is that the normalized impedance of the design line in (3.88) and (3.89) appears squared. Thus, implementing the sensor with $(2n + 1)\cdot\pi/2$ design lines exhibiting extreme impedance contrast (either high or low, depending on the specific case) suffices for sensitivity optimization.[14]

It is important to emphasize that since the sensitivity varies with the dielectric constant of the MUT, the sensitivity optimization process discussed before, obviously, corresponds to a certain value of the dielectric constant of the MUT (i.e. the one of the REF material). The information relative to such REF dielectric constant is actually introduced implicitly in the optimization process, as far as the phase conditions that optimize the sensitivity [$\phi_s = (2n + 1)\cdot\pi/2$ or $\phi_s = n\cdot\pi$, depending on the case] are applied to the sensing line loaded with the REF material (and, obviously, the design value of the impedance of the sensing line, Z_s, that appears in (3.88) and (3.89), applies also to the line covered by the REF sample). Thus, the sensitivities given by (3.88) and (3.89) are those corresponding to the optimum values that can be obtained for small dielectric constant variations in the vicinity of that of the REF material. For large variations in the dielectric constant of the MUT, the sensitivity is given by (3.22), with (3.20), (3.21), (3.71), and (3.72), with the values of Z_s and ϕ_s corresponding to considered value of ε_{MUT}.

The sensitivity of these reflective-mode sensors can be further enhanced by cascading additional $(2n + 1)\cdot\pi/2$ design lines with alternate high and low impedance. Indeed, the sensitivity can be made as high as required by simply designing the sensor with the sufficient number, N, of cascaded lines. Obviously, this increases sensor size, but not the sensing area, which is determined by the length and width of the sensing line. To gain further insight on the effects of N on sensitivity optimization, let us next calculate the sensitivity for a multistep-impedance transmission line-based sensor with an arbitrary number of cascaded design lines.

---

14 Obviously, for sensitivity optimization, it is also necessary to set the electrical length and characteristic impedance of the sensing line to the convenient values.

According to Fig. 3.23, and assuming that the electrical lengths of all the design lines is $\phi_i = (2n + 1)\cdot\pi/2$, the impedance seen from any step-impedance discontinuity looking at the open end of the structure, $Z_{in,i}$, can be expressed in terms of the impedance seen from the previous discontinuity as

$$Z_{in,i} = \frac{Z_i^2}{Z_{in,i-1}} \tag{3.91}$$

Using mathematical induction, the impedance seen from the input port can be expressed as

$$Z_{in,N} = Z_{in}^{(-1)^N} \cdot \prod_{i=1}^{N} \left\{ Z_i^{2\cdot(-1)^{i+N}} \right\} \tag{3.92}$$

where the symbol $\prod$ denotes the product operator, and Z_{in} is the impedance seen from the first discontinuity, given by (3.23). Introducing (3.23) in (3.92) gives

$$Z_{in,N} = (-jZ_s)^{(-1)^N} \cdot (\cot\phi_s)^{(-1)^N} \cdot \prod_{i=1}^{N} \left\{ Z_i^{2\cdot(-1)^{i+N}} \right\} \tag{3.93}$$

and the reflection coefficient seen from the input port is thus

$$\rho = \frac{j(-1)^{N+1}\cdot(Z_s\cot\phi_s)^{(-1)^N}\cdot\prod - Z_0}{j(-1)^{N+1}\cdot(Z_s\cot\phi_s)^{(-1)^N}\cdot\prod + Z_0} \tag{3.94}$$

The phase of the reflection coefficient is given by

$$\phi_\rho = 2\arctan\left(\frac{(-1)^N\cdot(Z_s\cot\phi_s)^{(-1)^N}\cdot\prod}{Z_0}\right) + \pi \tag{3.95}$$

Note that in (3.94) and (3.95) the argument of the product operator has been omitted for simplicity. According to (3.95), it is clear that for N odd, the phase ϕ_ρ is identical to the one for $N = 1$, with the exception of $\prod$, which depends on the number of stages. For N even, the following identity results:

$$\phi_\rho = 2\arctan\left(\frac{(Z_s\cot\phi_s)\cdot\prod}{Z_0}\right) + \pi = 2\arctan\left(\frac{-(Z_s\cot\phi_s)^{-1}\cdot Z_0}{\prod}\right) \tag{3.96}$$

Thus, except by the constant factor π, the phase for N even is also formally identical to the phase for $N = 1$, but replacing $\prod$ ($=Z^2$ for $N = 1$) with $Z_0^2/\prod$. Since $\prod$ does not depend on the phase ϕ_s (it is given by the impedances of the quarter-wavelength design lines), it is obvious that sensitivity optimization for N arbitrary is achieved for phases satisfying either $\phi_s = (2n + 1)\cdot\pi/2$ or $\phi_s = n\cdot\pi$, similar to the sensor with $N = 1$ (and the convenient values of the impedances). For these phases, and $\phi_i = (2n + 1)\cdot\pi/2$, it follows that $d\phi_\rho/dZ_s = 0$, and the sensitivity can be straightforwardly calculated by simply multiplying $S_{\phi_s} = d\phi_\rho/d\phi_s$ times (3.20).

After some simple (but tedious) algebra, S_{ϕ_s} is found to be

$$S_{\phi_s} = -\frac{2}{\dfrac{Z_0}{\prod\cdot(Z_s)^{(-1)^N}}\cdot\dfrac{(\sin\phi_s)^{(-1)^N+1}}{(\cos\phi_s)^{(-1)^N-1}} + \dfrac{\prod\cdot(Z_s)^{(-1)^N}}{Z_0}\cdot\dfrac{(\cos\phi_s)^{(-1)^N+1}}{(\sin\phi_s)^{(-1)^N-1}}} \tag{3.97}$$

and it takes a maximum or a minimum value, depending on the set of impedances Z_i, when the electrical length of the sensing line is either $\phi_s = (2n + 1)\cdot\pi/2$ or $\phi_s = n\cdot\pi$ (the phases that optimize the overall sensitivity). In order to evaluate (3.97) for these specific values of ϕ_s, it is also necessary to distinguish whether the number of sections, N, is even or odd. Consequently, four different cases appear, and, for each case, S_{ϕ_s} is found to be:

- Case A': $\phi_s = (2n + 1)\cdot\pi/2$ and N odd.

$$S_{\phi_s} = -\frac{2Z_sZ_0}{\prod\limits_{i=1}^{N}\left\{Z_i^{2\cdot(-1)^{i+N}}\right\}} \tag{3.98a}$$

- Case B': $\phi_s = n\cdot\pi$ and N odd.

$$S_{\phi_s} = -\frac{2\cdot\prod\limits_{i=1}^{N}\left\{Z_i^{2\cdot(-1)^{i+N}}\right\}}{Z_sZ_0} \tag{3.98b}$$

- Case C': $\phi_s = (2n + 1)\cdot\pi/2$ and N even.

$$S_{\phi_s} = -\frac{2Z_s\cdot\prod\limits_{i=1}^{N}\left\{Z_i^{2\cdot(-1)^{i+N}}\right\}}{Z_0} \tag{3.98c}$$

- Case D': $\phi_s = n\cdot\pi$ and N even.

$$S_{\phi_s} = -\frac{2Z_0}{Z_s\cdot\prod\limits_{i=1}^{N}\left\{Z_i^{2\cdot(-1)^{i+N}}\right\}} \tag{3.98d}$$

Note that for $N = 1$ (cases A' and B'), expressions (3.98a) and (3.98b) coincide with expressions (3.81) and (3.80), respectively, as expected.

Inspection of expressions (3.98) reveals that Z_s appears in the numerator for $\phi_s = (2n + 1)\cdot\pi/2$ and it appears in the denominator for $\phi_s = n\cdot\pi$. Thus, it can be concluded that for quarter-wavelength (or odd multiple) sensing lines, a high characteristic impedance is required for sensitivity optimization (regardless of the number of sections of the structure). Conversely, low impedance values are needed in half-wavelength (or multiple) sensing lines. To infer the effects of the characteristic impedances of the quarter-wavelength transmission line sections, Z_i (with $i = 1, 2 \ldots N$), on the sensitivity, we should analyze carefully the product operator that appears in expressions (3.98). For N odd (cases A' and B'), it follows that the characteristic impedance of a section with odd order (i.e. with i odd) appears as Z_i^2, whereas for an even-order section the corresponding term in the product is Z_i^{-2}. Under these circumstances, the requirement of a high or low value of Z_i for sensitivity optimization depends on whether the product operator is present either in the numerator or in the denominator in expressions (3.98). Particularly, for case A', with the product operator in the denominator, the odd-order transmission line sections must exhibit low impedance values, whereas high characteristic impedance sections are required for the even sections. The opposite conditions apply for case B', as far as the product operator appears in the numerator of (3.98b). For N even (cases C' and D') and i odd, the impedance is negative squared (Z_i^{-2}), whereas it appears as Z_i^2 for N even and i even. Consequently, for case C', with the product operator in the numerator of (3.98c), the odd sections must exhibit low characteristic impedance, and the line impedance must be high for the even sections. Finally, it is obvious that for case D' (half-wavelength sensing line), the sections that should exhibit high impedance for sensitivity optimization are those with odd index.

From the previous analysis, it can be concluded that for sensitivity optimization, regardless of the number of sections N, if the electrical length of the sensing line is $\phi_s = (2n + 1){\cdot}\pi/2$, the impedance of this line must be high. The impedance of the cascaded quarter-wavelength transmission line sections must alternatively exhibit low and high values (with a low impedance value for the section adjacent to the sensing line, i.e. the one with $i = 1$). For $\phi_s = n{\cdot}\pi$, the sensing line must be a low-impedance line, a high-impedance line is required for the first transmission line section ($i = 1$), a low-impedance line for the second section, and so on. Thus, it is clear that a non-uniform stepped-impedance transmission line based on quarter-wavelength sections, cascaded to an open-ended quarter- or half-wavelength sensing line, is a very useful structure for sensitivity enhancement in reflective-mode phase-variation sensors. Note that with the reported approach, the sensitivity can be as high as desired, without the need of increasing the length of the sensing line (as typically occurs in phase-variation sensors – see Section 3.2.1). For that purpose, it suffices to design the sensor with the necessary number of sections (with as much contrast of impedance as possible) to obtain the required (high or low) value of the product operator in (3.98).

The previous analysis does not consider the effects of losses on sensor performance (sensitivity). In [5] (see Appendix B), S_{ϕ_s} is also evaluated by considering the effects of losses in the MUT (to a first-order approximation, the ohmic losses of the step-impedance transmission line, as well as the dielectric losses of the substrate, can be neglected, provided the considered substrate is a low-loss material). The main conclusion is that losses do not alter so much the sensitivity, as compared to the lossless case, provided the low-loss approximation is satisfied (i.e. $\alpha l_s \ll 1$, where α is the attenuation constant of the sensing line). This conclusion also applies to the phase of the reflection coefficient, ϕ_ρ, the output variable. Therefore, ϕ_ρ can be used for the determination of the dielectric constant in lossy MUTs (provided losses are small). It is also concluded in the Appendix of [5] that from the measurement of the reflection coefficient at the operating frequency, including not only the phase, but also the modulus, it is possible to infer the attenuation constant of the sensing line, α. From the attenuation constant, it is possible to obtain the loss tangent of the MUT, e.g. from a calibration curve inferred from MUTs with well-known loss factor. The approach is similar to the one reported in Section 3.2.1.1 in reference to transmission-mode phase-variation sensors. Nevertheless, as it was mentioned in Section 3.2.1, in general, resonant sensing methods offer better accuracy for the determination of the loss tangent of the MUT.

3.2.2.2 Highly Sensitive Dielectric Constant Sensors

In order to validate the previous analysis and to point out the potential for sensitivity enhancement of reflective-mode phase-variation sensors based on a step-impedance configuration, let us report in this section the performance of two designed prototypes. The sensors are implemented in microstrip technology and consist of the sensing line cascaded to a single design line section (i.e. $N = 1$). In one sensor (designated as sensor A), the sensing line is an open-ended 90° line with high characteristic impedance ($Z_s = 150\,\Omega$), cascaded to a 90° line section with low characteristic impedance ($Z_1 = 25\,\Omega$). In the second prototype (sensor B), the electrical length of the sensing line is 180°, whereas the design line section is also a 90° line, as required for sensitivity optimization. In this second sensor, the line impedances are opposite, i.e. $Z_s = 25\,\Omega$ and $Z_1 = 150\,\Omega$. The considered operating frequency of the sensors is $f_0 = 2\,\text{GHz}$ and the devices have been implemented in the *Rogers RO4003C* substrate with dielectric constant $\varepsilon_r = 3.55$, thickness $h = 1.524\,\text{mm}$, and loss tangent $\tan \delta = 0.0022$. With this frequency value and the considered line impedances and substrate, the dimensions of the sensors (length and width of the line sections) can be easily inferred, e.g. by means of a transmission line calculator. It has been considered that the REF dielectric constant is the one of air ($\varepsilon_{REF} = 1$), that is, the sensitivity is optimized for small perturbations of the dielectric constant in the vicinity of $\varepsilon_{MUT} = \varepsilon_{REF} = 1$. The photograph of the fabricated devices is depicted in Fig. 3.25, where dimensions are indicated. For comparison purposes, four additional sensors, where the design line is omitted, have been fabricated in the same substrate. These sensors, designated as sensors C, D, E, and F, are also depicted in Fig. 3.25, where the electrical lengths and characteristic impedances of the constitutive sensing lines are indicated (see caption).

Figure 3.26 plots the measured phase variation experienced by the reflection coefficient of the different sensors, in reference to the one of the bare sensor, when the sensing area is loaded with different MUTs. The figure also plots the phase variation inferred through full-wave electromagnetic simulations corresponding not only to the MUTs used to experimentally validate the sensor performance, but also to hypothetical MUTs with dielectric constants in the range $\varepsilon_{MUT} = 1$ to $\varepsilon_{MUT} = 10.5$. It can be seen that the sensors with step-impedance discontinuity exhibit stronger variation of the phase of the reflection coefficient with the dielectric constant of the MUT in the vicinity $\varepsilon_{MUT} = 1$, the reference dielectric constant. By contrast, sensors E and F, based on a uniform matched line, exhibit the softer dependence. The sensitivity inferred by derivation of the simulated data points is also included in Fig. 3.26. The sensitivity in the low perturbation limit ($\varepsilon_{MUT} \approx 1$) is indicated in Fig. 3.26 for each sensor. The theoretical sensitivity for $\varepsilon_{MUT} = 1$, S_{th}, is calculated by means of expression (3.89) for sensor A and by means of (3.88) for sensor B, and using (3.31) for sensors C, D, E, and F. These theoretical sensitivities are also indicated in Fig. 3.26. As it can be seen, the agreement with the sensitivities inferred from the simulated data points, S, is good, thereby validating the sensitivity analysis developed in the preceding section. It is remarkable that

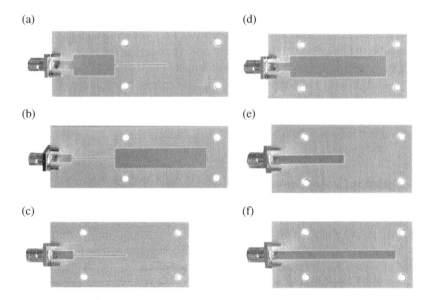

(a) (d)

(b) (e)

(c) (f)

Figure 3.25 Photographs of the fabricated reflective-mode phase-variation sensors. (a) Sensor A, with step-impedance discontinuity satisfying $Z < Z_0 < Z_s$; (b) sensor B, with step-impedance discontinuity satisfying $Z > Z_0 > Z_s$; (c) sensor C, with uniform mismatched 90° sensing line satisfying $Z_s = 150\,\Omega > Z_0$; (d) sensor D, with uniform mismatched 180° sensing line satisfying $Z_s = 25\,\Omega < Z_0$; (e) sensor E, based on a 90° uniform 50-Ω sensing line, and (f) sensor F, based on a 180° uniform 50-Ω sensing line. For sensor A, the width of the sensing and design lines is $W_s = 0.242$ mm and $W = 9.1$ mm, respectively, and the lengths of such line sections is $l_s = 23.65$ mm and $l = 18.3$ mm. For sensor B, dimensions are: $W_s = 9.1$ mm, $W = 0.242$ mm, $l_s = 42.1$ mm, and $l = 20.7$ mm. Note that 50-Ω access line (with length of 10 mm) have been added for connector soldering (the width of these 50-Ω lines is 3.42 mm). *Source:* Printed with permission from [5]; copyright 2020 IEEE.

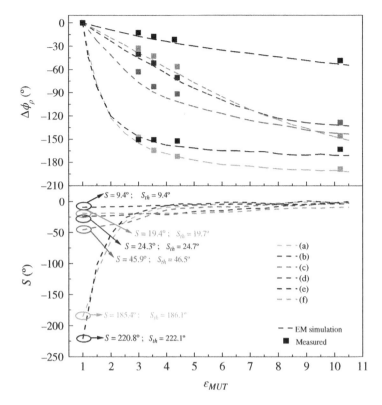

Figure 3.26 Measured and simulated phase of the reflection coefficient for the sensors of Fig. 3.25, and simulated sensitivity. (a) Sensor A; (b) sensor B; (c) sensor C; (d) sensor D; (e) sensor E, and (f) sensor F. The measured dielectric loads (MUTs) are 3-mm slabs of uncoated *PLA* (ε_{MUT} = 3), *Rogers RO4003C* (ε_{MUT} = 3.55), *FR4* (ε_{MUT} = 4.4), and *Rogers RO3010* (ε_{MUT} = 10.2) substrates. Such MUT thicknesses are achieved by stacking up two 1.5-mm samples. The sensitivities in the limit of small perturbations are given in absolute value. *Source:* Reprinted with permission from [5]; copyright 2020 IEEE.

the sensitivity for small perturbations is significantly enhanced by considering the step-impedance discontinuity. Note that for sensor A, the sensitivity is roughly 4 times larger than the one of sensor C, with identical sensing line, and 19.7 times larger than the one of sensor E. On the other hand, sensor B exhibits a sensitivity better than the one of sensors D and F by a factor of 9 and 11.4, respectively. In [5], a prototype sensor with $N = 2$ and a maximum sensitivity of 528.7° is reported. These sensitivities are very competitive, but at the expense of a limited linearity, i.e. the region where the sensor exhibits a linear response decreases as the sensitivity increases. Nevertheless, the main interest in these sensors are applications where small variations in the dielectric constant of the MUT should be detected (this includes not only the accurate measurement of dielectric constants within a certain limited range, but also other related variables, e.g. solute content in very diluted solutions, or the detection of defects in samples, among others).

To summarize, the phase-variation sensors subject of this section, based on a step-impedance open-ended configuration and devoted to dielectric constant measurements, exhibit a good combination of size (sensing region) and performance (sensitivity). These sensors operate at a single frequency, are based on a reflective-mode one-port structure, and, most important, the maximum sensitivity, the key sensor parameter, can be enhanced at wish, maintaining unaltered the region devoted to the MUT.

3.2.2.3 Displacement Sensors

The sensor design strategy for sensitivity optimization, based on step-impedance discontinuities and discussed before, can also be applied to displacement sensing [14]. Let us report in the present section a prototype example devoted to the measurement of linear displacements between the static part of the sensor, an open-ended step-impedance transmission line, and the movable part, a dielectric slab. The output variable is the phase of the reflection coefficient, whereas the input variable is the relative distance between the static and the movable part. The topology and a sketch in perspective view of the sensor is depicted in Fig. 3.27, where the sensitive region (open-ended sensing line) is indicated (sensor implementation in microstrip technology is considered). Specifically, the sensor is devoted to the measurement of a linear displacement of the slab in the direction of the line axis. Slab motion in this direction modifies the phase of the sensing line, thereby producing a variation in the phase of the reflection coefficient seen from the input port.

The sensing line can be considered to be formed by two cascaded sections of variable length, i.e. the uncovered and the covered region [see Fig. 3.27(a)]. Obviously, the effective dielectric constant of the sensing line is different (larger) in the region covered by the movable slab. Let us designate by l the total length of the sensing line and by l_d and l_a the lengths of the covered and uncovered regions, respectively (so that $l = l_d + l_a$). The phase constant and the characteristic impedance of the sensing line are also different in the covered and uncovered regions. Let us call these phase constants β_a and β_d, and the impedances Z_a and Z_d, where the subindexes a and d are used to

(a)

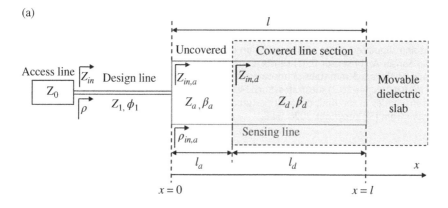

(b)

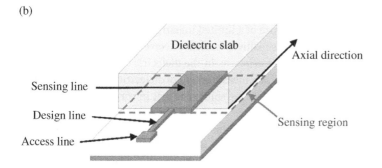

Figure 3.27 Topology (a) and sketch in perspective view (b) of the phase-variation displacement sensor implemented in microstrip technology. *Source:* Reprinted with permission from [14]; copyright 2020 IEEE.

distinguish whether the line is uncovered (air) or covered (dielectric). The impedance, $Z_{in,d}$, seen from the plane separating the covered and uncovered line sections, and looking at the open end (Fig. 3.27), can be expressed as

$$Z_{in,d} = -jZ_d \cot (\beta_d l_d) \tag{3.99}$$

This impedance is the load of the uncovered line section, with input impedance given by

$$Z_{in,a} = \frac{jZ_a(Z_a \tan (\beta_a(l-l_d)) - Z_d \cot (\beta_d l_d))}{Z_a + Z_d \cot (\beta_d l_d) \tan (\beta_a(l-l_d))} \tag{3.100}$$

Note that (3.100) has been expressed in terms of the considered input displacement variable, l_d.

Let us first consider the sensor as formed only by the sensing line. The reflection coefficient seen from the input port is thus $\rho = \rho_{in,a}$, i.e.

$$\rho_{in,a} = \frac{-Z_0(Z_a + Z_d \tan (\beta_a(l-l_d)) \cot (\beta_d l_d)) + jZ_a(Z_a \tan (\beta_a(l-l_d)) - Z_d \cot (\beta_d l_d))}{+Z_0(Z_a + Z_d \tan (\beta_a(l-l_d)) \cot (\beta_d l_d)) + jZ_a(Z_a \tan (\beta_a(l-l_d)) - Z_d \cot (\beta_d l_d))} \tag{3.101}$$

and the phase of the reflection coefficient is therefore

$$\phi_{\rho_{in,a}} = 2 \arctan \left(\frac{Z_a(Z_d \cot (\beta_d l_d) - Z_a \tan (\beta_a(l-l_d)))}{Z_0(Z_a + Z_d \tan (\beta_a(l-l_d)) \cot (\beta_d l_d))} \right) + \pi \tag{3.102}$$

where Z_0 is the reference impedance of the port (as usually designated). The sensitivity, S, given by the derivative of $\phi_{\rho_{in,a}}$ with l_d, can be expressed as

$$S \equiv \frac{d\phi_{\rho_{in,a}}}{dl_d} = \frac{2}{1 + \left(\dfrac{N}{D}\right)^2} \cdot \frac{D \cdot P - N \cdot Q}{D^2} \tag{3.103}$$

where N and D are the numerator and the denominator, respectively, of the argument of the arctan in (3.102), and P and Q are the corresponding derivatives, i.e.

$$P = -\frac{Z_a Z_d \beta_d}{\sin^2 (\beta_d l_d)} + \frac{Z_a^2 \beta_a}{\cos^2 (\beta_a(l-l_d))} \tag{3.104}$$

$$Q = -Z_0 Z_d \left\{ \frac{\beta_a \cot (\beta_d l_d)}{\cos^2 (\beta_a(l-l_d))} + \frac{\beta_d \tan (\beta_a(l-l_d))}{\sin^2 (\beta_d l_d)} \right\} \tag{3.105}$$

In the limit when l_d approaches l, or $x = 0$, the sensitivity is found to be

$$S_l \equiv S(l_d = l) = \frac{2\{-Z_d Z_0 \beta_d + Z_a Z_0 \beta_a \sin^2 (\beta_d l) + Z_0 Z_d^2 Z_a^{-1} \beta_a \cos^2 (\beta_d l)\}}{Z_0^2 \sin^2 (\beta_d l) + Z_d^2 \cos^2 (\beta_d l)} \tag{3.106}$$

In order to obtain the optimum line length, l, for sensitivity optimization, the derivative of (3.106) with l is obtained. After a straightforward (but tedious) calculation, the following expression results:

$$\frac{dS_l}{dl} = \frac{4Z_d Z_0 \beta_d \sin (\beta_d l) \cos (\beta_d l) \{(Z_a Z_d - Z_d Z_0^2 Z_a^{-1})\beta_a + (Z_0^2 - Z_d^2)\beta_d\}}{\{Z_0^2 \sin^2 (\beta_d l) + Z_d^2 \cos^2 (\beta_d l)\}^2} \tag{3.107}$$

The zeros of (3.107), where the sensitivity, S_l, is either a maximum or a minimum, are given by those values of line length satisfying $\beta_d l = n \cdot \pi$ and $\beta_d l = (2n + 1) \cdot \pi/2$ (with $n = 1, 2, 3 \ldots$). By introducing these values of $\beta_d l$ in (3.106), the corresponding sensitivities are found to be:

$$S_l = 2Z_0 \left\{ \frac{\beta_a}{Z_a} - \frac{\beta_d}{Z_d} \right\} \tag{3.108a}$$

for $\beta_d l = n \cdot \pi$, and

$$S_l = \frac{2}{Z_0} \{ Z_a \beta_a - Z_d \beta_d \} \tag{3.108b}$$

for $\beta_d l = (2n + 1) \cdot \pi/2$.

In order to determine if the sensitivity is either a maximum or a minimum for those line lengths providing such phases, the second derivative of (3.107), evaluated at $\beta_d l = n \cdot \pi$ and $\beta_d l = (2n + 1) \cdot \pi/2$, is calculated. The results are found to be:

$$\frac{d^2 S_l}{dl^2} = \frac{4Z_0 \beta_d^2}{Z_d^3} \left\{ (Z_a Z_d - Z_d Z_0^2 Z_a^{-1}) \beta_a + (Z_0^2 - Z_d^2) \beta_d \right\} \tag{3.109a}$$

for $\beta_d l = n \cdot \pi$, and

$$\frac{d^2 S_l}{dl^2} = -\frac{4Z_d \beta_d^2}{Z_0^3} \left\{ (Z_a Z_d - Z_d Z_0^2 Z_a^{-1}) \beta_a + (Z_0^2 - Z_d^2) \beta_d \right\} \tag{3.109b}$$

for $\beta_d l = (2n + 1) \cdot \pi/2$. Inspection of (3.109a) and (3.109b) reveals that the second derivative at the considered phase points exhibits different sign. Consequently, either (3.108a) is a maximum, (3.108b) being a minimum, or vice versa. In order to identify the maximum and the minimum, the sign of the term in brackets in (3.109) should be obtained. By dividing such term by Z_d, and rearranging the terms, one obtains the following positive result, i.e.

$$Z_0^2 \left\{ \frac{\beta_d}{Z_d} - \frac{\beta_a}{Z_a} \right\} + Z_a \beta_a - Z_d \beta_d > 0 \tag{3.110}$$

Expression (3.110) is positive since $Z_d < Z_a$, $\beta_d > \beta_a$, and $Z_a \beta_a - Z_d \beta_d = 0$ (this later equality is consequence of the fact that the product of the characteristic impedance times the phase constant of the line does not vary by covering the line with a dielectric material, see the Appendix in [14]). According to this, (3.109a) is positive and therefore (3.108a) is a minimum. However, since the sensitivity is negative, the absolute value of the sensitivity, the relevant parameter, exhibits a maximum for $\beta_d l = n \cdot \pi$, and a minimum for $\beta_d l = (2n + 1) \cdot \pi/2$ (indeed, the sensitivity is zero for this phase condition, since $Z_a \beta_a - Z_d \beta_d = 0$).

From expression (3.108a), one might erroneously conclude that the sensitivity increases by decreasing the characteristic impedance of the uncovered part of the sensing line, Z_a, and consequently Z_d. The reason is that both impedances appear in the denominator of the corresponding terms in the right-hand side member of (3.108a). However, a reduction in the line impedance is achieved by increasing the width of the line, and this makes both the impedance and the phase constant of the line less sensitive to the presence of a material on top of it. This seems to go against sensitivity optimization, according to (3.108a). Thus, to determine whether Z_a (or Z_d) should be high or low, it is convenient to carry out a numerical analysis of expression (3.108a). For that purpose, β_a and β_d are inferred from (3.1), with v_p given by (3.3), and the effective dielectric constant calculated according to (3.6), with the considered semi-infinite approximation for the movable dielectric slab. Concerning the characteristic impedance of the covered line section, Z_d, assuming that $t \ll h$, it can be approximated by [1]

$$Z_d = \frac{\eta_0}{\sqrt{\varepsilon_{eff,d}}} \left\{ \frac{W_s}{h} + 1.393 + 0.667 \ln \left(\frac{W_s}{h} + 1.444 \right) \right\}^{-1} \tag{3.111a}$$

for $W_s/h \geq 1$, where $\eta_0 = 120\pi\ \Omega$ is the characteristic impedance of vacuum, or by

$$Z_d = \frac{\eta_0}{2\pi\sqrt{\varepsilon_{eff,d}}} \ln\left(\frac{8h}{W_s} + \frac{W_s}{4h}\right) \tag{3.111b}$$

for $W_s/h < 1$. For Z_a, expressions (3.111) also apply, but $\varepsilon_{eff,d}$, the effective dielectric constant of the covered line, must be replaced with $\varepsilon_{eff,a}$, the effective dielectric constant of the uncovered line.

Figure 3.28 depicts the dependence of the sensitivity S_l, given by (3.106), with $\beta_d l$ for different values of the characteristic impedance of the uncovered and covered line sections. The substrate thickness is set to $h = 1.524$ mm, whereas different line widths are considered, so that different curves parametrized by the characteristic impedances of the covered and uncovered line sections are generated. According to Fig. 3.28, the maximum sensitivity (for $\beta_d l = n \cdot \pi$, and given by 3.108a) increases by decreasing the impedance Z_a, or Z_d.

From the previous analysis, it follows that for sensitivity optimization in the displacement sensor based uniquely on the sensing line, the phase of the covered line must be set to $\beta_d l = n \cdot \pi$, and the characteristic impedance, Z_d, or Z_a, must be low, as compared to the reference impedance of the ports. A movable slab with a relatively high dielectric constant is also convenient, since this generates an appreciable difference between the characteristic impedances and phase constants of the uncovered and covered line sections.

Let us now analyze the sensor by including a cascaded transmission line section with characteristic impedance Z_1 and electrical length $\phi_1 = \beta_1 l_1$ (β_1 and l_1 being the phase constant and the line length, respectively), as depicted in Fig. 3.27(a). This additional transmission line section contributes to sensitivity enhancement, provided the line parameters are adequately chosen. In particular, the electrical length of this line at the design frequency should be $\phi_1 = (2n + 1) \cdot \pi/2$, similar to the dielectric constant sensors based on a step-impedance transmission line configuration. The impedance seen from the input port, Z_{in}, is thus

$$Z_{in} = \frac{Z_1^2}{Z_{in,a}} \tag{3.112}$$

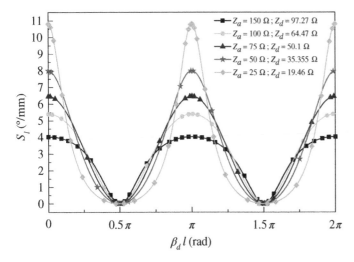

Figure 3.28 Sensitivity S_l as a function of $\beta_d l$ for different values of the characteristic impedance of the sensing line. *Source:* Reprinted with permission from [14]; copyright 2020 IEEE.

and the reflection coefficient is

$$\rho = \frac{Z_1^2/Z_{in,a} - Z_0}{Z_1^2/Z_{in,a} + Z_0} = \frac{Z_1^2/jX_{in,a} - Z_0}{Z_1^2/jX_{in,a} + Z_0} = \frac{Z_1^2 - jX_{in,a}Z_0}{Z_1^2 + jX_{in,a}Z_0} \tag{3.113}$$

where $X_{in,a}$ is the reactance of the sensing line, i.e. $Z_{in,a} = jX_{in,a}$. The phase of the reflection coefficient is thus

$$\phi_\rho = 2\arctan\left(-\frac{X_{in,a}}{\frac{Z_1^2}{Z_0}}\right) \tag{3.114}$$

On the other hand, the phase of the reflection coefficient seen from the sensing line is[15]

$$\phi_{\rho_{in,a}} = 2\arctan\left(-\frac{X_{in,a}}{Z_0}\right) + \pi \tag{3.115}$$

By comparing (3.114) and (3.115), it can be concluded that the sensitivity analysis carried out by considering only the sensing line can be applied to the sensor composed by the sensing line plus the quarter-wavelength ($\phi_1 = 90°$) transmission line. It suffices to replace Z_0 with Z_1^2/Z_0 in (3.108), given that (3.115) is identical to (3.114) with this change of variable (except by a constant phase, π, irrelevant to the sensitivity).

According to the previous paragraph, the displacement sensor based on the step-impedance transmission line structure with $\beta_d l = n \cdot \pi$ and $\phi_1 = \pi/2$ exhibits a sensitivity for small displacements given by

$$S_l = 2\frac{Z_1^2}{Z_0}\left\{\frac{\beta_a}{Z_a} - \frac{\beta_d}{Z_d}\right\} \tag{3.116}$$

Such sensitivity can be optimized by choosing the impedances according to $Z_1 > Z_0 > Z_a$ (or Z_d), with a high impedance contrast (i.e. $Z_1/Z_a \gg 1$).

Further increasing the sensitivity in the displacement sensor is possible by including additional quarter-wavelength transmission line sections with alternating high/low characteristic impedance. If N 90°-lines with characteristic impedance Z_i are cascaded to the sensing line, the input impedance can be expressed as

$$Z_{in,N} = Z_{in,a}^{(-1)^N} \cdot \prod_{i=1}^{N}\left\{Z_i^{2\cdot(-1)^{i+N}}\right\} \tag{3.117}$$

or, using $Z_{in,a} = jX_{in,a}$,

$$Z_{in,N} = j(-1)^N X_{in,a}^{(-1)^N} \cdot \prod_{i=1}^{N}\left\{Z_i^{2\cdot(-1)^{i+N}}\right\} \tag{3.118}$$

where Π is the product operator, as indicated before. The reflection coefficient is thus

$$\rho = \frac{j(-1)^N X_{in,a}^{(-1)^N} \cdot \prod_{i=1}^{N}\left\{Z_i^{2\cdot(-1)^{i+N}}\right\} - Z_0}{j(-1)^N X_{in,a}^{(-1)^N} \cdot \prod_{i=1}^{N}\left\{Z_i^{2\cdot(-1)^{i+N}}\right\} + Z_0} \tag{3.119}$$

---

15 This result is obtained by taking into account that the reflection coefficient seen from the sensing line can be expressed as $\rho = (jX_{in,\,a} - Z_0)/(jX_{in,\,a} + Z_0)$.

For N odd, the phase of the reflection coefficient is

$$\phi_\rho = 2\arctan\left(-\frac{\chi_{in,a}}{\dfrac{\prod\limits_{i=1}^{N}\left\{ z_i^{2\cdot(-1)^{i+N}} \right\}}{Z_0}} \right) \tag{3.120}$$

whereas for N even, the phase is found to be

$$\phi_\rho = 2\arctan\left(-\frac{\dfrac{\chi_{in,a}}{Z_0}}{\prod\limits_{i=1}^{N}\left\{ z_i^{2\cdot(-1)^{i+N}} \right\}} \right) + \pi \tag{3.121}$$

Thus, the sensitivity S_l is

$$S_l = 2\frac{\prod\limits_{i=1}^{N}\left\{ z_i^{2\cdot(-1)^{i+N}} \right\}}{Z_0}\left\{ \frac{\beta_a}{Z_a} - \frac{\beta_d}{Z_d} \right\} \tag{3.122a}$$

for N odd, whereas for N even, it can be expressed as

$$S_l = 2\frac{Z_0}{\prod\limits_{i=1}^{N}\left\{ z_i^{2\cdot(-1)^{i+N}} \right\}}\left\{ \frac{\beta_a}{Z_a} - \frac{\beta_d}{Z_d} \right\} \tag{3.122b}$$

with the phase of the sensing line set to $\beta_d l = n\cdot\pi$ (since for $\beta_d l = (2n+1)\cdot\pi/2$, the sensitivity is null, as discussed before). In summary, sensitivity optimization requires a low-impedance sensing line with the phase set to $\beta_d l = n\cdot\pi$, whereas the subsequent cascaded lines, if they are present, must exhibit a phase of $\phi_i = (2n+1)\cdot\pi/2$ and an alternately high and low characteristic impedance. By this means, there is a multiplicative effect, and the sensitivity can be substantially enhanced without the need of implementing the lines with extreme impedances.

Based on the previous analysis, the design and characterization of two prototype displacement sensors, for validation purposes, is reported next. In both sensors, the phase of the sensing line is set to 180° when it is covered with the movable dielectric slab (the operating frequency of the sensors is set to $f_0 = 2\,\text{GHz}$). The difference concerns the characteristic impedance of such line. Thus, in the prototype identified as sensor A, $Z_d = 97.27\,\Omega$ and $Z_a = 150\,\Omega$, whereas in sensor B, $Z_d = 19.46\,\Omega$ and $Z_a = 25\,\Omega$. In both sensors, a single 90°-line (i.e. $N = 1$) with high characteristic impedance ($Z_1 = 150\,\Omega$) is cascaded to the sensing line (the reference impedance of the port is $Z_0 = 50\,\Omega$). The sensors are implemented in the *Rogers RO4003C* substrate with dielectric constant $\varepsilon_r = 3.55$ and thickness $h = 1.524\,\text{mm}$. The movable dielectric slab is an uncladded piece of the *Rogers RO3010* substrate with dielectric constant $\varepsilon_{r,d} = 10.2$ and thickness $h_d = 3.81\,\text{mm}$ (with such dielectric slab, the above-indicated values of Z_d in both sensors result). Figure 3.29 depicts the photographs of both sensors (dimensions are indicated in the figure). Slab displacement over the sensing line was carried out by means of a linear stepper motor (model *THORLABS LTS300/M*), with reference (REF) position corresponding to the slab with its edge located in the step discontinuity (i.e. $l_a = 0$, or $l_d = l$, or $x = 0$). The picture of the experimental setup is shown in Fig. 3.30.

Figure 3.31 depicts the variation of the phase of the reflection coefficient as a function of the slab displacement, l_a, for both sensors. Actually, the phase is depicted in reference to the phase corresponding to the REF position. The figure includes the measured data points, the phase inferred

(a)

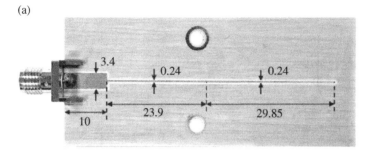

(b)

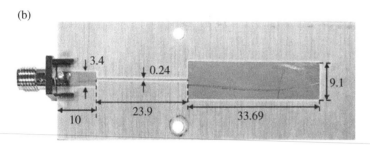

Figure 3.29 Photograph of the fabricated reflective-mode phase-variation displacement sensors. Sensors A (a) and B (b). Dimensions are given in mm. *Source:* Printed with permission from [14]; copyright 2020 IEEE.

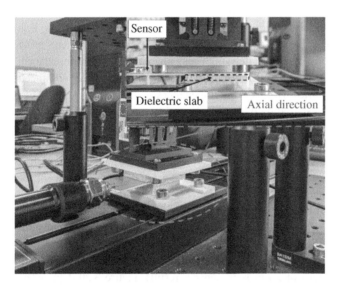

Figure 3.30 Photograph of the experimental setup for validation of the linear displacement sensors of Fig. 3.29. *Source:* Printed with permission from [14]; copyright 2020 IEEE.

from the *HFSS* commercial simulator, and the results predicted by the theory [expression (3.114)]. The sensitivity, inferred by simple derivation of the experimental data points, is also depicted in Fig. 3.31. The maximum sensitivity, for $l_d = l$, or S_l, is found to be 35.14 and 95.24°/mm for sensors A and B, respectively. Such maximum sensitivity (achieved when the slab is in the REF position) should be coherent with the value predicted by expression (3.116). Evaluation of this expression for both sensors gives $|S_l| = 36.44$°/mm and $|S_l| = 97.67$°/mm for sensors A and B, respectively. These values coincide with the maximum sensitivities of Fig. 3.31 to a good approximation. Therefore, these simple calculations validate the sensitivity analysis developed before.

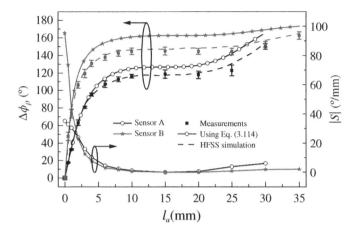

Figure 3.31 Dependence of the phase of the reflection coefficient with the slab displacement, and sensitivity, for sensors A (a) and B (b). For the measured data, error bars, corresponding to the standard deviation divided by the square root of the number of measurements that make up the mean (3 measurement points), are depicted. *Source:* Reprinted with permission from [14]; copyright 2020 IEEE.

To end this section, let us mention that inspection of expressions (3.108a), (3.116), and (3.122) reveals that the maximum sensitivity of these reflective-mode phase-variation displacement sensors, the one for the REF position, S_l, does not depend on the length of the sensing line (by contrast, the sensitivity of the phase-variation dielectric constant sensors studied in the previous section for the REF dielectric constant increases with the length of the line). Nevertheless, long sensing lines, with a length equal to a multiple of a half-wavelength (at the operating frequency and with the line covered with the movable slab), may be of interest in applications where a high input dynamic range is necessary.[16] It should also be mentioned that, by redesigning the sensing line and the movable slab with a circularly shaped geometry, these sensors could potentially be applied to the measurement of angular motion.

3.2.2.4 Reflective-Mode Differential Sensors

Similar to transmission-mode phase-variation sensors, reflective-mode phase-variation sensors operating differentially can also be implemented. Moreover, it is possible to transform the differential phase information (the usual output variable in such sensors) to magnitude information by means of a rat-race hybrid coupler. Let us report an example of a reflective-mode differential sensor based on a pair of identical step-impedance open-ended transmission lines, devoted to dielectric constant measurements. The input variable is the differential dielectric constant (between the REF and the MUT samples, each one located on top of the corresponding sensing line). Rather than the differential phase recorded at the differential port of the pair of step-impedance lines, the output variable is the magnitude of the transmission coefficient of the two-port structure that results by connecting the pair of lines to one of the pairs of isolated ports of the rat-race coupler.

Figure 3.32 depicts the sketch of the specific sensor configuration. The input port of the resulting two-port network is the Δ-port (port 1), whereas the output variable is recorded in the Σ-port (port 2). Let us designate by ρ_{REF} and ρ_{MUT} the reflection coefficients seen from port 3 and 4, respectively, of

---

16 Nevertheless, the main interest of these sensors is for small-range displacement measurements, where the sensor response is roughly linear, and the sensitivity is high.

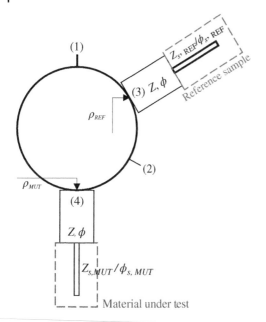

Figure 3.32 Sketch of the two-port reflective-mode differential sensor based on a pair of step-impedance open-ended lines and a rat-race hybrid coupler. In this schematic, high-impedance 90° open-ended sensing lines for the REF and MUT lines are considered, whereas a single ($N = 1$) low-impedance 90° line section (design line) is cascaded between port (3) of the coupler and the REF sensing line, and between port (4) of the coupler and the MUT sensing line. The sensing regions for the REF and MUT samples are indicated by the dashed rectangles.

the coupler (where the REF and MUT lines are connected). The transmission coefficient between the input and the output port, S_{21}, can be expressed in terms of these reflection coefficients as

$$S_{21} = -\frac{1}{2}(\rho_{REF} - \rho_{MUT}) = -\frac{1}{2}\left(e^{j\phi_{REF}} - e^{j\phi_{MUT}}\right) \tag{3.123}$$

where ϕ_{REF} and ϕ_{MUT} are the phases of the reflection coefficients. In (3.123), it has been considered that losses can be neglected, i.e. $|\rho_{REF}| = |\rho_{MUT}| \approx 1$.[17] From (3.123), the magnitude of the transmission coefficient can be easily calculated, i.e.

$$|S_{21}| = \frac{1}{2}\sqrt{2(1 - \cos\Delta\phi_\rho)} \tag{3.124}$$

where $\Delta\phi_\rho = \phi_{REF} - \phi_{MUT}$. Note that for identical REF and MUT samples $|S_{21}| = 0$ (since $\Delta\phi_\rho = 0$), whereas the output variable is a maximum ($|S_{21}| = 1$) for a combination of REF and MUT samples providing out-of-phase reflection coefficients (with $\Delta\phi_\rho = \pi$, or $\rho_{REF} = -\rho_{MUT}$).

The sensitivity, or derivative of the output variable, $|S_{21}|$, with the input variable, i.e. the differential dielectric constant, $\Delta\varepsilon$, can be expressed as

$$S = \frac{d|S_{21}|}{d\Delta\varepsilon} = \frac{d|S_{21}|}{d\Delta\phi_\rho} \cdot \left\{ \frac{d\Delta\phi_\rho}{d\Delta\phi_s} \cdot \frac{d\Delta\phi_s}{d\Delta\varepsilon} + \frac{d\Delta\phi_\rho}{d\Delta Z_s} \cdot \frac{d\Delta Z_s}{d\Delta\varepsilon} \right\} \tag{3.125}$$

---

17 The lossless approximation ($|\rho_{REF}| = |\rho_{MUT}| \approx 1$) is reasonable as far as the sensing structure, the REF sample, and the MUT sample are considered to be made of low-loss materials. Nevertheless, under some circumstances, radiation losses might not be negligible. Indeed, in the reported example (to be discussed later in this subsection), radiation losses are significant, and the magnitude of the transmission coefficient, as given by (3.124), is reduced.

with $\Delta\phi_s = \phi_{s,REF} - \phi_{s,MUT}$, where $\phi_{s,REF}$ and $\phi_{s,MUT}$ are the electrical lengths of the sensing line sections for the REF and MUT lines, respectively, and $\Delta Z_s = Z_{s,REF} - Z_{s,MUT}$, where $Z_{s,REF}$ and $Z_{s,MUT}$ are the impedances of those sections. The first term of the right-hand side member in (3.125) is

$$\frac{d|S_{21}|}{d\Delta\phi_\rho} = \frac{\sin\Delta\phi_\rho}{2\sqrt{2(1-\cos\Delta\phi_\rho)}} \tag{3.126}$$

The term $d\Delta\phi_\rho/d\Delta\phi_s$ is indeed the derivative S_{ϕ_s} defined in Section 3.2.2.1.[18] Such derivative is given by expression (3.71) for open-ended lines with a single step-impedance discontinuity ($N = 1$) and line sections (sensing line and design line) of arbitrary electrical length. If the design line is not present ($N = 0$), S_{ϕ_s} is given by the simplified expression (3.26), whereas for multiple step discontinuities ($N > 1$), expression (3.97) applies, but such expression considers the optimum electrical lengths for the design lines, i.e. odd multiples of 90°. Nevertheless, as it was demonstrated in Section 3.2.1.1, the optimized sensitivity S_{ϕ_s} in the limit of small perturbations, given by (3.80) or (3.81) for a single step-impedance discontinuity, and by expressions (3.98) for a line with multiple discontinuities, is achieved for specific electrical lengths of the sensing line (and quarter-wavelength, or odd multiple, design lines). As it was concluded in that section, such phases [$\phi_s = (2n + 1)\cdot\pi/2$ or $\phi_s = n\cdot\pi$, depending on the impedance of the sensing line] do also optimize the overall sensitivity, and null the term $d\phi_\rho/dZ_s$ in (3.22), see (3.72), which can be assimilated to $d\Delta\phi_\rho/d\Delta Z_s$ in (3.125). Therefore, the second summand in the right-hand side member of (3.125) can be neglected for the optimum phases. The term $d\Delta\phi_s/d\Delta\varepsilon$ is given by expression (3.20) for microstrip lines, where ε_{eff} is calculated according to (3.6) with ε_{MUT} replaced with ε_{REF} (for CPW lines, $d\Delta\phi_s/d\Delta\varepsilon$ is also given by (3.20) with $F = 0$, and ε_{eff} calculated using (3.10) with ε_{MUT} replaced with ε_{REF}). In the limit of small perturbations, $\Delta\phi_\rho \to 0$, (3.126) simplifies to

$$\frac{d|S_{21}|}{d\Delta\phi_\rho} = \frac{1}{2} \tag{3.127}$$

Thus, in summary, for the phases and impedances that optimize the sensitivity in the limit of small perturbations ($\phi_s = (2n + 1)\cdot\pi/2$ and high impedance sensing line, or $\phi_s = n\cdot\pi$ and low impedance sensing line), the sensitivity is determined by (3.125), with the first term given by (3.127), S_{ϕ_s} given by (3.80) or (3.81) for a line pair with a single step-impedance discontinuity, and by (3.98) for a pair of lines with multiple step-impedance discontinuities, and with $d\Delta\phi_s/d\Delta\varepsilon$ calculated by means of (3.20).

Figure 3.33 depicts a specific implementation of these reflective-mode differential sensors in microstrip technology [15]. The device is implemented in the *Rogers RO4003C* substrate with dielectric constant $\varepsilon_r = 3.55$, thickness $h = 1.52$ mm, and dissipation factor $\tan\delta = 0.0021$, and designed to operate at $f_0 = 2$ GHz. The step-impedance open-ended lines consist of a low-impedance ($Z_s = 23\,\Omega$) 180° sensing line, cascaded to a high-impedance ($Z = 157\,\Omega$) 90° transmission line section. The REF sample is a stacked pair of uncladded pieces of the *Rogers RO4003C* substrate considered for sensor implementation, i.e. with dielectric constant $\varepsilon_{REF} = 3.55$, thickness $h = 1.52$ mm, and dissipation factor $\tan\delta = 0.0021$ (thereby giving a total thickness of roughly 3 mm, i.e. enough to consider the REF sample semi-infinite in the vertical direction). Consequently, the physical length of the 180° sensing lines is inferred by considering the presence of such REF samples on top of them. These step-impedance open-ended lines are connected to the isolated ports (3) and (4) of the coupler, as shown in Fig. 3.33.

---

18 Identifying $d\Delta\phi_\rho/d\Delta\phi_s$ of the reflective-mode differential sensor with $S_{\phi s} = d\phi_\rho/d\phi_s$ of the reflective-mode single-ended sensor is perfectly licit, as far as $\Delta\phi_\rho = \phi_{REF} - \phi_{MUT}$ and $\Delta\phi_s = \phi_{s,REF} - \phi_{s,MUT}$, where ϕ_{REF} and $\phi_{s,REF}$ are constant.

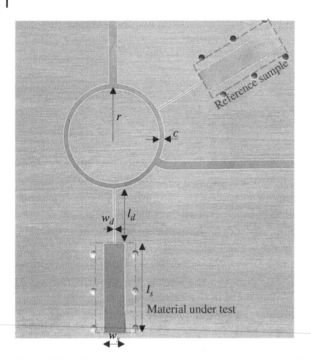

Figure 3.33 Photograph of the reported reflective-mode differential sensor implemented in microstrip technology. The sensing regions, including the one devoted to the MUT and the one devoted to the REF sample, are indicated by dashed rectangles. Dimensions (in mm) are: w_s = 9.07, l_s = 39.81, w_d = 0.20, l_d = 24.65, c = 1.88, and r = 21.15. *Source:* Reprinted with permission from [15]; copyright 2021 IEEE.

Figure 3.34(a) plots the measured and simulated frequency response of the sensor ($|S_{21}|$) for various MUTs. In all the cases, the REF sample is the one indicated in the previous paragraph. It can be seen that at the operating frequency ($f_0 = 2\,\text{GHz}$), the transmission coefficient is null for the *RO4003C* MUT, because it coincides with the REF sample. The measured magnitude of the transmission coefficient evaluated at f_0 for the different samples is depicted in Fig. 3.34(b). The figure includes also the simulated value of $|S_{21}|_{f=f_0}$ as a function of the dielectric constant of the MUT. Using this curve, the sensitivity was inferred from a simple derivative. The resulting (simulated) sensitivity is also depicted in Fig. 3.34(b), where it is compared with the theoretical sensitivity (inferred by means of 3.125). As it can be appreciated, the difference is substantial, and losses explain such discrepancy. Figure 3.34(c) depicts the simulated magnitude of the reflection coefficient of the step-impedance open-ended sensing line with the REF material on top of the sensing region, $|\rho_{REF}|$, by including and by excluding ohmic and dielectric losses. These results reveal that losses are significant (the modulus of the reflection coefficient at f_0 is 0.57) and that a significant portion of losses is due to radiation. However, since there is a quasi-perfect balance when the loads of the sensing lines are identical (the REF sample), the resulting transmission coefficient at f_0 is null to a very good approximation, as Fig. 3.34(a) reveals. Indeed, the loss level of $|\rho_{REF}|$ at f_0 for the sensing line loaded with the REF material can reasonably explain the difference in the sensitivity, at least in the limit of small perturbations. That is, if we assume that for MUTs with dielectric constants close to the one of the REF material, the loss level at f_0 does not vary significantly (i.e. $|\rho_{MUT}| \approx |\rho_{REF}|$), expression (3.123) should be modified by simply multiplying the right-hand side member by $|\rho_{REF}|$. Under this approximation, the sensitivity analysis carried out before is valid, with the exception of a multiplicative factor given by $|\rho_{REF}|$. Consequently, by correcting the lossless sensitivity [$|S| = 2.01$, see Fig. 3.34(b)] taking into account the value of $|\rho_{REF}| = 0.57$, the resulting

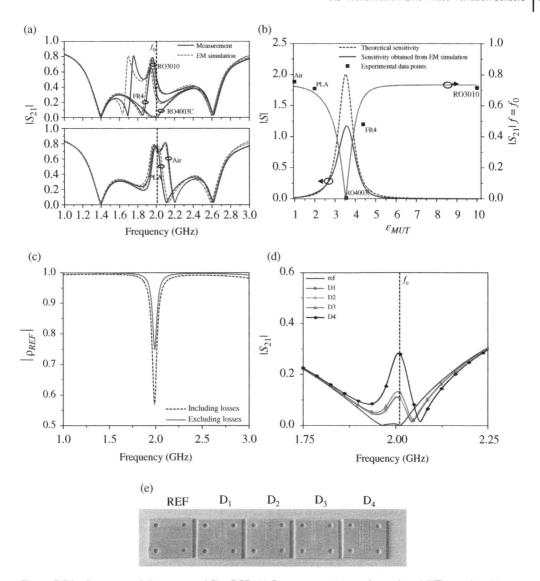

Figure 3.34 Response of the sensor of Fig. 3.33. (a) Frequency response for various MUT samples; (b) magnitude of the transmission coefficient at f_0 and sensitivity, as a function of the dielectric constant of the MUT, obtained by electromagnetic simulation, and experimental data points corresponding to the output variable for different MUTs; (c) reflection coefficient of the step-impedance open-ended sensing line seen from port (3) of the coupler; (d) magnitude of the transmission coefficient for different REF defected samples with different densities of holes across the substrate; (e) photograph of the defected samples. *Source:* Reprinted with permission from [15]; copyright 2021 IEEE.

sensitivity in the limit of small perturbations should be reasonably predicted, as it actually occurs. Note that $|S| \cdot |\rho_{REF}| = 1.17$, whereas the sensitivity resulting from the simulated data points, obviously by including losses, is 1.18, i.e. in excellent agreement. It is obvious that losses tend to degrade the sensitivity, but it does not mean that the sensor cannot be useful. Indeed, the results of Fig. 3.34 show the potential of the approach to determine the dielectric constant of the MUT from the measurement of the magnitude of the transmission coefficient. Moreover, the sensitivity can be enhanced, if needed, by adding further high-/low-impedance 90° line sections to the sensing line.

In order to evaluate the capacity of the sensor to detect small variations in the REF material (comparator functionality), a set of REF samples with arrays of holes of different densities across the substrate (similar to the samples of Section 3.2.1.2.4) were prepared (this is equivalent to slightly modifying the dielectric constant of the MUT sample in the vicinity of the one of the REF material). The measured frequency responses are depicted in Fig. 3.34(d), and point out the potential of this approach to detect small defects in samples [the samples are shown in Fig. 3.34(e)].

Although not reported, and similar to the transmission-mode differential sensors based on a pair of meandered lines (studied in Section 3.2.1.2 and designated as sensors based on differential-mode to common-mode conversion detectors), these reflective-mode differential sensors can be equipped with fluidic channels located on top of the sensing line sections. By these means, the sensors can be applied to the characterization of liquid samples.

As compared to the sensors based on differential-mode to common-mode conversion detectors, the sensors subject of this section are smaller (note that only one rat-race coupler is involved in the design) and, most important, can be designed to exhibit superior sensitivity. This later aspect is obvious by comparing (3.64) and (3.125). The first term of (3.125) is identical to the one of (3.64), provided the sensors are adequately designed [see expressions (3.62) and (3.127)]. By considering sensor implementation in microstrip technology, the derivatives $d\Delta\phi/d\Delta\varepsilon$ in (3.64) and $d\Delta\phi_s/d\Delta\varepsilon$ in (3.125) are both given by (3.20). However, (3.125) includes an extra term ($d\Delta\phi_\rho/d\Delta\phi_s$) that can be substantially high, depending on the number of sections of the step-impedance transmission lines and their impedance contrasts, thereby contributing significantly to sensitivity enhancement. Regardless of the specific implementation, i.e. single-ended or differential, reflective-mode transmission line-based sensors are very promising in applications requiring high sensitivities of the output variable with the dielectric constant of the MUT. Such potentially high sensitivities do also contribute to implement sensors with very good resolution. Thus, the differential-mode realizations are of high interest as real-time comparators able to discriminate subtle differences between the REF and MUT samples.

3.3 Resonant-Type Phase-Variation Sensors

In this section, phase-variation sensors based on planar resonant elements are discussed. Two types of sensors are considered: (i) dielectric constant sensors implemented by combining planar resonators (the sensing element) with step-impedance transmission lines (for sensitivity enhancement), and (ii) angular displacement sensors based on rotatable circular resonators. The working principle of the former sensors is very similar to the one of the reflective-mode phase-variation sensors studied in Section 3.2.2. The angular displacement sensors exploit the cross-polarization effects [16, 17] of the considered resonant elements.

3.3.1 Reflective-Mode Sensors Based on Resonant Sensing Elements

The reflective-mode phase-variation dielectric constant sensors studied in Section 3.2.2 can be implemented by replacing the open-ended sensing line with a planar resonant element [18].[19] By choosing resonant sensing elements exhibiting a significant phase dependence on the dielectric

---

19 Note that the open-ended sensing lines of the sensors of Section 3.2.2 are indeed quarter-wavelength or half-wavelength resonators (for sensitivity optimization). However, these sensors have not been categorized as being based on resonant elements since the resonance condition is not an *a priori* requirement, but a result of sensitivity optimization. Such sensors may operate at frequencies where the sensing lines are not a quarter- or a half-wavelength resonator.

constant of the surrounding medium, the intrinsic sensitivity of such sensors can be potentially very high. Moreover, the sensitivity can be enhanced by means of a step-impedance configuration, as discussed in Section 3.2.2. In that section, it was demonstrated that for sensitivity optimization, the length and characteristic impedance of the open-ended sensing line should not be arbitrarily chosen, i.e. either a high-impedance quarter-wavelength (or odd multiple) line or a low-impedance half-wavelength (or multiple) line is necessary for that purpose. It is well known that an open-ended transmission line can be modeled by means of a series resonator for frequencies close to the one providing a line length of a quarter wavelength [1]. Conversely, a shunt resonant tank describes the behavior for frequencies close to the one corresponding to a half-wavelength transmission line. According to these words, replacing the sensing lines with planar resonators that can be described by either a series or a parallel resonant tank seems to be an alternative strategy to implement reflective-mode phase-variation sensors with high sensitivity and small-sized sensing region.

To gain further insight on the previously mentioned equivalence between quarter- or half-wavelength open-ended lines and the corresponding resonators, let us first consider a $(2n + 1)\cdot\pi/2$ open-ended line and a series resonant tank, both perturbed by a certain MUT. Particularly, let Z_s and $\phi_s = (2n + 1)\cdot\pi/2$ be the impedance and phase, respectively, of the line when it is loaded with an MUT, which is considered to be the REF material, at the angular frequency designated as $\omega_{0,REF}$. With these loading conditions and frequency, the input reactance is null, due to the impedance inversion of the open-ended line with phase $\phi_s = (2n + 1)\cdot\pi/2$. However, if the dielectric constant of the MUT (the input variable) is slightly perturbed, and losses are neglected, the input reactance changes to [1]

$$Z_{in,line} = -j(Z_s + \Delta Z_s) \cot(\phi_s + \Delta\phi_s) = j(Z_s + \Delta Z_s) \tan \Delta\phi_s \approx jZ_s\Delta\phi_s \tag{3.128}$$

where ΔZ_s and $\Delta\phi_s$ are the variations generated in the impedance and phase, respectively, of the line by the small perturbation in the input variable.

On the other hand, let us designate by L and C the inductance and capacitance, respectively, of the unperturbed series resonator (i.e. surrounded by the REF material), and let us consider that the MUT modifies the value of the resonator capacitance by a small quantity ΔC. The impedance of the resonator can be expressed as

$$Z_{in,res} = jL\omega\left[1 - \frac{\omega_{0,REF}^2}{\omega^2\left(1 + \frac{\Delta C}{C}\right)}\right] \tag{3.129}$$

where $\omega_{0,REF} = 1/\sqrt{LC}$ is the (angular) resonance frequency of the unperturbed resonator (and also the angular frequency providing a quarter-wavelength line length, or an odd multiple, to the unperturbed line, i.e. the operating frequency). At such frequency, the impedance of the resonator can be expressed as

$$Z_{in,res} = jL\omega_{0,REF}\frac{\Delta C}{C + \Delta C} \approx jL\omega_{0,REF}\frac{\Delta C}{C} \tag{3.130}$$

since $\Delta C \ll C$. Expression (3.130) is identical to (3.128), provided the following mapping is satisfied:

$$Z_s\Delta\phi_s = L\omega_{0,REF}\frac{\Delta C}{C} \tag{3.131}$$

The phase variation in the line can be expressed as

$$\Delta\phi_s = \omega_{0,REF}\left(\sqrt{L_l(C_l + \Delta C_l)} - \sqrt{L_lC_l}\right) \tag{3.132}$$

where L_l and C_l are the line inductance and capacitance, respectively, corresponding to the considered unperturbed line section [of electrical length $(2n + 1) \cdot \pi/2$], and ΔC_l is the variation of the line capacitance caused by the perturbation. Since for small perturbations ΔC_l is small, (3.132) can be expressed as

$$\Delta \phi_s = \omega_{0,REF} \sqrt{L_l C_l} \left(\sqrt{1 + \frac{\Delta C_l}{C_l}} - 1 \right) \approx \omega_{0,REF} \sqrt{L_l C_l} \left(\frac{\Delta C_l}{2 C_l} \right) = \frac{(2n + 1)\pi}{2} \left(\frac{\Delta C_l}{2 C_l} \right) \quad (3.133)$$

If $\Delta C_l/C_l$ is identified with $\Delta C/C$, introducing (3.133) in (3.131) yields

$$Z_s = \frac{4 L \omega_{0,REF}}{(2n + 1)\pi} = \frac{4}{(2n + 1)\pi} \sqrt{\frac{L}{C}} \quad (3.134)$$

Thus, by choosing the elements of the series resonator as

$$L = \frac{(2n + 1)\pi Z_s}{4 \omega_{0,REF}} ; C = \frac{4}{(2n + 1)\pi Z_s \omega_{0,REF}} \quad (3.135)$$

an identical behavior to the one of the $(2n + 1) \cdot \pi/2$ sensing line in the limit of small perturbations is expected. Note that for sensitivity optimization in sensing lines with such electrical lengths, the impedance Z_s must be high (see Section 3.2.2). Consequently, for sensitivity enhancement in reflective-mode sensors based on series resonators, it is necessary to choose L high and C low, or, in other words, a high-Q resonator. Nevertheless, the same conclusion results by calculating the phase of the reflection coefficient for the impedance given by (3.130), that is

$$\phi_{\rho,res} = 2 \arctan \left(- \frac{L \omega_{0,REF} \Delta C}{C Z_0} \right) + \pi \quad (3.136)$$

where Z_0 is the reference impedance of the port. From (3.136), the sensitivity of the phase of the reflection coefficient with ΔC in the limit of small perturbations ($\Delta C \rightarrow 0$) is found to be

$$\left. \frac{d\phi_{\rho,res}}{d\Delta C} \right|_{\Delta C = 0} = \frac{-2 L \omega_{0,REF}}{C Z_0} \quad (3.137)$$

and it increases with the ratio L/C. Nevertheless, note that, rather than the resonator's capacitance, the input variable in the considered sensors is the dielectric constant of the MUT. Therefore, the sensitivity of interest is actually

$$S = \frac{d\phi_{\rho,res}}{d\varepsilon_{MUT}} = \frac{d\phi_{\rho,res}}{d\Delta C} \cdot \frac{d\Delta C}{d\varepsilon_{MUT}} \quad (3.138)$$

and, in the limit of small perturbations, the first term is given by (3.137), whereas the second term is proportional to the capacitance, C, of the unperturbed resonator. Namely, as it was shown in Chapter 2, the capacitance of the resonator loaded with a MUT with an arbitrary dielectric constant is

$$C + \Delta C = C \left(\frac{\varepsilon_r + \varepsilon_{MUT}}{\varepsilon_r + \varepsilon_{REF}} \right) \quad (3.139)$$

and therefore

$$\frac{d\Delta C}{d\varepsilon_{MUT}} = \frac{C}{\varepsilon_r + \varepsilon_{REF}} \quad (3.140)$$

Thus, the sensitivity in that limit, for operation at $\omega_{0,REF}$, can be finally expressed as[20]

$$S\big|_{\Delta C = 0, \omega_{0,REF}} \equiv \frac{d\phi_{\rho,res}}{d\varepsilon_{MUT}}\bigg|_{\Delta C = 0, \omega_{0,REF}} = -\frac{2\omega_{0,REF}L}{Z_0(\varepsilon_r + \varepsilon_{REF})} \tag{3.141}$$

and it increases by increasing L and by decreasing C, provided the product of these reactive elements is constant and given by the resonance frequency of the unperturbed resonator, i.e. $\omega_{0,REF} = 1/\sqrt{LC}$.

The sensitivity (3.141) is the one corresponding solely to the sensing series resonator. If N quarter-wavelength (or odd multiple) transmission line sections with alternating high/low impedance, Z_i, are cascaded to the resonant element, the sensitivity can be enhanced, thanks to the multiplicative effect generated by the impedance contrast of such lines, as follows:

$$S\big|_{\Delta C = 0, \omega_{0,REF}} = -\frac{2\omega_{0,REF}L}{(\varepsilon_r + \varepsilon_{REF})} \cdot \frac{\prod\limits_{i=1}^{N}\left\{Z_i^{2\cdot(-1)^{i+N}}\right\}}{Z_0} \quad (N \text{ even}) \tag{3.142a}$$

$$S\big|_{\Delta C = 0, \omega_{0,REF}} = -\frac{2\omega_{0,REF}L}{(\varepsilon_r + \varepsilon_{REF})} \cdot \frac{Z_0}{\prod\limits_{i=1}^{N}\left\{Z_i^{2\cdot(-1)^{i+N}}\right\}} \quad (N \text{ odd}) \tag{3.142b}$$

where the line section adjacent to the sensing resonator must exhibit low impedance, and the impedance of the subsequent cascaded lines must alternately be high and low.

It can be demonstrated, using a similar procedure, that, in the limit of small perturbations, the behavior of a $n\cdot\pi$ open-ended line is identical to the one of a parallel LC resonator. In this case, the relationship between the characteristic impedance of the sensing line and the elements of the resonant tank should be

$$Z_s = \frac{n\pi L\omega_{0,REF}}{2} = \frac{n\pi}{2}\sqrt{\frac{L}{C}} \tag{3.143}$$

or

$$L = \frac{2Z_s}{n\pi\omega_{0,REF}}; C = \frac{n\pi}{2Z_s\omega_{0,REF}} \tag{3.144}$$

Sensitivity optimization requires small and high value of the inductance and capacitance, respectively, of the sensing resonator, provided Z_s must be low for $n\cdot\pi$ open-ended sensing lines. Therefore, high-Q parallel resonators are an alternative to series resonators for the implementation of highly sensitive reflective-mode phase-variation sensors based on resonant elements. Nevertheless, let us obtain analytically the sensitivity of a parallel resonator with inductance L and unperturbed capacitance C at the resonance frequency, i.e. $\omega_{0,REF} = 1/\sqrt{LC}$. The impedance at such frequency is

$$Z_{in,res} = -j\omega_{0,REF}L\frac{C}{\Delta C} \tag{3.145}$$

and the phase of the reflection coefficient is thus

---

20 Expressions (3.139) and (3.141) are valid by considering that the MUT present on top of the planar sensing resonator is semi-infinite in the vertical direction, a common approximation adopted in this book. Nevertheless, the need for a high inductance and low capacitance for sensitivity enhancement in sensors based on series sensing resonators holds regardless of the fulfillment of such approximation.

$$\phi_{\rho,res} = 2\arctan\left(\frac{\omega_{0,REF}LC}{Z_0\Delta C}\right) + \pi \tag{3.146}$$

Calculation of the derivative of the phase of the reflection coefficient with ΔC in the limit of small perturbations ($\Delta C \to 0$) gives

$$\left.\frac{d\phi_{\rho,res}}{d\Delta C}\right|_{\Delta C=0} = -2Z_0\omega_{0,REF} \tag{3.147}$$

Finally, using (3.138), with (3.147) and (3.140), the sensitivity in the limit of small perturbations evaluated at $\omega_{0,\,REF}$ is found to be

$$S\big|_{\Delta C=0,\omega_{0,REF}} \equiv \left.\frac{d\phi_{\rho,res}}{d\varepsilon_{MUT}}\right|_{\Delta C=0,\omega_{0,REF}} = -\frac{2Z_0\omega_{0,REF}C}{\varepsilon_r + \varepsilon_{REF}} \tag{3.148}$$

and it increases with C (and consequently it decreases with L, provided $\omega_{0,\,REF}$ is set to a fixed value).

By adding N quarter-wavelength (or odd multiple) transmission line sections with alternating high/low impedance, the sensitivity (3.148) changes to

$$S\big|_{\Delta C=0,\omega_{0,REF}} = \frac{-2\omega_{0,REF}C}{\varepsilon_r + \varepsilon_{REF}} \cdot \frac{Z_0}{\displaystyle\prod_{i=1}^{N}\left\{Z_i^{2\cdot(-1)^{i+N}}\right\}} \quad (N \text{ even}) \tag{3.149a}$$

$$S\big|_{\Delta C=0,\omega_{0,REF}} = \frac{-2\omega_{0,REF}C}{\varepsilon_r + \varepsilon_{REF}} \cdot \frac{\displaystyle\prod_{i=1}^{N}\left\{Z_i^{2\cdot(-1)^{i+N}}\right\}}{Z_0} \quad (N \text{ odd}) \tag{3.149b}$$

and sensitivity enhancement requires that the characteristic impedance of the line cascaded to the sensing resonator exhibits a high impedance value.

Resonant-type reflective-mode phase-variation sensors have been applied to permittivity measurements [18], as well as to the measurement of solute content in liquid solutions [19]. In [18], various single-ended sensors were implemented in CPW technology by cascading an open complementary split ring resonator (OCSRR), the sensing element, to a step-impedance line (with different number of high/low impedance quarter-wavelength line sections). The OCSRR behaves as a parallel resonator [20].[21] Consequently, for sensitivity optimization, the ratio L/C should be small, as justified before. Moreover, since this resonant element is equivalent to a $n \cdot \pi$ sensing line, the 90° transmission line section (of the step-impedance CPW) adjacent to the OCSRR must be designed with high impedance, and the subsequent cascaded line sections (if they are present) with alternating low and high impedance. These design guidelines were applied to the prototype sensors reported in [18], depicted in Fig. 3.35. The number of cascaded 90° line sections is $N = 0$ (sensor A), $N = 1$ (sensor B), and $N = 2$ (sensor C). These sensors were implemented in the *Rogers RO3010* with dielectric constant $\varepsilon_r = 10.2$, thickness $h = 1.27$ mm, and loss tangent tan $\delta = 0.0022$, and the considered REF dielectric constant was the one of air, i.e. $\varepsilon_{REF} = 1$. In sensor B, the characteristic impedance of the high-impedance 90° line section was set to $Z_1 = 70\,\Omega$. In sensor C, an identical impedance for the high-impedance 90° line adjacent to the sensing region was considered, whereas the impedance of the cascaded low-impedance 90° line section was set to $Z_2 = 35.35\,\Omega$. The inductance and unperturbed capacitance, extracted by means of a parameter extraction procedure detailed in [18], were found to be $L = 2.48$ nH and $C = 4.90$ pF. It should be

---

21 Actually, for the accurate modeling of an OCSRR-terminated CPW, an additional inductance, in series with the LC parallel resonant tank, must be included [18]. However, such inductance has negligible effect on the sensitivity for sensor operation at the intrinsic resonance frequency of the unperturbed OCSRR, $\omega_{0,REF}$.

mentioned that for sensors B and C, the length of the high-impedance 90° line adjacent to the OCSRR was adjusted (reduced as compared to the nominal value) in order to compensate for the phase shift generated by the parasitic capacitance that appears in the contact plane between such line and the OCSRR (see further details in [18]). The dimensions of the sensors are indicated in the caption of Fig. 3.35.

The phase of the reflection coefficient for sensors A, B, and C as a function of the dielectric constant of the (semi-infinite) MUT was inferred by electromagnetic simulation by means of the *CST Microwave Studio* commercial software (Fig. 3.36). The operating frequency was set to

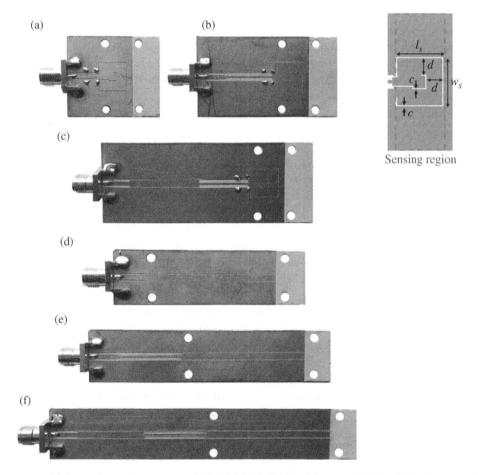

Figure 3.35 Photographs of sensors A (a), B (b), C (c), A′ (d), B′ (e), and C′ (f). The dimensions of the OCSRR (in mm) are: $w_s = 10$, $l_s = 10$, $c = 0.2$, and $d = 3.3$. The dimensions of the high/low step-impedance CPW line sections (in mm) are: strip width $w_1 = 0.72$, slot width $G_1 = 1.04$, and length $l_1 = 22.40$ for sensors B′ and C′, whereas $l_1 = 16.15$ for sensors B and C; $w_2 = 2.25$, $G_2 = 0.28$, $l_2 = 22.53$. The dimensions of the low-impedance 180° open-ended sensing line are (in mm): $w_s = 2.25$, $G_s = 0.28$, $l_s = 44.78$. In all the cases, the dimensions of the access line (50 Ω) are (in mm): $w_0 = 1.39$, $G_0 = 0.71$, and $l_0 = 10$. In sensors A, B and C, the ground plane regions of the CPW are short-circuited by means of vias and strips etched in the back-substrate side, in order to avoid the appearance of the parasitic slot mode (this is necessary since the OCSRRs do not exhibit axial symmetry). The subindexes 1, 2, s, and 0 that appear in the dimension variables of the line sections refer to the high-impedance 90° line (with $i = 1$), the low-impedance 90° line (with $i = 2$), the 180° sensing line, and the access line, respectively.

$f_{0,REF} = \omega_{0,REF}/2\pi = 1.442\,\text{GHz}$, the intrinsic resonance frequency of the unperturbed OCSRR. By numerically obtaining the derivative of such phases with ε_{MUT}, the sensitivities were inferred (the results are also included in Fig. 3.36). It can be appreciated that the sensitivities in the limit of small perturbations (i.e. for $\Delta C = 0\,\text{pF}$, or $\varepsilon_{MUT} = 1$) are in reasonable good agreement with the theoretical predictions, given by expression (3.148) or (3.149), and indicated in the figure.

The distributed counterparts of these sensors, i.e. implemented by replacing the OCSRR with a 180° open-end sensing line were also fabricated in [18] (see Fig. 3.35). These sensors were designated as sensors A′, B′, and C′, and the impedance of the sensing line was set to the value given by (3.143) with $n = 1$. With this impedance value, identical sensitivity in the limit of small perturbations is expected between the sensors based on OCSRRs and their distributed counterparts.[22] To demonstrate this equivalence, the phase of the reflection coefficient at the operating frequency as a function of ε_{MUT} was inferred by electromagnetic simulation. Such phase curves as well as the corresponding sensitivities are also included in Fig. 3.36 for comparison purposes. As it can be seen, the agreement between the responses of the OCSRR-based and fully distributed sensors is very good, not only in the limit of small perturbations, but also for the whole considered input dynamic range.

The OCSRR-based sensors are therefore equivalent to the sensors based on 180° sensing lines, provided the latter are adequately designed, i.e. with the impedance calculated according to (3.143). However, the OCSRR-based sensors exhibit a substantially smaller size of the sensing region, and an improved shape factor of that region, as compared to those of the distributed sensors. A figure of merit (FoM) in phase-variation sensors is the ratio between the sensitivity and the size of the sensing region expressed in terms of the squared-guided wavelength, λ^2. For sensors C and C′, the FoM, calculated for the maximum sensitivity, was found to be FoM = $5643°/\lambda^2$ and FoM = $801°/\lambda^2$, respectively, i.e. substantially higher in sensor C. The FoM for sensor C is very competitive thanks to the use of an electrically small sensing element (the OCSRR), and, obviously, by virtue of the multiplicative effect of the step-impedance configuration.

Experimental validation of the sensors of Fig. 3.35 was achieved in [18] by covering the sensing regions with several commercially available (uncladded) microwave substrates, and with a MUT sample (PLA) fabricated with a 3D printer (the *Ultimaker 3 Extended*). In all the cases, the thickness of the MUT samples was roughly 3 mm (achieved by stacking two 1.5-mm-thick samples), so that such samples can be considered to be roughly semi-infinite in the vertical direction. The measured phases at the operating frequency for such MUTs, inferred by means of the *Keysight 85072A* vector network analyzer, are also included in Fig. 3.36. Actually, the measurements were performed three times, in order to ensure repeatability of the results (the error bars are included in Fig. 3.36). The good agreement with the simulated data points validates experimentally the reported sensors.

3.3.2 Angular Displacement Sensors

The angular displacement sensors subject of this section rely on the phase difference of the reflection coefficients seen from the ports of a transmission line loaded with a rotatable circular resonator. Such phase difference is related to symmetry disruption caused by rotation, which in turn produces mixed coupling (electric and magnetic) between the resonant element and the host transmission line (also known as cross-polarization [16, 17]). Thus, let us first briefly review the origin of cross-polarization in SRR- and CSRR-loaded lines, since these are the structures used for the implementation of the sensors to be reported later.

---

22 Indeed, such identical sensitivities are consequence of the fact that both sensors exhibit the same dependence of the variable capacitance (the resonator's capacitance in OCSRR and the line capacitance in the 180° sensing line) with the dielectric constant of the MUT (see further details in [18]).

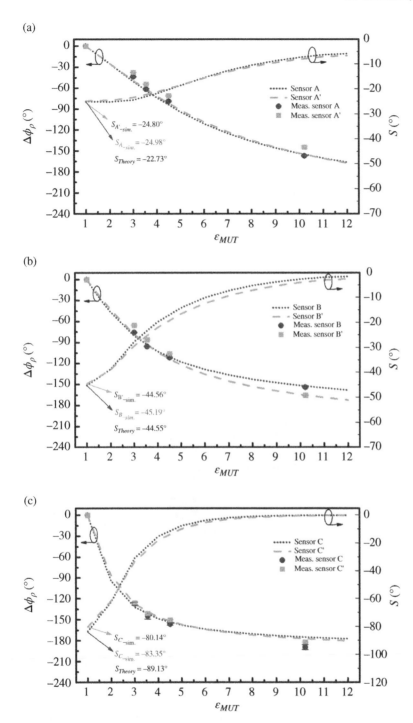

Figure 3.36 Simulated differential phase ($\Delta\phi_\rho = \phi_\rho - \phi_{\rho,\varepsilon_{MUT}=1}$) at $f_{0,REF}$, and sensitivity, for the designed and fabricated sensors of Fig. 3.35. Measured data points for specific values of ε_{MUT} are also included. (a) Sensors A and A′; (b) sensors B and B′; (c) sensors C and C′. *Source:* Reprinted with permission from [18]; copyright 2021 IEEE.

3.3.2.1 Cross-Polarization in Split Ring Resonator (SRR) and Complementary SRR (CSRR) Loaded Lines

The typical topologies of a circularly shaped split ring resonator (SRR) and complementary SRR (CSRR) are depicted in Fig. 3.37. The SRR is an electrically small planar resonator that can be excited by means of an axial time-varying magnetic field, or by means of an electric field applied parallel to the plane of the particle, in the direction orthogonal to its symmetry plane [21–23].[23] Thus, when a SRR is the loading element of a transmission line, depending on the relative orientation between the line and the SRR, mixed coupling (i.e. through the electric and magnetic field simultaneously) may arise. This mixed coupling is also known as cross-polarization, and is useful for sensing purposes. Let us consider a CPW line section loaded with a pair of SRRs, as depicted in Fig. 3.38(a).[24] With this relative orientation between the line and the resonators, the magnetic field generated by the line in the slot regions exhibits a significant component in the axial direction of the SRRs, thereby being able to excite the particle (magnetic coupling). However, the electric field generated by the line does not exhibit any component in the direction of the line axis (i.e. the orthogonal direction to the symmetry plane of the SRRs). Consequently, the orientation illustrated in Fig. 3.38(a) precludes the appearance of electric coupling between the line and the resonator. The situation changes by SRR rotation, since the symmetry plane of the SRR is no longer orthogonal

(a) (b)

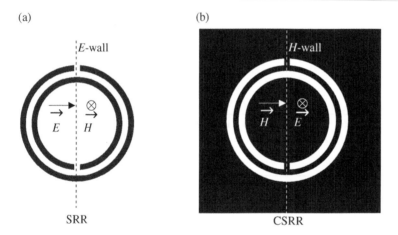

Figure 3.37 Topologies of a circularly shaped SRR (a) and CSRR (b), with indication of the driving mechanisms, an electric field, $\vec{E}$, or a magnetic field, $\vec{H}$.

---

23 The SRR is electrically small at the fundamental resonance frequency. Such frequency is smaller than the fundamental resonance frequencies of the isolated open-loop resonators, by virtue of the coupling between both elements. The closer the concentric rings of the SRR, the larger their mutual coupling and the smaller the fundamental resonance. An electrically small-sized SRR contributes to reduce the dimensions of the sensor. Nevertheless, the analysis of the present paragraph is also valid for SRRs with small inter-ring coupling, and even for SRRs with a single ring (usually designated as open-loop resonators). Note also that the symmetry plane of the SRR (or open-loop resonator) is an electric wall at the fundamental resonance. For this reason, this resonant particle can be excited at this frequency by means of an electric field applied in the direction orthogonal to the symmetry plane.

24 CPW transmission lines are typically loaded with a pair of SRRs (rather than one), etched in the backside of the substrate beneath the CPW slots and oriented symmetrically with regard to the axial symmetry plane of the line, in order to avoid the appearance of the CPW slot mode.

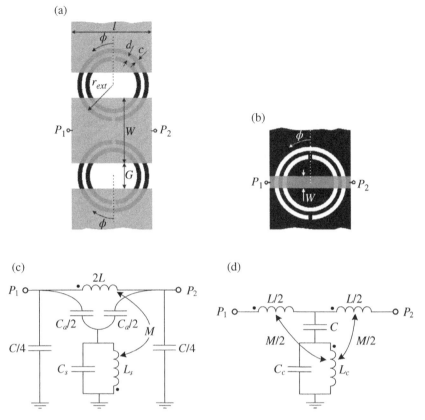

Figure 3.38 SRR-loaded CPW (a), CSRR-loaded microstrip line (b), and circuit models of the SRR-loaded line (c), and CSRR-loaded line (d) with arbitrary resonator orientation. In the circuit of the SRR-loaded line, the magnetic wall concept is applied, namely, the circuit accounts for one of the halves of the structure (with regard to the axial symmetry plane of the CPW). In (a) and (b), the upper metal level is depicted in gray, whereas the metal in the back-substrate side is depicted in black. *Source:* Reprinted with permission from [17]; copyright 2013 IEEE.

to the line axis. This symmetry disruption by rotation activates the electric coupling between the CPW line and the SRR, and therefore the resonator is driven through mixed coupling [17]. Note that the electric coupling is maximized when the rotation angle (see Fig. 3.38) is $\phi = 90°$. Moreover, for a rotation angle of $\phi = 180°$, the structure is again symmetric with respect to the bisecting plane between the input and the output ports, and line-to-resonator coupling is purely magnetic.

The CSRR is the negative (or dual) counterpart of the SRR [24, 25], i.e. the metallic rings of the SRR are replaced with slots, etched in a metallic film, in the CSRR. From duality arguments, it can be concluded that the CSRR can be excited by means of a time-varying axial electric field, or by means of a magnetic field applied in the plane of the particle, in the direction orthogonal to the symmetry plane (a magnetic wall at the fundamental resonance of the CSRR) [26].[25] Let us consider the CSRR-loaded microstrip line depicted in Fig. 3.38(b). Particle excitation through the electric

---

25 According to the Babinet principle, the solution of the electromagnetic field in a planar structure consisting of a metallic film with apertures (or slots) is the same as the one of its dual counterpart (i.e., with apertures replaced with metals and vice versa), but with the electric and magnetic field lines interchanged. The symmetry plane of the CSRR is therefore a magnetic wall at the fundamental resonance (see Section 2.2.1.2 for further details).

field generated by the line (essentially axial to the CSRR) does not depend on CSRR orientation. However, magnetic coupling between the line and the resonator is only possible if $\phi \neq 0°$ and $\phi \neq 180°$, this being soft for angles close to $\phi = 0°$ or $\phi = 180°$, and appreciable for angles in the vicinity of $\phi = 90°$ or odd multiples. Thus, except for $\phi = 0°$ or $\phi = 180°$, the CSRR-loaded microstrip line exhibits cross-polarization [17].

The transmission coefficient in the SRR- or CSRR-loaded lines of Fig. 3.38 is identical in both directions regardless of the rotation angle, since these structures are reciprocal. However, the phase of the reflection coefficients seen from both ports is unequal (except for $\phi = 0°$ or $\phi = 180°$) and the phase difference depends on the rotation angle. The circuit models of the SRR- and CSRR-loaded lines with arbitrary resonator orientation are depicted in Fig. 3.38(c) and (d), respectively [17]. Note that the asymmetry with regard to the bisecting plane between the ports (in case $\phi \neq 0°$ and $\phi \neq 180°$) is accounted for in the circuit models by the presence of the capacitive (electrical) and inductive (magnetic) coupling between the line and the resonant elements. In the circuit of the SRR-loaded CPW, C_a and M provide the capacitive and inductive coupling, respectively, between the line (modeled by the inductance L and the capacitance C) and the SRRs (described by the resonant tank L_s–C_s). Thus, C_a strongly depends on the rotation angle, and $C_a = 0$ if $\phi = 0°$ or $\phi = 180°$ (in this case, the circuit model is symmetric with regard to the ports). In the circuit of the CSRR-loaded line, the capacitive and inductive coupling between the line and the CSRR are modeled by the capacitor C and by the inductance M, respectively, and $M = 0$ if $\phi = 0°$ or $\phi = 180°$ (note that the circuit model is symmetric if $M = 0$). In the circuit of Fig. 3.38(d), L is the line inductance and the L_c–C_c resonator describes the CSRR. Figure 3.39 depicts the responses of the structures of Fig. 3.38 for a rotation angle of $\phi = 90°$, where the significant phase difference between the reflection coefficients measured at both ports can be appreciated. Therefore, this phase difference can be considered to be the output variable in angular displacement sensors based on either SRRs or CSRRs.

3.3.2.2 Slot-Line/SRR Configuration

In order to prevent the appearance of the undesired slot mode, SRR-loaded CPWs are typically implemented by means of a pair of SRRs symmetrically oriented with regard to the axial symmetry plane of the line (as it was mentioned in the preceding section). However, this structure is not of practical use for angular displacement sensors based on the rotation of a unique SRR, the rotor. In [27], a slot-line/SRR configuration, with a single SRR, was alternatively proposed for rotation sensing. Indeed, the structure was equipped with a pair of slot-line to microstrip transitions, so that sensor feeding proceeds by means of a microstrip line, as depicted in Fig. 3.40. Note that the rotation angle in this figure, designated as θ, has its REF position ($\theta = 0°$) for the indicated orientation (with maximum cross-polarization). It is obvious from Fig. 3.40 that the input dynamic range of the sensor is comprised between $\theta = 0°$ and $\theta = 180°$. Namely, the phase difference between the ports is undistinguishable for angles θ_1 and θ_2 satisfying $\theta_1 + \theta_2 = 2\pi$, as it can be appreciated in Fig. 3.41.

By adding a flag resonator to the structure of Fig. 3.40, the input dynamic range of the sensor is extended to 360°. Figure 3.42 shows the top view of the sensor with extended dynamic range, where the flag resonator in the form of a half ring in the rotor is visible. The half ring is concentric with the SRR but its symmetry plane is 90° rotated with respect to the symmetry plane of the SRR. With this configuration, the phase difference between the reflection coefficients at the two ports of the structure around the resonance frequency of the half ring (roughly $f_2 = 4.5$ GHz) can be used as a flag. The fabricated stator and rotor are depicted in Fig. 3.43, whereas Fig. 3.44 plots the simulated and measured phase difference at $f_1 = 3.5$ GHz and $f_2 = 4.5$ GHz. The presence of the flag resonator allows us to resolve the ambiguity in determining rotation angles greater than 180°. For that purpose, it suffices to compare the phase difference at the two frequencies f_1 and f_2. The small discrepancies between simulation and measurement and the slight asymmetry observed in measurement

(a)

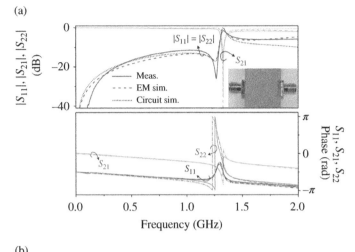

(b)

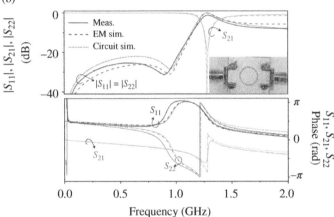

Figure 3.39 Frequency response of the SRR-loaded CPW (a) and CSRR-loaded microstrip line (b) for $\phi = 90°$. The substrate (in both cases) is the *Rogers RO3010* with dielectric constant $\varepsilon_r = 11.2$ and thickness $h = 1.27$ mm. For the structure (a), dimensions (in reference to Fig. 3.38a) are $W = 10.4$ mm, $G = 1.6$ mm, $l = 10.4$ mm, $r_{ext} = 5$ mm, and $c = d = 0.2$ mm. For the structure (b), dimensions are $W = 1.15$ mm, $l = 10.4$ mm, $r_{ext} = 5$ mm, and $c = d = 0.2$ mm. The fabricated structures (view of the resonators) are shown in the insets. *Source:* Reprinted with permission from [17]; copyright 2013 IEEE.

are attributed to certain misalignment of the two substrates and to a small precession of the rotor. The average sensitivity is found to be 0.63 (note that such sensitivity is dimensionless, provided it gives the variation of the phase difference of the reflection coefficients of the ports, in degrees, with the rotation angle, also in degrees). The linearity of this sensor is reasonably good within the whole sensor range.

3.3.2.3 Microstrip-Line/CSRR Configuration

A rotary sensor based on a microstrip line loaded with a movable CSRR was reported in [28]. The output variable of this sensor is also the difference in the phase of the reflection coefficients seen from both ports. The fabricated stator and rotor are depicted in Fig. 3.45 (actually, sensor functionality by considering several rotors, as depicted in the figure, was demonstrated in [28]). Note that the considered resonators of the rotors exhibit a modified topology as compared to the one of the

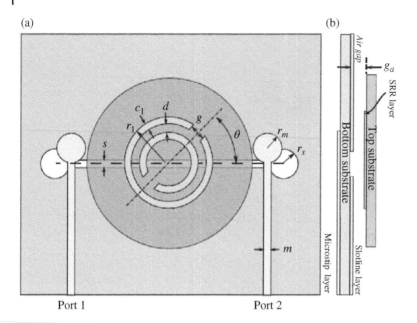

Figure 3.40 Layout of the phase-variation rotation sensor based on a slot-line/SRR configuration feed by means of microstrip lines (a) and side view (b). The stator (slot-line and slot-line to microstrip transitions) as well as the rotor (SRR) are implemented on the *Rogers RO3010* substrate with dielectric constant ε_r = 10.2, thickness h = 0.635 mm, and loss tangent tan δ = 0.0035. Dimensions are: r_m = 2 mm, r_s = 1.6 mm, m = 0.58 mm, s = 0.2 mm, r_1 = 3.2 mm, c_1 = 0.6 mm, d = 0.6 m, and g = 0.4 m. The dimensions of the SRR provide a resonance frequency of 3.5 GHz. The vertical distance between the rotor and stator, or air gap, is considered to be g_a = 0.6 mm. *Source:* Reprinted with permission from [27]; copyright 2020 IEEE.

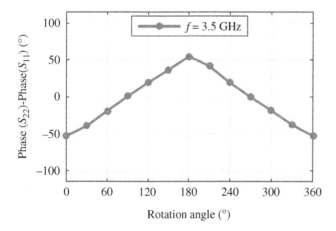

Figure 3.41 Simulated phase difference between the reflection coefficients at the two ports of the sensor for different rotation angles from 0° to 360°, at the fixed frequency f = 3.5 GHz. *Source:* Reprinted with permission from [27]; copyright 2020 IEEE.

CSRR shown in Fig. 3.37. This modified CSRR (MCSRR) was used in [28] in order to enhance the magnetic coupling with the line, thereby enhancing the cross-polarization effects, with direct impact on sensitivity optimization. Figure 3.46 depicts the phase response of the sensor with the rotation angle for the different considered rotors (note that, in this case, the reference, $\theta = 0°$,

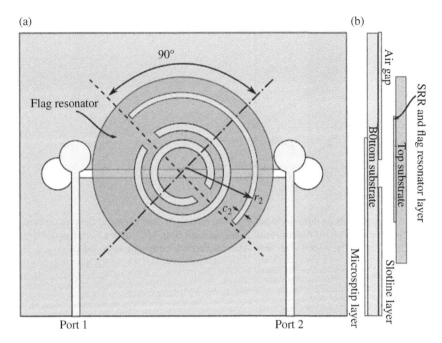

Figure 3.42 Layout of the rotation sensor based on a slot-line/SRR configuration with flag resonator (a) and side view (b). The outer radius of the half ring is r_2 = 5 mm and its width is c_2 = 0.1 mm (the other dimensions are those given in the caption of Fig. 3.40). *Source:* Reprinted with permission from [27]; copyright 2020 IEEE.

Figure 3.43 Top (a) and bottom (b) view of the fabricated stator, photograph of the rotor (c) and sensor in its housing with rotatable knob (d). *Source:* Printed with permission from [27]; copyright 2020 IEEE.

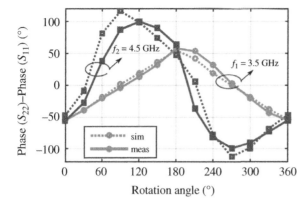

Figure 3.44 Simulated and measured phase difference between the reflection coefficients at the two ports of the sensor with expanded dynamic range, for different rotation angles from 0° to 360°, at the fixed frequencies of f = 3.5 GHz and f_2 = 4.5 GHz. *Source:* Reprinted with permission from [27]; copyright 2020 IEEE.

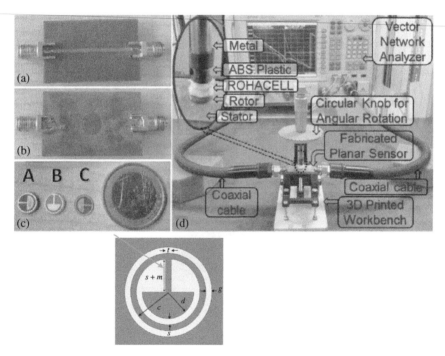

Figure 3.45 Photographs of the phase-variation sensor system based on a microstrip line loaded with a rotatable MCSRR. (a) Bottom and (b) top views of the stator, fabricated on the 0.5-mm-thick *RT5880* substrate with dielectric constant ε_r = 2.2, and dissipation factor tan δ = 0.009; (c) rotor A (1.5-mm-thick *RT5880* laminate, ε_r = 2.2, tan δ = 0.009), rotor B (1-mm-thick *FR4* laminate, ε_r = 4.3, tan δ = 0.025), and rotor C (0.5-mm-thick *RT5880* laminate); (d) measurement setup for directional angular rotation. The microstrip line length and width are 40 and 1.49 mm. The dimensions of the MCSRR, in regard to the zoom view topology, are: c = 3.5 mm, d = 1.5 mm, s = 0.5 mm, t = 0.5 mm, and g = 0.5 mm. *Source:* Reprinted with permission from [28]; copyright 2020 IEEE.

corresponds to the symmetric orientation of the rotor, contrary to the example reported in the previous section).

As compared to the angular displacement sensor of the previous section, the sensor of Fig. 3.45 exhibits smaller input dynamic range $[-90°, +90°]$ and larger output dynamic range. Therefore, the sensitivity of the reported sensor based on the microstrip/CSRR configuration is better than the one of the rotation sensor implemented by means of a slot-line/SRR combination. In particular, the maximum sensitivities achieved with rotors A, B, and C (see Fig. 3.45) are 3.2, 2.9, and 3.1, respectively. It is worth mentioning that the sign of the differential phase (the output variable) indicates if the angular displacement with regard to the reference angular position (symmetric orientation of the rotor, or $\theta = 0°$) proceeds clockwise or counterclockwise (see Fig. 3.46). This capability to detect the motion direction applies also to the sensor of Fig. 3.43 (note, however, that, according to the angle definition in that sensor, the reference angle is $\theta = 90°$).

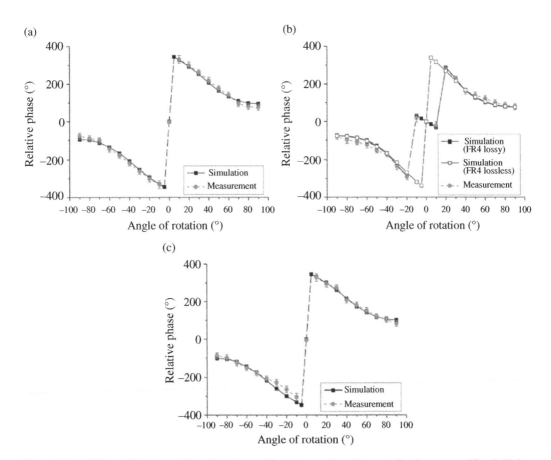

Figure 3.46 Differential phase of the reflection coefficients seen from the ports for the sensor of Fig. 3.45, for (a) rotor A, (b) rotor B, and (c) rotor C. *Source:* Reprinted with permission from [28]; copyright 2020 IEEE.

3.4 Phase-Variation Sensors Based on Artificial Transmission Lines

Artificial transmission lines are guided-wave propagating structures consisting of a host line loaded with reactive elements, and/or with different types of inclusions or defects, or exhibiting a variation (typically periodic) in their cross-sectional geometry [29]. The presence of such additional elements increases the degrees of freedom of such lines. Thus, by replacing the constitutive ordinary lines of distributed microwave components with such artificial lines, the implementation of RF and microwave devices with reduced size, with superior performance, and/or exhibiting novel functionalities is possible. In particular, these artificial lines are applicable to the design of highly sensitive phase-variation dielectric constant sensors, by virtue of the controllability of their dispersion diagram.

As it was indicated at the beginning of Section 3.2, the sensitivity of a phase-variation sensor implemented by means of an ordinary line increases by lengthening the line [see expression (3.2)]. Reducing the phase velocity, v_p, by increasing the substrate dielectric constant, ε_r, boosts up the first term of the last member in (3.2). However, the sensitivity of v_p with the dielectric constant of the MUT, ε_{MUT}, is degraded by using high dielectric constant substrate materials. The overall effect of an increase in ε_r is a degradation of the sensitivity, as it was demonstrated in Section 3.2 for microstrip and CPW transmission lines [see expressions (3.9) and (3.11)].

In ordinary lines, increasing ε_r is the single procedure to reduce the phase velocity. However, there are alternatives to ordinary lines to implement planar (and open) guided-wave propagating media exhibiting small phase velocity. The most canonical (but not exclusive) example is the so-called slow-wave transmission line [30, 31], a type of artificial line that can be implemented by periodically loading a host line, e.g. with shunt-connected capacitors, or with series-connected inductors (or with both elements simultaneously) [32–38].[26] The presence of such reactive elements enhances the per-unit-length effective capacitance or inductance of the line, thereby decreasing the phase velocity. This slow-wave effect can be achieved without the need of increasing the dielectric constant of the substrate. Indeed, the effective dielectric constant of the line (the one that determines the phase velocity) is artificially engineered (magnified) by means of the shunt-connected capacitors. Similarly, the presence of series inductors in the line increases the effective magnetic permeability of the line, thus contributing to decrease the phase velocity.

Figure 3.47 depicts a typical dispersion diagram of a slow-wave transmission line, compared to the one of an ordinary line (in dashed line) with phase velocity v_{p0}.[27] The figure includes also a typical topology (with three unit cells), as well as the circuit schematic (unit cell), of a slow-wave transmission line based on capacitive loading (achieved by means of capacitive patches). Regardless of the specific line type, the phase velocity can be calculated from the dispersion diagram as

---

26 These reactive elements (shunt-connected capacitors or series-connected inductors) can be implemented by means of surface-mount technology (*smd*) components, or through semi-lumped (i.e. electrically small) planar components. Patch capacitors and omega inductors are typical topologies used for the implementation of slow-wave transmission lines in fully planar form. It should also be mentioned that semi-lumped resonators are also useful to reduce the phase velocity. For example, dumbbell-shaped defect ground structure (DB-DGS) resonators [39], etched in the ground plane of a microstrip line, behave as series-connected parallel resonant tanks, and step-impedance shunt stubs (SISS) can be described by means of shunt-connected series resonators [40]. In both cases, below the resonance frequency, the resonator exhibits a reactance of the convenient sign for phase velocity reduction (i.e. positive in the case of the DB-DGS, and negative for the SISS).

27 The dispersion is represented in a reduced Brillouin diagram, with frequency in the vertical axis, and the electrical length (or phase, $\phi = \beta l$) corresponding to the physical length, l, of the unit cell in the horizontal axis (l is also the length of the considered ordinary line section). Note that the angular frequency is proportional to the phase constant of the host (ordinary) line, i.e. $k = \omega/v_{p0}$. The phase velocity of the ordinary line (constant with frequency) is distinguished from the one of the slow-wave transmission line by means of the sub-index "0". Note also that the slow-wave effect occurs in the first band, with frequencies satisfying $kl/\pi < 1$.

(a)

(b)

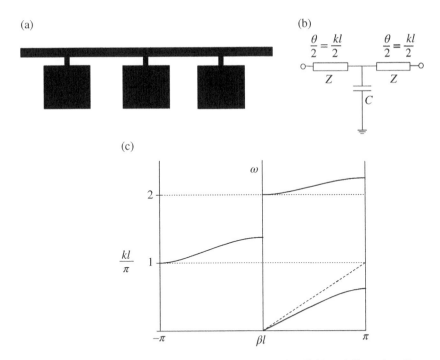

(c)

Figure 3.47 Typical topology (a), circuit model of the unit cell (b), and dispersion diagram (c) of a capacitively loaded slow-wave transmission line. The phase constant of the host line, proportional to frequency, is designated as k, in order to distinguish it from the phase constant of the loaded line, β. Thus, $\theta = kl$ is the electrical length of the host line, whereas $\phi = \beta l$ is the electrical length of the loaded line, the measurable output variable. In both cases, l is the physical length of the unit cell. Z is the characteristic impedance of the host line, different to the characteristic impedance of the loaded line (usually referred to as Bloch impedance, or image impedance, and designated as Z_B). Note that Z_B is frequency dependent, contrary to Z, which is constant with frequency, and determined by the transverse geometry of the host line, substrate dielectric constant, and dielectric constant of the MUT. The loading shunt-connected capacitance is designated as C.

$$v_p = \frac{\omega}{\phi} l \tag{3.150}$$

namely, it is proportional to the tangent of the angle given by the imaginary line crossing the origin and the point in the dispersion diagram where v_p is calculated. In view of the diagram of Fig. 3.47, it can be concluded that the phase velocity of the slow-wave transmission line, v_p, progressively decreases as frequency increases. Moreover, v_p is smaller than v_{p0} from DC up to the first cutoff frequency. Note that the dispersion curve is indicative of the slow-wave effect from the fact that such curve is situated below the straight line corresponding to the $\omega-\phi$ diagram of the (dispersionless) ordinary line (with phase velocity v_{p0} at, ideally, all frequencies). With such a dispersion diagram, it is clear that a slow-wave transmission line exhibits a strong dependence of the phase (or electrical length) with the phase velocity, $d\phi/dv_p$, at least as compared to the dependence achievable in an ordinary line with identical physical length [first term of the right-hand side member in expression (3.2)].[28] Thus, the sensitivity of ϕ with ε_{MUT} (the one of interest) can be potentially

---

28 The reason is the smaller phase velocity in the slow-wave transmission line.

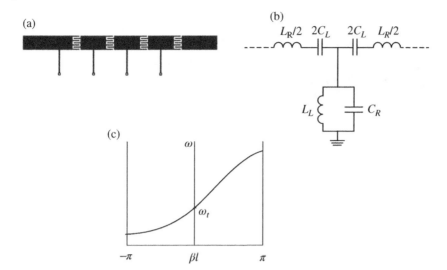

Figure 3.48 Typical topology (a), circuit model of the unit cell (b), and dispersion diagram (c) of a CRLH transmission line. Note that the considered unit cell is a microstrip host line loaded with series interdigital capacitors (C_L) and shunt-connected inductors (L_L). The host line is described by the line inductance, L_R, and by the line capacitance, C_R. The depicted dispersion is the one corresponding to a balanced CRLH line, which satisfies $\omega_s = \omega_p \equiv \omega_t$, where $\omega_s = (L_R C_L)^{-1/2}$ is the resonance frequency of the series branch, and $\omega_p = (L_L C_R)^{-1/2}$ is the resonance frequency of the shunt branch. The frequency that delimits the left-handed and the right-handed bands in a balanced CRLH line, ω_t, is designated as transition frequency.

improved in a slow-wave transmission line, as far as the last term contributing to the sensitivity ($dv_p/d\varepsilon_{MUT}$) is not degraded as a consequence of an increase in the dielectric constant of the substrate.[29] Nevertheless, this potential for sensitivity improvement in guided-wave propagating media exhibiting a flat dispersion diagram, like the one of slow-wave transmission lines, must be studied in detail on the basis of the specific structure and sensitive element. For example, in capacitively loaded slow-wave transmission lines, the sensing element can be either the host line or the shunt-connected (patch) capacitor (or both elements simultaneously).

There are further artificial lines exhibiting a flat (high) dispersion, in particular CRLH lines [41, 42] and magnetoinductive-wave (MIW) [43–47], or EIW [48], transmission lines. The typical dispersion diagram, topology, and circuit model (unit cell) of such lines are depicted in Figs. 3.48 (CRLH lines) and 3.49 (EIW lines).[30] CRLH lines exhibit backward (or left-handed) wave propagation at low frequencies, and forward (or right-handed) wave propagation at high frequencies, and the transition between the left-handed and the right-handed band is continuous if a certain condition is satisfied (balanced CRLH line [41, 42], see the caption of Fig. 3.48). By contrast, EIW and MIW lines exhibit only a backward transmission band. Both CRLH and EIW/MIW transmission lines do not support wave propagation at low frequencies. However, their dispersion

---

29 The dielectric constant of the substrate in a slow-wave transmission line can be maintained at low values since the phase velocity is reduced by the presence of the reactive elements.

30 The dispersion diagram of MIW lines is identical to the one of EIW transmission lines. For this reason, and because only phase-variation sensors based on EIW lines (but not on MIW lines) are later reported, neither the topology, nor the circuit model and dispersion diagram of MIW transmission lines are included in this book. Nevertheless, for the interested readers, a detailed analysis of these artificial lines can be found in [29, 43–47].

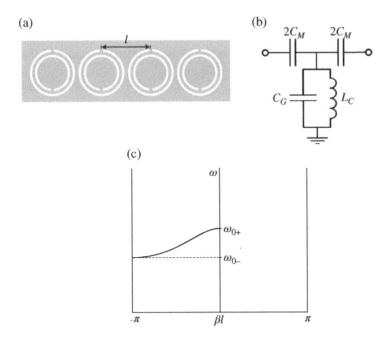

Figure 3.49 Typical topology (a), circuit model of the unit cell (b), and dispersion diagram (c) of an EIW transmission line based on coupled CSRRs. The isolated CSRRs are described by the resonant tank L_c–C_c, whereas C_M accounts for the mutual coupling between adjacent CSRRs, and $C_G = C_c - 2C_M$. Source: Printed with permission from [26].

diagram can be engineered (to some extent) in order to achieve a flat shape, thereby favoring sensitivity, as discussed before.

The periodic structures of Figs. 3.47, 3.48, and 3.49 can be analyzed from the transfer matrix method [29], which provides the electrical length of the unit cell, $\phi = \beta l$, and the equivalent to the characteristic impedance, usually referred to as Bloch impedance, Z_B. Note that l is the physical length of the unit cell (not the full length of the artificial line) and β is the phase constant. It is obvious that β, or ϕ, is not proportional to frequency, as occurs in ordinary lines, due to dispersion. Moreover, Z_B is frequency dependent, contrary to the characteristic impedance in conventional lines. It should also be mentioned that, due to periodicity, the artificial lines depicted in Figs. 3.47, 3.48, and 3.49 exhibit stop bands, where wave propagation is not allowed [29, 49]. In those bands, the evanescent waves are characterized by the attenuation constant, α, the phase constant being null ($\beta = 0$). By contrast, $\beta \neq 0$ and $\alpha = 0$ in the allowed bands.[31] The analytical expressions providing the electrical length of the unit cell in the allowed bands and the Bloch impedance are [29]

$$\cos \phi = A \tag{3.151}$$

and

$$Z_B = \frac{B}{\sqrt{A^2 - 1}} \tag{3.152}$$

---

31 This is valid if losses are neglected (lossless approximation).

respectively, where A and B are the elements of the first row of the ABCD matrix of the unit cell. Expressions (3.151) and (3.152) are valid if $A = D$, that is, if the unit cell of the considered artificial transmission line exhibits symmetry with regard to the bisecting plane between the input and the output port, the usual case (D is the last element of the ABCD matrix).

It is obvious that (3.151) is a relevant expression for the design of phase-variation sensors based on artificial transmission lines. Nevertheless, the characteristic impedance of the line may play also an important role on sensor performance, as it has been shown in several sensor implementations. In the next sections, further analysis and examples of phase-variation sensors implemented by means of artificial lines, devoted to dielectric characterization, are presented.

3.4.1 Sensors Based on Slow-Wave Transmission Lines

Let us consider a slow-wave artificial line like the one depicted in Fig. 3.47. Calculating the ABCD matrix of the unit cell, and introducing the parameters A and B in (3.151) and (3.152), the phase of the unit cell (at the operating frequency, $f_0 = \omega_0/2\pi$) and the Bloch impedance are found to be [29, 50]

$$\cos\phi = \cos\theta - \frac{\omega_0 CZ}{2}\sin\theta \qquad (3.153)$$

$$Z_B = Z\frac{\sin\theta - \omega_0 CZ(\sin\theta/2)^2}{\sin\phi} \qquad (3.154)$$

where Z is the impedance of the host line. The usual design (input) parameters of a capacitively loaded slow-wave transmission line devoted to the implementation of a certain microwave component with reduced size (the most canonical application) are ϕ and Z_B. Circuit specifications dictate such parameters [29, 50].[32] However, from the assigned values to these parameters, it is not possible to univocally determine θ, Z, and C by inverting (3.153) and (3.154). An additional condition is required for that purpose. Such condition is usually the so-called slow-wave ratio (*swr*), or relation between the phase velocities of the slow-wave transmission line and ordinary (host) line, given by

$$swr = \frac{v_p}{v_{p0}} = \frac{\omega/\beta}{\omega/k} = \frac{\theta}{\phi} \qquad (3.155)$$

Application of the slow-wave structure of Fig. 3.47 to dielectric constant measurements requires the definition of the sensing region. One possibility is to consider the sensing region constituted by the area enclosing the host line, excluding the patch capacitors. Complementarily, the sensing region can be defined as the region including only the patch capacitors (but not the host line).[33] Let us next analyze the former situation, and then the slow-wave transmission line sensor based on sensing capacitive patches. Prototype sensors corresponding to both cases are also reported.

3.4.1.1 Sensing Through the Host Line

A variation in the dielectric constant of the MUT, placed on top of the sensing area (the host line), is expected to modify the phase velocity, v_{p0}, and the characteristic impedance, Z, of the host line. In turn, a variation in the phase velocity of the host line modifies the phase of the unit cell of such line, θ. According to expression (3.153), θ and Z contribute to the phase (unit cell) of the capacitively

---

32 For instance, $\phi = 90°$ and $Z_B = 70.71\,\Omega$ in slow-wave-based impedance inverters devoted to replace the pair of ordinary inverters of a Wilkinson power divider.

33 Obviously, both the host line and the patch capacitances may be part of the sensing region, but in this case, the analysis is not simple.

loaded structure, ϕ. Thus, the sensitivity, defined as the derivative of the phase of the reactively loaded line, $N\phi$, with respect to ε_{MUT}, can be expressed as

$$S = N\frac{d\phi}{d\varepsilon_{MUT}} = N\left(\frac{d\phi}{d\theta} \cdot \frac{d\theta}{dv_{p0}} \cdot \frac{dv_{p0}}{d\varepsilon_{MUT}} + \frac{d\phi}{dZ} \cdot \frac{dZ}{d\varepsilon_{MUT}}\right) \tag{3.156}$$

N being the number of cells. The terms in (3.156) are

$$\frac{d\phi}{d\theta} = \frac{\sin\theta + \frac{\omega_0 CZ}{2}\cos\theta}{\sin\left(\frac{\theta}{swr}\right)} \tag{3.157a}$$

$$\frac{d\theta}{dv_{p0}} = -\frac{\omega_0 l}{v_{p0}^2} \tag{3.157b}$$

$$\frac{dv_{p0}}{d\varepsilon_{MUT}} = -\frac{v_{p0}}{4\varepsilon_{eff}}(1-F) \tag{3.157c}$$

$$\frac{d\phi}{dZ} = \frac{\frac{\omega_0 C}{2}\sin\theta}{\sin\left(\frac{\theta}{swr}\right)} \tag{3.157d}$$

$$\frac{dZ}{d\varepsilon_{MUT}} = -\frac{Z}{4\varepsilon_{eff}}(1-F) \tag{3.157e}$$

where implementation in microstrip technology is considered. Introducing expressions (3.157) in (3.156) gives

$$S = N\frac{d\phi}{d\varepsilon_{MUT}} = N\frac{1-F}{4\varepsilon_{eff}}\left\{\theta \cdot \frac{\sin\theta + \frac{\omega_0 CZ}{2}\cos\theta}{\sin\left(\frac{\theta}{swr}\right)} - Z \cdot \frac{\frac{\omega_0 C}{2}\sin\theta}{\sin\left(\frac{\theta}{swr}\right)}\right\} \tag{3.158}$$

Note that the output variable in the sensitivity given by (3.158) is actually the phase of the slow-wave artificial line, ϕ, rather than the phase of the transmission coefficient, the usual measurable phase variable. However, by designing the line with $Z_B = Z_0$ (the reference impedance of the ports, typically 50 Ω) for the REF dielectric constant, as required for matching purposes in the limit of small perturbations, the sensitivity (3.158) coincides with the variation of the phase of the transmission coefficient with ε_{MUT}, the actual sensitivity, in the limit of small perturbations [i.e. a variation in Z_B has negligible effect on the phase of the transmission coefficient for $Z_B = Z_0$, see expression (3.18)[34]].

If (3.156) is compared with (3.2), the sensitivity of the phase of an ordinary line with the dielectric constant of the MUT, it follows that the first product of the right-hand side member of (3.156) is indeed (3.2) multiplied by $d\phi/d\theta$.[35] Thus, such term can be enhanced, as compared to (3.2), as far as $d\phi/d\theta > 1$. Calculation of this derivative, using (3.153), gives (3.157a). As discussed in [51], for $Z_B = Z_0$, and for reasonable values of the *swr* and phase of the unit cell, ϕ, it follows that $d\phi/d\theta > 1$. However, this term cannot be made significantly larger than unity unless ϕ is set to values close to π, and this gives unrealizable (very high) values of Z, as inferred from inversion of (3.153)–(3.155) [51]. Moreover, the second summand contributing to the sensitivity in (3.156), not present in an ordinary line, is of opposite sign to the first term, as can be appreciated in the developed formula (3.158). Thus, sensing through the host line in slow-wave transmission lines does not seem to be an efficient method for sensitivity improvement, as compared to ordinary lines with identical physical length (at least for $Z_B = Z_0$ and implementable values of the host line impedance, Z, and shunt capacitance,

---

34 In (3.18), Z_s is the impedance of the sensing line, equivalent to Z_B in the artificial slow-wave transmission line.
35 Note, however, that in (3.2), the phase of the ordinary line is designated as ϕ, rather than θ, and the phase velocity as v_p, rather than v_{p0}.

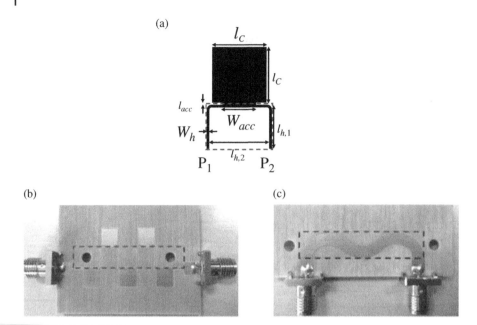

Figure 3.50 Layout (unit cell) (a) and photograph of the capacitively loaded transmission line-based sensor (b), and photograph of the meander line-based sensor (c). Dimensions (in mm) are: l_c = 5.0, l_{acc} = 0.2, W_{acc} = 3.0, $l_{h,1}$ = 3.19, $l_{h,2}$ = 5.35, and W_h = 0.23. The width of the ordinary meandered line gives a characteristic impedance of 50 Ω in the considered substrate. *Source:* Reprinted with permission from [51]; copyright 2020 EuMA.

C). Nevertheless, $Z > Z_B$ in capacitively loaded slow-wave transmission lines, and therefore the host line is narrower than the equivalent ordinary line with identical impedance. This typically facilitates host line meandering, an interesting strategy to achieve reasonable shape factors, as discussed before in this chapter. Indeed, the sensitivity of the designed and fabricated slow-wave transmission line-based sensor with meandered host line reported in [51] is better than the one of the meandered ordinary matched line accommodated in the same sensing area (substantially wider than the host line of the slow-wave sensor). However, the improvement is limited and mainly due to the reduced length of the ordinary line.[36]

Figure 3.50 depicts the slow-wave transmission line-based sensor with meandered host line (with $N = 5$ unit cells), as well as the sensor based on a meandered ordinary line (with identical sensing region), that were reported in [51]. The sensors were fabricated in the *Rogers RO4003C* substrate with dielectric constant $\varepsilon_r = 3.55$ and thickness $h = 1.524$ mm. For the capacitively loaded line sensor, the impedance of the structure, the phase of the unit cell, and the slow-wave ratio were set to $Z_B = 50 \,\Omega$, $\phi = 120°$, and $swr = 0.5$, respectively. From inversion of equations (3.153)–(3.155), the host line impedance, the electrical length of the host line (unit cell), and the capacitance value are found to be $Z = 150 \,\Omega$, $\theta = 60°$, and $C = 0.817$ pF, respectively (the frequency of operation was set to $f_0 = 3$ GHz). Figure 3.51 shows the variation of the measured differential phase of the transmission coefficient at $f_0 = 3$ GHz (i.e. considering as reference the phase of the bare sensors) with the dielectric constant of the MUT. Such data have been inferred by loading the sensing area with

---

36 Another cause of sensitivity degradation in ordinary lines implemented in microstrip technology is line width, since, as it increases, the form factor F approaches unity, thereby reducing the sensitivity, see (3.9). By contrast, in capacitively loaded slow-wave transmission lines, with narrow host lines, the terms (3.157c) and (3.157e) can be made substantially high, but such terms have opposite contribution to the sensitivity.

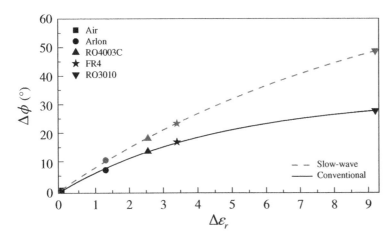

Figure 3.51 Dependence of the differential phase of the measured transmission coefficient at the operating frequency with the dielectric constant of the MUT. The differential phase is defined as $\Delta\phi = N\cdot\phi_{MUT} - N\cdot\phi_{air}$, where $N\cdot\phi_{MUT}$ and $N\cdot\phi_{air}$ are the measured phases obtained with MUT and air, respectively, and $\Delta\varepsilon_r = \varepsilon_{r,MUT} - \varepsilon_{r,air}$. *Source:* Reprinted with permission from [51]; copyright 2020 EuMA.

materials of different dielectric constants. The phase experiences a stronger variation with the dielectric constant for the slow-wave transmission line phase-variation sensor, but, as anticipated, the sensitivity improvement, as compared to the one of the ordinary line, is limited.

3.4.1.2 Sensing Through the Patch Capacitors

The analysis of Section 3.4.1.1 explains the limited capacity for sensitivity enhancement in slow-wave capacitively loaded structures with sensing based exclusively on the host line. Moreover, excluding the capacitive patches from the sensing region has another drawback, i.e. it is not possible to extend the sensitive region (or the presence of the MUT) far enough from the host line strip in its transverse direction (at least in the positions of the line where the patches are shunt connected).[37] Under these conditions, sensitivity is degraded, and the effects of the dielectric constant of the MUT on the variation of the phase velocity and impedance of the host line cannot be accurately predicted by (3.157c) and (3.157e), as far as these expressions are valid, provided the electric field generated by the line is confined within the substrate and MUT.

The previous limitative aspects can be solved by considering an alternative approach, i.e. placing the MUT on top of the patches (excluding the host line) [52]. It is not expected that the MUTs have strong influence on the (broad-side) capacitance of the patches. However, by closely spacing the patches, it is expected that a coupling (edge) capacitance between them, magnified by the presence of the MUT, arises. Thus, in the sensor structure analyzed and reported in this section, the patches are tiny separated, and the circuit model of the resulting capacitively loaded slow-wave transmission line includes an additional capacitor, C_c, in order to account for such coupling. Dielectric constant sensing is based on the effects that a variation in C_c (caused by changes in ε_{MUT}) produce on the electrical length of the line.

The typical topology and circuit model of these capacitively loaded slow-wave transmission lines with inter-capacitor coupling are depicted in Fig. 3.52. The sensing region is indicated by the dashed rectangle. Since the circuit model is not identical to the one of Fig. 3.47, the dispersion relation and

---

37 This is apparent in view of Fig. 3.50b, where the dashed rectangle delimiting the sensing region is roughly in contact with the host line in the regions where the patches are connected.

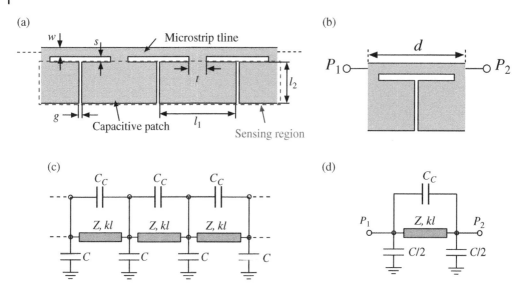

Figure 3.52 Typical topology of the slow-wave transmission line based on coupled loading capacitors (a), unit cell (b), circuit model (c), and circuit model of the unit cell (d). Relevant dimensions are indicated.

the Bloch impedance are no longer given by expressions (3.153) and (3.154). Calculation of the ABCD matrix for the unit cell of Fig. 3.52, and application of (3.151) and (3.152) gives [52]

$$\cos \phi = \frac{\cos \theta - \omega_0 Z (0.5C + C_c) \sin \theta}{1 - \omega_0 Z C_c \sin \theta} \tag{3.159}$$

$$Z_B = Z \frac{\sin \theta}{\sin \phi (1 - \omega_0 Z C_c \sin \theta)} \tag{3.160}$$

Note that if $C_c = 0$, the dispersion relation (3.159) coincides with (3.153). However, the Bloch impedance (3.160) is different than (3.154). This discrepancy is explained by the fact that the unit cell in the circuit of Fig. 3.47 is delimited by the bisecting planes of the adjacent transmission line sections sandwiching the shunt capacitor. By contrast, in the circuit of Fig. 3.52, the unit cell consists of the line section of length $\theta = kl$, in parallel with the capacitor C_c, sandwiched between the two halves of the patch capacitances. Given a periodic structure, line dispersion does not depend on how the unit cell is selected. However, this does not apply to the Bloch impedance, as it is demonstrated in [29].

The sensitivity of the sensor of Fig. 3.52 can be expressed as

$$S = N \frac{d\phi}{d\varepsilon_{MUT}} = N \frac{d\phi}{dC_c} \cdot \frac{dC_c}{d\varepsilon_{MUT}} \tag{3.161}$$

and sensitivity optimization requires the minimum possible gap distance, g, between adjacent patches. Thus, for sensor design, g should be set to the minimum value allowed by the technology in use. It should be mentioned, however, that, contrary to the slow-wave transmission line of Section 3.4.1.1, in the structure of Fig. 3.52, the line parameters $(Z, C, C_c, \text{and } \theta)$ are not univocally determined by the three design parameters $(Z_B, \phi, \text{and } swr)$. Setting C_c to a certain value compatible with the minimum gap distance seems to be a convenient procedure to solve this indeterminacy.

However, this univocally provides the patch length, l_2 (see Fig. 3.52).[38] Thus, with the resulting value of C, obtained by solving (3.159), (3.160), and (3.155), the width of the patches, l_1, is also determined (since C is given by the patch area, $l_1 \times l_2$). This sets the period of the structure to $l = l_1 + g$, a value not necessarily compatible with the electrical length of the host line of the unit cell, $\theta = kl$, also inferred from expressions (3.159), (3.160), and (3.155), unless meandering is applied to the host line.[39]

One procedure to solve the inconsistency mentioned in the preceding paragraph is to vary the value of C_c, until the geometry and line parameters are consistent (and obviously satisfy the specifications). For that purpose, it is convenient first to ignore C_c, and determine the remaining line parameters (Z, C, and θ) from the specifications (Z_B, ϕ, and swr). Then, with the resulting values of C and θ, and g set to the minimum separation, determine l_1 and l_2 (note that $l_1 = l - g$, and, once l_1 is found, l_2 is determined by C). From l_2, the coupling capacitance C_c can be obtained (provided g is known). This procedure provides the first tentative value of C_c. The next step is to iteratively vary C_c around such value, as indicated above. However, it is expected that C_c does not vary excessively from that value. Therefore, from a practical viewpoint, and for the sake of agility and simplicity, C_c is set to the value given by the procedure explained before. With this value, the other line parameters are calculated (by solving the design equations), and, finally, the dimensions of the structure are fine-tuned, if necessary, to satisfy the required specifications.

Let us next detail the design process for sensitivity optimization. The Bloch impedance is set to the reference impedance of the ports ($Z_B = Z_0$, typically 50 Ω), since this value minimizes the insertion loss. The combination of swr and ϕ providing the maximum sensitivity, yet keeping implementable values of Z and C, should then be found. For that purpose, swr is first set to a certain value, and ϕ is varied in appropriate steps. For each value of ϕ, the line parameters Z, C, and θ are calculated. For this value of swr, the best option for ϕ, i.e. the one providing the maximum derivative $d\phi/dC_c$ keeping implementable value of the host line impedance, Z, is obtained. The procedure is then repeated for different values of swr.

Using the above procedure, the values of C, Z, and $d\phi/dC_c$ plotted in Fig. 3.53 as a function of $\phi = \beta l$, and parametrized by swr, are obtained. The curves in Fig. 3.53 can be used in designing dielectric constant sensors based on slow-wave transmission lines loaded with coupled capacitor patches.[40] According to Fig. 3.53, by increasing ϕ for a specific swr value, the required value of C increases, whereas Z decreases. In addition, a larger electrical length of the unit cell results in a higher sensitivity. However, note that $d\phi/dC_c$ does not depend on swr [52]. In the sensor example reported in this section, the design parameters are set to $swr = 0.5$, $\phi = 90°$, and $Z_B = 50$ Ω. With these values, the design curves in Fig. 3.53 provide $\theta = 45°$, $C = 2.22$ pF, and $Z = 68.5$ Ω. The geometrical parameters of the artificial line that result by considering the dielectric substrate used for the determination of C_c in Fig. 3.53 are: $d = 11.3$ mm, $g = 0.2$ mm, $l_1 = 11.1$ mm, $l_2 = 3.55$ mm, $s = 0.5$ mm, $t = 0.75$ mm, and $w = 0.9$ mm. Figure 3.54 depicts the frequency response (unit cell) of the designed structure, as well as the dispersion relation and Bloch impedance. The figure points out the good agreement between the results inferred from circuit and electromagnetic simulation. Moreover, it is verified that $Z_B = 50$ Ω and $\phi = 90°$ at $f_0 = 2$ GHz.

---

38 C_c depends on the separation, g, and length, l_2, of the gap, and, obviously, on the dielectric constant of the substrate material and MUT (if it is present).

39 Although θ depends also on the phase constant of the host line, k, this parameter cannot be tailored, since it is given by the substrate dielectric constant and by the host line width, and hence by Z, also determined from the solution of expressions (3.159), (3.160) and (3.155).

40 Obviously, these curves are valid for a specific operating frequency ($f_0 = 2$ GHz in this case) and substrate (in this case, the *Rogers RO4350* with thickness $h = 0.762$ mm and relative permittivity of $\varepsilon_r = 3.66$).

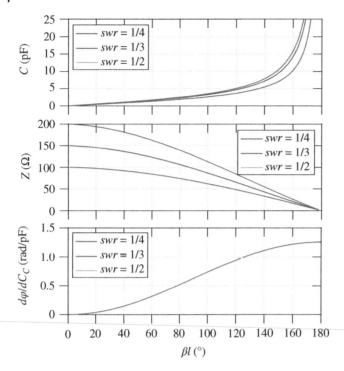

Figure 3.53 Calculated values of *C*, *Z*, and *dφ/dC_c* as a function of *φ* for three different values of *swr* and by assuming C_c = 0.04 pF. *Source:* Reprinted with permission from [52]; copyright 2021 IEEE.

The prototype sensor that results by cascading four ($N = 4$) unit cells is depicted in Fig. 3.55. The phase variation of S_{21} at $f_0 = 2$ GHz for various dielectric loadings (MUTs) is depicted in Fig. 3.56(a), where it is compared with the phase variation of a 50-Ω meandered microstrip line (with approximately the same sensing area) implemented in the same substrate (the meandered line-based sensor is also depicted in Fig. 3.55). As it can be seen in Fig. 3.56(b), the slow-wave transmission line exhibits significantly better sensitivity, in comparison with the meandered line. Moreover, contrary to most phase-variation sensors, the sensitivity increases with the dielectric constant of the MUT in the reported slow-wave transmission line-based sensor of Fig. 3.55(a) (as it is justified in [52]).

3.4.2 Sensors Based on Composite Right-/Left-Handed (CRLH) Lines

The dispersion relation and the Bloch impedance of a CRLH transmission line described by the model of Fig. 3.48(b) are given by [29, 41]

$$\cos \phi = 1 - \frac{\omega_0^2}{2\omega_R^2}\left(1 - \frac{\omega_s^2}{\omega_0^2}\right)\left(1 - \frac{\omega_p^2}{\omega_0^2}\right) \tag{3.162}$$

$$Z_B = \sqrt{\frac{L_R\left(1 - \frac{\omega_s^2}{\omega_0^2}\right)}{C_R\left(1 - \frac{\omega_p^2}{\omega_0^2}\right)} - \frac{L_R^2\omega_0^2}{4}\left(1 - \frac{\omega_s^2}{\omega_0^2}\right)^2} \tag{3.163}$$

(a)

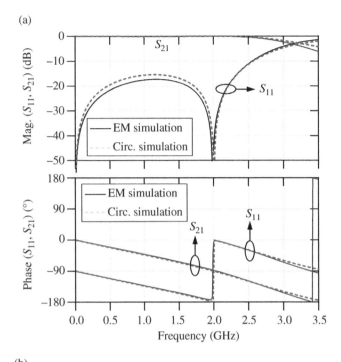

(b)

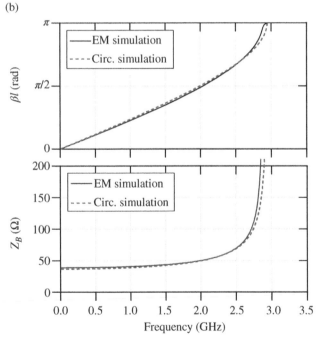

Figure 3.54 (a) Magnitude and phase response of the unit cell of the designed slow-wave structure based on coupled capacitors; (b) dispersion relation and Bloch impedance. *Source:* Reprinted with permission from [52]; copyright 2021 IEEE.

(a)

(b)

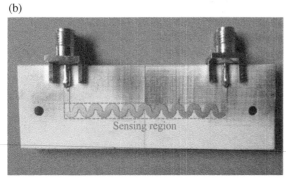

Figure 3.55 Photograph of the designed phase-variation sensor based on a slow-wave transmission line with coupled capacitors (a) and photograph of the phase-variation sensor based on a meander line with similar sensing area for the MUT (b). *Source:* Reprinted with permission from [52]; copyright 2021 IEEE.

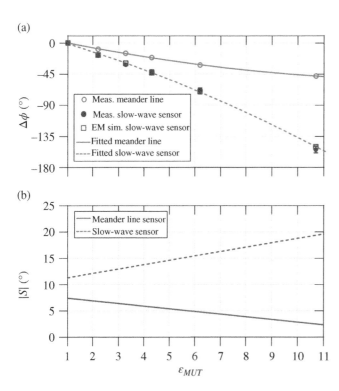

Figure 3.56 Variation of the phase of the transmission coefficient at f_0 = 2 GHz for the slow-wave artificial line-based sensor and for the sensor based on a meandered line (a) and sensitivity (b). *Source:* Reprinted with permission from [52]; copyright 2021 IEEE.

respectively, where ω_0 is the operating (angular) frequency.[41] In (3.162) and (3.163), the following variables

$$\omega_R = \frac{1}{\sqrt{L_R C_R}} \tag{3.164a}$$

$$\omega_L = \frac{1}{\sqrt{L_L C_L}} \tag{3.164b}$$

$$\omega_s = \frac{1}{\sqrt{L_R C_L}} \tag{3.164c}$$

$$\omega_p = \frac{1}{\sqrt{L_L C_R}} \tag{3.164d}$$

have been introduced, in order to simplify the expressions. Regardless of whether the CRLH line is balanced ($\omega_s = \omega_p = \omega_t$) or not ($\omega_s \neq \omega_p$), there are portions of the dispersion diagram lying below the straight line corresponding to the ω–ϕ diagram of a (dispersionless) ordinary line. Therefore, CRLH lines are potentially useful for the implementation of dielectric constant sensors based on phase variation, as anticipated before.

Similar to the section dedicated to phase-variation sensors based on slow-wave transmission lines, the objective in this section is to determine the optimum conditions for sensitivity enhancement. To this end, the operating frequency and the Bloch impedance should be set to certain values (the latter is typically forced to coincide with the reference impedance of the ports). However, dealing with expressions (3.162) and (3.163) is not straightforward, as far as these expressions depend on many parameters. Nevertheless, for sensitivity optimization, the CRLH should preferably operate in a ϕ–ω point within the dispersion diagram close to the extremes of the Brillouin zone. In these regions, the dispersion can be approximated either to the one of a purely left-handed line, where the parameters modeling the host line (C_R and L_R) can be neglected, or to the dispersion of a right-handed line based on a cascade of series inductances and shunt capacitances. In this latter case, the neglected parameters are C_L and L_L. Let us next consider the operation of the CRLH line in the lower frequency region of the dispersion diagram, where the structure resembles a purely left-handed line. Neglecting C_R and L_R, the dispersion diagram and the Bloch impedance can be expressed as

$$\cos\phi = 1 - \frac{1}{2L_L C_L \omega_0^2} \tag{3.165}$$

$$Z_B = \sqrt{\frac{L_L}{C_L}\left(1 - \frac{\omega_c^2}{\omega_0^2}\right)} \tag{3.166}$$

where $\omega_c = 1/2(L_L C_L)^{1/2}$ is the cutoff frequency.

The simplification that represents considering the purely left-handed line approximation, i.e. neglecting L_R and C_R, inherently assumes that the effects of the MUT on the phase of the host line

---

41 Do not confuse ω_0, the angular frequency of operation, with the transition frequency of a balanced CRLH line, usually referred to as ω_0 in many textbooks [29, 41]. The balance condition of a CRLH line, or continuous transition between the left-handed and the right-handed band, requires identical series and shunt resonance frequencies ($\omega_s = \omega_p$), and such resonance frequency is the transition frequency, designated as ω_t in this book (i.e. $\omega_t = \omega_s = \omega_p$ for a balanced CRLH line).

are also neglected. This is obvious as far as C_R (the capacitance of the host line) is the single parameter accounting for the effects of the MUT on the host line.[42] Thus, within this approximation, any variation in the dielectric constant of the MUT modifies uniquely the loading capacitance C_L of the line (it is assumed that such capacitance is implemented in planar form, e.g. by means of interdigital capacitors or series gaps). The sensitivity (or variation of the phase of the line with the dielectric constant of the MUT) can thus be written as

$$S = N \frac{d\phi}{d\varepsilon_{MUT}} = N \frac{d\phi}{dC_L} \cdot \frac{dC_L}{d\varepsilon_{MUT}} \tag{3.167}$$

where the first term is the key parameter for sensitivity optimization, N being the number of stages. Using (3.165), such term is found to be

$$\frac{d\phi}{dC_L} = -\frac{1}{C_L \sqrt{4L_L C_L \omega_0^2 - 1}} \tag{3.168}$$

Isolating L_L from (3.166) and introducing the result in (3.168) gives

$$\frac{d\phi}{dC_L} = -\frac{1}{2 C_L^2 \omega_0 Z_B} \tag{3.169}$$

From this result, it follows that decreasing C_L increases the sensitivity of ϕ with C_L. Note, however, that a reduction in C_L (with Z_B fixed) means also an increase in L_L.[43] Therefore, the maximum achievable sensitivity is related to the maximum value of L_L compatible with an implementable inductance. For simplification purposes, let us designate the left-hand side member of (3.169) as ϕ'. Isolating C_L from (3.169) and introducing the result in (3.166), the inductance L_L is found to be

$$L_L = \frac{Z_B(2\omega_0 Z_B + |\phi'|)}{2\omega_0 \sqrt{2\omega_0 Z_B |\phi'|}} \tag{3.170}$$

Finally, let us express the dispersion relation in terms of the design parameters. Using L_L and C_L as inferred from (3.170) and (3.169), respectively, and introducing the results in (3.165) gives

$$\cos\phi = 1 - \frac{2|\phi'|}{2\omega_0 Z_B + |\phi'|} \tag{3.171}$$

Note that for $|\phi'| \to \infty$ (high sensitivity), the phase of the unit cell tends to $\phi = -\pi$, pointing out (and verifying) that for sensitivity enhancement, operation in the flat region of the dispersion diagram is necessary.

It is worth mentioning that, despite the fact that reducing C_L increases the first term of the right-hand side of (3.167), a drop in C_L degrades the second term $(dC_L/d\varepsilon_{MUT})$. However, (3.169) is inversely proportional to the square of C_L, whereas $dC_L/d\varepsilon_{MUT}$ is roughly proportional to C_L.[44]

---

42 The dielectric constant and thickness of the MUT determine the effective dielectric constant of the host line and the per-unit-cell capacitance C_R.

43 Actually this is true for values of the capacitance satisfying $C_L < (2Z_B\omega_0)^{-1}$ (the case of interest), which provides a value of L_L giving a flat dispersion, i.e. a high value of ϕ', at the operating frequency. For values of $C_L > (2Z_B\omega_0)^{-1}$, the resulting cutoff frequency is substantially below the operating frequency (it is easy to demonstrate that for $C_L = (2Z_B\omega_0)^{-1}$ the cutoff frequency is $\omega_c = \omega_0/\sqrt{2}$). In the limit of high C_L values ($C_L \gg (2Z_B\omega_0)^{-1}$), with Z_B and ω_0 fixed (as input parameters), L_L and C_L are related by $L_L = C_L Z_B^2$, and the line operates within the so-called long-wavelength approximation, with a frequency-independent Bloch impedance.

44 To a first-order approximation, by considering a semi-infinite MUT, the capacitance C_L when a MUT is on top of the CRLH line can be expressed as $C_L = C_{L,air}\left(\frac{\varepsilon_r + \varepsilon_{MUT}}{\varepsilon_r + 1}\right)$, where $C_{L,air}$ is the capacitance C_L of the bare CRLH line and ε_r is the substrate dielectric constant. Consequently, it follows that $dC_L/d\varepsilon_{MUT} = C_{L,\,air}/(\varepsilon_r + 1)$.

Therefore, the net effect is an increase of the overall sensitivity (3.167), the one of interest, as C_L decreases. A potential issue of these sensors is related to the fact that working in the flat region of the dispersion diagram, as required to obtain very high sensitivity, limits the robustness of the device against fabrication-related tolerances. However, relaxing this aspect (i.e. limiting the sensitivity per unit cell) may be compensated by increasing the number of stages, as far as the unit cell size of CRLH lines is small (note that the sensitivity of the sensor is proportional to the number of stages).

An illustrative example of a phase-variation sensor based on a CRLH line is reported in [53]. It is a differential-mode sensor based on a pair of identical CRLH lines, a Wilkinson power divider and a combiner, as depicted in Fig. 3.57(a). The input signal is divided between the two CRLH lines. However, the lines connecting the divider/combiner with the feeding points of the CRLH lines are different. Particularly, the feeding lines of the sensing CRLH line are half-wavelength longer than the ones of the reference CRLH line. Therefore, the signals at the input ports of the combiner are out of phase. Because of this, the magnitude of the transmission coefficient of the resulting two-port structure, or the amplitude of the output signal, is expected to be roughly null. However, loading the sensing line with a MUT alters the perfect imbalance (180° phase difference) between the signals

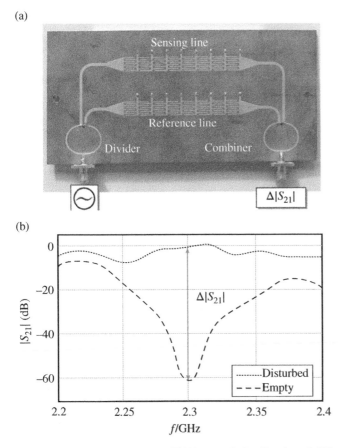

Figure 3.57 Photograph of the CRLH transmission line-based differential-mode phase-variation sensor (a) and comparison between the transmission coefficients for the empty case (unloaded sensing line) and for the sensing line loaded with a MUT (b). *Source:* Reprinted with permission from [53]; copyright 2009 IEEE.

present at the input ports of the combiner. The reason is the phase variation experienced by the CRLH sensing line due to the presence of the MUT on top of it. The consequence is the presence of a certain signal level at the output port of the structure (or $S_{21} \neq 0$), related to ε_{MUT}. Figure 3.57(b) depicts the frequency response of the differential-mode phase-variation sensor based on the pair of CRLH lines for two cases, i.e. with unloaded (empty) sensing line, and with the sensing line loaded with a MUT [disturbed case, in reference to Fig. 3.57(b)]. The combination of CRLH line characteristics and ε_{MUT} has been determined in order to obtain the maximum possible transmission coefficient when the MUT is on top of the sensing line (this occurs provided the signals at the input ports of the combiner are in-phase). The output dynamic range at the operating frequency, $f_0 = 2.3$ GHz, is roughly 60 dB. Further details on this structure are out of the scope of this example, with merely illustrative purpose, but can be found in [53].

3.4.3 Sensors Based on Electro-Inductive Wave (EIW) Transmission Lines

The dispersion relation and the Bloch impedance of the EIW transmission line depicted in Fig. 3.49 are given by [29]

$$1 - \frac{2C_M}{C_c}\cos\phi = \frac{\omega_r^2}{\omega_0^2} \tag{3.172}$$

$$Z_B = \sqrt{\frac{L_c}{C_M} \cdot \frac{L_c\omega_0^2(C_c + 2C_M) - 1}{4L_cC_M\omega_0^2(1 - L_cC_G\omega_0^2)}} \tag{3.173}$$

where ω_0 is the angular frequency of operation, $C_c = C_G + 2C_M$ is the capacitance of the isolated CSRR, and

$$\omega_r = \frac{1}{\sqrt{L_c(C_G + 2C_M)}} = \frac{1}{\sqrt{L_cC_c}} \tag{3.174}$$

The allowed (left-handed) band is restricted to frequencies delimited by

$$\omega_{0\pm} = \frac{\omega_r}{\sqrt{1 \mp \frac{2C_M}{C_c})}} \tag{3.175}$$

At the lower frequency of the interval, ω_{0-}, the electrical length of the line (unit cell) and the Bloch impedance are $\phi = -\pi$ and $Z_B = 0$, respectively, whereas in the upper limit, ω_{0+}, evaluation of (3.172) and (3.173) gives $\phi = 0$ and $Z_B = \infty$.

According to (3.175), the bandwidth is determined by the coupling capacitance between the resonant elements, C_M, typically small in EIW transmission lines.[45] Thus, EIW lines exhibit very flat dispersion diagrams and are highly dispersive. Thus, it is expected that the phase of these lines experiences strong variations by loading it with the MUT (which should be located on top of the CSRRs). However, because the bandwidth of EIW transmission lines is very narrow, the effects of the MUT on the response of the device are actually twofold. Namely, the variation in the phase at ω_0 (caused by an overall shift in the dispersion diagram) is accompanied by a significant excursion of the magnitude of the transmission coefficient. In other words, the peaked magnitude response of the weakly coupled EIW transmission lines experiences significant shifts by varying the dielectric

---

45 EIW transmission lines based on CSRRs exhibit weak coupling because the resonant elements are coplanar (see Fig. 3.49a). Consequently, even by closely spacing the CSRRs to the minimum limit dictated by the technology in use, the mutual capacitance C_M cannot take values comparable to those of C_c, i.e. $C_M \ll C_c$ (weak coupling condition).

constant of the MUT (as it will be shown later). Therefore, these sensors can also be used as frequency-variation sensors (studied in Chapter 2).

In EIW transmission lines based on CSRRs, the presence of the MUT on top of the CSRRs (sensing region) alters both the coupling capacitance, C_M, and the capacitance of the resonators, C_c (and consequently C_G). Thus, a sensitivity analysis similar to the one carried out in Sections 3.4.1 and 3.4.2 in reference to slow-wave and CRLH transmission line phase-variation sensors does not seem to be straightforward in this case. However, a reasonable approximation simplifies the analysis. It considers that the relative variation in C_c, caused by the effects of the MUT, is identical in C_M. Under this approximation, the ratio $2C_M/C_c$ that appears in the dispersion relation (3.172) is constant. Therefore, any variation in the electrical length of the unit cell, ϕ, caused by a change in ε_{MUT}, can be accounted for through the effects on C_c only.[46] According to these words, the sensitivity for a N-cell EIW transmission line-based sensor can be expressed as

$$S = N\frac{d\phi}{d\varepsilon_{MUT}} = N\frac{d\phi}{dC_c} \cdot \frac{dC_c}{d\varepsilon_{MUT}} \tag{3.176}$$

Let us designate the capacitance ratio appearing in (3.172) as $2C_M/C_c = F$. Isolating ϕ from (3.172) and taking the derivative with respect to C_c provides the first term of the right-hand side member of (3.176). After some simple algebra, the following result is obtained:

$$\frac{d\phi}{dC_c} = -\frac{1}{C_c\sqrt{F^2\frac{\omega_0^4}{\omega_r^4} - \left(\frac{\omega_0^2}{\omega_r^2} - 1\right)^2}} \tag{3.177}$$

Calculating analytically the values of the element parameters of the EIW unit cell model [Fig. 3.49 (b)] that are necessary to set the sensitivity (3.176) and the Bloch impedance (3.173), the design parameters, to certain desired values is not simple. Obviously, such calculation can be carried out numerically. However, the physical implementation of the resulting values and, most important, ensuring that the design parameters are satisfied in practice, is not apparent. This second issue is due to the high dispersion of these artificial lines (i.e. small variations in the line parameters may cause significant changes in the phase and magnitude response of the line). Nevertheless, let us calculate the derivative (3.177), the relevant one contributing to the sensitivity, for the operating (angular) frequency set to roughly the central frequency of the allowed band, i.e. for $\omega_0 = \omega_r$. The following expression is obtained

$$\frac{d\phi}{dC_c}\bigg|_{\omega_0 = \omega_r} \equiv \phi' = -\frac{1}{C_c F} = -\frac{1}{2C_M} \tag{3.178}$$

i.e. a result that confirms that weakly coupled EIW lines are highly sensitive. From (3.173), it follows that the Bloch impedance evaluated at $\omega_0 = \omega_r$ is

$$Z_B = \frac{1}{2}\sqrt{\frac{L_c C_c}{C_M^2}} = \sqrt{\frac{L_c}{2C_M F}} \tag{3.179}$$

and, finally,

$$\frac{L_c}{F} = \frac{Z_B^2}{|\phi'|} \tag{3.180}$$

---

46 Note that the right-hand side member of (3.172) depends on ω_r, and consequently on C_c, according to (3.174).

as inferred by introducing $1/(2C_M) = |\phi'|$ in (3.179). Note, however, that using $F = 2C_M/C_c = 1/(|\phi'| \cdot C_c)$ and combining (3.180) and (3.174) yields

$$\omega_0^2 = \omega_r^2 = \frac{|\phi'|^2}{Z_B^2} \tag{3.181}$$

This means that for the specific case of sensor operation at $\omega_0 = \omega_r$, the design parameters ϕ' and Z_B dictate the operating frequency. For sensor design, the procedure consists in first determining a reasonable combination of ϕ' and ω_0 satisfying (3.181), provided Z_B is typically set to the reference impedance of the ports for matching purposes. The coupling capacitance C_M is obtained by means of (3.178), whereas (3.174) determines the product L_cC_c, i.e. the values of L_c and C_c are not univocally determined. In practice, a certain CSRR topology providing the required resonance frequency, $\omega_r = \omega_0$, must be selected (note that this is the frequency of the isolated resonator[47]). Then, by adjusting the distance between adjacent CSRRs, the capacitance C_M is set to the desired value.

The previous approach is very simple, but it imposes a severe constraint concerning the frequency of operation, given by (3.181). Indeed, there is not an a priori reason to set the operating frequency to $\omega_0 = \omega_r$. Thus, let us relax this condition, and let us calculate the frequency, within the allowed band, where (3.177) is a minimum, $\omega_{0,min}$. For that purpose, the derivative of (3.177) with respect to ω_0 is calculated and forced to be null. This gives

$$\omega_{0,min} = \frac{\omega_r}{\sqrt{1 - F^2}} \approx \omega_r \tag{3.182}$$

and (3.177) at this frequency is found to be

$$\left. \frac{d\phi}{dC_c} \right|_{\omega_0 = \omega_{0,min}} = -\frac{1}{C_cF}\sqrt{1 - F^2} \approx -\frac{1}{2C_M} \tag{3.183}$$

Expression (3.177) is minimized at $\omega_0 = \omega_{0,min}$ since its derivative is null at this frequency, and (3.177) progressively increases as the frequency approaches the limits of the allowed band, as inferred by evaluating (3.177) at $\omega_0 = \omega_{0\pm}$. From (3.182), it follows that the frequency where (3.177) is a minimum roughly coincides with ω_r, provided F is very small. Consequently, the minimum value of (3.177), given by (3.183), coincides approximately with (3.178), and it increases as C_M decreases. Thus, by weakly coupling the CSRRs of the EIW transmission line, the structure is highly sensitive regardless of the specific frequency of operation within the allowed band. Since the excursion experienced by the Bloch impedance goes from $Z_B = 0$ to $Z_B = \infty$ as the frequency varies between the limits of the band, there is a frequency within that band where the line is matched to the ports. Thus, by setting the operation frequency to this value, both line matching and high sensitivity are achieved simultaneously.

From the previous analysis, it follows that the key aspect for sensitivity optimization is to weakly couple the resonant elements of the EIW line (as anticipated before). However, as the coupling decreases, tuning the operating frequency to the convenient value (for line matching) is increasingly more difficult, since the variation of the Bloch impedance with frequency also increases. Moreover, very weakly coupled EIW lines exhibit narrow bandwidths, with peaked magnitude responses severely sensitive to the effects of losses (not considered in the present analysis). This

---

47 By isolated resonator, the authors mean a CSRR without the presence of further adjacent resonators (or far enough to neglect their mutual coupling). CSRR layout generation from the reactive element values, L_c and C_c, is out of the scope of this book. Nevertheless, the authors recommend to the interested readers the book [29] and papers [54, 55], where this aspect is discussed.

means that, in designing the sensor, a trade-off is necessary. Thus, a moderately coupled EIW line typically suffices to achieve high sensitivity combined with reasonable insertion losses (intimately related to the bandwidth), as it is demonstrated next.

Concerning the second term contributing to the overall sensitivity [$dC_c/d\varepsilon_{MUT}$, see expression (3.176)], such term is proportional to $C_{c,air}$ (the capacitance of the bare CSRR), as far as the capacitance of the CSRR can be expressed as $C_c = C_{c,\,air}(\varepsilon_r + \varepsilon_{MUT})/(\varepsilon_r + 1)$. Such expression is valid if the MUT can be considered to be semi-infinite in the vertical direction (see Chapter 2 for further details and Section 3.4.2, where a similar approach for the capacitance C_L of the CRLH line was adopted).

It should be mentioned that the previous sensitivity analysis is meaningful for small perturbations of the dielectric constant of the MUT. Namely, it has been indicated that high sensitivity (considering the phase as the output variable) is achieved if the resonant elements are weakly coupled (the usual situation in EIWs), and the device should be tuned to the frequency providing matching to the ports. However, due to the high dispersion of EIW transmission lines, relatively small variations of the dielectric constant of the MUT with regard to the reference value are expected to generate strong mismatch at the operating frequency, or even to significantly shift the pass band of the structure and leave the operational frequency in the stop band region. Under these conditions, sensor functionality based on the variation of the phase of the transmission coefficient with the dielectric constant of the MUT is not the best option (this is indeed the case of the example reported next). However, the phase of EIW transmission lines is very dependent on the dielectric constant of the material surrounding it. For this reason, the canonical output variable in these EIW-based sensors is the phase of the lines (or phase of the transmission coefficient), and this justifies the inclusion of these sensors in this chapter (similar to CRLH- and slow-wave-based sensors).

Let us complete this section by reporting an illustrative example of an EIW transmission line-based sensor [56]. The layout of the device, consisting of a chain of three moderately coupled CSRRs fed by means of a pair of 50-Ω microstrip lines, is depicted in Fig. 3.58(a), where dimensions are indicated (see the caption). The complete circuit model of the structure, based on the unit cell model of Fig. 3.49(b), and including the coupling of the first and third CSRR to the corresponding feed line (through the capacitance C), as well as these lines and their extensions, is depicted in Fig. 3.58(b).[48] The parameters of the circuit model, extracted according to the method detailed in [29, 57], provide a circuit response in very good agreement with the electromagnetic simulation, as it can be appreciated in Fig. 3.59. This figure points out the strong dependence of the phase response with frequency, a requirement to achieve high sensitivity in the limit of small perturbations.

As indicated before, due to their peaked response, these sensors can operate also as frequency-variation sensors, with the output variable (the peak frequency) strongly dependent on the dielectric constant of the MUT. Figure 3.60 depicts the photograph of the fabricated device, as well as the frequency response (magnitude) for different MUTs on top of the sensing region (indicated in the caption). The variation of the peak frequency with the dielectric constant is significant. Nevertheless, the main interest of this sensor would be to exploit its functionality as phase-variation sensor. The natural output variable for this mode of operation is, logically, the phase of the transmission coefficient at the operating frequency (as mentioned before, the device must be tuned close to the resonance frequency of the isolated CSRR). However, such phase variation is not exempt of magnitude variation, as well (as indicated before). If significant perturbations in the dielectric constant of the MUT are going to be considered, the overall transmission coefficient is preferred. However,

---

48 In the model, L is the inductance of the line sections on top of the CSRRs. Note that in this model, the coupling capacitance between adjacent CSRRs is called C_R, rather than C_M, adopting the nomenclature of the original source [56].

(a)

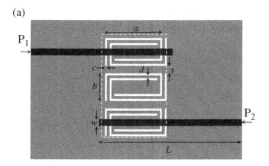

(b)

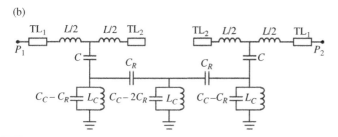

Figure 3.58 Topology (a) and circuit model (b) of the EIW transmission line phase-variation sensor. The microstrip-fed lines (in black) are etched in the upper side of the substrate, whereas the CSRRs are etched in the ground plane (in gray). Dimensions (in mm) are: $a = 4.90$, $b = 2.30$, $c = 0.20$, $d = 0.20$, $s = 0.60$, $w = 0.56$, and $L = 11.86$. The length of the 50-Ω access lines is 6.31 mm, and the line extension of the access lines beyond the CSRR limits is 0.49 mm. The sensitive area (ground plane) is indicated by the dashed rectangle. The considered substrate is the *Rogers RO3010* with dielectric constant $\varepsilon_r = 10.2$, thickness $h = 0.635$ mm, and dissipation factor $\tan \delta = 0.0023$. *Source:* Reprinted with permission from [56]; copyright 2020 IEEE.

the authors in [56] considered a differential-mode scheme, where the transmission coefficient of the line loaded with a certain MUT is recorded, and then it is subtracted from the transmission coefficient of the line loaded with the REF sample (the uncladded *Rogers RO4003C* substrate, see details in the caption of Fig. 3.60). This is equivalent to calculate the cross-mode transmission coefficient, S_{21}^{DC}, of a real differential-mode device based on a pair of EIW transmission lines.[49] The results are depicted in Fig. 3.61, where it can be appreciated the high sensitivity of the sensor to variations in the dielectric constant of the MUT (as compared to the one of the REF sample). It should be clarified that such high sensitivity is not only due to the effects of the phase variation (between the line loaded with the REF and the MUT samples), but also to the magnitude variation. The sensitivity at the operating frequency (4.2 GHz) is so high ($S = 25.33$ dB[50]), that for the *FR4* sample (with dielectric constant very similar to the one of the REF sample, see caption of Fig. 3.60), a saturation

---

49 The cross-mode transmission coefficient for uncoupled line pairs is given by $S_{21}^{DC} = \frac{1}{2}\left(S_{21}^{REF} - S_{21}^{MUT}\right)$, where S_{21}^{REF} and S_{21}^{MUT} are the transmission coefficients of the REF and MUT lines, respectively. A single EIW line is used in the reported example. Thus, S_{21}^{DC} evaluation proceeds by first loading the line with the REF sample, and measuring the corresponding transmission coefficient, S_{21}^{REF}, and subsequently by loading the line with the considered MUTs (measuring also the transmission coefficient, S_{21}^{MUT}, for each case). The cross-mode transmission coefficient is the considered output variable in most differential sensors reported in Chapter 6.

50 Note that the units of the sensitivity are dB since the considered output variable is the cross-mode transmission coefficient, as defined in the text.

(a)

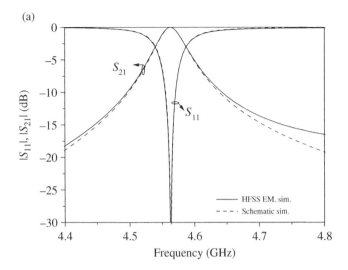

(b)

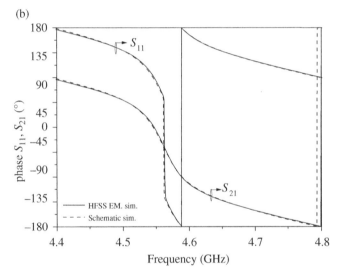

Figure 3.59 Lossless electromagnetic and circuit (schematic) simulation of the microstrip-fed EIW transmission line of Fig. 3.58. (a) Magnitude response; (b) phase response. The circuit parameters are: L = 3.467 nH, C = 0.749 pF, L_c = 0.524 nH, C_c = 2.269 pF, and C_R = 0.086 pF. *Source:* Reprinted with permission from [56]; copyright 2020 IEEE.

effect is observed. For that reason, measurements with MUT samples consisting of the REF substrate with sparse arrays of drilled holes are considered in [56] (see also the results in Fig. 3.61). The device is able to detect the presence of holes despite the fact that their size and density does not represent an appreciable change in the effective dielectric constant of the REF sample.

To summarize this section, let us mention that the high sensitivity of the EIW transmission line-based sensor prototype of Fig. 3.60(a) has been achieved by limiting the sensing area (and overall sensor dimensions) to a small size. Comparison of this sensor with other phase-variation sensors is carried out in the next section, where the advantages and drawbacks of the different implementations considered in this chapter are pointed out.

(a)

(b)

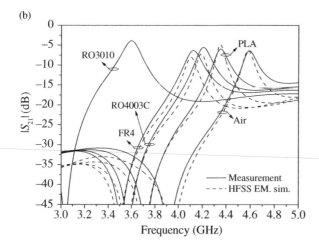

(c)

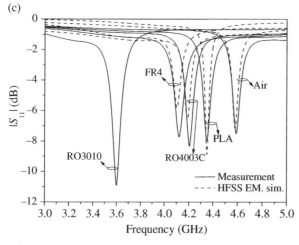

Figure 3.60 Photographs of the top (left) and bottom (center) views of the EIW transmission line-based sensor, including a picture with a MUT (right) in the sensing region (a), and transmission (b) and reflection (c) coefficients without dielectric loading (air) and with MUTs corresponding to the indicated samples. The electromagnetic simulations have been carried out by including losses. The specific samples are: a piece of uncladded *Rogers 4003C* substrate with ε_r = 3.55, h = 1.52 mm, and tan δ = 0.0027; a piece of uncladded *FR4* with ε_r = 4.4, h = 1.52 mm, and tan δ = 0.02; a piece of uncladded *Rogers RO3010* substrate with ε_r = 10.2, h = 1.52 mm, and tan δ = 0.0023; and a slab of *PLA (Polylactic acid)* with ε_r = 2.7, h = 1.5 mm, and tan δ = 0.02. *Source:* Reprinted with permission from [56]; copyright 2020 IEEE.

(a)

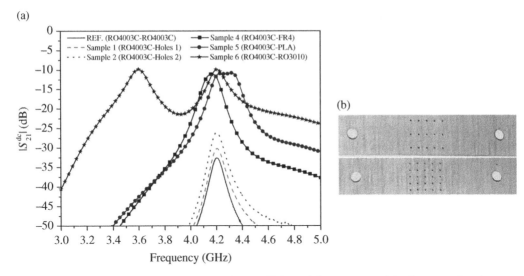

(b)

Figure 3.61 Measured cross-mode transmission coefficient of the sensor resulting by comparing the transmission coefficient of the EIW line loaded with various MUT samples with the transmission coefficient of the line loaded with the REF sample, a piece of *RO4003C* substrate (a), and photograph of the REF sample with generated defects consisting of square arrays of holes drilled across the substrate (b). *Source:* Reprinted with permission from [56]; copyright 2020 IEEE.

3.5 Advantages and Drawbacks of Phase-Variation Sensors

Several approaches for the design of sensors whose working principle relies on the variation of the phase of a transmission line-based structure or resonant element have been reviewed in this chapter. One of the main advantages of these sensors, as compared to other microwave sensors based on different principles (and discussed throughout this book), is operation at a single frequency. This reduces the cost of the associated electronics, particularly for the generation of the feeding signal, necessary for measurement in a real scenario. Note that sensor feeding with a harmonic signal suffices, as far as the phase information (the typical, although not exclusive, output variable) should be provided at a unique frequency, the one that optimizes the sensitivity. Several sensor implementations where the phase is transformed into magnitude information have also been reported in the chapter. Examples are the differential-mode phase-variation sensors, either working in transmission or operating in reflection, where inclusion of rat-race hybrid couplers in the sensor topology has been demonstrated to be an efficient approach to obtain highly sensitive two-port sensors (the output variable in such sensors is the magnitude of the transmission coefficient).

The most canonical application of phase-variation sensors is dielectric constant measurements, since the phase is, in general, very sensitive to the relative permittivity of the MUT. It is obvious that these sensors are, therefore, of interest for the measurement of other variables related to the dielectric constant, e.g. solute concentration in liquid solutions, or for the analysis of materials (including determination of their composition, defect detection, etc.). However, for the measurement of the complex dielectric constant of the MUT, including the imaginary part (or the loss tangent), resonant methods are preferred. The reason is that resonant elements are very sensitive to the effects of losses of the MUT. Note that the resonant phase-variation sensors reported in this chapter are mainly devoted to the measurement of angular displacements.[51] These phase-variation angular displacement sensors are simple and small, but in terms of performance are not at the same level as some of the coupling-modulation angular displacement and velocity sensors discussed in Chapter 4, particularly the rotary encoders (see Section 4.3.2.2).

Concerning sensor size, transmission line phase-variation sensors operating either in transmission or reflection are typically larger than other sensors reported in this book (particularly, the coupling-modulation or frequency-splitting sensors, based on one and two electrically small sensing resonators, respectively). Moreover, the sensitivity of transmission line-based sensors (including those sensor implementations based on artificial lines, see Section 3.4) increases with the dimensions of the sensing line. Nevertheless, several strategies to enhance the sensitivity, simultaneously maintaining the sensing region within very reasonable limits, have been reported and discussed throughout this chapter (see, for instance, the reflective-mode sensors based on step-impedance open-ended lines, as well as the artificial line-based sensors). It is remarkable that the sensitivity of the step-impedance open-ended line-based sensors can be made extremely high by simply cascading few quarter-wavelength sections (with appreciable impedance contrast) to the open-ended (90° or 180°) sensing line.

Comparing the studied phase-variation sensors in terms of size, sensor performance, design and fabrication simplicity, etc., is not an easy task, since the operation frequencies, substrate materials, output variables, and, even, the considered measurands for the various reported implementations (including those available in the literature, and not reported in this chapter) are different. Journal papers devoted to sensors typically report comparison tables, but the required conditions for a valid and significant comparison are not always satisfied. Nevertheless, from the analysis and examples provided in the chapter, some general conclusions, of interest to the readers, can be extracted.

Concerning size, the resonant sensors reported in Section 3.3.2, devoted to rotation measurements, are competitive, but such sensors do not exhibit the performance of other angular displacement and velocity sensors. The artificial lines considered in Section 3.4 for the implementation of the reported phase-variation sensors are intrinsically small. Therefore, the corresponding sensors are also competitive in terms of dimensions. By contrast, the phase-variation sensors based on ordinary lines (meandered, step-impedance), operating either in reflection or transmission, are typically larger (moreover, the whole dimensions of the sensor layout increase by including additional elements devoted to phase-to-magnitude conversion, e.g. hybrid couplers). Nevertheless, it is important to highlight that in the reflective-mode phase-variation sensors reported in Section 3.2.2, based on step-impedance open-ended lines, the sensitivity can be enhanced by maintaining unaltered the size of the sensing region, thereby providing a very competitive FoM, or ratio between the maximum sensitivity and the size of the sensing region expressed in terms of the squared wavelength.

With regard to performance, the potential of reflective-mode phase-variation sensors based on step-impedance discontinuities is very high, as demonstrated in Section 3.2.2. The sensitivity of these sensors can be made unprecedentedly high by virtue of the impedance contrast between the different cascaded line sections. Thanks to such high sensitivity, these sensors are able to resolve tiny variations in the dielectric constant of the MUT. Among the sensors based on artificial lines,

---

51 The exception is the reflective-mode phase-variation sensor reported in Section 3.3.1, where the sensing element is a resonator, and sensor application is dielectric constant measurements. However, for the measurement of the loss factor of the MUT, magnitude measurements, rather than phase measurement, are required (see [18]). Obviously, the magnitude of the reflection coefficient in such sensors is related to the loss tangent of the MUT, but transmission-mode sensors based on transmission lines loaded with resonant elements typically exhibit better accuracy in the determination of the loss factor of the MUT. Examples of such sensors can be found in Chapters 2, 5, and 6.

Table 3.1 Qualitative comparison of the phase-variation sensors.

Category	Transmission line-based sensors		Resonant-type sensors		Artificial transmission line-based sensors		
Type/mode	Transmission-mode	Reflective-mode	Reflective-mode	Rotation[a]	Slow-wave	CRLH	EIW
Size	Large	Moderate	Moderate	Very small	Small	Small	Very small
Sensitivity	Moderate	Very good	Very good	Good	Moderate	Very good	Very good
Design	Simple	Simple	Complex	Complex	Moderate	Complex	Complex

[a] These sensors are resonant-type reflective-mode sensors devoted to rotation (discussed in Section 3.3.2), contrary to the resonant-type reflective-mode sensors of the immediate adjacent column to the left (analyzed in Section 3.3.1), which are permittivity sensors.

those based on EIW transmission lines have also been found to exhibit very competitive sensitivity, yet keeping small sensing regions.

An important aspect in microwave sensors is the design burden.[52] This is especially simple in transmission line phase-variation sensors. Such sensors merely use ordinary lines (meandered or step-impedance lines), and eventually include simple and well-known microwave components (e.g. couplers). Thus, designing such sensors from the required specifications by using the reported design guidelines and formulas is easy. By contrast, in the sensors based on artificial lines or resonant elements, the design process is not so simple. The main reason is the lack of available expressions, at least in most cases, accurately linking the geometrical and circuit parameters. Moreover, these sensors are less robust against fabrication-related tolerances, as compared to transmission line-based sensors.

To finalize this section, let us qualitatively summarize in Table 3.1 the main characteristics of the sensors reviewed in this chapter, highlighted in the preceding paragraphs.

References

1 D. M. Pozar, *Microwave Engineering*, 4th ed., John Wiley, Hoboken, NJ, USA, 2011.

2 F. J. Ferrández-Pastor, J. M. García-Chamizo, and M. Nieto-Hidalgo, "Electromagnetic differential measuring method: application in microstrip sensors developing," *Sensors*, vol. 17, p. 1650, 2017.

3 J. Muñoz-Enano, P. Vélez, M. Gil, and F. Martín, "An analytical method to implement high sensitivity transmission line differential sensors for dielectric constant measurements," *IEEE Sensors J.*, vol. 20, pp. 178–184, Jan. 2020.

4 L. Su, J. Muñoz-Enano, P. Vélez, P. Casacuberta, M. Gil, and F. Martín, "Phase-variation microwave sensor for permittivity measurements based on a high-impedance half-wavelength transmission line," *IEEE Sensors J.*, vol. 21, no. 9, pp. 10647–10656, May 2021.

5 J. Muñoz-Enano, P. Vélez, L. Su, M. Gil, P. Casacuberta, and F. Martín, "On the sensitivity of reflective-mode phase variation sensors based on open-ended stepped-impedance transmission lines:

---

52 Many advanced planar microwave components exhibit high performance and small size, but at the expense of complex design processes that require the experience of the design engineers.

theoretical analysis and experimental validation," *IEEE Trans. Microw. Theory Techn.*, vol. 69, no. 1, pp. 308–324, Jan. 2021.

6 J. Muñoz-Enano, P. Casacuberta, L. Su, P. Vélez, M. Gil, and F. Martín, "Open-ended-line reflective-mode phase-variation sensors for dielectric constant measurements," *IEEE Sensors 2020*, Rotterdam, The Netherlands, Oct. 2020.

7 J. Muñoz-Enano, P. Vélez, M. Gil Barba, J. Mata-Contreras, and F. Martín, "Differential-mode to common-mode conversion detector based on rat-race hybrid couplers: analysis and application to differential sensors and comparators," *IEEE Trans. Microw. Theory Techn.*, vol. 68, no. 4, pp. 1312–1325, Apr. 2020.

8 D. E. Bockelman and W. R. Eisenstadt, "Combined differential and common-mode scattering parameters: theory and simulation," *IEEE Trans. Microw. Theory Techn.*, vol. 43, no. 7, pp. 1530–1539, Jul. 1995.

9 W. R. Eisenstadt, B. Stengel, and B. M. Thompson, *Microwave Differential Circuit Design Using Mixed-Mode S-Parameters*, Artech House, Norwood, MA, USA, 2006.

10 F. Martín, L. Zhu, J. S. Hong, and F. Medina, *Balanced Microwave Filters*, Wiley/IEEE Press, Hoboken, NJ, USA, 2018.

11 J. Muñoz-Enano, P. Vélez, and F. Martín, "Signal balancing in unbalanced transmission lines," *IEEE Trans. Microw. Theory Techn.*, vol. 67, no. 8, pp. 3339–3349, Aug. 2019.

12 P. Vélez, J. Muñoz-Enano, K. Grenier, J. Mata-Contreras, D. Dubuc, and F. Martín, "Split ring resonator (SRR) based microwave fluidic sensor for electrolyte concentration measurements," *IEEE Sensors J.*, vol. 19, no. 7, pp. 2562–2569, Apr. 2019.

13 G. Galindo-Romera, F. Javier Herraiz-Martínez, M. Gil, J. J. Martínez-Martínez, and D. Segovia-Vargas, "Submersible printed split-ring resonator-based sensor for thin-film detection and permittivity characterization," *IEEE Sensors J.*, vol. 16, no. 10, pp. 3587–3596, May 2016.

14 J. Muñoz-Enano, P. Vélez, L. Su, M. Gil, and F. Martín, "A reflective-mode phase-variation displacement sensor," *IEEE Access*, vol. 8, pp. 189565–189575, Oct. 2020.

15 C. Herrojo, P. Vélez, J. Muñoz-Enano, L. Su, P. Casacuberta, M. Gil, and F. Martín, "Highly sensitive defect detectors and comparators exploiting port imbalance in rat-race couplers loaded with step-impedance open-ended transmission lines," *IEEE Sensors J.*, vol. 21, pp. 26731–26745, Dec. 2021.

16 R. Marqués, F. Medina, and R. Rafii-El-Idrissi, "Role of bianisotropy in negative permeability and left-handed metamaterials," *Phys. Rev. B*, vol. 65, p. 144440, 2002.

17 J. Naqui, M. Durán-Sindreu, and F. Martín, "Modeling split ring resonator (SRR) and complementary split ring resonator (CSRR) loaded transmission lines exhibiting cross polarization effects," *IEEE Ant. Wireless Propag. Lett.*, vol. 12, pp. 178–181, 2013.

18 L. Su, J Muñoz-Enano, P. Vélez, M. Gil, P. Casacuberta, and F. Martín, "Highly sensitive reflective-mode phase-variation permittivity sensor based on a coplanar waveguide (CPW) terminated with an open complementary split ring resonator (OCSRR)," *IEEE Access*, vol. 9, pp. 27928–27944, 2021.

19 J. Muñoz-Enano, P. Vélez, M. Gil, and F. Martín, "Microfluidic reflective-mode differential sensor based on open split ring resonators (OSRRs)," *Int. J. Microw. Wireless Technol.*, vol. 12, pp. 588–597, Sep. 2020.

20 A. Vélez, F. Aznar, J. Bonache, M. C. Velázquez-Ahumada, J. Martel, and F. Martín, "Open complementary split ring resonators (OCSRRs) and their application to wideband CPW band pass filters," *IEEE Microw. Wireless Compon. Lett.*, vol. 19, pp. 197–199, Apr. 2009.

21 R. Marqués, F. Martín, and M. Sorolla, *Metamaterials with Negative Parameters: Theory*, Design and Microwave Applications, John Wiley, Hoboken, NJ, USA, 2008.

22 J. B. Pendry, A. J. Holden, D. J. Robbins, and W. J. Stewart, "Magnetism from conductors and enhanced nonlinear phenomena," *IEEE Trans. Microw. Theory Tech.*, vol. 47, no. 11, pp. 2075–2084, Nov. 1999.

23 F. Martín, F. Falcone, J. Bonache, R. Marqués, and M. Sorolla, "Split ring resonator based left handed coplanar waveguide," *Appl. Phys. Lett.*, vol. 83, pp. 4652–4654, Dec. 2003.

24 F. Falcone, T. Lopetegi, J. D. Baena, R. Marqués, F. Martín, and M. Sorolla, "Effective negative-ε stopband microstrip lines based on complementary split ring resonators," *IEEE Microw. Wireless Compon. Lett.*, vol. 14, pp. 280–282, Jun. 2004.

25 J. D. Baena, J. Bonache, F. Martín, R. Marqués, F. Falcone, T. Lopetegi, M. A. G. Laso, J. García, I Gil, M. Flores-Portillo, and M. Sorolla, "Equivalent circuit models for split ring resonators and complementary split rings resonators coupled to planar transmission lines," *IEEE Trans. Microw. Theory Techn.*, vol. 53, pp. 1451–1461, Apr. 2005.

26 F. Falcone, T. Lopetegi, M. A. G. Laso, J. D. Baena, J. Bonache, R. Marqués, F. Martín, and M. Sorolla, "Babinet principle applied to the design of metasurfaces and metamaterials," *Phys. Rev. Lett.*, vol. 93, paper 197401, Nov. 2004.

27 A. K. Horestani, Z. Shaterian, and F. Martín, "Rotation sensor based on the cross-polarized excitation of split ring resonators (SRRs)," *IEEE Sensors J.*, vol 20, pp. 9706–9714, Sep. 2020.

28 A. K. Jha, A. Lamecki, M. Mrozowski, and M. Bozzi, "A highly sensitive planar microwave sensor for detecting direction and angle of rotation," *IEEE Trans. Microw. Theory Techn.*, vol. 68, no. 4, pp. 1598–1609, Apr. 2020.

29 F. Martín, *Artificial Transmission Lines for RF and Microwave Applications*, John Wiley, Hoboken, NJ, USA, 2015.

30 K. Wu, "Slow wave structures," in: Webster, J. G. (ed) *Encyclopedia of Electrical and Electronics Engineering*, vol. 19, Wiley, New York, 1999, pp. 366–381.

31 M. C. Scardelletti, G. E. Ponchak, and T. M. Weller, "Miniaturized wilkinson power dividers utilizing capacitive loading," *IEEE Microw. Wirel. Compon. Lett.*, vol. 12, pp. 6–8, Jan. 2002.

32 K. W. Eccleston and S. H. M. Ong, "Compact planar microstripline branch-line and rat-race couplers," *IEEE Trans. Microw. Theory Techn.*, vol. 51, pp 2119–2125, Oct. 2003.

33 J. García-García, J. Bonache, and F. Martín, "Application of electromagnetic bandgaps (EBGs) to the design of ultra wide band pass filters (UWBPFs) with good out-of-band performance," *IEEE Trans. Microw. Theory Techn.*, vol. 54, pp. 4136–4140, Dec. 2006.

34 M. Orellana, J. Selga, P. Vélez, A. Rodríguez, V. Boria, and F. Martín, "Design of capacitively-loaded coupled line bandpass filters with compact size and spurious suppression," *IEEE Trans. Microw. Theory. Techn.*, vol. 65, pp. 1235–1248, Jan. 2017.

35 L. Zhu, "Guided-wave characteristics of periodic microstrip lines with inductive loading: slow-wave and bandstop behaviors," *Microw. Opt. Technol. Lett.*, vol. 41, pp. 77–79, Mar. 2004.

36 P. Vélez, J. Selga, J. Bonache, and F. Martín, "Slow-wave inductively-loaded electromagnetic bandgap (EBG) coplanar waveguide (CPW) transmission lines and application to compact power dividers," *2016 46th Europ. Microw. Conf. (EuMC)*, London, UK, Oct. 2016.

37 F. Aznar, J. Selga, A. Fernández-Prieto, J. Coromina, P. Vélez, J. Bonache, and F. Martín, "Slow wave coplanar waveguides based on inductive and capacitive loading and application to compact and harmonic suppressed power splitters," *Int. J. Microw. Wireless Technol.*, vol. 10, no. 5/6, pp. 530–537, Jun. 2018.

38 J. Selga, P. Vélez, J. Coromina, A. Fernández-Prieto, J. Bonache, and F. Martín, "Harmonic suppression in branch-line couplers based on slow-wave transmission lines with simultaneous inductive and capacitive loading," *Microw. Opt. Technol. Lett.*, vol. 60, pp. 2374–2384, Oct. 2018.

39 D. Ahn, J. S. Park, C. S. Kim, J. Kim, Y. Qian, and T. Itoh, "A design of the low-pass filter using the novel microstrip defected ground structure," *IEEE Trans. Microw. Theory Tech.*, vol. 49, pp. 86–93, 2001.

40 J. Naqui, M. Durán-Sindreu, J. Bonache, and F. Martín, "Implementation of shunt connected series resonators through stepped-impedance shunt stubs: analysis and limitations," *IET Microw. Ant. Propag.*, vol. 5, pp. 1336–1342, Aug. 2011.

41 C. Caloz and T. Itoh, *Electromagnetic Metamaterials: Transmission Line Theory and Microwave Applications*, Wiley/IEEE Press, Hoboken, NJ, USA, 2005.

42 A. Lai, T. Itoh, and C. Caloz, "Composite right/left-handed transmission line metamaterials," *IEEE Microw. Mag.*, vol. 5, no. 3, pp. 34–50, Sep. 2004.

43 E. Shamonina, V. A. Kalinin, K. H. Ringhofer, and L. Solymar, "Magneto-inductive waveguide," *Electron. Lett.*, vol. 38, no. 8, pp. 371–373, Apr. 2002.

44 E. Shamonina, V. A. Kalinin, K. H. Ringhofer, and L. Solymar, "Magnetoinductive waves in one, two, and three dimensions," *J. Appl. Phys.*, vol. 92, no. 10, pp. 6252–6261, 2002.

45 R. R. A. Syms, E. Shamonina, V. Kalinin, and L. Solymar, "A theory of metamaterials based on periodically loaded transmission lines: interaction between magnetoinductive and electromagnetic waves," *J. Appl. Phys.*, vol. 97, no. 6, Art. no. 064909, 2005.

46 M. J. Freire, R. Marqués, F. Medina, M. A. G. Laso, and F. Martín, "Planar magnetoinductive wave transducers: theory and applications," *Appl. Phys. Lett.*, vol. 85, no. 19, pp. 4439–4441, 2004.

47 F. J. Herraiz-Martínez, F. Paredes, G. Z. Gonzalez, F. Martín, and J. Bonache, "Printed magnetoinductive-wave (MIW) delay lines for chipless RFID applications," *IEEE Trans. Antennas Propag.*, vol. 60, no. 11, pp. 5075–5082, Nov. 2012.

48 M. Beruete, F. Falcone, M. J. Freire, R. Marqués, and J. D. Baena, "Electroinductive waves in chains of complementary metamaterial elements," *Appl. Phys. Lett.*, vol. 88, Art. no. 083503, Jan. 2006.

49 M. A. G. Laso, T. Lopetegi, M. J. Erro, D. Benito, M. J. Garde, and M. Sorolla, "Multiple-frequency-tuned photonic bandgap microstrip structures," *IEEE Microw. Guided Wave Lett.*, vol. 10, no. 6, pp. 220–222, Jun. 2000.

50 J. Coromina, P. Vélez, J. Bonache, and F. Martín, "Branch line couplers with small size and harmonic suppression based on non-periodic step impedance shunt stub (SISS) loaded lines," *IEEE Access*, vol. 8, pp. 67310–67320, 2020.

51 J. Coromina, J. Muñoz-Enano, P. Vélez, A. Ebrahimi, J. Scott, K. Ghorbani, and F. Martín, "Capacitively-loaded slow-wave transmission lines for sensitivity improvement in phase-variation permittivity sensors," *50th Europ. Microw. Conf.*, Utrecht, The Netherlands, Sep. 2020 (held as virtual conference in Jan. 2021).

52 A. Ebrahimi, J. Coromina, J. Muñoz-Enano, P. Vélez, J. Scott, K. Ghorbani, and F. Martín, "Highly sensitive phase-variation dielectric constant sensor based on a capacitively-loaded slow-wave transmission line," *IEEE Trans. Circ. Syst. I: Regular Papers*, vol. 68, no. 7, pp. 2787–2799, July 2021.

53 C. Damm, M. Schübler, M. Puentes, H. Maune, M. Maasch, and R. Jakoby, "Artificial transmission lines for high sensitive microwave sensors," *2009 IEEE Sensors*, Christchurch, New Zealand, 25–28 Oct. 2009, pp. 755–758.

54 J. Selga, A. Rodríguez, M. Orellana, V. Boria, and F. Martín, "Automated synthesis of transmission lines loaded with complementary split ring resonators (CSRRs) and open complementary split ring resonators (OCSRRs) through aggressive space mapping," *Appl. Phys. A Mater. Sci. Process.*, vol. 117, pp. 557–565, Nov. 2014.

55 J. Selga, A. Rodríguez, V. E. Boria, and F. Martín, "Synthesis of split rings based artificial transmission lines through a new two-step, fast converging, and robust aggressive space mapping (ASM) algorithm," *IEEE Trans. Microw. Theory Techn.*, vol. 61, pp. 2295–2308, Jun. 2013.

56 M. Gil, P. Vélez, F. Aznar, J. Muñoz-Enano, and F. Martín, "Differential sensor based on electro-inductive wave (EIW) transmission lines for dielectric constant measurements and defect detection," *IEEE Trans. Ant. Propag.*, vol. 68, pp. 1876–1886, Mar. 2020.

57 J. Bonache, M. Gil, I. Gil, J. Garcia-García, and F. Martín, "On the electrical characteristics of complementary metamaterial resonators," *IEEE Microw. Wireless Compon. Lett.*, vol. 16, pp. 543–545, Oct. 2006.

4

Coupling-Modulation Sensors

This chapter is devoted to the analysis and applications of a type of planar microwave sensors where the measurand modifies the transmission coefficient of a transmission line loaded with at least one resonator, the sensing element.[1] However, unlike frequency-variation sensors, the typical output variable of the sensors studied in the present chapter is the magnitude of the transmission coefficient of the line at the design frequency.[2] Thus, like phase-variation sensors, the sensors subject of this chapter do not need wideband signals for measurement purposes. Indeed, in most cases, a single-tone harmonic signal suffices, contrary to frequency-variation sensors, where the canonical output variable is the resonance frequency of the sensing element, thereby requiring wideband signals in order to obtain the information relative to the input variable. Almost all the sensors reported in this chapter are based on a resonator (or a set of resonators) in relative motion (linear or angular) to a static transmission line (the stator). Therefore, the canonical application of these sensors is the measurement of linear or angular displacements and velocities. Moreover, the general working principle of these sensors is the variation of the coupling level between the line and the resonant elements caused by the input variable (typically a linear or angular displacement, as indicated). For this reason, these sensors are designated as coupling-modulation sensors. Nevertheless, it is also shown in this chapter that it is possible to design sensors for dielectric characterization exploiting the variation of the magnitude level (voltage amplitude, for instance) of a resonator-loaded line at a single (operating) frequency. The key aspect in the design of coupling-modulation sensors and related measuring devices is to achieve the maximum possible excursion of the output variable (the magnitude of the transmission coefficient or the output voltage level) when the measurand varies within the input dynamic range. Several strategies can be considered for that purpose. One of such strategies exploits electromagnetic symmetry properties in transmission lines loaded with a single symmetric resonator. For this reason, the next section is focused on the general analysis of such symmetry properties. After that analysis, the chapter focuses on the working principle of coupling-modulation sensors. Examples of such sensors for motion control applications are reported in Section 4.3, which constitutes the main part of the chapter. Nevertheless, a section dedicated to sensors for dielectric characterization is also included in the chapter (Section 4.4). The chapter ends by highlighting the main advantages and drawbacks of the studied coupling-modulation sensors.

---

1 Nevertheless, an example of a device operating in reflection is also included in this chapter (see Section 4.3.1).
2 Alternatively, the output variable can be the amplitude of the voltage wave generated at the output port of the line, when it is fed by means of a harmonic signal tuned to the design frequency.

Planar Microwave Sensors, First Edition. Ferran Martín, Paris Vélez, Jonathan Muñoz-Enano, and Lijuan Su.
© 2023 John Wiley & Sons, Inc. Published 2023 by John Wiley & Sons, Inc.

4.1 Symmetry Properties in Transmission Lines Loaded with Single Symmetric Resonators

Let us consider a symmetric transmission line loaded with a symmetric planar resonant element.[3] Under these conditions, electromagnetic coupling between the resonator and the line is possible, and, indeed, it is usually manifested as a notch in the transmission coefficient of the line at resonance. However, depending on the resonator and line combination and on their relative orientation or position, such coupling can be prevented. Specifically, the electromagnetic coupling between the resonator and the line cancels if the symmetry planes of the line and resonant element are of different electromagnetic sort (i.e. one a magnetic wall and the other one an electric wall) and such planes are perfectly aligned [1, 2]. Under these conditions, the electromagnetic field generated by the line is not able to excite the resonant element, since neither a net magnetic field nor a net electric field in the resonator region arises due to symmetry.

To gain further insight on this aspect, let us consider the coplanar waveguide (CPW) transmission line of Fig. 4.1, loaded with a split ring resonator (SRR). For the fundamental CPW mode, the symmetry plane of the line is a magnetic wall, since this mode is even. By contrast, the symmetry plane of the SRR is an electric wall at the fundamental resonance [3]. Thus, by symmetrically loading the CPW transmission line with the SRR, as Fig. 4.1(a) illustrates, the electromagnetic coupling between the line and the SRR is prevented. Although the inner region of the SRR extends beyond the slots of the CPW, where the magnetic field flux is significant, there is a perfect cancellation of the magnetic field lines in the area enclosed by the SRR if the structure exhibits perfect symmetry. Obviously, this occurs because the magnetic field in the slots of the CPW flows in opposite directions. Thus, the SRR cannot be driven by means of a net axial magnetic field, the most usual procedure for SRR excitation. Note that due to the presence of an electric wall in the SRR at the fundamental resonance, an alternative method to excite the SRR at that frequency is to apply a net electric field in the direction orthogonal to its symmetry plane. However, for perfect symmetry, the electric field generated by the CPW transmission line does not satisfy the previous requirement, i.e. there is not a net electric field across the symmetry plane of the SRR, and therefore the SRR can neither be excited by the electric field generated by the CPW transmission line.

This coupling cancellation under perfect symmetry conditions applies to other line and resonator combinations. For the most usual lines, microstrip and CPW, the resonant element should exhibit an electric wall at the resonance frequency of interest, as far as the indicated lines exhibit a magnetic wall at their axial symmetry plane. Nevertheless, for balanced lines, such as differential lines or slot lines, where the axial symmetry plane is an electric wall, coupling is prevented as far as the line is loaded with a resonator exhibiting a magnetic wall at the resonance frequency of interest.[4]

Regardless of the specific line and resonator combination, it is expected that symmetry disruption activates the electromagnetic coupling between the line and the resonator, as illustrated in Fig. 4.1

---

3 In the context of this chapter, a resonator-loaded line is a transmission line with a planar resonant element in close proximity to it (but not in contact with it), so that the resonator is within the region of influence of the electromagnetic field generated by the line. The exception is the structure considered in Section 4.4, devoted to dielectric constant measurements, where a microstrip transmission line is loaded with a shunt-connected step impedance resonator (SIR), in turn loaded with a complementary split ring resonator (CSRR).

4 For example, the complementary split ring resonator (CSRR) [4] exhibits a magnetic wall at its fundamental resonance. This particle is not excited by the electromagnetic field generated by a differential line pair operating in the differential mode (the one of interest in such lines), provided the structure is symmetric. Consequently, the line is transparent for that mode despite the presence of the loading resonator. However, the CSRR can be excited by a common-mode signal tuned to the fundamental resonance of the CSRR, since the line exhibits a magnetic wall for this mode. Therefore, CSRR-loaded differential lines are of interest for common-mode noise suppression, since CSRR excitation prevents the propagation of that mode in the line (obviously, in the vicinity of the fundamental resonance frequency) [5].

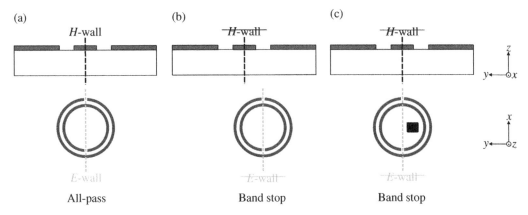

Figure 4.1 Sketch showing the effects of symmetry disruption in a CPW transmission line loaded with a SRR etched in the backside of the substrate. (a) Perfect symmetry, preventing line-to-resonator coupling; (b) symmetry cancelation by means of a lateral displacement of the SRR, thereby activating coupling; (c) symmetry perturbation by means of an asymmetric dielectric loading, also activating coupling. The CPW and the SRR (etched in the back-substrate side) are depicted in cross-sectional and bottom views, respectively, for better understanding.

(b) and (c), thereby generating a stopband (notched) response in the transmission coefficient. Obviously, the notch depth is determined by the coupling level, in turn related to the magnitude of the perturbation. Thus, symmetry disruption in these structures can be exploited for sensing purposes, as it is discussed in the next section.

4.2 Working Principle of Coupling-Modulation Sensors

The general working principle of coupling-modulation sensors is the variation of the transmission coefficient of a resonator-loaded line at a certain (operating) frequency, caused by a variation (or modulation) in the coupling level between the line and the resonator [6]. Symmetry truncation from the unperturbed state (perfect symmetry) in transmission lines loaded with a single symmetric resonator can be exploited for sensing, as anticipated in the previous section. Particularly, by considering a movable resonator, linear or angular displacements can be detected and measured (note that any relative lateral shift between the resonator and the line, or a rotation of the resonant element, represents a symmetry perturbation). Despite the fact that the canonical applications of these sensors are motion-related measurements, symmetry can also be altered by asymmetrically loading the resonant element, as Fig. 4.1(c) illustrates. Thus, coupling-modulation sensors implemented by means of transmission lines loaded with a symmetric resonator are also applicable to dielectric characterization.

The output variable in coupling-modulation sensors is, typically, the magnitude of the transmission coefficient of the line at the operating frequency, which should be set to the resonance frequency of the loading element or to a frequency in close proximity to it. However, since any variation in the level of electromagnetic coupling between the line and the resonator has the effect of modulating the amplitude of the signal generated at the output port in response to a harmonic signal injected to the input port, such amplitude level can be considered to be the output variable, as well.

There is, however, a different procedure to modify the magnitude of the transmission coefficient of a transmission line at a given frequency, also based on the modulation of the electromagnetic coupling between the line and a resonant element, but in this case without the need to consider a symmetric resonator. It merely consists of displacing an arbitrary (planar) resonant element at short

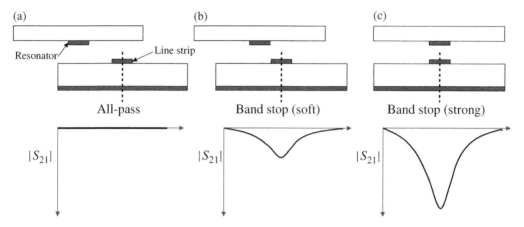

Figure 4.2 Effects of a displacement of an arbitrary resonator over a transmission line (microstrip), across the line axis. As the resonator approaches the line axis, coupling progressively intensifies (as revealed by the increasing notch depth). Cross-sectional schematic views for three situations: (a) resonator sufficiently separated from the line axis (all-pass response without line-to-resonator coupling), (b) resonator in close proximity to the line axis (stopband response with a soft notch), and (c) resonator on top of the line axis (stopband response with maximum coupling and the deepest notch).

distance across the line axis (Fig. 4.2). The single requirement is that the considered element is excited when it is sufficiently close to the line. In this case, electromagnetic coupling is modulated by the relative distance between the resonator and the line. For short-range displacements, i.e. comparable to the transverse dimensions of the considered line and resonant element, coupling-modulation sensors based on symmetry truncation are preferred over those based on line-to-resonator proximity. The reason is that the former sensors exhibit, in general, superior sensitivity. However, for high input dynamic ranges (long-range displacements), a single movable resonator is not useful, since the region of influence of the electromagnetic field generated by the line is very limited.

For long-range displacements, a possible solution is to consider a linear periodic chain of closely spaced resonant elements (Fig. 4.3). If we assume that the line is fed with a harmonic signal conveniently tuned, the displacement of the linear chain across the line axis is expected to generate an amplitude modulated (AM) signal at the output port of the line, with peaks, or dips, in the envelope function appearing each time a resonator is on top of the line axis. In this case, linear displacement measurement proceeds by cumulative pulse counting (peaks or dips), and the resolution is clearly given by the period of the chain. Moreover, the instantaneous velocity can be easily inferred from the time distance between adjacent peaks or dips. Actually, these coupling-modulation sensors based on periodic chains of resonant elements can also be applied to the measurement of angular displacements and velocities [7]. For that purpose, the resonators' chain must be circularly shaped along the edge of the so-called rotor, or circular rotatable element of the sensor. The working principle of these sensors based on resonators' chains (either devoted to linear or angular motion measurements) is very similar to the one of optical encoders,[5] i.e. pulse counting. For this reason, such sensors are designated as electromagnetic (linear or rotary) encoders [7].

---

5 In optical rotary encoders, a metallic disc (rotor) with apertures is situated between the optical source and the detector. When an aperture is located in the optical path between the source and the detector, a pulse is generated. Thus, the apertures in optical encoders are the equivalent of the resonant elements in electromagnetic encoders. The working principle of both encoder types is indeed very similar, the difference being related to the type of signal used for sensing, i.e. optical or electromagnetic. A figure of merit in optical rotary encoders, intimately related to sensor resolution, is the number of pulses per revolution, PPR. Optical encoders exhibiting thousands of pulses per revolution are commercially available. Electromagnetic rotary encoders cannot compete against optical encoders in terms of PPR, but they exhibit other advantageous aspects, as it will be discussed later in this chapter.

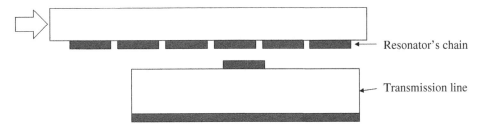

Figure 4.3 Sketch (cross-sectional view) of a transmission (microstrip) line with a movable chain of resonant elements on top of it. The time-varying coupling between the line and the resonator's chain, due to chain motion, periodically modulates the transmission coefficient of the line. By conveniently tuning the (carrier) frequency of a harmonic signal injected to the input port of the line, an amplitude modulated (AM) signal is generated at the output port of the line.

Electromagnetic encoders can also be implemented by means of chains of metallic strips [8] or patches [9], and even by means of chains of dielectric inclusions [10] (e.g. apertures on a dielectric substrate, or inclusions made of a high, or low, dielectric constant material as compared to the one of the host substrate). Like resonant elements, all these inclusions do also modulate the transmission coefficient of the line. However, several advantages, as compared to encoder implementation by means of chains of metallic resonant elements, are obtained when such inclusions' chains are used. In particular, it has been demonstrated that linear metallic strips transversally oriented to the chain axis constitute a very interesting approach for reducing the chain period, thereby improving the linear or angular resolution of the encoders [10]. Metallic patches are robust against aging effects and potential damage caused by mechanical wearing and friction, as far as their functionality is not jeopardized by the presence of chafing, scratches, and cracks (an aspect demonstrated in [11]), similarly to dielectric inclusions.

One of the main advantages of coupling-modulation sensors based on electromagnetic symmetry properties over frequency-variation or phase-variation sensors is their inherent robustness against the effects of cross sensitivities caused by environmental changes (e.g. temperature, pressure, and humidity). The reason is that variations in ambient conditions occur at a scale much larger than the typical dimensions of the sensors. Therefore, such variations are seen as common-mode stimuli by the sensors, not able to disrupt symmetry.[6] Although electromagnetic encoders are not based on symmetry properties, such sensors are also tolerant to variations in ambient conditions. Namely, despite the fact that such changes may slightly modify the resonance frequency of the resonators of the chain, their effect on the modulation index of the AM signal generated at the output port of the line, the key parameter in these sensors, is not expected to be significant.[7]

4.3 Displacement and Velocity Coupling-Modulation Sensors

This section reviews several implementations of coupling-modulation sensors for sensing linear and angular displacements and velocities.

---

6 Concerning their ability to suppress the adverse effects of ambient-related cross sensitivities, coupling-modulation sensors based on electromagnetic symmetry properties are similar to differential-mode sensors. However, as far as coupling-modulation sensors use a single sensing element (a resonator-loaded line), they cannot be considered to be true differential sensors.

7 In electromagnetic encoders, the most important aspect for sensor functionality is to keep the air gap (or vertical distance between the line and the inclusions chain), as well as any potential lateral shift of the chain, e.g. caused by mechanical vibrations, within certain tolerance limits. The tolerances of such sensors against air gap variations and lateral displacements are discussed later in this chapter.

4.3.1 One-Dimensional and Two-Dimensional Linear Displacement Sensors

The working principle and a preliminary proof-of-concept demonstrator of a coupling-modulation sensor based on symmetry disruption, and devoted to short-range one-dimensional displacement measurements, were first reported in [6]. The considered line was a CPW, whereas the symmetric resonator was a rectangular SRR with the longer side oriented in the direction of the line axis. In [12], the same idea was applied to the implementation of a two-dimensional displacement sensor proof-of-concept prototype. For that purpose, the CPW was bended 90°, and it was loaded with two pairs of SRRs, one pair (SRR$_{\Delta x}$ and SRR$_{\Delta y}$) for detecting and measuring lateral displacements (in both the x- and y-directions), and the other one (SRR$_{\pm x}$ and SRR$_{\pm y}$) for determining the motion direction (Fig. 4.4). The four resonators were tuned to different frequencies, and thereby exhibit different dimensions. When the displacement sensing resonators (SRR$_{\Delta x}$ or SRR$_{\Delta y}$) are misaligned with regard to the line axis, the corresponding resonance emerges as a notch in the transmission coefficient. The resonators designated as SRR$_{\pm x}$ and SRR$_{\pm y}$ allow us to discern whether the displacement proceeds in the $\pm x$- or $\pm y$-direction of motion, since the corresponding resonance frequency (notch) is "activated" only for positive displacement in the x- or y direction. For negative displacements, the resonators SRR$_{\pm x}$ and SRR$_{\pm y}$ are out of the region of influence of the CPW and are not coupled to it. Note that the resonators SRR$_{\pm x}$ and SRR$_{\pm y}$ act as flags that are activated for positive displacements in their respective directions.

In the proof-of-concept demonstrator reported in [12], several samples with SRRs etched on the back-substrate side of the CPW but at different locations (emulating different displacements), were

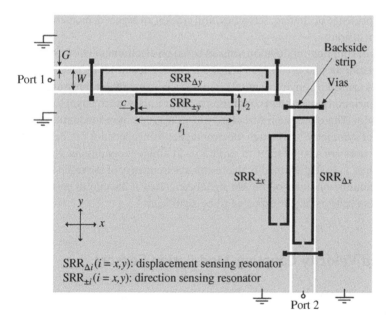

Figure 4.4 Topology of the two-dimensional coupling-modulation displacement sensor based on a bended CPW transmission line loaded with two pairs of SRRs. CPW dimensions are W = 1.67 mm and G = 0.2 mm. The dimensions of the SRRs are: l_1(SRR$_{\Delta x}$) = 9.95 mm, l_1(SRR$_{\pm x}$) = 7.05 mm, l_1(SRR$_{\Delta y}$) = 13.4 mm, l_1(SRR$_{\pm y}$) = 7.8 mm, l_2 = 1.67 mm, and c = 0.2 mm. The considered substrate is the *Rogers RO3010* with relative permittivity ε_r = 10.2, thickness h = 127 μm, and loss tangent tanδ = 0.0023. *Source:* Reprinted with permission from [12]; copyright 2012 MDPI.

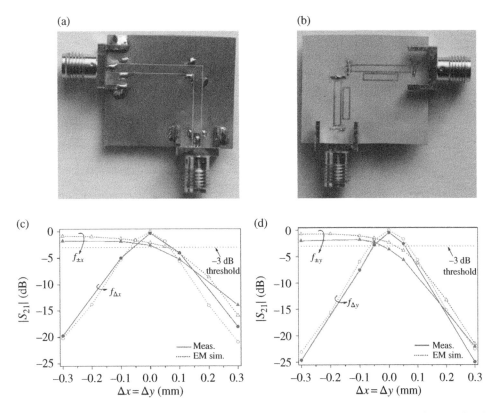

Figure 4.5 Photograph of the top (a) and bottom (b) faces of the proof-of-concept coupling-modulation two-dimensional displacement sensor for the aligned position (unperturbed state), and notch magnitude of the transmission coefficient, S_{21}, at the indicated frequencies for displacement in the $x = y$ direction. *Source:* Reprinted with permission from [12]; copyright 2012 MDPI. The results for x- and y-axis position sensing are indicated in (c) and (d), respectively.

actually fabricated.[8] The device with the SRRs etched in the positions corresponding to the unperturbed state is depicted in Fig. 4.5. The figure includes the notch magnitudes of the transmission coefficient at the frequencies of the different resonators for a linear displacement satisfying $x = y$. For positive displacements, the $SRR_{\pm x}$ and $SRR_{\pm y}$ resonators are activated as revealed by a clear increase in the notch at $f_{\pm x}$ and $f_{\pm y}$, whereas the threshold level (-3 dB) is not exceeded for negative displacements (indicating that the shift is in the negative direction).

The linearity in the structure of Fig. 4.5 is good, but the reported dynamic range is limited (actually the range of displacements in the x- and y-directions in Fig. 4.5 is restricted to the region where the response of the sensor is roughly linear). The dynamic range and linearity can be improved by increasing the transverse dimensions of the CPW and resonant element, and by modifying the shape of the resonator. Namely, by considering rectangular resonators, as those of Fig. 4.5, a quasi-saturation effect arises when the SRR intercepts the slot of the CPW. Beyond this point, as demonstrated by simulation and measurement results in [6, 12], the displacement sensor goes into saturation, and only a minor increase in the notch magnitude can be observed. This limitation

---

8 In [12], sensor implementation by etching the SRRs on an independent movable substrate, with regard to the line, was not considered.

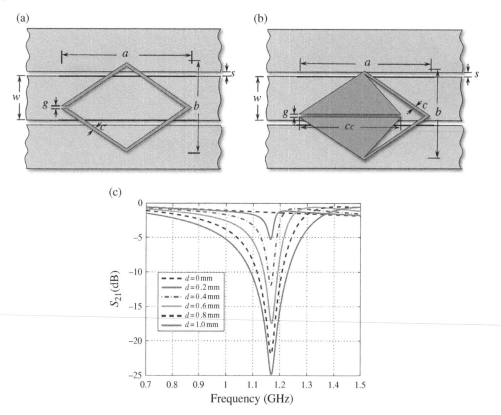

Figure 4.6 Topology of the coupling-modulation one-dimensional displacement sensor based on a diamond-shaped SRR (a) and modified topology to make the notch frequency invariant to SRR displacements (b). The simulated frequency responses of the sensor with modified SRR topology for different displacements, *d*, of the resonant element are depicted in (c). The considered substrate is the *Rogers RO3010* with dielectric constant ε_r = 10.2 and thickness *h* = 0.127 mm. CPW dimensions are *w* = 3.1 mm and *s* = 0.3 mm, corresponding to a 50-Ω characteristic impedance. SRR dimensions are *a* = 10 mm, *b* = 6 mm, and *c* = *g* = 0.2 mm. *Source:* Reprinted with permission from [13]; copyright 2013 IEEE.

can be addressed by considering a diamond-shape geometry [Fig. 4.6(a)], since for this SRR geometry, the magnetic flux through the SRR surface is maximized when the SRR horizontal diagonal is aligned with the CPW slot. Thus, in contrast to the rectangular SRR geometry, the angle between the sides of the diamond-shaped SRR and the slots of the CPW makes the spatial dynamic range of the diamond-shaped sensor dependent on the width of the CPW's signal strip (such dynamic range is approximately half the width of the signal strip). Therefore, a diamond-shaped SRR coupled to a wide CPW can be used to increase the dynamic range and linearity of the displacement sensor, as it was pointed out in [13].

It should be mentioned, however, that there is another limitative aspect in the sensor structure of Fig. 4.5, i.e. the resonance frequency, or notch position, varies slightly with the lateral displacement of the resonant element. This also occurs in the diamond-shaped topology of Fig. 4.6(a). To solve this issue, the authors in [13] proposed a modification of the diamond-shape SRR geometry by including triangular patches [see Fig. 4.6(b)]. By this means, the SRR inductance is reduced, whereas its capacitance is increased, and the resulting resonator is less sensitive to the effects of the displacement on the resonance frequency [13]. Figure 4.6(c) depicts the simulated frequency response of the sensor for different displacements, where it can be appreciated that the notch

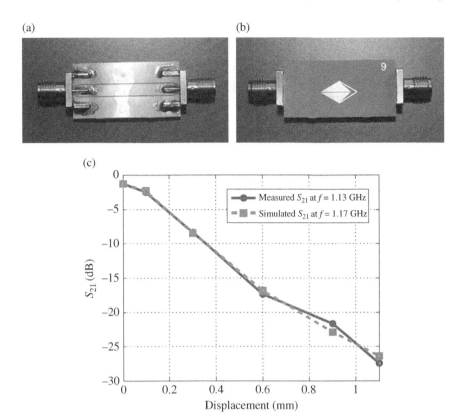

Figure 4.7 Photograph of the top (a) and bottom (b) faces of the fabricated proof-of-concept coupling-modulation one-dimensional displacement sensor based on a diamond-shaped SRR, for the aligned position (unperturbed state), and magnitude of the transmission coefficient measured at the notch frequency (1.13 GHz) (c). The response inferred from simulation at 1.17 GHz (see Fig. 4.6) is also depicted (the notch position varies slightly between simulation and measurement due to fabrication-related tolerances). *Source:* Reprinted with permission from [13]; copyright 2013 IEEE.

position does not depend on the SRR displacement. Such notch position determines the frequency of the feeding signal, necessary for measuring purposes. Figure 4.7 shows the photograph of the fabricated device with the SRR symmetrically etched in the back-substrate side,[9] as well as the sensor response (notch magnitude as a function of the SRR position). The improved dynamic range, as compared to the sensor of Fig. 4.5, and the good linearity over the considered input dynamic range are apparent.

The last example reported in this section is a two-dimensional displacement sensor operating in reflection [14]. The structure consists of a pair of orthogonally oriented microstrip lines each one loaded with a SRR at the open end (see Fig. 4.8). Unlike the sensor proof-of-concept prototypes of Figs. 4.5 and 4.7, in this sensor, the SRRs are etched in a movable substrate. If the relative position between the SRRs and the lines is the one depicted in Fig. 4.8, with the symmetry planes of the lines perfectly aligned with those of the corresponding SRRs, electromagnetic coupling between the lines and the resonators is not possible. Consequently, the injected signals at both lines are totally reflected for such relative position between the line and the SRRs. By contrast, any lateral shift of the SRRs with regard to the corresponding line truncates the symmetry and activates

---

9 Like the structure of Fig. 4.5, in the sensor proof-of-concept based on the diamond-shaped SRR, a set of CPW lines with the SRR etched in the back-substrate side at different lateral positions (rather than etched on a movable substrate) emulates the displacement.

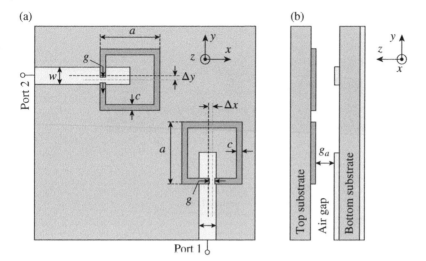

Figure 4.8 Top (a) and side (b) views of the coupling-modulation reflective-mode two-dimensional displacement sensor based on a pair of SRR-loaded microstrip lines. *Source:* Reprinted with permission from [14]; copyright 2014 IET.

line-to-resonator coupling. The consequence is that part of the injected power at the input port of the lines is absorbed, due to SRR losses, and part of it is radiated, provided the injected signal is tuned to the resonance frequency of the SRRs.[10] Therefore, the reflection coefficients at both ports should exhibit a notch when the symmetry condition is not perfectly fulfilled, with notch depth determined by the magnitude of the lateral displacement of the SRRs.

The photograph of the fabricated device is shown in Fig. 4.9. The figure depicts also the reflection coefficient measured from any of the ports for different lateral displacements (seen from the corresponding port). One important feature of the proposed sensor is that displacement affects only the depth of the notch, and leaves the resonance frequency roughly unaltered. As mentioned before, the invariability of the notch frequency is important since it enables the proposed sensor to operate at a fixed frequency. Figure 4.9 includes the simulated and measured reflection coefficient at a fixed frequency of 4.253 GHz against Δx, while $\Delta y = 0$ mm. The figure clearly shows that the movement in the x-direction can be sensed from $|S_{11}|$, whereas $|S_{22}|$ remains unaffected, thereby indicating the perfect alignment in the y-direction. Thus, any displacement in the x-direction has no effect on the depth of the notch in $|S_{22}|$ (nor displacement in the y-direction onto $|S_{11}|$). Thus, misalignment in the x- and y-directions can be independently sensed from the depth of the notches in $|S_{11}|$ and $|S_{22}|$, respectively. With the sensor configuration of Fig. 4.8, the direction of motion cannot be determined. Nevertheless, additional resonant elements, tuned at different frequencies, can be added to the movable substrate, in order to distinguish the motion direction for each axis, similarly to the strategy of the sensor of Fig. 4.4.

The input dynamic range of the sensor of Fig. 4.9 is comparable to that of the sensor of Fig. 4.7, and both are superior to the input dynamic range of the sensor of Fig. 4.5. However, the linearity and sensitivity for small displacements is worst in the sensor of Fig. 4.9, according to the responses depicted in Figs. 4.5, 4.7, and 4.9. Nevertheless, it should be taken into account that in the sensor prototype of Fig. 4.9, an actual movable structure necessarily separated from the static part to avoid friction has been considered, contrarily to the proof-of-concept demonstrators of Figs. 4.5 and 4.7.

---

10 In this sensor, it is not necessary to tune the SRRs to different frequencies, i.e. both SRRs can be identical.

(a)

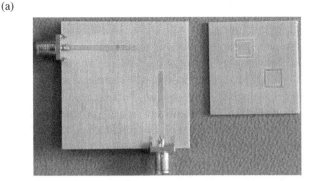

(b)

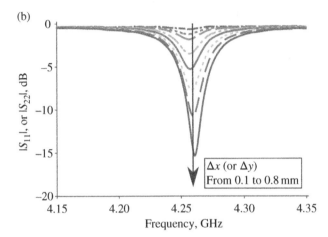

(c)

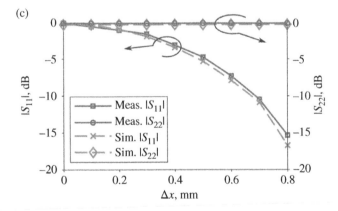

Figure 4.9 (a) Photograph of the sensor of Fig. 4.8; (b) measured $|S_{11}|$ (or $|S_{22}|$) for different values of displacement Δ_x (or Δ_y) from 0.1 to 0.8 mm in steps of 0.1 mm; (c) comparison between measured and simulated $|S_{11}|$ and $|S_{22}|$ at a fixed frequency of f = 4.253 GHz as a function of Δx, for Δy = 0 mm. Owing to the symmetry of the proposed sensor, identical comparison is valid between $|S_{11}|$ and $|S_{22}|$ against Δy. Dimensions (in reference to Fig. 4.8) are: w = 1.84 mm, a = 7 mm, g = 0.5 mm, and c = 0.5 mm. The nominal air gap between the two substrates is g_a = 0.76 mm. The structure was fabricated on the *Rogers RO4003* substrate (for both the static and the movable parts) with thickness h = 0.81 mm, dielectric constant ε_r = 3.38, and loss tangent tanδ = 0.0022. The control of the relative position between the static and the movable parts of the sensor, including the air gap separation, was carried out by means of three pairs of micrometer actuators. *Source:* Reprinted with permission from [14]; copyright 2014 IET.

4.3.2 Angular Displacement and Velocity Sensors

It was demonstrated in [6] that a CPW transmission line loaded with a movable SRR can be used for sensing angular displacements through symmetry disruption. For that purpose, the axis of the rotor (the movable part) should be contained in the symmetry planes of the SRR and line. Rotating the SRR from the orientation corresponding to perfect alignment between the symmetry planes of the line and resonator activates the electromagnetic coupling between both elements, and the rotation angle determines the notch depth of the transmission coefficient, the output variable. In [15], a solution to minimize the dependence of the notch frequency with the rotation angle, consisting of replacing the rectangular SRR with a horn-shaped SRR, was proposed. Nevertheless, the angular sensors reported in [6, 15] exhibit a very limited input dynamic range (of few degrees), not useful in many applications. The main reason explaining such limited capability for the measurement of large angles is the specific shape of the SRR (either rectangular or horn-shaped) and CPW transmission line (uniform in the implementations reported in [6, 15]).

In the next section, it is shown that the input dynamic range can be significantly improved by using a circularly shaped resonant element with its axis perfectly aligned with the axis of the rotor (axial configuration). In particular, the considered resonator is an electric-LC (ELC) resonator [16], a bisymmetric particle that exhibits both an electric wall and a magnetic wall (orthogonally oriented) at the fundamental resonance frequency. By using such resonant element, etched in the rotor, and a nonuniform (circularly shaped) CPW transmission line (stator), the input dynamic range can be extended up to 90°, as demonstrated in [17]. Nevertheless, the input dynamic range of such axial configuration can be by far extended by considering the so-called edge configuration, where the rotor is made of a circular dielectric disc with a circular chain of hundreds of resonant elements etched along its edge (microwave rotary encoder). Both configurations are discussed in detail next. The reported prototype sensors are useful not only for measuring angular displacements, but also for providing rotation speeds.

4.3.2.1 Axial Configuration and Analysis

Since the axial configuration is based on a rotor implemented by means of an ELC resonator, let us first analyze this particle, in particular when it is coupled to a CPW transmission line. As mentioned before, the ELC of the rotor is circular in order to obtain a high dynamic range with as much linearity as possible. Nevertheless, the qualitative behavior of the particle does not depend on the specific shape, and, for this reason, it can be explained based on a square-shaped ELC geometry (as done in [17]), depicted in Fig. 4.10. The orthogonally oriented magnetic and electric walls of the particle at the fundamental resonance frequency are indicated in the figure. At such resonance, the currents in both loops flow in opposite directions, one clockwise and the other one counterclockwise. Therefore, the ELC cannot be excited by means of a uniform time-varying magnetic field applied in the direction orthogonal to the plane of the particle (the usual driving mechanism in SRRs). However, the ELC resonator can be electrically excited through application of a uniform electric field in the direction orthogonal to the electric wall. It can also be driven by application of counter magnetic fields in both loops of the particle. Consequently, the ELC resonator is a very useful resonant element to control the transmission coefficient of a CPW transmission line at the ELC fundamental resonance through the relative angular orientation between both elements, the rotor (ELC) and the stator (CPW). From these words, it follows that the ELC/CPW combination is very convenient (although not exclusive) for the implementation of coupling-modulation angular displacement and velocity sensors with high dynamic range.

To gain further insight on the orientation-modulated coupling between the CPW and the ELC resonator, let us consider the two relative orientations depicted in Fig. 4.11. Both orientations

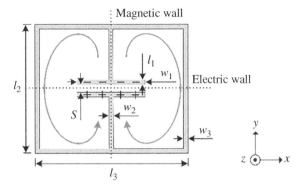

Figure 4.10 Typical topology of a square-shaped ELC resonator. The electric and magnetic walls at the fundamental resonance, as well as a sketch of the distribution of charges and currents, are indicated. *Source:* Reprinted with permission from [17]; copyright 2013 IEEE.

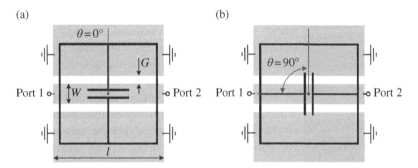

Figure 4.11 CPW (in gray) symmetrically loaded with an ELC resonator (in black) etched in the backside of the substrate. The orientation between the CPW and the resonator is determined by the rotation angle, θ. (a) Line axis perfectly aligned with the resonator electric wall ($\theta = 0°$); (b) line axis perfectly aligned with the resonator magnetic wall ($\theta = 90°$). *Source:* Reprinted with permission from [17]; copyright 2013 IEEE.

exhibit perfect symmetry, but in one case, the line axis is aligned with the electric wall of the ELC resonator [Fig. 4.11(a)], and in the other case with the magnetic wall [Fig. 4.11(b)]. For the fundamental CPW mode (i.e. the even mode), the symmetry plane of the line along its axis is a magnetic wall. Hence, the ELC resonator cannot be excited if its electric wall is aligned with the line axis [configuration of Fig 4.11(a)], and the line is transparent. In this situation, perfect cancellation of the axial magnetic field lines through the individual loops occurs (i.e. there is not a net magnetic flux in each ELC resonator loop). Conversely, the resonator is magnetically driven if its magnetic wall is aligned with the line axis [Fig. 4.11(b)], thereby generating a notch in the transmission coefficient. In this case, a net magnetic flux generated by the line penetrates axially and oppositely each loop. This is a sufficient condition for ELC resonator excitation. Obviously, for intermediate orientations, specified with the angle θ (see Fig. 4.11), the line is able to drive the resonator, but the coupling level decreases as the angle between the line and the ELC resonator approaches $\theta = 0°$. It is thus expected that the magnitude and bandwidth of the notch increase as the relative orientation approaches the angle $\theta = 90°$, corresponding to maximum magnetic coupling.

The previous behavior can be explained in terms of the circuit model of the ELC-loaded CPW transmission line, depicted in Fig. 4.12(a). The ELC resonator is described by the inductances

(a)

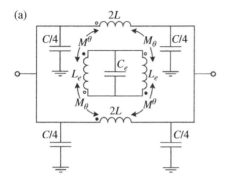

(b)

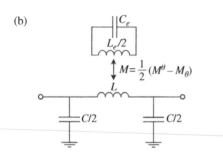

(c)

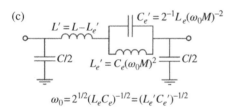

$$\omega_0 = 2^{1/2}(L_e C_e)^{-1/2} = (L_e' C_e')^{-1/2}$$

Figure 4.12 (a) Lossless equivalent circuit model of a CPW loaded with an ELC resonator, including the different magnetic coupling mechanisms, accounted for through the mutual inductances M_θ and M^θ; (b) simplified circuit model; (c) transformed simplified circuit model. *Source:* Reprinted with permission from [17]; copyright 2013 IEEE.

L_e and the capacitance C_e, modeling the inductive loops and the capacitive gap, respectively. The CPW is modeled by its inductance L and capacitance C, and it is divided into two identical halves for convenience. Finally, each CPW half is magnetically coupled to each loop through the mutual inductances M_θ and M^θ, both being dependent on θ (different dots are used to distinguish the magnetic coupling sign associated to each of the halves). Therefore, the frequency response of the circuit of Fig. 4.12(a) directly depends on the angle θ. When $\theta = 0°$, due to symmetry, $M_\theta = M^\theta \neq 0$, which means that the currents flowing in the CPW induce a pair of equal and antiphase voltages in the loops; that is, no net voltage is induced in the resonator due to an absolute cancellation. In other words, there is not a net magnetic coupling, the resonator cannot be magnetically driven, and the resulting model is the one of a conventional transmission line section. On the other hand, as θ increases, the magnetic coupling is complementarily distributed; M^θ increases at the expense of a decrease in M_θ. Hence, for $\theta > 0°$, a net induced voltage arises and the line is indeed capable of magnetically exciting the resonator. The larger the angle, the higher the induced voltage. Thus, at the upper limit, $\theta = 90°$, M^θ is maximum while M_θ completely vanishes, and the resonator is expected to be tightly coupled to the line.

The physical understanding of magnetic coupling modulation as a function of the relative orientation between the CPW line and the ELC resonator is well illustrated by this model. An equivalent and simplified circuit model is the one depicted in Fig. 4.12(b), where an effective mutual inductance M is defined. Such a model is equivalent to that of an SRR magnetically coupled to a CPW transmission line, which can be transformed to the circuit model of Fig. 4.12(c) [18].

From such model, it is clear that a transmission zero at ω_0 arises, as long as $M \neq 0$. Moreover, the rejection bandwidth broadens with M (the reason is that the susceptance slope of the parallel resonator decreases as the ratio L_e'/C_e' increases).

Let us consider for circuit validation a circular-shaped topology for both the resonator and the line (see Fig. 4.13). This structure is preferred for sensor implementation, as indicated previously. Moreover, the mutual inductance as a function of the orientation of the ELC resonator can be easily obtained. The large (and circular) patch capacitor is justified, as it contributes to reduce the electrical size of the ELC at resonance.[11] The frequency responses inferred by electromagnetic simulation for different resonator orientations are plotted in Fig. 4.14 (for model validation, it has been considered that the ELC resonator is etched in the backside of the substrate). It can be seen

---

11 The model of Fig. 4.12 accurately predicts the behavior of the ELC resonator over a wide band in the vicinity of the fundamental resonance frequency for electrically small particles.

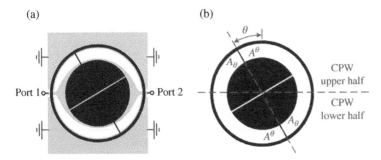

Figure 4.13 (a) Layout of a circular-shaped ELC-loaded CPW transmission line; (b) illustration of the areas A_θ and A^θ (for a specific angle of 30°). The considered substrate is *Rogers RO3010* with thickness $h = 1.27$ mm and dielectric constant $\varepsilon_r = 11.2$. The dimensions are: for the line, W and G are tapered such that the characteristic impedance is 50 Ω; for the ELC resonator: mean radius $r_0 = 8.05$ mm, capacitor outer radius $r_1 = 5.6$ mm, and, in reference to Fig. 4.10, $w_2 = s = 0.2$ mm, and $w_3 = 0.5$ mm. *Source:* Reprinted with permission from [17]; copyright 2013 IEEE.

Figure 4.14 (a) Magnitude and (b) phase of the reflection and transmission coefficients given by the lossless electromagnetic simulation for the structure of Fig. 4.13 and by the circuit simulation for the models of Fig. 4.12. The circuit parameters are those given in Table 4.1. *Source:* Reprinted with permission from [17]; copyright 2013 IEEE.

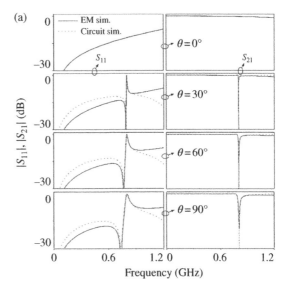

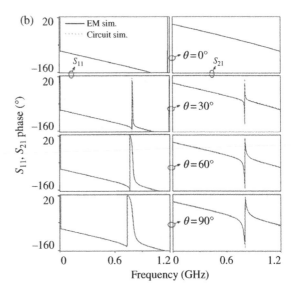

Table 4.1 Extracted lumped element values of the circuit of Fig. 4.12(b) for the structure of Fig. 4.13.

θ (°)	C (pF)	L (nH)	C_e (pF)	L_e (nH)	M (nH)
30	5.86	5.95	3.09	25.6	0.94
60	5.73	6.22	3.09	25.6	1.91
90	5.59	6.40	3.07	25.6	2.72

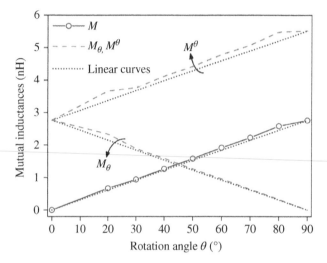

Figure 4.15 Estimated mutual inductances of the models of Fig. 4.12 for the structure of Fig. 4.13 versus the rotation angle. The ideal linear curves are also shown for comparison purposes. *Source:* Reprinted with permission from [17]; copyright 2013 IEEE.

that the transmission zero frequency (notch position) is situated at $f_0 = \omega_0/2\pi = 0.80$ GHz, and it does not vary with the angle θ.[12] The extracted parameters for different orientations, inferred from the method reported in [1, 18], are listed in Table 4.1. With the exception of the mutual inductance, all the circuit parameters are roughly invariant with the rotation angle. The reflection and transmission coefficients inferred from circuit simulation with the extracted parameters are also depicted in Fig. 4.14. The good agreement between the circuit and electromagnetic simulations in the vicinity of the notch frequency validates the proposed circuits with arbitrary ELC orientation.

To gain more insight on the effective mutual inductance, Fig. 4.15 shows the extracted value of M (as well as M_θ and M^θ) for the structure of Fig. 4.13 as a function of the rotation angle. Note that M varies continuously and quasi-linearly (the ideal curve is also plotted) as a consequence of the circular shape of the line and ELC resonator. This phenomenology can be explained from the definition of the mutual inductance between two circuits

---

12 According to the circuit model, regardless of the value of M, the notch frequency f_0 is invariant with the ELC orientation. Indeed, in the structure of Fig. 4.13, the notch frequency is roughly constant with the angle (the maximum frequency deviation is 0.48%) and this is because the parameters L_e and C_e are nearly invariant with the angular orientation. Nevertheless, it should be clarified that, in the proposed circuit model, C_e is the effective capacitance of the resonator with the presence of the CPW. Therefore, electric interaction between the resonator and the CPW may actually exist and change the effective capacitances of them.

$$M_{ij} = \frac{\Phi_{ij}}{I_j} \tag{4.1}$$

where Φ_{ij} is the magnetic flux through circuit i generated by a current I_j flowing on a circuit j. The magnetic flux through any surface S can be computed as the surface integral

$$\Phi_{ij} = \int_S B_n \cdot ds \tag{4.2}$$

where B_n is the normal component to S of the magnetic flux density generated on the circuit j. If B_n is assumed to be uniform over the entire surface, then (4.2) yields

$$\Phi_{ij} = B_n \cdot A_s \tag{4.3}$$

the magnetic flux being proportional to the area A_s of the surface over which the flux is being quantified. In our case, we are interested in the magnetic flux generated by the CPW linking either loop of the resonator. According to the considered circular resonator (Fig. 4.13), the area of each loop is $A_l = \pi[(r_0 - w_3/2)^2 - r_1^2]/2$. If A_s is chosen to be that portion of the loop area that lies beneath half of the CPW (as illustrated in Fig. 4.13), the resulting complementary areas are then

$$A_\theta = A_l \left(\frac{1}{2} - \frac{\theta}{\pi} \right) = A_l - A^\theta \tag{4.4a}$$

$$A^\theta = A_l \left(\frac{1}{2} + \frac{\theta}{\pi} \right) = A_l - A_\theta \tag{4.4b}$$

which are linearly dependent on the angle θ.

Regarding the magnetic flux density generated by a CPW, it is not spatially uniform. However, for an electrically small and uniform (i.e. the per-unit length inductance is uniform) CPW transmission line section, like the one shown in Fig. 4.11, it seems reasonable to assume a longitudinally uniform magnetic flux density (i.e. a uniform per-unit length magnetic flux density). Analogously, for a small circular-shaped CPW as the considered one, the magnetic flux density can be assumed to be radially uniform with θ.[13] Therefore, assuming a uniform B_n over the areas (4.4a) and (4.4b) along θ, the mutual inductance (4.1) can be written as

$$M_\theta = \frac{B_n}{I_{2L}} A_l \left(\frac{1}{2} - \frac{\theta}{\pi} \right) \tag{4.5a}$$

$$M^\theta = \frac{B_n}{I_{2L}} A_l \left(\frac{1}{2} + \frac{\theta}{\pi} \right) \tag{4.5b}$$

where I_{2L} is the current through one of the halves of the CPW. With regard to the effective mutual inductance, since the effective area is found to be

$$A = A^\theta - A_\theta = A_l \frac{2\theta}{\pi} \tag{4.6}$$

and the current flowing on the circuit in this case is

$$I_L = 2I_{2L} \tag{4.7}$$

---

13 Although in a nonuniform CPW the per-unit length inductance is not strictly uniform, this approximation is reasonable.

(a)

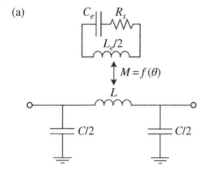

(b)

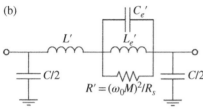

(c)

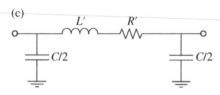

Figure 4.16 (a) Equivalent circuit model of a CPW loaded with an ELC resonator, including losses in the resonator; (b) transformed circuit model; and (c) transformed circuit model at the notch frequency. *Source:* Reprinted with permission from [17]; copyright 2013 IEEE.

the resulting effective mutual inductance is

$$M = \frac{B_n}{I_L} A_l \frac{2\theta}{\pi} = \frac{1}{2}\left(M^\theta - M_\theta\right) \tag{4.8}$$

Equation (4.8) indicates that, to a first-order approximation, for a circularly shaped ELC-loaded CPW, the net mutual inductance modeling the magnetic coupling between the line and the resonator depends linearly on the loading angle θ. The reason is the linear dependence of the net magnetic flux penetrating the loops. Note that the assumed uniform magnetic flux density, B_n, in (4.5) and (4.8) depends on the spatial location of the resonator, and it decreases with the physical distance between the coupled elements (given by the substrate thickness, h, since the ELC is etched in the backside of the substrate in the present study). In the limiting case $\theta = \pi/2$, and $h \to 0$, (4.8) can be expressed as $M = L \cdot f$ [19], where f is the fraction of the loops that lies beneath the slots. Since in our geometry $f = 1$, in this hypothetic case, the magnetic flux linkage would be maximum, i.e. $M \approx L$.

As far as the typical output variable in the sensors under study is the notch depth, it is convenient to analyze the effect of losses, since losses prevent the attenuation from going to infinity.[14] In addition, losses round off the stopband characteristic [20], thereby enhancing the rejection bandwidth around the notch frequency. The loss mechanisms in CPWs loaded with planar resonators are ohmic (conductor) losses, dielectric losses in the substrate, and radiation losses. Losses in CPWs use to be negligible; conversely, losses in planar resonant elements may be critical, and hence, dominate the power loss.

Losses in an ELC-loaded CPW can thus be modeled, to a good approximation, by means of a series resistance R_s representing the losses in the resonator, as illustrated in Fig. 4.16(a). This is equivalent to introduce a parallel resistance R' in the transformed model, as depicted in Fig. 4.16(b) [1, 21]. In the absence of losses, R_s vanishes while R' is infinite. It is also important to emphasize that the equivalent resistance R' indicates that minimum losses in the equivalent circuit model are achieved by maximizing M, and obviously, by minimizing R_s. On one hand, M is proportional to the area of the loops A_l according to (4.8), and it is also influenced by the distance between the coupled elements. On the other hand, in general, the larger the resonator, the higher the radiation losses, but the smaller the conductor losses (there is a trade-off) [20]. Since commonly the radiation efficiency of a small loop antenna is small [22], the conductor loss of the ELC resonator is likely to be larger than the radiation loss. To sum up, what is to be expected is that the notch depth at a given frequency increases with the electrical size of the resonator and with the proximity between the line and the resonator.

The series resistance R_s can be extracted by curve fitting the lossy circuit simulation to the lossy electromagnetic simulation. The equivalent parallel resistance can then be estimated from the transformation equation given in Fig. 4.16(b) by using the extracted mutual inductance of Fig. 4.15.

---

14 The notch magnitude in the transmission coefficient of an ideally lossless notch filter is infinite.

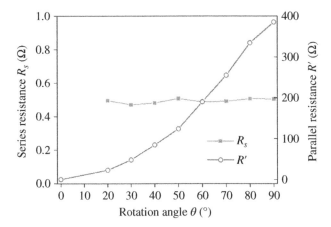

Figure 4.17 Extracted series and parallel resistance of the circuit models of Fig. 4.16 for the structure of Fig. 4.13 versus the rotation angle. *Source:* Reprinted with permission from [17]; copyright 2013 IEEE.

The extracted resistances for the structure of Fig. 4.13 are shown in Fig. 4.17 (for the lossy simulations, the metallization layer is considered to be copper, with conductivity $\sigma = 5.8 \cdot 10^7$ S/m and thickness $t = 35\,\mu$m, and the loss tangent of the substrate material is set to $\tan\delta = 0.0023$). As can be seen, R_s is roughly constant while R' increases strongly with the rotation angle due to its quadratic dependence on M, that also increases with the angle. Figure 4.18 depicts the transmission coefficient of the lossy electromagnetic and circuit simulations. As expected, the higher the parallel resistance R', the deeper the notch magnitude. This is intimately related to the fact that an increase in M simultaneously broadens the stopband bandwidth (because L'_e/C'_e increases). Therefore, the higher the M, the higher the stopband attenuation (depth and bandwidth). The previous results are in accordance with the estimated maximum stopband attenuation in bandstop filters designed from the low-pass prototypes, since the attenuation increases with the stopband bandwidth and the Q-factor of the resonators [20].

The notch magnitudes observed in the transmission coefficients shown in Fig. 4.18 are plotted in Fig. 4.19. The magnitude of the notch in logarithmic scale is roughly linearly dependent with the

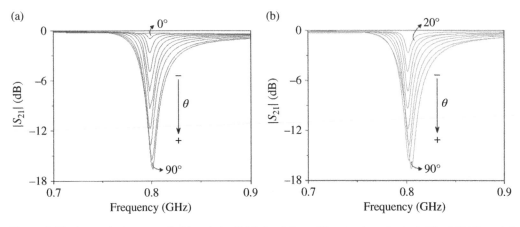

Figure 4.18 Lossy electromagnetic (a) and circuit (b) simulation of the structure shown in Fig. 4.13. The notch depth and the rejection bandwidth increase with the rotation angle θ from 0° to 90°, incremented in steps of 10°. The circuit parameters for some of the orientations are given in Table 4.1, and $R_s = 0.49\,\Omega$. *Source:* Reprinted with permission from [17]; copyright 2013 IEEE.

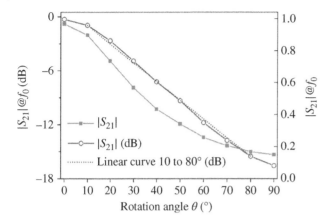

Figure 4.19 Notch magnitude at the notch frequency f_0 extracted from the lossy simulations of Fig. 4.18. *Source:* Reprinted with permission from [17]; copyright 2013 IEEE.

rotation angle (i.e. the sensitivity is nearly constant). Indeed, from 10° to 80°, the linearity is quite good (for comparison purposes, the ideal linear curve is also shown). To gain insight on the linearity, let us now focus on the circuit model at the notch frequency. At that frequency, the resonator L'_e-C'_e opens and the resulting equivalent circuit is the one of a conventional transmission line with R' in the series branch, as depicted in Fig. 4.16(c). The series inductance is $L' \approx L$ since L'_e is typically negligible as compared with L. Given that the line parameters L and C are roughly constant with the rotation, the sensitivity is then an exclusive function of R', which depends on θ through M. By the calculation of the transmission ABCD matrix of that circuit model, the analytical transmission coefficient is found to be [23]

$$S_{21} = \frac{2}{(Z_L + R')(3Y_C + 1/Z_0) + 3} \tag{4.9}$$

where Z_L is the impedance of the line inductance, Y_C is the admittance of the line capacitance, and Z_0 is the port impedance. It has been verified that the analytical solution of (4.9) coincides with the value obtained through circuit simulation. Despite the fact that the variation of the notch magnitude (in logarithmic scale) with θ may be obtained in closed form, the resulting expression is cumbersome. Nevertheless, in the light of numerical solutions (Fig. 4.19) and experimental data (to be shown later), the sensitivity is approximately constant.

The previous analysis has been conducted by considering a nonuniform circular CPW and a circular-shaped ELC resonator, etched in the backside of the substrate. Nevertheless, the general ideas in regard to the dependence of the notch depth with the rotation angle θ, derived from such analysis, can be applied to a real system with the sensing resonant element etched on an independent movable substrate (rotor). In the next two sections, two different prototypes of coupling-modulation angular displacement and velocity sensors based on the axial configuration are reported. The first prototype is indeed a CPW/ELC-based device similar to the one of Fig. 4.13(a) [17]. In the second prototype, the stator is based on a circular microstrip line configuration, the rotor being identical to the one of the first prototype [24]. It should also be mentioned that in [25], an angular velocity sensor similar to the one of [17], but replacing the ELC resonator of the rotor with an S-shaped split ring resonator (S-SRR) [26–28], was reported. Despite the fact that the S-SRR does not exhibit any symmetry plane, the excitation of such resonant particle by means of a CPW transmission line is very similar to the excitation of an ELC resonator [25]. However, the electrical size of the S-SRR is much smaller than the one of the ELC resonator. Thus, a performance comparable to the one of the

ELC-based sensor, but with significantly smaller rotor size (provided the operation frequency is the same), is achievable by means of an S-SRR (see [25] for further details).

4.3.2.1.1 Coplanar Waveguide (CPW) Stator

This section is devoted to the implementation of an angular displacement and velocity sensor based on an ELC-loaded CPW structure similar to the one considered in the previous analysis. For angle measurements, it was previously indicated that the sensor is able to measure angles in the range $0°$ $< \theta < 90°$ (input dynamic range). The performance of the sensor is mainly determined by two parameters, the linearity and the sensitivity, the latter being related to the output dynamic range. It was shown before that good linearity is achievable by considering circularly shaped topologies for both sensor elements, the stator (CPW) and the rotor (ELC resonator). The notch depth for the $90°$ orientation (i.e. the one providing maximum coupling between the line and the resonator) determines the output dynamic range. The smaller the R_s and the higher the M values, the deeper the notch. On one hand, R_s decreases and M increases with the electrical size of the particle. For this main reason, it is convenient to design the ELC resonator with a small capacitance [17]. On the other hand, the smaller the distance between the CPW and the resonator, the higher the M. However, for very tiny distances, the increase in the output dynamic range may be at the expense of some insertion loss for $\theta = 0°$, caused by a strong influence of the resonator to the line parameters, L and C (this produces line mismatching). Thus, a trade-off is necessary, in order to obtain a good balance between the insertion loss at $0°$ and the dynamic range.

The first prototype of this type of axial sensor was first reported in [29] (the layout is depicted in Fig. 4.20). Since access lines were added for connectors soldering, vias and backside strips were introduced at each side of the resonator in order to prevent mode conversion (from the fundamental CPW mode to the slot mode) for nonsymmetric orientations (i.e. $0° < \theta < 90°$). In [29], the CPW and the ELC resonator were etched on different substrates, parallel oriented with an air layer in between, and the resonator substrate (rotor) was suspended (by a suspending substrate attached to the top of the CPW substrate). The rotor was manually moved, so that the angular orientation was inferred by the use of a graded angular grid. This experimental setup offered robustness and reliability to the measurement, and the approach was validated successfully since the sensor

Figure 4.20 Layout of the angular displacement sensor reported in [17, 29] for an angle of $\theta = 30°$. The dimensions are those given in the caption of Fig. 4.13 with the exception of $l_1 = 0.2$ mm and $w_1 = 6$ mm (see Fig. 4.10). Vias and backside strips at each side of the resonator are used to suppress the parasitic slot mode. *Source:* Reprinted with permission from [17]; copyright 2013 IEEE.

(a)

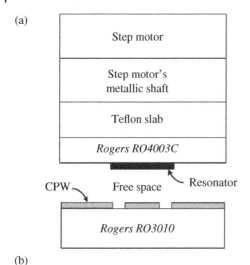

(b)

(c)

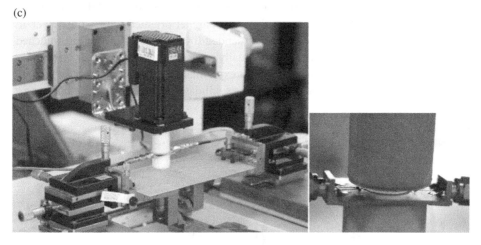

Figure 4.21 Setup for the angular displacement measurement. (a) Layer cross section, (b) photograph of the fabricated CPW and ELC resonator, and (c) photographs of the experimental setup with positioners and a step motor *STM 23Q-3AN*. The parameters of the substrates are: *Rogers RO3010*, ε_r = 11.2, h = 1.27 mm, and tanδ = 0.0023; *Rogers RO4003C*, ε_r = 3.55, h = 0.8128 mm, and tanδ = 0.0021; Teflon, ε_r = 2.08, h = 3.5 mm; the in-between air layer (air gap) is h = 1.27 mm. *Source:* Reprinted with permission from [17]; copyright 2013 IEEE.

operated as expected. By contrast, in [17], a more realistic experimental setup was used (Fig. 4.21). The resonator substrate was attached to the metallic shaft of a step motor by means of a Teflon cylindrical slab, used to avoid close proximity between the resonator and the metallic shaft (thereby ensuring that the electrical performance of the sensor does not change). The step motor (model *STM 23Q-3AN* in the setup of [17]) was software controlled from a host computer.

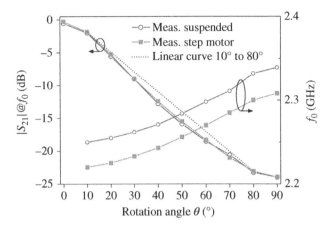

Figure 4.22 Notch magnitude at the notch frequency f_0 extracted from the measurement of the transmission coefficient as a function of the rotation angle in the structure of Fig. 4.21. The measurement with the suspended substrate was first reported in [29]. *Source:* Reprinted with permission from [17]; copyright 2013 IEEE.

The measured notch magnitude and frequency as a function of the rotation angle, obtained after a 3-D spatial calibration between the CPW and the resonator by means of positioners, is plotted in Fig. 4.22. The results are very similar to those with the suspended substrate [29], also depicted in the figure. The slight discrepancy (30 MHz) in the notch frequency is due to the Teflon slab. The variation in the notch depth between both sets of experiments is negligible. These results indicate that the notch magnitude exhibits not only quite robustness, but also less sensitivity than frequency-based sensors to variations or tolerances on the properties of the substrate.

According to Fig. 4.22, the output dynamic range is 23.7 dB, with an average sensitivity of 0.26 dB/°. Such high output dynamic range was achieved while preserving a low insertion loss for small angles (0.29 dB for $\theta = 0°$), although the measured linearity was not as good as that of the structure shown in Fig. 4.13 (see Fig. 4.19). It should be mentioned that for the same stator and resonant element, the notch position shifts upward when the resonant element is etched in an independent (movable) substrate (rotor), as compared to the case with the resonator attached to the backside of the stator substrate. Moreover, the thinner the rotor substrate and the lower its dielectric constant, the higher the resonance frequency, the wider the bandwidth, and the deeper the notch. For this reason, the resonator substrate was chosen in [17] to be relatively thin and with low dielectric constant, resulting in an enhancement of the notch depth. Note that the notch frequency is shifted from 2.22 GHz ($\theta = 10°$) to 2.308 GHz ($\theta = 90°$), see Fig. 4.22. This frequency variation is attributed to the fact that resonator parameters are somehow dependent on the rotation angle. By contrast, in the structure of Fig. 4.13, the notch frequency variation is roughly negligible, probably because the ELC resonator in that preliminary prototype is electrically smaller.

The setup of Fig. 4.21 can be easily modified in order to use the coupling-modulation sensor under study as an angular velocity sensor. In this application, the stator line must be fed by means of a harmonic signal conveniently tuned. In an ideal situation where the notch frequency f_0 does not vary with the angle, the frequency of the feeding signal should be f_0. Note that, due to rotation, the transmission coefficient at f_0 varies. Therefore, the amplitude of the injected signal is modulated at the output port. By assuming that the rotation speed is constant, such angular velocity can be deduced from the time lapse between two adjacent peaks, or minima, in the envelope function of such AM signal. Indeed, for constant angular velocity, the envelope function is periodic and it exhibits two peaks, or minima, per cycle. Thus, if the time distance between two adjacent peaks,

(a)

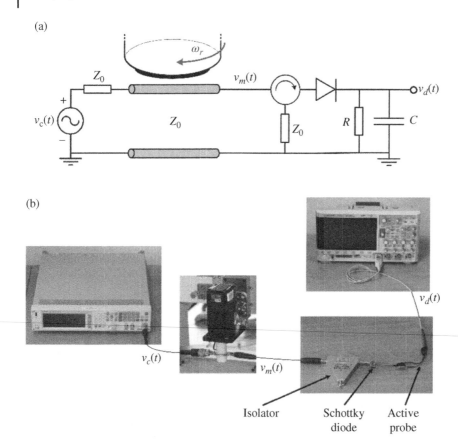

(b)

Isolator   Schottky   Active
                  diode   probe

Figure 4.23 Setup for the angular velocity measurement. (a) Schematic and (b) photographs of the experimental implementation. The *RC* filter is implemented by means of an active probe. The source impedance, the CPW characteristic impedance, and the circulator impedance is $Z_0 = 50\,\Omega$. The harmonic signal is injected by means of the *Agilent N5182A MXG* signal generator, the isolator is implemented by means of the circulator *ATM ATc1-2* model, and the envelope detector uses the *Avago Technologies HSMS-2860* diode and the active probe *Agilent N2795A*, with resistance $R = 1\,\mathrm{M\Omega}$ and capacitance $C = 1\,\mathrm{pF}$. The oscilloscope *Agilent 3054A* visualizes the envelope signal in time domain. *Source:* Reprinted with permission from [17]; copyright 2013 IEEE.

or period of the envelope function (a measurable quantity), is designated as T_m, the angular velocity is simply calculated as

$$\omega_r = \frac{2\pi}{T_r} = \frac{\pi}{T_m} \tag{4.10}$$

since the rotation period is $T_r = 2T_m$. In order to obtain the envelope function, an envelope detector, consisting of a diode and a RC low-pass filter, should be cascaded to the output port of the stator line. Nevertheless, since diodes are highly nonlinear devices, an isolator sandwiched between the output port of the stator line and the envelope detector is recommended (by this means, mismatching reflections back to the sensor are prevented).[15] The system functionality is thus similar to the

---

15 An isolator is a two-port device able to transmit signals in one direction but not in the opposite direction. Such device can be implemented by means of a circulator, a nonreciprocal three-port device, with one of the ports terminated with a matched load [23]. It should be mentioned that, alternatively, the envelope function can be obtained by means of an integrated AM detector.

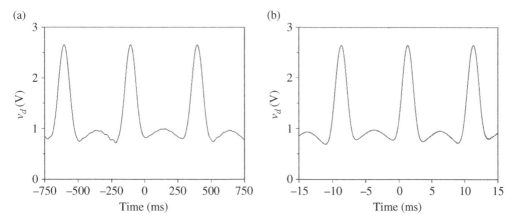

Figure 4.24 Measured envelope signals using the experimental setup of Fig. 4.23. (a) f_r = 1 Hz and (b) f_r = 50 Hz. *Source:* Reprinted with permission from [17]; copyright 2013 IEEE.

one of an AM modulator with carrier frequency tuned to $f_c = f_0$ [see Fig. 4.23(a)]. In a real system, where the notch frequency slightly varies during the rotation cycle, it suffices to tune the carrier frequency in the vicinity of the value of f_0 corresponding to maximum attenuation (90° orientation). The photograph of the experimental setup used for validation purposes in [17] is depicted in Fig. 4.23(b) (the specific component and equipment models are indicated in the caption).

In [17], the stator used for angular velocity measurements was the one depicted in Fig. 4.21, but the rotor was the one of Fig. 4.13, since with this rotor the notch frequency is more stable. By contrast, the dynamic range of attenuation with this rotor is poorer, but this is not crucial in this application. The important aspect is to obtain a well-defined period in the envelope function. The specific carrier frequency of the feeding signal was set to f_c = 1.515 GHz in [17], corresponding to the notch frequency for a rotation angle of 10°. The validity of the approach was demonstrated by configuring the step motor with a uniform angular velocity of $\omega_r = 2\pi f_r$ (with $f_r = 1/T_r$) for $f_r = 1$ Hz and f_r = 50 Hz (60 and 3000 rpm, respectively, the latter being the maximum angular velocity of the motor). Figure 4.24 depicts the measured envelope functions. The measured time periods between consecutive peaks are T_m = 501 ms (f_r = 1 Hz) and T_m = 10.05 ms (f_r = 50 Hz). The measured angular velocities are then f_r = 0.998 Hz and f_r = 49.751 Hz, values very close to the nominal velocities specified in the step motor. To enhance precision, the measurement can be done between very distant nonconsecutive peaks, although the rotation velocity should be constant in that case.

As long as the carrier frequency is much higher than the frequency of rotation, the measurable range of velocities is theoretically unlimited. However, the sensor is not able to provide instantaneous velocities, i.e. angular velocity variations within the time lapse between two adjacent peaks. Actually, this sensor provides the average angular velocity corresponding to such time lapse. This is an intrinsic limitation of these angular velocity sensors, able to generate only two pulses (peaks in the envelope function) per revolution (PPR). In order to discern potential variations in the angular velocity within a revolution cycle, or even to measure angular accelerations, it is necessary to enhance significantly the PPR. For that purpose, the edge configuration (microwave rotary encoders), discussed in Section 4.3.2.2, is the solution. Moreover, as it will be shown, microwave rotary encoders are able to determine the motion direction, an ability not present in the rotary coupling-modulation sensors based on the axial configuration, at least as they are presented in the different reported implementations [17, 24, 25, 29]). Nevertheless, before studying and analyzing the microwave rotary encoders, let us present an alternative realization of an axial rotary sensor based on a microstrip line stator [24].

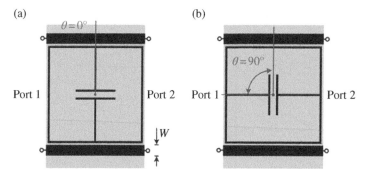

Figure 4.25 Parallel microstrip lines loaded with an ELC resonator for the two extreme cases of angular orientation: (a) $\theta = 0°$ and (b) $\theta = 90°$. *Source:* Reprinted with permission from [24]; copyright 2014 IEEE.

4.3.2.1.2 Microstrip Stator

The orientation-dependent coupling in an ELC resonator can also be achieved by means of a pair of microstrip lines (see Fig. 4.25) [24]. For common mode signals, the symmetry plane is a magnetic wall. Therefore, for the configuration of Fig. 4.25(a), the ELC is not excited. By contrast, significant coupling between the resonator and the line pair is expected for the configuration of Fig. 4.25(b), since in this case, identical electromagnetic walls (magnetic walls) are perfectly aligned. For sensor implementation, the ELC resonator is circularly shaped (for the reasons explained in the previous section), and the four-port line pair structure is replaced with a two-port network consisting of a circular splitter/combiner configuration (stator). The specific topology and photograph of the stator and rotor are depicted in Fig. 4.26. Note that with this stator configuration, the axial symmetry

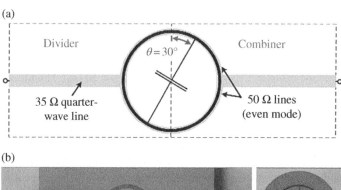

Figure 4.26 Angular displacement sensor based on a microstrip-based stator and ELC-based rotor. (a) Layout and (b) photograph. The substrates are *Rogers RO3010* with dielectric constant $\varepsilon_r = 11.2$, thickness $h = 1.27$ mm (microstrip line) and $h = 0.635$ mm (resonator), and loss tangent $\tan\delta = 0.0023$. The line widths are 2.06 mm (35.35-Ω quarter-wavelength line) and 1.04 mm (50-Ω circular-shaped line). ELC mean radius $r_0 = 8.05$ mm, and, in reference to Fig. 4.10, $w_1 = 6$ mm, $w_2 = l_1 = s = 0.2$ mm, and $w_3 = 0.5$ mm. *Source:* Reprinted with permission from [24]; copyright 2014 IEEE.

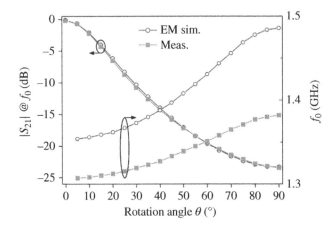

Figure 4.27 Notch magnitude at the notch frequency f_0 extracted from the measurement of the transmission coefficient as a function of the rotation angle in the structure of Fig. 4.26. *Source:* Reprinted with permission from [24]; copyright 2014 IEEE.

plane of the stator is a magnetic wall (provided the ELC resonator is symmetrically oriented). The divider acts as a single-ended to common-mode signal transition. Conversely, the combiner converts the common-mode signal into a single-ended signal. The circularly shaped lines exhibit 50 Ω even-mode characteristic impedance, and the combiner/divider is implemented with 35.35 Ω impedance inverters to achieve matching to 50-Ω reference ports.

The divider/combiner suppresses any differential mode signal eventually generated by nonsymmetric orientations. However, if the circularly shaped lines are electrically long, mode mixing is generated along them. In order to reduce electrically the size of these lines at resonance, the electrical size of the resonator was decreased by using a high dielectric constant substrate ($\varepsilon_r = 11.2$). The displacement measurement was performed in [24] following the procedure in [17], and detailed in the precedent section. The air gap between the ELC and the microstrip lines was set to 0.254 mm. As shown in Fig. 4.27, the rotation angle can be sensed from the notch magnitude, being the dependence reasonably linear. For angular velocity measurements, the setup is the one detailed in the previous section. The carrier frequency of the feeding harmonic signal was tuned to the notch frequency corresponding to $\theta = 10°$ ($f_c = 1.308$ GHz). The step motor was configured with $f_r = 1$ Hz and $f_r = 50$ Hz. The corresponding envelope signals are depicted in Fig. 4.28, and the

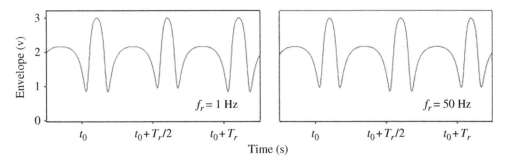

Figure 4.28 Measured envelope signals for the sensor of Fig. 4.26, inferred by means of the experimental setup of Fig. 4.23 corresponding to $f_r = 1$ Hz and $f_r = 50$ Hz. *Source:* Reprinted with permission from [24]; copyright 2014 IEEE.

velocities, derived from the time difference between consecutive transmission peaks, are $f_t = 0.998$ Hz and $f_r = 50.251$ Hz, very close to the nominal values.

4.3.2.2 Edge Configuration. Electromagnetic Rotary Encoders

The main limitation of the axial rotation sensors of the previous section is the limited number of pulses per revolution (PPR = 2). This prevents from measuring instantaneous velocities, as discussed. On the other hand, despite the fact that the achieved sensitivity in the previous sensors is reasonably good (the average sensitivity in the sensor of Fig. 4.21 is 0.26 dB/°), such sensors do not constitute the best option in applications requiring high accuracy or resolution in the measurement of angular displacements. This is because the output variable for such application is the notch depth, susceptible to the effects of electromagnetic interference or noise. The electromagnetic, or microwave, rotary encoders are an alternative to the previous sensors. In such sensing systems, the rotor is a dielectric disc with a circular chain of metallic inclusions (typically, but not necessarily resonators) etched along its edge. Similar to optical rotary encoders, electromagnetic encoders are based on pulse counting. That is, the stator, an element able to detect the crossing of the chain inclusions through a certain position, provides the angular displacement from the recorded number of cumulative pulses. According to this, the angle resolution, θ_{res}, is intimately related to the number of PPR. Specifically,

$$\theta_{res} = \frac{2\pi}{PPR} \tag{4.11}$$

and for this reason a figure of merit in these sensors is the number of pulses per cycle, PPR.

The working principle of electromagnetic and optical rotary encoders is essentially the same, i.e. pulse counting. The main difference is the type of signal used for sensing, an RF/microwave signal in electromagnetic encoders, and an optical signal in the optical counterparts. Optical encoders use a light source, which generates an optical beam directed toward a photodetector. The system uses an opaque disc (made of metal or glass) with a grid of apertures, the rotor. When any aperture lies in the optical path between the source and the detector, an optical pulse reaches the photodetector, which converts it to an electronic signal (the working principle is illustrated in Fig. 4.29). According to this working principle, the angular displacement from a well-known reference is determined from the cumulative number of pulses, and the information relative to the angular velocity and acceleration can be inferred from the time between adjacent pulses. These encoders are designated as incremental-type encoders. However, it should be mentioned that there is a different type of

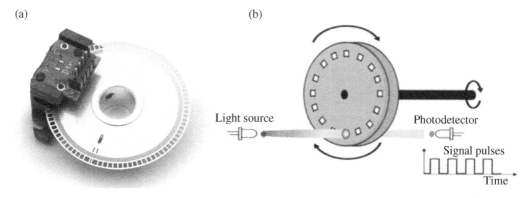

Figure 4.29 Photograph of a typical optical rotary encoder (a) and sketch illustrating their working principle (b). *Source:* Reprinted with permission from [30]; copyright 2020 Springer.

optical rotary encoders, the so-called absolute encoders, which are able to provide the angular position without the need to predefine a reference position. The electromagnetic rotary encoders that are studied in this section are similar to incremental optical encoders.

Before studying in detail the strategy to detect the presence of metallic inclusions in electromagnetic encoders using microwaves, let us first briefly discuss the advantages and drawbacks of such devices, as compared to optical encoders. In terms of performance, it should be taken into account that the main relevant parameter in rotary encoders is the angle resolution. In this regard, microwave encoders cannot compete against their optical counterparts, since the achievable PPR is by far superior in optical encoders (optical encoders with thousands of PPR are commercially available). However, microwave encoders are cheaper, and can be used in scenarios subjected to harsh conditions (e.g. radiation, extreme temperatures, and space) or contaminants (dirt, dust, grease, etc.). In [31], the potential use of electromagnetic rotary encoders for attitude control in space vehicles is discussed.

The working principle of electromagnetic rotary encoders was first introduced in [7], where a preliminary prototype was presented, and patented in [32]. Then, a detailed analysis of these rotary sensors was reported in [31], where two prototypes were presented, one with a single chain of resonant elements in the rotor, and the other one with two concentric chains. With this latter rotary encoder, resolution was improved, as far as a very competitive number of pulses was achieved (PPR = 1200). In a subsequent paper [33], it was demonstrated that by adding a nonperiodic extra chain, the motion direction could be discerned. These encoders are studied in detail in a book devoted to chipless-RFID systems based on near-field coupling and sequential bit reading [30], due to their similarity with such systems. Indeed, the main difference is that such chipless-RFID systems are based on linear (rather than circular) chains of inclusions, and not all the inclusions are present at their predefined positions in the chain, in order to provide a unique ID code to the chain (an overview of these chipless-RFID systems is given in [30, 34] and references therein).[16] Nevertheless, the working principle is identical in both types of systems: inclusions' detection through near-field coupling with a transmission line (acting as a stator in rotary sensors and as a reader in chipless-RFID systems). For that purpose, the transmission line of the stator/reader is fed by means of a harmonic (carrier) signal conveniently tuned, similar to the angular velocity sensors based on the axial configuration, reported in the previous section. As the inclusions of the chain are displaced (due to rotor or tag motion) over the stator/reader transmission line, the amplitude of the signal at the output port is AM modulated, and the envelope function contains as many peaks, or dips, as inclusions in the chain. For rotary sensors, the total number of inclusions in the circular chain, N, provides the number of peaks, or dips, in the envelope function per cycle, or PPR. Nevertheless, by properly tuning the frequency of the input signal, the PPR can be doubled, as it will be shown. The sketch of a single-chain electromagnetic rotary encoder showing the working principle is illustrated in Fig. 4.30. The considered stator line in that figure is a CPW, and the inclusions are rectangular split ring resonators (SRRs). However, different combinations of lines and inclusions can be used for sensor implementation. In the next section, single- and double-chain encoders based on a CPW stator are presented. The following section reports an encoder system where the stator is implemented by means of a microstrip line.

---

16 Moreover, to reduce their cost, chipless-RFID tags based on chains of metallic inclusions are typically implemented on plastic [35] or paper substrates [36], rather than on rigid microwave substrates. An additional important property of such chipless-RFID tags is their programing capability [37]. This results in a significant reduction of manufacturing costs, as far as all-identical tags can be fabricated (e.g. by means of large-scale printing processes, such as screen-printing, or offset, among others), and later programmed by means of low-cost printing processes (e.g. inkjet-printing).

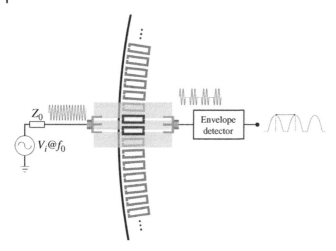

Figure 4.30 Sketch of the microwave rotary encoder, showing the working principle. *Source:* Reprinted with permission from [31]; copyright 2017 IEEE.

4.3.2.2.1 CPW Stator

Let us first present the electromagnetic rotary encoder corresponding to the sketch of Fig. 4.30. Actually, the stator is not as simple as a mere CPW transmission line. Namely, a pair of SRRs identical to those of the rotor, with identical separation, are etched in the back-substrate side of the CPW line, as depicted in Fig. 4.31. However, the two SRRs of the stator are oppositely oriented to the SRRs of the rotor (a zoom view of a part of the rotor layout and a cross-sectional view of the stator/rotor are also included in Fig. 4.31). The presence of the two SRRs in the stator is justified by the need to avoid multiple couplings between the line and the SRRs of the rotor, as well as

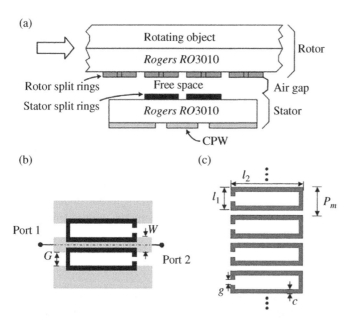

Figure 4.31 Sensor cross section (a), and layout of the (b) stator and (c) rotor (zoom view). Dimensions (in mm) are: $W = 1.3$, $G = 0.9$, $P_m = 2$, $l_1 = 1.6$, $l_2 = 6.2$, $c = 0.4$, and $g = 0.2$. The substrates have relative permittivity $\varepsilon_r = 11.2$, thickness $h = 0.635$ mm (stator) and $h = 1.27$ mm (rotor), and loss tangent $\tan\delta = 0.0023$.

inter-resonator coupling in the rotor [7, 31]. By orienting the SRRs of the stator in opposition to those of the rotor, significant coupling between the two SRRs of the stator and a pair of SRRs of the rotor arises when there is perfect alignment between both pairs of resonators (a situation illustrated in the cross-sectional view of Fig. 4.31). Indeed, each pair of oppositely oriented face-to-face resonators (one belonging to the rotor and the other one to the stator) can be considered to form a single resonant particle, designated in the literature as broadside-coupled SRR (BC-SRR) [3] (see Chapter 2, Section 2.2.1.1). Such particle, with perfectly aligned SRRs, exhibits a resonance frequency significantly smaller than the one of the individual SRRs. Thus, at such frequency, the coupling between the SRRs of the stator and those of the rotor occurs only for those SRRs of the rotor aligned with the SRRs of the stator. Thus, by tuning the carrier signal to such frequency, multiple couplings between the stator and the rotor, as well as inter-resonator coupling in the rotor, are avoided. Moreover, a notch in the transmission coefficient at this frequency is expected when a pair of SRRs of the rotor is perfectly aligned, i.e. face-to-face, with the SRRs of the stator (for convenience, let us call this relative displacement between the stator and the rotor REF position).

The transmission coefficient for the REF position is depicted in Fig. 4.32. The figure includes also the response for different relative incremental/decremental displacements between the stator and the rotor, expressed in terms of the period of the rotor, P_m. As the misalignment between the stator and rotor increases, the notch frequency shifts upward. It can also be appreciated that for the notch frequency corresponding to perfect alignment, f_0, the attenuation decreases as the rotor moves from the REF position. In other words, the transmission coefficient at f_0 is modulated by rotor motion. However, there is not an a priori reason for tuning the carrier frequency to f_0, as far as the transmission coefficient is modulated by rotor motion over a wide frequency range. Indeed, if the carrier frequency is set to a value above f_0, two attenuation peaks, or dips, per period, rather than one, are expected by rotor motion. This is interesting since, by this means, the number of PPR, the key figure of merit of these sensors, can be set to PPR = 2N, twice the number of chain inclusions.

Figure 4.33 depicts the transmission coefficient as a function of the displacement (for one entire period) for three different frequencies. As it can be seen, a single dip appears at the REF position (perfect alignment) when the carrier frequency is tuned to f_0. However, for the frequencies

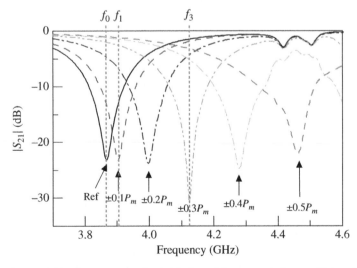

Figure 4.32 Electromagnetic simulation of the transmission coefficient for different relative displacements between the stator and the rotor. The considered air gap, or distance between the stator and rotor, is 0.5 mm. *Source:* Reprinted with permission from [31]; copyright 2017 IEEE.

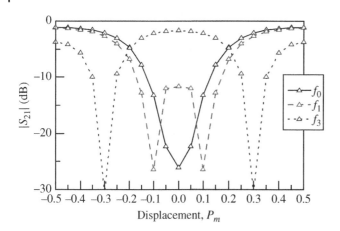

Figure 4.33 Attenuation as a function of the rotor displacement at the indicated frequencies. *Source:* Reprinted with permission from [31]; copyright 2017 IEEE.

designated as f_1 and f_3 in Fig. 4.32, two attenuation peaks arise. It can be then concluded that for carrier frequencies satisfying $f_c > f_0$, the number of pulses per cycle is PPR $= 2N$. It should be taken into account, however, that as the carrier frequency approaches f_0, the attenuation peaks get closer, and they merge in the limit when $f_c \rightarrow f_0$. Ideally, tuning the carrier frequency close to f_3 is a good choice, as far as a similar level of transmission between attenuation peaks is obtained at that frequency. It is important to mention that for the automatic determination of the angular displacement and velocity of the rotor, a high modulation index is necessary, and it is intimately related to the excursion of the attenuation level of the transmission coefficient at the operating (carrier) frequency.

An important aspect in the sensors under consideration is the influence of the vertical distance between stator and rotor (air gap) on sensor performance and robustness. The reason is that a real system is subjected to mechanical vibrations and to rotor precession, which continuously modify the air gap separation. This means that the system must be tolerant to air gap variations. To analyze this aspect, Fig. 4.34 depicts the transmission coefficient as a function of frequency for different relative displacements parametrized by the air gap, corresponding to the structure of Fig. 4.31. According to this figure, if the carrier signal is tuned to a frequency close to f_3, see Fig. 4.32, the system should be functional under air gap variations comprised, at least, between 0.3 and 0.7 mm. However, if the carrier frequency is set to f_0 (the resonance frequency for the REF position when the air gap is 0.5 mm, see Fig. 4.32), system functionality is not preserved for an air gap of 0.7 mm. This is obvious in view of Fig. 4.34(c), showing that the transmission coefficient at $f_0 = 3.86$ GHz does not vary with the displacement of the rotor. Consequently, the injected signal cannot be AM modulated if it is tuned to this frequency. This analysis provides a further reason to tune the carrier signal to a frequency above f_0. To conclude this analysis relative to the effects of the air gap, let us mention that if the air gap varies during a cycle (within the limits of system tolerance), the modulation index of the output signal also changes with time. However, the time position of the peaks/dips in the envelope function are not affected by such air gap variations (unavoidable in practice), and, therefore, the system functionality is preserved.

Experimental validation of the electromagnetic rotary encoders was carried out in [31] by considering a setup similar to the one used for the measurements of angular displacements and velocities in the axial sensors of the previous section. Nevertheless, the system was somehow modified in order to accommodate the large-sized rotor (Fig. 4.35). Such rotor (also depicted in Fig. 4.35)

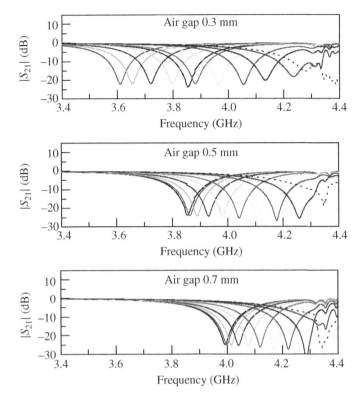

Figure 4.34 Electromagnetic simulation of the transmission coefficient for different relative displacements (from 0 to 1 mm in steps of 0.1 mm) between the stator and the rotor, parametrized by the air gap separation. Note that such relative displacements correspond to variation between $0P_m$ and $P_m/2$ in steps of $0.05P_m$. Dashed lines indicate the curves corresponding to $P_m/2$. *Source:* Reprinted with permission from [31]; copyright 2017 IEEE.

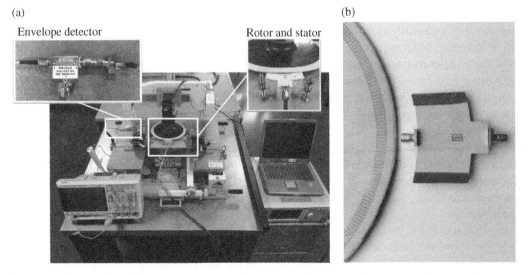

Figure 4.35 Experimental setup for measuring the angular velocity and displacement by means of rotors based on the edge configuration (a), and photographs of a designed and fabricated rotor, based on a single chain of SRRs, and stator (b).

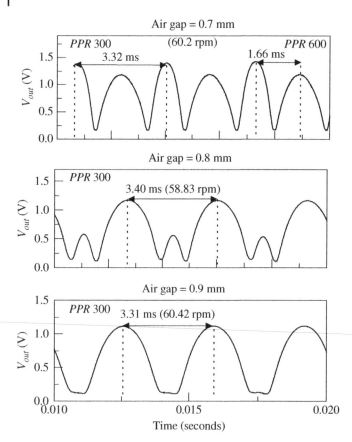

Figure 4.36 Envelope of the modulated carrier signal for different air gaps, corresponding to the rotor with a single chain of SRRs. For the smaller air gap, the amplitude of the pulses is similar, and the angular velocity can be obtained from the time distance between adjacent pulses, or between higher amplitude pulses, as indicated. *Source:* Reprinted with permission from [31]; copyright 2017 IEEE.

contains $N = 300$ equally spaced SRRs along its perimeter (this explains the relatively large diameter of the rotor, 101.6 mm). Figure 4.36 depicts the envelope functions for different nominal air gaps, for a carrier frequency tuned to $f_c = 4.2$ GHz. This frequency is slightly above f_3 in Fig. 4.32. With this choice, the system can operate under a relatively wide air gap span, and the nominal air gaps can be slightly superior to those of Fig. 4.34.[17] The rotation speed of the rotor was set to 60 rpm. For the smaller air gap (0.7 mm), two pulses per period of the encoder chain are visible, although the amplitude of such pulses is not identical. The difference in the amplitude of the pulses increases as the air gap increases (see Fig. 4.36, corresponding to an air gap of 0.8 mm), and the pulses merge when the air gap is sufficiently high (0.9 mm in Fig. 4.36). Note that, depending on the air gap, the number of PPR is either PPR $= N = 300$ or PPR $= 2N = 600$. The rotation speed inferred from the envelope functions, indicated in Fig. 4.36, coincides with the nominal value to a good approximation.

The results of Fig. 4.36 validate the functionality of the microwave rotary encoders. The PPR can be further increased, keeping unaltered the rotor diameter, by including an additional SRR chain in the rotor and a pair of SRRs in the stator, as depicted in Fig. 4.37 [31]. Note that the symmetry planes

---

17 It is convenient to avoid small nominal air gaps in order to avoid mechanical friction by rotor precession.

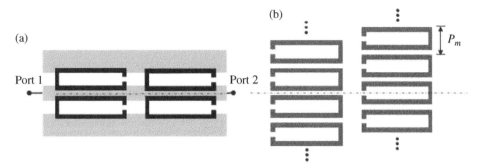

Figure 4.37 Stator (a) and zoom view of the rotor (b) for the rotary encoder based on a rotor with two SRR chains.

of the SRRs of one chain coincide with the intermediate planes between adjacent SRRs of the other chain. With this configuration, the PPR can be twice the PPR resulting when a single rotor chain is considered. The effects of the air gap on the number of pulses of the envelope function per chain period do also apply in this case. Therefore, the PPR achieved with a rotor based on two chains of $N = 300$ SRRs is PPR = 1200 [31]. Essentially, the single- and the double-chain rotor sensors work similarly, i.e. the amplitude of the signal at the output port of the stator CPW varies at a rate dictated by the rotor speed and the whole number of SRRs. However, using two chains is an efficient procedure to enhance the PPR without the need to increase the diameter of the rotor (a trivial way to accommodate further SRRs in a single chain).

Determining the motion direction is possible by including an additional SRR chain, designated in [33] as direction chain. Such chain must be nonperiodic and the SRRs must be tuned to a frequency different from that of the velocity chain/s. Particularly, by progressively increasing, or decreasing, the separation between adjacent SRRs in the direction chain as the angular position of the SRRs varies, e.g. clockwise, it is possible to discern the motion direction. It depends on whether the time lapse between adjacent pulses in the corresponding envelope function progressively increases or decreases. Two envelope functions are thus needed in this case, and for this reason, the SRRs of the direction chain must be tuned to a different frequency. This means that the system needs two different carrier signals, one for detecting the angular velocity (and position) and the other one to determine the motion direction. Such signals must be combined and injected to the input port of the CPW line of the stator. The generated AM signals, containing the information relative to the angular velocity and motion direction are separated by means of a diplexer. The schematic of the complete system is depicted in Fig. 4.38. Further details of this sensor system, e.g. the design of the diplexer and combiner, can be found in [33]. Nevertheless, it should be mentioned that the direction chain does not require a pair of SRRs in the stator, since the SRRs of such chain are separated enough and the aforementioned multiple couplings, as well as inter-resonator coupling, do not appear in this case. The photograph of the rotor with direction detection capability is depicted in Fig. 4.39, whereas Fig. 4.40 plots the envelope functions inferred from the experimental setup used for determining the angular velocity in the previous prototype.[18] The number of PPR is 1200 and the angular velocity is correctly estimated, as it can be seen in Fig. 4.40(a), i.e. the envelope function providing the angular velocity (the nominal rotation speed was set to 60 rpm). Figure 4.40(b) shows a progressively increasing time distance between adjacent pairs of pulses, thereby indicating that

---

18 The only difference between the experimental setups of [7, 33] concerns the mechanical system, more robust against precession and vibrations in [33]. For this reason, the envelope functions are more stable in the system reported in [33].

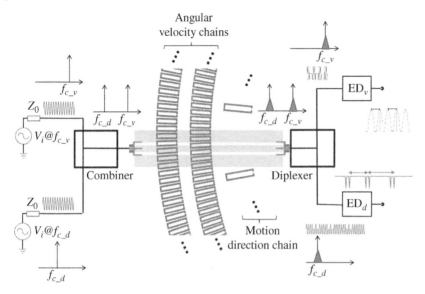

Figure 4.38 Sketch showing the working principle for the independent measurement of the angular velocity and rotation direction. The two carrier signals, tuned to f_{c_v} and f_{c_d}, can be injected to the CPW by means of a combiner. *Source:* Reprinted with permission from [33]; copyright 2018 IEEE.

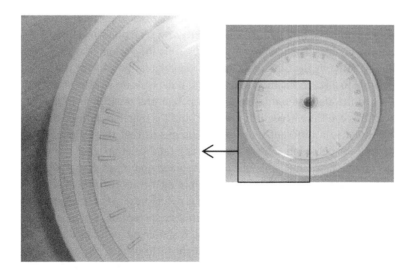

Figure 4.39 Fabricated rotor with an outer double-SRR chain for velocity measurement and an inner unequally spaced SRR chain for direction detection. The radius of the rotor is 101.6 mm. For the outer velocity sensor chain, SRR dimensions are those given in Fig. 4.31. For the inner velocity chain, SRR dimensions (in mm) are: $P_m = 1.81$, $l_1 = 1.45$, $l_2 = 6.35$, $c = 0.36$, and $g = 0.2$ (see Fig. 4.31, where the spatial variables are defined). For the SRRs of the direction chain, dimensions (in mm) are: $l_1 = 2.0$, $l_2 = 8.4$, $c = 0.4$, and $g = 0.5$. The slightly different dimensions of the inner and outer velocity chains are necessary to accommodate the same number of SRRs ($N = 300$) within a smaller perimeter for the inner velocity chain. Nevertheless, such change does not substantially modify the resonance frequency of the pair of vertically aligned rings of the stator and rotor.

(a)

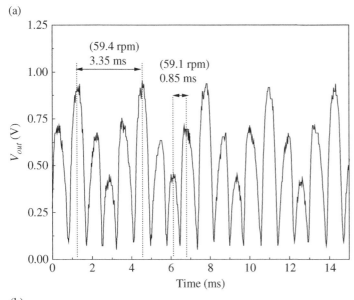

(b)

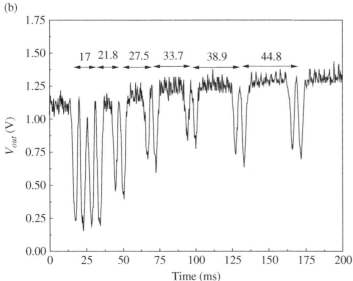

Figure 4.40 Envelope function providing the angular velocity (a) and motion direction (b). *Source:* Reprinted with permission from [33]; copyright 2018 IEEE.

the motion direction is the one corresponding to a progressively increasing time interval between crossings of adjacent SRR pairs through the CPW axis. In [33], a post-processing system for the automatic determination of the angular velocity and motion direction is reported, but this is out of the scope of this book.

4.3.2.2.2 *Microstrip Stator*

One potential limitation of CPW stators is the lack of backside isolation. For that purpose, micro-strip-based stators represent an interesting solution [38]. Figure 4.41 depicts the layout of the stator and rotor. The chain inclusions of the rotor are SRRs, oppositely oriented to the SRR present in the

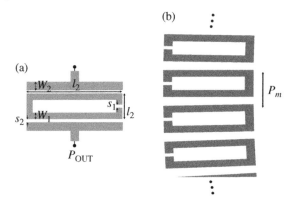

Figure 4.41 Layout of the stator (a) and rotor (b) corresponding to the angular displacement and velocity sensor based on a microstrip stator configuration. Dimensions (in mm) are: $l_1 = 6.2$, $l_2 = 1.6$, $s_1 = 0.2$, $s_2 = 0.2$, $W_1 = 0.4$, $W_2 = 0.535$, and $P_m = 2.0$.

stator. This microstrip stator configuration provides a bandpass response (contrary to the stopband behavior of the CPW-based stator of the previous section), since the structure is essentially an order-1 coupled resonator bandpass filter. The responses of the stator for different relative displacements between the stator and the rotor over one entire period of the rotor are depicted in Fig. 4.42. The response exhibiting the transmission peak at the lower frequency, $f_{BC\text{-}SRR}$, is the one corresponding to a perfect alignment between the SRR of the line and one of the SRRs of the rotor. The reason is that for this perfect alignment, the composite resonator formed by the SRRs of line and rotor exhibits the minimum possible frequency (the one of the so-called BC-SRR, as discussed before). The opposite situation occurs for the response with the peak at f_{stator}, where the stator and rotor SRRs are completely misaligned. Under these conditions, the resonance frequency of the resonator "seen" by the line (i.e. coupled to the input and output ports of the line) is essentially the one of the isolated resonator of the stator. The transmission coefficient at the indicated frequencies experiences a significant excursion. Indeed, such excursion can be better appreciated by representing the transmission coefficient at the indicated frequencies as a function of the displacement (Fig. 4.43). Clearly, two peaks per period of the resonator chain are expected for the frequency designated as f_{int}. The air gap in the simulation results of Figs. 4.42 and 4.43 is 0.5 mm. Varying the air gap modifies the results, as expected, but it has been found that system functionality is guaranteed for air gaps comprised between 0.3 and 0.7 mm [38]. Indeed, as discussed in [30], the

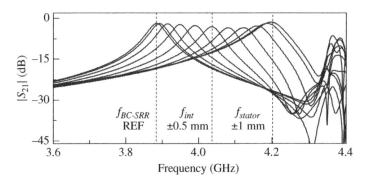

Figure 4.42 Simulated frequency responses corresponding to different relative displacements between the stator and the rotor. The considered substrate for the stator is the *Rogers RO3010* with dielectric constant $\varepsilon_r = 10.2$ and thickness $h = 0.635$ mm. The considered dielectric constant and thickness for the rotor are $\varepsilon_r = 10.2$ and $h = 1.27$ mm, respectively. The distance (air gap) between the stator and rotor is 0.5 mm. *Source:* Reprinted with permission from [38]; copyright 2018 IEEE.

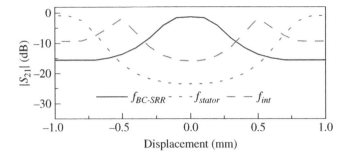

Figure 4.43 Variation of the transmission coefficient with the displacement for the indicated frequencies, for an air gap of 0.5 mm. These results have been inferred by electromagnetic simulation. *Source:* Reprinted with permission from [38]; copyright 2018 IEEE.

microstrip-based stator seems to be more robust against the effects of air gap variations than the CPW counterpart.

Figure 4.44 shows the photographs of the stator and rotor (with $N = 300$ SRRs) corresponding to the microstrip rotary encoder of Fig. 4.41. The experimental setup for measuring the angular velocities is identical to the one in reference to the CPW-based encoders. The envelope functions corresponding to the three frequencies of Fig. 4.42, for an angular velocity of 60 rpm, are depicted in Fig. 4.45. For the extreme frequencies, $f_{BC\text{-}SRR}$ and f_{stator}, a single pulse per resonant element is achieved, in coherence with Fig. 4.43. However, for the intermediate frequency, f_{int}, two pulses per resonant element appear. The distances between adjacent (equidistant) minima provide the angular velocities indicated in the figure. Such velocities are in very good agreement with the nominal angular velocity (i.e. the differences are within the typical tolerance errors of the considered stepped motor, around 1–2%).

It should be mentioned that, although not yet experimentally verified, microstrip rotary encoders with further number of pulses per resonant element of the rotor could be envisaged. However, rather than adding a chain of resonators in the rotor, the strategy in this case is to design the stator with a pair of parallel-connected SRR-loaded microstrip lines in bandpass configuration. By ensuring that when one of the SRRs of the rotor is perfectly aligned with one of the SRRs of the stator, the other SRR of the stator is completely misaligned with the SRRs of the rotor, up to four pulses per resonant element of the rotor can be potentially generated. This pulse-enhancing strategy is indeed

(a)   (b)

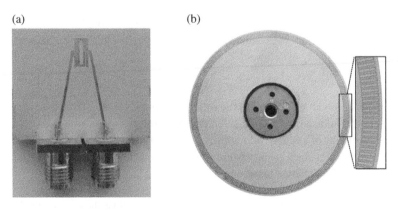

Figure 4.44 Photograph of the microstrip stator (a) and rotor (b).

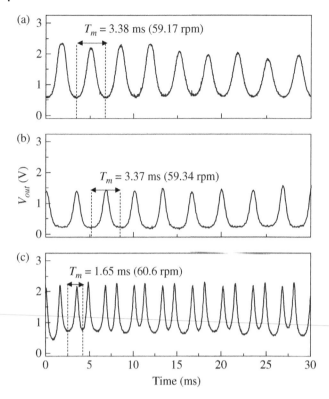

(a)

$T_m = 3.38$ ms (59.17 rpm)

(b)

V_{out} (V)

$T_m = 3.37$ ms (59.34 rpm)

(c)

$T_m = 1.65$ ms (60.6 rpm)

Time (ms)

Figure 4.45 Measured envelope functions for 60 rpm speed and carrier frequencies set to $f_{BC\text{-}SRR} = 3.88$ GHz (a), $f_{stator} = 4.20$ GHz (b), and $f_{int} = 4.03$ GHz (c). The estimated air gap is 0.5 mm. *Source:* Reprinted with permission from [38]; copyright 2018 IEEE.

equivalent to the one of the CPW-based stator, consisting of using two chains of resonant elements in the rotor. The difference is that in the microstrip stator system, the proposal is to double the sensing elements of the stator (SRR-loaded lines). This is necessary since the microstrip-based encoder system operates in bandpass configuration.

4.3.2.2.3 Resolution and Accuracy

Concerning sensor resolution in the measurement of angular displacements, it is given by the PPR, see expression (4.11). For the encoder of Fig. 4.39, based on two rotor chains devoted to the measurement of the angular velocity and position, the resolution is $\theta_{res} = 0.3°$. The angular displacement can be obtained from the angular velocity as follows:[19]

$$\Delta\theta = \omega_r \cdot \Delta t \tag{4.12}$$

where Δt is the considered time interval. Counting up the number of pulses is an alternative procedure to determine the angular displacement from a reference position.

The resolution in the measurement of angular velocities is given by time resolution, t_{res}, or the minimum time interval that can be resolved (time resolution is the inverse of the sampling frequency). Taking into account that the distance between adjacent pulses is given by

---

19 This expression is valid for constant angular velocity. If the rotation speed varies with time, the integral form is needed.

$$T_m = \frac{T_r}{PPR} \tag{4.13}$$

and that the resolution is t_{res}, the rotation frequency can be expressed as

$$f_r \equiv \frac{1}{T_r} = \frac{1}{PPR \cdot (T_m \pm t_{res})} \tag{4.14}$$

which can be approximated by

$$fr \approx \frac{1}{PPR \cdot T_m} \left(1 \mp \frac{t_{res}}{T_m}\right) \tag{4.15}$$

provided $t_{res} \ll T_m$. The time T_r in (4.13) is the rotation period, see expression (4.10). From (4.15), the resolution in the rotation frequency is

$$f_{r,res} = \frac{t_{res}}{PPR \cdot T_m^2} = PPR \frac{t_{res}}{T_r^2} \tag{4.16}$$

Angular velocity resolution depends on the number of PPR and on the rotation speed, as well. For a given rotation speed, the resolution increases (is worst) with PPR. This is due to the fact that T_m decreases. However, by increasing the number of PPR, variations of instantaneous rotation speed within a revolution can be better detected in time. Note that variations in time lapses smaller than the time between adjacent pulses cannot be detected. If the sampling frequency is, e.g. 1 MHz (the one in the measurements of Fig. 4.40), then the time resolution is $t_{res} = \pm 1\,\mu s$. By considering a rotation speed of 60 rpm (corresponding to $T_r = 1$ second) and PPR = 1200, the resolution in the rotation frequency is found to be 1.2 mHz, corresponding to a resolution of 0.072 rpm. Note that this very good resolution is achieved due to the high sampling frequency. Therefore, the sampling frequency can be relaxed, yet keeping the resolution within good limits. For a significantly higher rotation speed, i.e. 6000 rpm, the resolution is found to be 12 Hz, or 720 rpm. In this case, resolution can be improved by simply increasing the sampling frequency (reducing t_{res}).

Concerning the error in the determination of the instantaneous angular velocity, it depends on the width of the pulses and distance between them. Narrower pulses provide smaller error. Nevertheless, the accuracy can be improved either by averaging over several pulses, or by considering distant pulses. In both cases, however, the penalty is the degradation in the ability of the sensor to detect speed variations in small time intervals. To give a reasonable (very conservative) estimation of the sensor error, if we consider the width of the pulse as given by the value at 98% of the pulse maximum, for the double SRR chain rotor and four pulses per chain period (Fig. 4.40), the ratio between the pulse width and the distance between adjacent pulses is 0.1. This represents a 10% error in time (approximately $\pm 5\%$), but, as mentioned, this value can be reduced by 10 times, i.e. $\pm 0.5\%$, if the considered pulses are separated 10 positions. In terms of error in rotation frequency, these values transform to roughly ± 5 and $\pm 0.5\%$, respectively.

4.3.3 Electromagnetic Linear Encoders

Linear velocities and displacements can be measured by means of systems similar to microwave rotary encoders. The unique difference is that in such systems the circular chains of inclusions must be replaced with linear chains. Therefore, these linear displacement and velocity sensors can be designated as electromagnetic, or microwave, linear encoders. The working principle is identical to the one of rotary encoders: the detection of chain inclusions by near field, through microwaves. Moreover, as it was mentioned in Section 4.3.2.2, by etching or printing the inclusions on low-cost substrates, including paper or plastic, and by providing an identification (ID) code to the chain, these encoders are useful as chipless-RFID tags [30]. The ID code is implemented by etching or printing only certain inclusions of the chain at their predefined positions, i.e. those associated to the logic state "1." Therefore, the ID code is inferred sequentially, by displacing the tag (encoder) over the reader (stator) at short distance. Obviously, the ID code is inferred from the envelope function, where the number of peaks, or dips, coincides with the number of functional inclusions in the tag (corresponding to the "1" state). For the implementation of linear velocity sensors, it is convenient that all the inclusions of the encoder are present at their predefined positions. In this case, the envelope function is periodic (assuming that the velocity is constant), the velocity is given by the time lapse between adjacent pulses, and the displacement is determined by the cumulative number of pulses, provided the period of the chain is well known. In summary, these linear encoders can be used, indistinctly, as velocity/displacement sensors, or as near-field chipless-RFID tags.[20] In both applications, it is convenient to reduce the period of the chain as much as possible, since this improves the resolution (sensor functionality) and the data density per unit length, DPL (tag functionality).

Several examples of electromagnetic linear encoders based on chains of resonators have been reported in the literature [30, 34–37, 39]. To reduce the chain period, metallic inclusions consisting of narrow linear strips transversally oriented to the chain axis have been reported [8, 41]. Competitive chain periods have also been achieved by considering all-dielectric encoders based on permittivity contrast (an idea first reported in [10]), where the inclusions are either apertures (similar to optical encoders), or strips made of a high, or low, dielectric constant material, as compared to the one of the host substrate [42]. In the paper [41], very competitive linear encoders, with a period as small as 0.6 mm were reported. Such period was achieved by virtue of the narrow strips considered (with a width of 0.2 mm). Moreover, the length of the inclusions is only 6.4 mm. Consequently, the data density per length (DPL) and the data density per surface (DPS), in the functionality of the encoders as chipless tags, are as high as DPL = 16.7 bit/cm and DPS = 26.04 bit/cm^2, i.e. very competitive values.

Let us next briefly review the main aspects of the linear encoder system proposed in [41]. Since the encoders consist of transversally oriented narrow metallic strips, the challenge was to design a reader/stator able to detect such narrow and tiny separated strips by proximity. Despite the fact that the inclusions are linear strips, such strips do not behave as half-wavelength resonators. In this linear encoder system, the strips simply modify (perturb) the coupling level between a pair of resonant stubs loading a host transmission line. Such stubs are oriented face-to-face, as depicted in Fig. 4.46, and constitute the essential part of the reader. Specifically, system functionality relies on the enhancement of the capacitive coupling between the stubs when a strip of the encoder aligns with the open-end terminations of the stubs. Obviously, such coupling enhancement requires a small air gap. For the bare reader (without strip on top of the stub terminations), the coupling between the stubs is very small. However, the presence of a relative short strip on top of the stub terminations enhances substantially the inter-stub coupling. Indeed, by considering identical uncoupled stubs, a single transmission zero at the frequency where the stub length is a quarter-wavelength, and at the

---

20 It should be emphasized that, as chipless-RFID tags, the encoders (equipped with a certain ID code) are read by proximity, through near-field coupling with the reader (stator). The tag must be displaced over the reader at short distance, and with proper alignment. Therefore, these chipless-RFID systems are not useful in certain applications, where long- or moderate-range reading is required. However, there are applications where the reading distance is not a key factor, e.g. authentication. Indeed, reading by proximity in this case may be even preferred, in order to avoid eavesdropping or spying. Secure paper and authentication of premium products are canonical applications of these chipless-RFID systems based on near-field and sequential bit reading [39]. The main advantage over other chipless-RFID systems, e.g. those based on the frequency domain [40], is the highly achievable data capacity, only limited by tag (encoder) size. In [36], 80-bit chipless-RFID tags implemented on ordinary paper are reported.

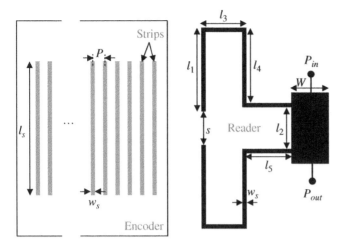

Figure 4.46 Topology of the double-stub reader and encoder, and relevant dimensions (in mm). $W = 1.81$, $l_0 = l_2 + w_s = 2.2$, $l_1 = 3.8$, $l_3 = 3.3$, $l_4 = 2.1$, $l_5 = 2.1$, $s = 1.60$, $P = 0.60$, $w_s = 0.20$, and $l_S = 6.4$. *Source:* Reprinted with permission from [41]; copyright 2019 IEEE.

odd harmonics, is expected. However, stub coupling, by the presence of a strip on top of the stub open-ends, splits the transmission zeros. Additionally, the response of the reader exhibits a pole that depends on the distance between the contact points of the stubs to the host line (l_2), and such pole also shifts down when inter-stub coupling increases. Thus, by appropriately designing the reader, it is possible to achieve a large excursion in the transmission coefficient for a certain frequency. The convenient strategy is to force the pole of the bare reader, f_0, to be identical to the first transmission zero of the reader loaded with a perfectly aligned strip (with the stub terminations), and to set the carrier signal, f_c, to that frequency. The specific design procedure is explained in detail in [41]. Figure 4.47 depicts the photograph of the fabricated reader and encoder. The frequency response of the reader with and without strip on top of the sensitive region is shown in Fig. 4.48, which indicates the carrier frequency of the feeding signal necessary for measuring purposes.

The linear encoder of Fig. 4.47 contains 100 linear strips. System validation was carried out by means of an experimental setup similar to the one of Fig. 4.23 or 4.35, but replacing the rotor shaft with a system able to provide linear motion of the encoder over the reader. The envelope function inferred by considering a carrier frequency set to $f_c = f_0 = 3.88$ GHz (see Fig. 4.48) is depicted in Fig. 4.49. The figure also depicts the envelope function corresponding to two coded chains, where encoding was achieved from the original chain by merely cutting certain inclusions (i.e. those corresponding to the logic state "0").[21] Strip cutting (representing encoder programming) is equivalent to strip absence, since inter-stub coupling requires a continuous strip on top of the open ends. The results indicate that the system is able to correctly read the encoders. It should be clarified that the voltage level for the "1" and "0" states is low and high, respectively, in coherence with the transmission coefficient at f_0. Nevertheless, this aspect is not relevant for the use of the system as displacement and velocity sensor, where pulse counting simply needs a different voltage level when the strip is perfectly aligned with the open-ended terminations of the stubs. The relative velocity between the encoder and the reader, inferred from the information of Fig. 4.49(c) by averaging over five periods, is 1.46 cm/s. The relative displacement between the reader and the encoder from a reference position is simply given by the cumulative number of peaks, or dips, in the

---

21 Note that the original chain is the one with code "1111 ... 1" in Fig. 4.49.

(a)

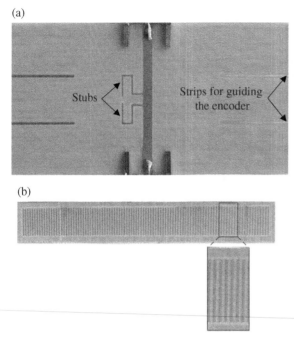

(b)

Figure 4.47 Photograph of the fabricated reader (a) and 100-inclusion encoder (b). The reader has been fabricated on the *Rogers RO4003C* with thickness h = 0.81 mm, dielectric constant ε_r = 3.55, and dissipation factor tanδ = 0.0021. For the encoder, the considered substrate is the same as the reader but with thickness h = 0.203 mm. The space occupied by the encoder is 60 mm × 6.4 mm.

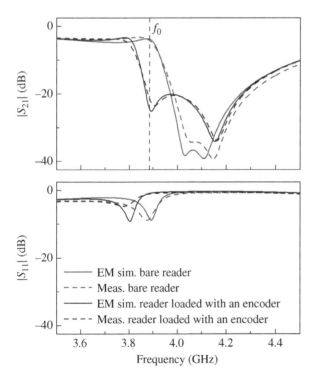

Figure 4.48 Measured and simulated responses of the bare reader and reader loaded with an encoder strip on top of it. The considered air gap is 0.2 mm. *Source:* Reprinted with permission from [41]; copyright 2019 IEEE.

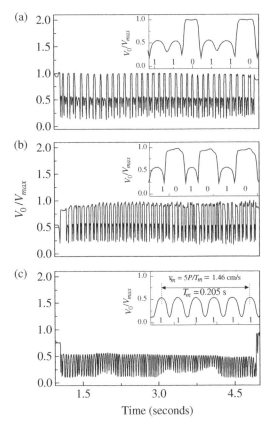

Figure 4.49 Measured normalized envelope function for the linear encoder system of Fig. 4.47, corresponding to the indicated codes. (a) code '110110...'; (b) code '101010...1'; (c) code '111111...'. *Source:* Reprinted with permission from [41]; copyright 2019 IEEE.

envelope function from that position, as it has been indicated before. A tolerance analysis carried out in [41] reveals that the air gap should be comprised between 0.17 and 0.22 mm. This range is somehow restrictive if the mechanical displacement system is not robust enough. Nevertheless, the air gap span can be expanded by slightly increasing the overall dimensions of the reader and the encoders.

4.3.3.1 Strategy for Synchronous Reading. Quasi-Absolute Encoders

In the electromagnetic encoders considered so far (either linear or angular), the displacement, from a reference position, is determined by the cumulative number of pulses. This approach is not always acceptable, as far as, for certain applications, providing the absolute position is a system requirement.[22] Absolute encoders should be able to determine the position of moving object regardless of their previous stages. For that purpose, a unique ID code should be assigned to the different discrete positions of the encoder. The required number of bits, N_b, depends on the total number of discrete positions, in turn given by the length, L, of the encoder and by the required resolution, p. Thus, the minimum number of bits should satisfy

$$N_b \geq \log_2\left(\frac{L}{p}\right) \tag{4.17}$$

---

22 In motion control applications, if a reset occurs in a system based on cumulative pulse counting, it must be initiated from the reference position, and this is not always possible, or convenient.

Although the absolute resolution of the encoder is p, the key factor from the point of view of encoder performance is the total number of discrete points L/p, or N_b (see 4.17), which should be significant in high-resolution encoders. To give an illustrative example, for an encoder intended to measure up to $L = 1$ m displacements, it is reasonable to consider a resolution of $p = 1$ mm. Therefore, application of (4.17) indicates that at least $N_b = 10$ bits are necessary for unequivocally providing the position. It has been indicated before that encoding can be achieved by means of the presence/absence of a functional inclusion in the encoder chain. However, this would require as many parallel encoder chains as number of bits, which is not reasonable in microwave encoders if the number of required bits is relatively high (e.g. $N_b = 10$). There is, however, an alternative approach based on the use of a single encoded chain [9]. As the encoder moves, the absolute position is determined by simply reading the bit corresponding to the predefined position of the inclusion (either present or absent) crossing the sensitive part of the reader. Such bit plus the previous $N_b - 1$ bits provide a unique N_b-bit subcode that univocally identifies the encoder position. For that purpose, the complete ID code of the chain along the whole encoder must be chosen according to the De Bruijn sequence [43]. This guarantees that any N_b-bit subcode does not repeat for the total set of different positions of the encoder. Note that the system must incorporate a table with the position assigned to the different N_b-bit subcode sequences, and this is necessary for both motion directions. For this reason, it is necessary to provide a means to discriminate the encoder direction. Note that after a system reset, or after a change in the direction of motion of the encoder chain, it must displace N_b positions in order to read the N_b bits of the subcode necessary to identify the absolute position from the table. This "after-reset uncertainty" can be tolerated in many applications. However, strictly speaking, a system like this cannot be categorized as absolute encoder. The designation as "quasi-absolute encoder" seems to be more reasonable.

In order to read the chain encoded following the De Bruijn sequence (position chain), a clock signal providing the instants of time for reading is necessary. Such clock signal can be generated by means of a parallel chain with all the inclusions present at their predefined positions. Note that such chain, designated as clock chain, provides also the instantaneous velocity (therefore, velocity chain is a reasonable alternative designation). Obviously, the predefined inclusion locations of the clock and position chains must be situated at the same axial positions of the encoder, a necessary condition for synchronously reading the position chain. It should also be mentioned that, by properly designing the reader (to be discussed next), the clock chain can also be used to determine the motion direction, thereby avoiding the use of a third chain for that purpose.

Let us report a prototype example of a microwave linear quasi-absolute encoder. The layouts of the encoder and reader are depicted in Fig. 4.50 [9], where the relevant dimensions are indicated. The encoder inclusions are metallic rectangular patches transversally oriented to the encoder axis. As indicated in the previous paragraph, the patches of both chains are located in parallel positions, as needed for synchronous reading. For this reason, the length of the patches in the axial direction, w, is identical in both chains. However, the length of the patches in the transverse direction is unequal for the clock and position chains. This is justified by the need of using different carrier signals for reading both chains, thereby requiring different sensing elements (with different size) in the reader.

The reader is a microstrip line loaded with three rectangular complementary split ring resonators (CSRRs) [4, 44] etched in the ground plane and tuned to different frequencies. The resonator designated as CSRR_p is used to determine the ID code (thus the position chain should be displaced above this resonator). For reading the ID code, a harmonic (carrier) signal tuned to the resonance frequency of the CSRR_p, $f_{0,p}$, should be injected to the line. When a patch is on top of the CSRR_p, the resonator is detuned and, consequently, the transmission coefficient at $f_{0,p}$ changes drastically (increases). Therefore, the input signal is AM modulated at the output port, and the envelope

(a)

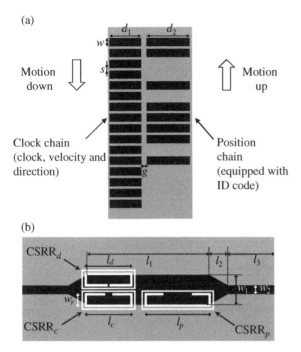

Motion
down

Motion
up

Clock chain
(clock, velocity and
direction)

Position
chain
(equipped with
ID code)

(b)

$CSRR_d$

$CSRR_c$

$CSRR_p$

Figure 4.50 Topology of the linear quasi-absolute encoder (a) and reader (b). The total area of the encoder is 30×70 mm^2, with the following metallic patch dimensions: $d_1 = 11.5$ mm; $d_2 = 15.9$ mm, $w = 3$ mm, $s = 1$ mm, and $g = 1.9$ mm. Reader dimensions are: $l_1 = 26.6$ mm; $l_2 = 3.8$ mm, $l_3 = 10$ mm, $w_1 = 6.4$ mm, $w_2 = 1.9$ mm, $l_c = l_d = 10.5$ mm; $l_p = 14.5$ mm, $w_r = 2.9$ mm. CSRR slots width is $c = 0.5$ mm, and ring splits are $s_d = 0.4$ mm, $s_c = 1.6$ mm, and $s_p = 6.2$ mm.

function contains the ID code (with as many peaks as patches in the position-chain). The other two resonators, designated as $CSRR_c$ and $CSRR_d$, are both aligned with the clock chain. When a patch of that chain is located below the $CSRR_c$ (tuned to $f_{0,c}$), the transmission coefficient at $f_{0,c}$ varies. Thus, by injecting a harmonic signal tuned to that frequency, such signal should be AM modulated at the output port, the envelope function being the clock signal necessary for synchronously obtaining the ID code. The clock signal also provides the encoder velocity, from the time lapse between adjacent peaks, since the clock-chain period is well known. To determine the motion direction, a redundant signal is generated from the AM modulation of a third injected harmonic signal tuned to the frequency of the resonator $CSRR_d$, $f_{0,d}$. By encoder motion, the clock-chain generates an AM signal at the output port of the line with a period (envelope function) identical to the one of the clock signal. The only difference is a lag or a lead depending on whether the clock-chain patches first cross the $CSRR_c$ or the $CSRR_d$. Thus, from this redundant signal, the motion direction can be deduced.

From a practical viewpoint, the three required harmonic signals should be injected to the three input ports of a 3 : 1 combiner, with the output port of such combiner connected to the input port of the CSRR-loaded line (reader). To independently obtain the envelope functions (containing the ID code, the clock signal – and velocity – and the redundant signal), a triplexer scheme, followed by envelope detectors, can be implemented (this strategy is similar to the one used for reading the rotary encoder of Fig. 4.39). The complete scheme of this approach, to separately obtain the ID code, the velocity, and motion direction, is depicted in Fig. 4.51. Alternatively, to avoid the use of three envelope detectors, a switching scheme based on a microcontroller can be considered [45] (a system of this type is discussed in the next section).

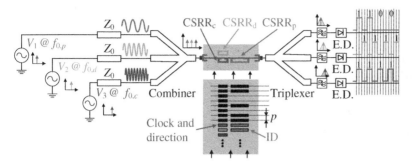

Figure 4.51 Sketch of the quasi-absolute encoder system with synchronous reading.

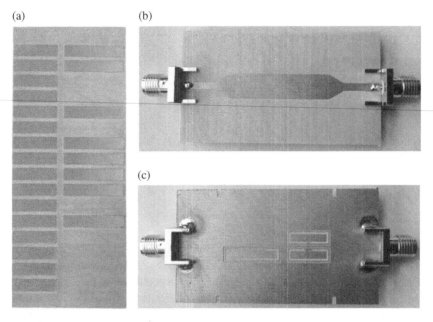

Figure 4.52 Photograph of the fabricated encoder (a), and top (b) and bottom (c) views of the reader. The ID code of the encoder is "1100-1011-1101-0000." The encoder was fabricated in the *Rogers RO4003C* substrate with thickness $h = 0.81$ mm, dielectric constant $\varepsilon_r = 3.38$, and loss factor $\tan\delta = 0.0022$. The considered substrate for the fabrication of the reader is identical to the one of the encoder, except the thickness, $h = 0.2$ mm.

System validation was carried out in [9] by considering an encoder with 16 different discrete positions, exhibiting a De Bruijn ID code corresponding to $N_b = 4$. The fabricated encoder and reader are depicted in Fig. 4.52 (dimensions are those indicated in the caption of Fig. 4.50). The dimensions of the CSRRs provide roughly equidistant resonance frequencies in the range 4–4.6 GHz, as Fig. 4.53 illustrates. According to this figure, the specific (measured) resonance frequencies for the CSRR$_p$, CSRR$_c$, and CSRR$_d$ are $f_{0,p} = 4.030$ GHz, $f_{0,d} = 4.270$ GHz, and $f_{0,c} = 4.540$ GHz, respectively. In [9], rather than using a combiner and a triplexer (or, alternatively, a switching scheme), the authors opted (as a preliminary proof-of-concept demonstration) for independently injecting the three interrogation signals (tuned to $f_{0,p}, f_{0,c}$, and $f_{0,d}$) to the input port of the reader line. Then, the envelope function of the generated AM modulated signal for each case was obtained. The experimental setup was similar to those used in previous encoders. Figure 4.54 depicts the measured envelope

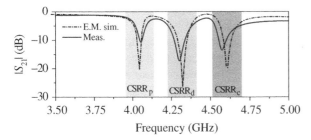

Figure 4.53 Simulated and measured frequency response (transmission coefficient) of the bare reader. *Source:* Reprinted with permission from [9]; copyright 2020 IEEE.

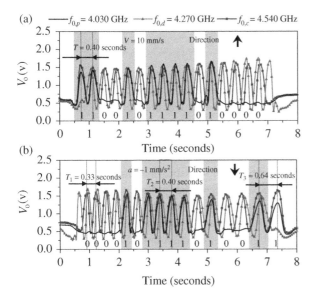

Figure 4.54 Measured envelope functions for (a) encoder motion up and $v = 10$ mm/s; (b) encoder motion down with constant acceleration of $a = -1$ mm/s^2. The considered vertical distance (air gap) between the encoder and the reader is 1 mm in both cases. *Source:* Reprinted with permission from [9]; copyright 2020 IEEE.

functions generated from the three carrier signals for two different cases. In one case, the encoder is displaced upward (according to the scheme of Fig. 4.50) at a constant velocity of $v = 10$ mm/s [Fig. 4.54(a)]. In the second case, the encoder is displaced in the downward direction with a nominal encoder acceleration of $a = -1$ mm/s^2 [Fig. 4.54(b)]. The ID codes are perfectly reproduced (the "1" logic states are revealed as peaks in the envelope function). For displacement in the upward direction, the clock-chain patches first cross the CSRR$_c$ and then the CSRR$_d$. Therefore, the redundant envelope function should be delayed with regard to the clock signal, as it actually occurs (i.e. the motion direction is correctly predicted). From the distance between adjacent pulses in the clock signal ($T_m = 0.40$ seconds), and taking into account the encoder period ($p = 0.4$ cm), the encoder velocity is found to be 10 mm/s, i.e. in perfect agreement with the nominal value (10 mm/s).

Concerning Fig. 4.54(b), with encoder displacement in the downward direction, the redundant envelope function is advanced with regard to the clock signal (the clock-chain patches first cross the CSRR$_d$ and then the CSRR$_c$). The measured instantaneous velocities corresponding to the indicated time lapses between different pairs of adjacent pulses (i.e. $T_1 = 0.33$ seconds, $T_2 = 0.40$ seconds, and

$T_3 = 0.64$ seconds) are $v_1 = 12.12$ mm/s, $v_2 = 10$ mm/s, and $v_3 = 6.25$ mm/s. This gives an acceleration of -1.01 mm/s^2, in very good agreement with the nominal value.

The previous results validate the functionality of the system as quasi-absolute microwave encoder with synchronous reading, able to provide the position, the velocity (and acceleration if required), and the direction of motion. An interesting prototype of a microwave linear encoder with synchronous reading is reported in [11]. In that system, the encoder consists of a single chain of unequal rectangular patches etched at periodic positions, and the reader is a microstrip line loaded with a pair of CSRRs, one inside the other. All the patches, regardless of their size, excite the smaller CSRR. Thus, the carrier signal tuned to the frequency of this CSRR, AM modulated at the output port of the line, provides the clock signal and the encoder velocity. By contrast, only the larger patches are able to drive the larger CSRR. Thus, a second harmonic signal tuned to the frequency of such CSRR is used to read the ID code (the logic state is determined by patch size). Despite the fact that this synchronous system uses a single encoder chain, it is not able to discriminate the direction of motion.

4.3.3.2 Application to Motion Control

Industrial applications of the quasi-absolute encoders reported in the previous section can be envisaged. Encoders, either linear or angular, are essential components in motion control, as far as the precise control of a moving object requires the accurate measurement of its position. For a precise motion control, closed-loop systems are typically used. In such closed-loop motion control systems, besides the motion controller and actuator (e.g. a step motor), position and velocity sensors are key components to provide the necessary feedback signal to the controller (which may eventually compensate for any potential error). There are a multitude of industrial systems where an accurate control of the position and velocity of their moving parts is essential for their correct functionality. Examples include conveyor belts, elevators, servomotors, and positioning systems, among others. Motion control is also fundamental in sectors as diverse as space, automotive and aeronautic industry, robotics, textile industry, packaging, and medical instrumentation, among others.

In particular, the quasi-absolute encoder system of Fig. 4.52 can prospectively be applied in conveyor belts and elevator systems. The accurate control of cabin position and velocity in elevators is typically carried out by means of optical rotary encoders. As it was previously mentioned, optical encoders provide very good resolution. However, the zone of the elevator system where the encoder is located (pulley) is subjected to contaminants, grease, and pollution, thereby requiring the regular service of the maintenance staff (otherwise, the functionality of the encoders may be jeopardized). Such harsh environments are also encountered in many industrial systems where conveyor belts are used. Thus, the application of electromagnetic encoders for the determination of the cabin position and velocity in elevators, as well as belt displacement in conveyor systems might be of interest. Modern elevators use rubber belts with a steel core that provides the necessary mechanical strength to support the tensile stress generated by the weight of the cabin. On the other hand, many conveyor belts are also made of rubber (sometimes combined with other materials), since rubber provides mechanical flexibility. Besides its mechanical properties, rubber is a good dielectric. Therefore, the rubber belts of elevators and conveyors, i.e. the moving target, can potentially be used as the substrate material for the electromagnetic encoder implementation. That is, rather than using an encoder, made of a dedicated dielectric material, attached to the moving object, the idea is to implement the encoder directly on the moving target, provided it is made of a dielectric material (rubber in the considered cases). For the implementation of the encoders (based on metallic patches, similar to those of Fig. 4.52) on rubber belts, additive processes such as screen-printing can be used. The use of commercial rubber elevator belts for encoder implementation was explored in [46], where the functionality of a quasi-absolute encoder displacement and velocity sensor system was demonstrated, despite the presence of the metallic core (carcase) in the belts (the carcase represents a substantial part of the cross section of the belt, but sufficiently separated from the belt edges).

The quasi-absolute encoder system reported in [46] is based on the same stator depicted in Fig. 4.52. The difference is the encoder, implemented on a commercial rubber elevator belt by screen-printing the two chains of patch inclusions. The size and separation of the patches are identical to those of the patches of the encoder of Fig. 4.52. The length, width, and thickness of the belt are 20 cm, 3 cm, and 3 mm, respectively. With such length and chain period, the clock chain contains 48 patches, i.e. 48 different positions can be discerned, and the number of bits of the subcode sequence necessary to univocally determine the encoder position is $N_b = 6$. Thus, the position chain was codified by means of the De Bruijn sequence, where any subset of adjacent $N_b = 6$ bits does not repeat. The metallic inclusions were screen-printed using the *Norcote ELG* conductive silver ink. Then the ink was UV cured by a halogen light (5 seconds at 500 W), and the printed belt was finally introduced in an air flow oven (10 minutes at 130 °C) to make thermal curing. Figure 4.55 depicts the photograph of the fabricated encoder.

Apart from the encoder, an advanced system architecture as compared to the one of Fig. 4.51 was implemented in the system reported in [46] [see Fig. 4.56(a)]. Let us review the different components next. For the generation of the harmonic signal, the VCO *HMC391LP4* was used. The control voltage of such component was managed by means of the *ATmega328* microcontroller (*Arduino* development platform). As shown in Fig. 4.56, the microcontroller output was connected to the 12-bit *MCP4725* DAC (digital/analogic converter), followed by an amplifier stage in order to cover the full operation range of the VCO. A linear displacement system (model *STM 23Q-3AN*) was employed to displace the encoder over the stator (the separation between the stator and encoder, or air gap, was set to 1 mm). Finally, the *ADL5511* commercial envelope detector was connected between the output port of the stator and the input port of the microcontroller. The photograph of the experimental setup is depicted in Fig. 4.56(b). The procedure to generate the signals and process the data is as follows. First, the microcontroller sets a specific control voltage V_1 (so that the VCO generates a tone at $f_{0,p}$, the carrier frequency of the position chain) and reads the output signal of the envelope (AM) detector. After the sample period, $t_s = 1.25$ ms, the control voltage is changed to V_2 (corresponding to the unmodulated tone $f_{0,d}$, carrier frequency of the direction chain) and the procedure is repeated (the microcontroller retrieves the signal given by the AM detector) until the end of the sample period, $2t_s$. Finally, at $2t_s$, the voltage is set to V_3 (with the VCO giving the tone at $f_{0,c}$, carrier frequency of the clock chain) and this voltage is kept until $t = 3t_s$. This sequence is constantly repeated while the encoder is in motion, and the three AM modulated signals (envelope functions) are obtained. With this system, the combiner and the triplexer of Fig. 4.51 are avoided, and a single AM detector suffices.

To validate the system, the encoder was displaced at a velocity of $v = 40$ mm/s in the upward direction. The measured signal at the output port of the envelope (AM) detector, which contains the three AM demodulated signals (envelope functions), is depicted in Fig. 4.57(a), whereas Fig. 4.57 (b) shows the separated envelope functions (achieved by the microcontroller). There is a good synchronism between the clock and the ID code signals, and the motion direction signal is delayed with

Figure 4.55 Photograph of the fabricated encoder, with patches screen-printed on a piece of commercial elevator belt.

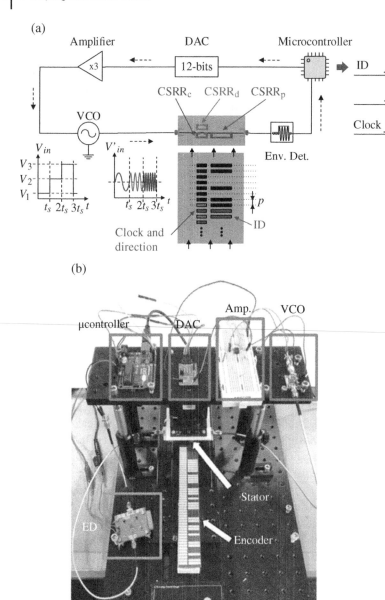

Figure 4.56 Sketch (a) and photograph (b) of the quasi-absolute encoder system with synchronous reading based on an advanced architecture with signal generation and processing by means of a microcontroller.

regard to the clock signal (this indicates that the encoder moves upward). The separation between peaks is 100 ms, in coherence with the nominal encoder velocity of 40 mm/s (the period of the metallic inclusions is $p = 4$ mm). Finally, the peaks in the ID code signal perfectly correlate with the presence of rectangular patches in the position chain (see Fig. 4.55).

The previous results validate system functionality with the quasi-absolute encoder implemented in a commercial elevator belt. Nevertheless, sensor functionality in a real scenario (either an elevator or conveyor system) has not been yet demonstrated. Let us mention that the screen-printed

(a)

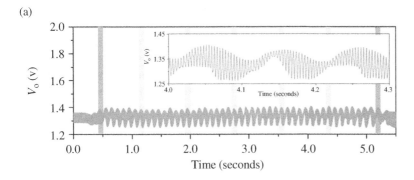

(b)

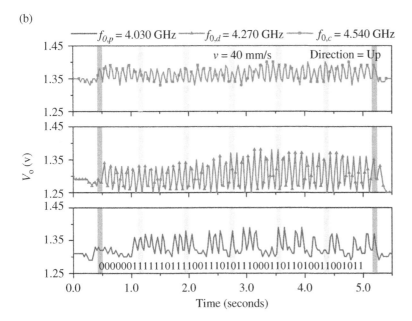

Figure 4.57 (a) Measured signal at the output port of the envelope (AM) detector; (b) measured envelope functions for the clock, direction, and position signals. *Source:* Reprinted with permission from [46]; copyright 2022 MDPI.

rubber encoder of Fig. 4.55 supports significant tensile stress and torsion, but tolerance against wear out, e.g. as consequence of pulley action in elevator systems, still needs to be demonstrated. Let us also indicate that due to mechanical vibration of the belts during motion, the reader/stator should be closely located to one of the pulleys that the considered systems typically require. The belt movement in the radial direction of the pulley is negligible, and, by this means, the air gap, or distance between the reader and the encoder, can be maintained within very restrictive limits, an important aspect for the correct functionality of the sensor.

4.4 Coupling-Modulation Sensors for Dielectric Characterization

Coupling-modulation sensors based on magnitude-level measurements at a single (operating) frequency are especially suited for displacement and velocity sensing (the previous sections report several examples). However, exploiting the magnitude variation of the transmission coefficient of a resonator-loaded line can also be applied to dielectric characterization. In the sensor implementations studied so far in the present chapter, a movable resonant element, or a set of inclusions (including also resonant elements), perturb the transmission coefficient of a transmission line-based structure. The strategy for sensitivity optimization was to achieve the maximum possible excursion of the transmission coefficient at the operating frequency. In certain implementations, the relative orientation between the sensor line and the movable element modulates the depth of the generated notch (see, e.g. Fig. 4.18), and the carrier frequency must be tuned to such notch frequency. In other prototypes, the movable element generates an overall shift in the transmission coefficient (including poles and zeros if they are present). In these cases, a significant excursion of the transmission coefficient is achieved in the vicinity of the transmission zeros, provided they are present.[23] Hence, notched responses are of interest not only for the implementation of frequency-variation sensors (as discussed in detail in Chapter 2), but also for the design of magnitude modulation sensors operating at a single frequency. The design strategy is to achieve the maximum possible excursion of the output variable (the magnitude of the transmission coefficient or the output voltage level) when the measurand varies within the input dynamic range. In sensors where the measurand shifts the transmission zeros and poles, considering responses exhibiting closely spaced transmission zeros and poles is convenient. An example of such a sensor, devoted to dielectric characterization is presented next.

According to the Foster's reactance Theorem [47], the reactance (or susceptance) of a passive and lossless two-terminal (one-port) network always exhibits a positive slope (derivative) with frequency. Consequently, if the network exhibits zeros and poles, they must alternate. The number of reactive elements of the network determines the total number of poles and zeros. For instance, the reactance of a capacitor and an inductor exhibits a pole and a zero at DC, respectively. A parallel LC resonant tank exhibits a zero in the reactance at DC, and a pole at the resonance frequency. Conversely, the role of the zeros and poles is interchanged in a series LC resonator. For a one-port network parallel connected to a matched transmission line, the zeros in the reactance of the network are transmission zeros (notches) of the composite structure, whereas total transmission (reflection zeros) occurs at the poles of the parallel-connected network. If the interest is to generate a response with at least one transmission zero and one reflection zero at finite (and closely spaced) frequencies by loading a line by means of a parallel one-port network, at least three reactive elements for that network are necessary. Moreover, for the implementation of a dielectric characterization sensor, the network should represent an implementable structure sensitive to variations in the dielectric constant of the surrounding medium. Thus, such network should describe at least a planar resonator.

In [48], a one-port network that combines a step impedance resonator (SIR) and a complementary split ring resonator (CSRR) was presented and used for the implementation of a dielectric constant sensor based on the variation of the magnitude of the transmission coefficient at a specific frequency. The topology of the whole sensor consists of a matched microstrip transmission line loaded with a shunt-connected SIR (also known as step impedance shunt stub [49]), with a CSRR etched in the ground plane, beneath the SIR capacitive patch. Figure 4.58 depicts the typical topology and the equivalent circuit model of this structure. The SIR is modeled by means of a shunt-connected series resonator, whereas the CSRR is modeled by means of a grounded parallel resonant tank. Thus, the lumped element equivalent circuit model (by excluding losses) of the composite resonator, designated as CSRR-loaded SIR in [48], is the one depicted in Fig. 4.58(b). In such model,

---

23 A clear example is the linear encoder of Fig. 4.47. In such encoder, the reader exhibits a transmission zero that splits when the strips of the encoder lie on top of the sensitive region, and the carrier frequency is tuned to one of the split zeros (see Fig. 4.48).

L_c and C_c are the inductance and capacitance, respectively, of the CSRR, L is the inductance of the narrow strip of the SIR, and C is the coupling capacitance between the SIR and the CSRR (i.e. the patch capacitance of the SIR). It is important to mention that the validity of this circuit model is subjected to the fact that CSRR dimensions should be larger than the dimensions of the SIR patch. If this condition is not satisfied, part of the electric field lines generated by the SIR patch is not expected to be circumscribed to the inner metallic region of the CSRR, and, under these conditions, an additional capacitance, C_x, should be included in the model [indicated in dashed line in Fig. 4.58 (b)]. Z_0 and kl are the characteristic impedance and electrical length, respectively, of the transmission line sections.

Note that, by excluding the parasitic capacitance C_x, the parallel network is described by means of four reactive elements. Therefore, the response of the whole structure should exhibit two poles and two zeros. One of the poles is at DC, where the resonator opens due to the effect of the patch capacitor C. The other pole, where total transmission is expected, is given by the resonance frequency of the CSRR, i.e.

$$\omega_p = \frac{1}{\sqrt{L_c C_c}} \tag{4.18}$$

Thus, one of the zeros should be located to the left of ω_p, whereas the other one should appear at a higher frequency. The specific transmission zero frequencies are given by the solutions of $Z = 0$, where the impedance of the CSRR-loaded SIR is

$$Z = j\left(L\omega - \frac{1}{C\omega} - \frac{L_c\omega}{L_c C_c \omega^2 - 1}\right) \tag{4.19}$$

Forcing (4.19) to be zero, the two transmission zero (angular) frequencies are found to be given by the positive solutions of

$$\omega_z = \sqrt{\frac{LC + L_c C_c + L_c C \pm \sqrt{(LC + L_c C_c + L_c C)^2 - 4LCL_c C_c}}{2LCL_c C_c}} \tag{4.20}$$

For model validation, a specific topology for the CSRR-loaded SIR resonator and substrate parameters were considered in [48] (see caption of Fig. 4.58). The characteristic impedance of the transmission line is identical to the reference impedance of the ports ($Z_0 = 50\,\Omega$). The frequency response of this structure inferred by electromagnetic simulation by excluding losses, and depicted in Fig. 4.59, reveals that there is a pole between the two transmission zeros, as expected. Concerning parameter extraction, four conditions are needed to univocally determine the four lumped elements of the circuit model. Equation (4.18), the two positive solutions of (4.20), plus the susceptance slope[24] at the angular frequency of the pole, ω_p, are such four conditions. The extracted element values are indicated in the caption of Fig. 4.59, which includes the circuit simulation with such element values. The agreement between the lossless electromagnetic and circuit simulation is very good for both the magnitude and the phase of the transmission and reflection coefficients up to frequencies beyond the second transmission zero. Therefore, these comparative results validate the circuit model of the composite CSRR-loaded SIR resonator.

---

24 The susceptance slope is found to be $b = \frac{\omega_p}{2}\frac{dB}{d\omega}\big|_{\omega_p} = \omega_p C_c$, i.e. identical to the one of the $L_c C_c$ parallel resonant tank. It can be inferred from the simulated S-parameters of the composite CSRR-loaded SIR resonator with the ports directly connected to it (i.e. by excluding the access lines). Particularly, the admittance of the resonator ($Y = jB = Z^{-1}$) can be expressed in terms of the reflection coefficient, S_{11}, according to $Y = -2Y_0 S_{11}/(1 + S_{11})$, where $Y_0 = 1/Z_0$.

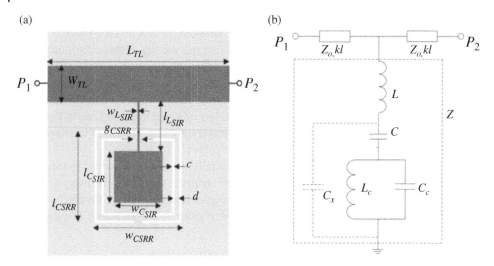

Figure 4.58 Typical topology of the CSRR-loaded SIR resonator loading a microstrip line (a) and equivalent circuit model by excluding losses (b). Relevant dimensions are indicated. The specific dimensions for the structure considered for validation purposes (in mm) are: l_{LSIR} = 4.6, l_{CSIR} = w_{CSIR} = 5.1, w_{LSIR} = 0.2, l_{CSRR} = 8.6, w_{CSRR} = 10.6, c = 0.2, d = 0.5, g_{CSRR} = 0.8, L_{TL} = 18, and W_{TL} = 3.4 (giving Z_0 = 50 Ω). Considered substrate parameters for validation are those of the *Rogers 4003C* substrate with dielectric constant ε_r = 3.55 and thickness h = 1.524 mm. *Source:* Reprinted with permission from [48]; copyright 2020 IEEE.

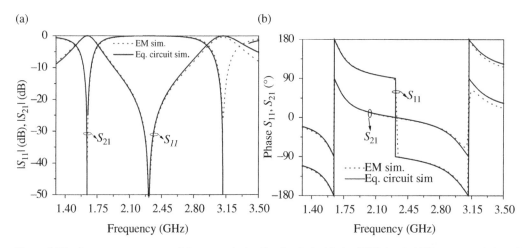

Figure 4.59 Frequency response of the transmission line loaded with the CSRR-loaded SIR resonator inferred from electromagnetic and circuit simulation by excluding losses. (a) Magnitude of the transmission and reflection coefficient; (b) phase of the transmission and reflection coefficient by excluding the access lines. The extracted lumped element values are L = 6.07 nH, C = 0.85 pF, L_c = 2.45 nH, and C_c = 1.92 pF. *Source:* Reprinted with permission from [48]; copyright 2020 IEEE.

The sensing principle of the proposed sensor is the variation of the transmission coefficient caused by the presence of a MUT (either solid or liquid) in contact with the CSRR, the sensitive element. As compared to the response of the bare sensor, the frequency response shifts down when a material is in contact with the CSRR. Thus, the device can operate as a frequency-variation sensor, where the output variable may be, e.g. the upper transmission zero frequency, or the pole frequency (the first transmission zero is less sensitive to the effects of the MUT [50]). However, a wideband interrogation signal is necessary for sensing under this mode of operation, as discussed in Chapter 2. Thus, the considered output variable is the magnitude of the transmission coefficient at a certain frequency, f_0. In particular, by choosing $f_0 = f_p$, the pole frequency of the bare sensor ($f_p = \omega_p/2\pi$), the transmission coefficient is expected to experience a significant variation by loading the CSRR with a MUT. Moreover, the transmission coefficient will be close to 0 dB for the bare sensor, and it will progressively decrease by increasing the dielectric constant of the MUT. Under these conditions, alternatively to the transmission coefficient, it is reasonable to consider the amplitude of the harmonic signal at the output port (in response to a harmonic interrogation signal tuned to f_p) as the output signal of the sensor. Such signal can be inferred by means of an envelope detector, following a scheme similar to those reported in previous sections in reference to microwave encoders.

In order to enhance the sensitivity, it is convenient to locate the second transmission zero (the most sensitive to variations in the dielectric constant) close to the pole. By doing this, the variation experienced by the transmission coefficient at f_0 with a change in the dielectric constant of the MUT, is expected to be stronger. Nevertheless, it should be mentioned that optimization of the sensitivity reduces the input dynamic range. For a proper sensor design, it is convenient to know the approximate range of variation of the dielectric constants of the considered MUTs. The reported prototype example is devoted to the dielectric characterization of solid samples, where the maximum dielectric constant is 10.2 (corresponding to a well-known commercially available microwave substrate) [48]. The sensor was designed in order to optimize the sensitivity and simultaneously achieve an input dynamic range of roughly 10.2 (the maximum considered dielectric constant value). For that purpose, the pole of the bare sensor, f_p, was set to the second transmission zero of the response of the sensor loaded with the MUT with $\varepsilon_{MUT} = 10.2$. The photograph of the designed sensor is depicted in Fig. 4.60, where the relevant dimensions are indicated. The simulated and measured responses for different materials (all with similar thickness, i.e. roughly 1.5 mm) are shown in Fig. 4.61, where it can be appreciated that f_p is very close to the second transmission zero of the response with $\varepsilon_{MUT} = 10.2$. It can also be seen that the excursion experienced by the transmission coefficient at $f_0 = f_p$, when ε_{MUT} varies between 1 (air) and 10.2 (the higher dielectric constant substrate), is roughly 17 dB, with the maximum value (for $\varepsilon_{MUT} = 1$) at 0.8 dB.

(a) (b)

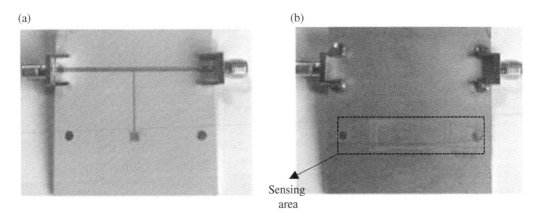

Sensing area

Figure 4.60 Photograph of the top (a) and bottom (b) views of the fabricated sensor based on a CSRR-loaded SIR resonator. Dimensions (in mm), in reference to Fig. 4.58(a), are: $l_{LSIR} = 18.2$, $l_{CSIR} = w_{CSIR} = 2.5$, $w_{LSIR} = 0.2$, $l_{CSRR} = 8.6$, $w_{CSRR} = 24.6$, $c = 0.2$, $d = 1.5$, $g_{CSRR} = 12.4$, $L_{TL} = 50$, and $W_{TL} = 1.28$ (50-Ω line). The sensor was implemented in the *Rogers RO3010* substrate with thickness $h = 1.27$ mm and dielectric constant $\varepsilon_r = 10.2$.

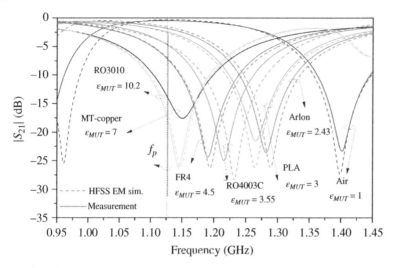

Figure 4.61 Simulated (using the *Ansys HFSS* electromagnetic solver) and measured responses of the sensor for the indicated MUTs (the dielectric constant is indicated). *Source:* Reprinted with permission from [48]; copyright 2020 IEEE.

In [50], it was demonstrated that by including the necessary electronics, and equipment, the output signal in the sensor of Fig. 4.60 can be the magnitude of the voltage signal present at the output port (in response to a harmonic signal tuned to f_p). An AM detector and an isolator, identical to those used in the microwave rotary encoders of Section 4.3.2.2, were added to the sensor structure of Fig. 4.60 for that purpose. The feeding harmonic signal was generated in [50] by means of a vector network analyzer (model *PNA N5221A*). Figure 4.62 depicts the amplitude of the output signal, visualized in an oscilloscope (model *Agilent MSO-X 3104A*), as a function of the dielectric constant of the MUT. The calibration curve inferred from these experimental data (included in Fig. 4.62) can be used to determine the dielectric constant of any unknown MUT from the measurement of the amplitude voltage level. It should be mentioned that for each sample, three measurements were carried out in order to ensure repeatability of the results (the averaged value for each sample as

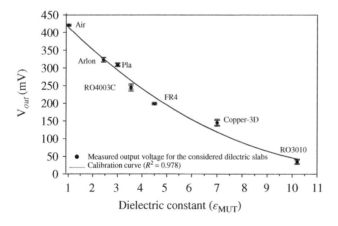

Figure 4.62 Variation of the output signal of the sensor of Fig. 4.60 (amplitude level of the harmonic signal at the output port) as a function of the dielectric constant of the MUT, ε_{MUT}. The considered samples are uncladded microwave substrates (*Arlon, RO4003C, FR4,* and *RO3010*) with well-known dielectric constant, or 3D-printed samples (PLA and Copper-3D) with the dielectric constant estimated by means of a commercial resonant cavity. *Source:* Reprinted with permission from [50]; copyright 2021 IEEE.

well as the obtained error bars are included in Fig. 4.62). The maximum sensitivity, defined as the absolute value of the derivative of the output voltage with the dielectric constant of the MUT, is found to be $S_{max} = 116.36$ mV. By adding the AM detector (plus the isolator), dielectric constant sensing involves a simple measurement of the amplitude of the output voltage signal, the interrogation signal being a single tone (harmonic) signal. Naturally, in a real sensor configuration, the vector network analyzer used to generate the harmonic signal can be replaced with a low-cost harmonic oscillator (or a VCO), and the output signal can be inferred by means of a post-processing stage (rather than being visualized in an oscilloscope).

Further details of the design of the sensor of Fig. 4.60 are given in [48, 50]. In [50], the sensor, conveniently redesigned, and including fluidic channels on top of the CSRRs, is also applied to the determination of solute content in liquid mixtures (particularly the concentration of isopropanol in DI water).

4.5 Advantages and Drawbacks of Coupling-Modulation Sensors

Like phase-variation sensors, probably the most interesting aspect of coupling-modulation sensors is device operation at a single frequency. This contributes to cost reduction of the associated electronics needed for the generation of the interrogation signal (a harmonic signal in these sensors). However, as compared to phase-variation sensors, coupling-modulation sensors are less robust to the effects of electromagnetic interference and noise, as far as magnitude measurements, rather than phase or frequency measurements, are involved in coupling-modulation sensors (nevertheless, in certain phase-variation sensors, the phase information is transformed to magnitude information, i.e. the magnitude of a transmission, or reflection, coefficient is the output variable in such sensors, since magnitude measurements are, in general, simple). As it has been shown throughout this chapter, an AM detector can be cascaded to the coupling-modulation sensors in order to obtain the magnitude of the output voltage (the output variable) in response to a harmonic feeding signal tuned to the convenient frequency. With this approach, the use of (high-cost) network analyzers to perform the measurements is avoided.

With the exception of the sensors reported in Section 4.4, focused on the measurement of the dielectric constant of materials and related variables, the other coupling-modulation sensors studied and discussed in this chapter are devoted to the measurement of linear or angular displacements and velocities. The reason is that the electromagnetic coupling between a transmission line and a sensing resonator (or a set of sensing resonators) is canonically controlled by the relative position and orientation between the line and the resonator/s. For short-range linear and angular displacement measurements (low input dynamic range), sensors based on a transmission line loaded with a single (symmetric) resonator (and based on symmetry disruption) constitute an interesting solution. Sensor size is the key advantageous aspect of such sensors, by virtue of the electrically small resonators typically used for sensing. However, size reduction of the sensing resonator decreases the input dynamic range (limited to the region of influence of the electromagnetic field generated by the line).

For the measurement of linear or angular displacements (and velocities) requiring higher input dynamic ranges, sensor solutions based on circular or linear chains of inclusions (typically, although not exclusively, metallic inclusions), in relative motion to the sensing structure (the stator), constitute a good option. These sensors, designated as electromagnetic encoders, cannot compete in terms of spatial resolution against their optical counterparts, which typically exhibit high densities of pulses per unit length, or angle, as compared to those achievable with electromagnetic encoders (nevertheless, rotary encoders with up to 1200 PPR have been reported in this chapter). Obviously, the encoder size for electromagnetic linear encoders is dictated by the required input dynamic range. For electromagnetic rotary encoders, the challenge is to accommodate the highest possible number of inclusions in the circular chain/s, with the minimum possible diameter of the rotor (one of the electromagnetic rotary encoders of Section 4.3.2.2, with two chains of 300 resonators and 1200 PPR, has been implemented in a dielectric disc with a diameter of 203.2 mm). Concerning the stator size, it is dictated by the considered transmission line and sensing element, and it is comparable to that of the displacement sensors based on symmetry disruption (reported in Sections 4.3.1 and 4.3.2.1).

Electromagnetic encoders (linear or angular) probably constitute the main contribution of this chapter. These devices are especially suited for motion control applications in harsh environments, typical of industry (with pollution, grease, dirtiness, etc.), or in ambient conditions subjected to extreme temperature, radiation, humidity, etc., where the functionality of their optical counterparts might be jeopardized. Electromagnetic linear encoders can also be applied to the implementation of near-field chipless-RFID systems with unprecedented data density and capacity (but this aspect is out of the scope of this book).

References

1 F. Martín, *Artificial Transmission Lines for RF and Microwave Applications*, John Wiley, Hoboken, NJ, USA, 2015.

2 J. Naqui, *Symmetry Properties in Transmission Lines Loaded with Electrically Small Resonators: Circuit Modeling and Applications*, Springer, Heidelberg, Germany, 2016.

3 R. Marqués, F. Medina, and R. Rafii-El-Idrissi, "Role of bianisotropy in negative permeability and left-handed metamaterials," *Phys. Rev. B*, vol. 65, p. 144440, 2002.

4 F. Falcone, T. Lopetegi, J. D. Baena, R. Marqués, F. Martín, and M. Sorolla, "Effective negative-ε stopband microstrip lines based on complementary split ring resonators," *IEEE Microw. Wireless Compon. Lett.*, vol. 14, pp. 280–282, Jun. 2004.

5 F. Martin, J. Naqui, A. Fernández-Prieto, P. Vélez, J. Bonache, J. Martel, and F. Medina, "The beauty of symmetry: common-mode rejection filters for high-speed interconnects and balanced microwave circuits," *IEEE Microw. Mag.*, vol. 18, pp. 42–55, Jan./Feb. 2017.

6 J. Naqui, M. Durán-Sindreu, and F. Martín, "Novel sensors based on the symmetry properties of split ring resonators (SRRs)," *Sensors*, vol. 11, pp. 7545–7553, 2011.

7 J. Naqui and F. Martín, "Application of broadside-coupled split ring resonator (BC-SRR) loaded transmission lines to the design of rotary encoders for space applications," *2016 IEEE MTT-S Int. Microw. Symp. (IMS'16)*, San Francisco, CA, USA, May 2016.

8 C. Herrojo, F. Muela, J. Mata-Contreras, F. Paredes, and F. Martín, "High-density microwave encoders for motion control and near-field chipless-RFID," *IEEE Sensors J.*, vol. 19, pp. 3673–3682, May 2019.

9 F. Paredes, C. Herrojo, and F. Martín, "Microwave encoders with synchronous reading and direction detection for motion control applications," *2020 IEEE-MTT-S Int. Microw. Symp. (IMS'20)*, Los Angeles, CA, USA, Jun. 2020.

10 C. Herrojo, P. Vélez, F. Paredes, J. Mata-Contreras, and F. Martín, "All-dielectric electromagnetic encoders based on permittivity contrast for displacement/velocity sensors and chipless-RFID tags," *2019 IEEE-MTT-S Int. Microw. Symp. (IMS'19)*, Boston, MA, USA, Jun. 2019.

11 F. Paredes, C. Herrojo, R. Escudé, E. Ramon, and F. Martín, "High data density near-field chipless-RFID tags with synchronous reading," *IEEE J. RFID*, vol. 4, no. 4, pp. 517–524, Dec. 2020.

12 J. Naqui, M. Durán-Sindreu, and F. Martín, "Alignment and position sensors based on split ring resonators," *Sensors*, vol. 12, pp. 11790–11797, 2012.

13 A. Karami-Horestani, C. Fumeaux, S. F. Al-Sarawi, and D. Abbott, "Displacement sensor based on diamond-shaped tapered split ring resonator," *IEEE Sensors J.*, vol. 13, no. 4, pp. 1153–1160, Apr. 2013.

14 A. K. Horestani, J. Naqui, D. Abbott, C. Fumeaux, and F. Martín, "Two-dimensional displacement and alignment sensor based on reflection coefficients of open microstrip lines loaded with split ring resonators," *Electron. Lett.*, vol. 50, no. 8, pp. 620–622, Apr. 2014.

15 A. K. Horestani, D. Abbott, and C. Fumeaux, "Rotation sensor based on horn-shaped split ring resonator," *IEEE Sensors J.*, vol. 13, pp. 3014–3015, 2013.

16 D. Schurig, J. J. Mock, and D. R. Smith, "Electric-field-coupled resonators for negative permittivity metamaterials," *Appl. Phys. Lett.*, vol. 88, Article ID 041109, 2006.

17 J. Naqui and F. Martín, "Transmission lines loaded with bisymmetric resonators and their application to angular displacement and velocity sensors," *IEEE Trans. Microw. Theory Tech.*, vol. 61, no. 12, pp. 4700–4713, Dec. 2013.

18 F. Aznar, M. Gil, J. Bonache, L. Jelinek, J. D. Baena, R. Marqués, and F. Martín, "Characterization of miniaturized metamaterial resonators coupled to planar transmission lines through parameter extraction," *J. Appl. Phys.*, vol. 104, Article ID 114501, Dec. 2008.

19 F. Martín, F. Falcone, J. Bonache, R. Marqués, and M. Sorolla, "Split ring resonator based left handed coplanar waveguide," *Appl. Phys. Lett.*, vol. 83, pp. 4652–4654, Dec. 2003.

20 J. S. Hong and M. J. Lancaster, *Microstrip Filters for RF/Microwave Applications*, John Wiley, New York, NY, USA, 2001.

21 R. Marqués, F. Martín, and M. Sorolla, *Metamaterials with Negative Parameters: Theory, Design and Microwave Applications*, John Wiley, Hoboken, NJ, USA, 2008.

22 C. Balanis, *Antenna Theory: Analysis and Design*, 2nd ed., John Wiley, New York, NY, USA, 1997.

23 D. M. Pozar, *Microwave Engineering*, 4th ed., John Wiley, Hoboken, NJ, USA, 2011.

24 J. Naqui and F. Martín, "Angular displacement and velocity sensors based on electric-LC (ELC) loaded microstrip lines," *IEEE Sensors J.*, vol. 14, pp. 939–940, Apr. 2014.

25 J. Naqui, J. Coromina, A. Karami-Horestani, C. Fumeaux, and F. Martín, "Angular displacement and velocity sensors based on coplanar waveguides (CPWs) loaded with S-shaped split ring resonator (S-SRR)," *Sensors*, vol. 15, pp. 9628–9650, 2015.

26 H. Chen, L. Ran, J. Huangfu, X. Zhang, K. Chen, T. M. Grzegorczyk, and J. A. Kong, "Left-handed materials composed of only S-shaped resonators," *Phys. Rev. E*, vol. 70, paper ID 057605, 2004.

27 H. Chen, L. Ran, J. Huangfu, X. Zhang, K. Chen, T. M. Grzegorczyk, and J. A. Kong, "Negative refraction of a combined double S-shaped metamaterial," *Appl. Phys. Lett.*, vol. 86, paper ID 151909, 2005.

28 H. Chen, L. Ran, J. Huangfu, X. Zhang, K. Chen, T. M. Grzegorczyk, and J. A. Kong, "Magnetic properties of S-shaped split ring resonators," *Prog. Electromagn. Res.*, vol. 51, pp. 231–247, 2005.

29 J. Naqui, M. Durán-Sindreu, and F. Martín, "Transmission lines loaded with bisymmetric resonators and applications," *IEEE MTT-S Int. Microw. Symp. Dig.*, Seattle, WA, USA, Jun. 2013.

30 F. Martin, C. Herrojo, J. Mata-Contreras, and F. Paredes, *Time-Domain Signature Barcodes for Chipless-RFID and Sensing Applications*, Springer, Heidelberg, Germany, 2020.

31 J. Mata-Contreras, C. Herrojo, and F. Martín, "Application of split ring resonator (SRR) loaded transmission lines to the design of angular displacement and velocity sensors for space applications," *IEEE Trans. Microw. Theory Tech.*, vol. 65, no. 11, pp. 4450–4460, Nov. 2017.

32 F. Martín and J. Naqui, "A contactless displacement and velocity measurement system," Patent, application WO-2017017181-A1, Feb. 2017.

33 J. Mata-Contreras, C. Herrojo, and F. Martín, "Detecting the rotation direction in contactless angular velocity sensors implemented with rotors loaded with multiple chains of split ring resonators (SRRs)," *IEEE Sensors J.*, vol. 18, no. 17, pp. 7055–7065, Sep. 2018.

34 C. Herrojo, F. Paredes, J. Mata-Contreras, E. Ramon, A. Núñez, and F. Martín, "Time-domain signature barcodes: near-field chipless-RFID systems with high data capacity," *IEEE Microw. Mag.*, vol. 20, no. 12, pp. 87–101, Dec. 2019.

35 C. Herrojo, J. Mata-Contreras, F. Paredes, A. Núñez, E. Ramón, and F. Martín, "Near-field chipless-RFID tags with sequential bit reading implemented in plastic substrates," *Int. J. Mag. Magnet. Mater.*, vol. 459, pp. 322–327, 2018.

36 C. Herrojo, M. Moras, F. Paredes, A. Núñez, E. Ramón, J. Mata-Contreras, and F. Martín, "Very low-cost 80-bit chipless-RFID tags inkjet printed on ordinary paper," *Technologies*, vol. 6, p. 52, 2018.

37 C. Herrojo, J. Mata-Contreras, F. Paredes, A. Núñez, E. Ramon, and F. Martín, "Near-field chipless-RFID system with erasable/programmable 40-bit tags inkjet printed on paper substrates," *IEEE Microw. Wireless Compon. Lett.*, vol. 28, pp. 272–274, Mar. 2018.

38 J. Mata-Contreras, C. Herrojo, and F. Martín, "Electromagnetic rotary encoders based on split ring resonators (SRR) loaded microstrip lines," *2018 IEEE MTT-S Int. Microw. Symp. (IMS'18)*, Philadelphia, PA, USA, Jun. 2018.

39 C. Herrojo, J. Mata-Contreras, F. Paredes, and F. Martín, "Near-field chipless RFID system with high data capacity for security and authentication applications," *IEEE Trans. Microw. Theory Tech.*, vol. 65, pp. 5298–5308, Dec. 2017.

40 S. Preradovic and N. C. Karmakar, *Multiresonator-Based Chipless RFID: Barcode of the Future*, Springer, Heidelberg, Germany, 2012.

41 C. Herrojo, F. Paredes, and F. Martín, "Double-stub loaded microstrip line reader for very high data density microwave encoders," *IEEE Trans. Microw. Theory Tech.*, vol. 67, no. 9, pp. 3527–3536, Sep. 2019.

42 C. Herrojo, F. Paredes, and F. Martín, "3D-printed high data-density electromagnetic encoders based on permittivity contrast for motion control and chipless-RFID," *IEEE Trans. Microw. Theory Tech.*, vol. 68, no. 5, pp. 1839–1850, May 2020.

43 N. G. de Bruijn, "Acknowledgement of priority to C. Flye Sainte-Marie on the counting of circular arrangements of 2n zeros and ones that show each n-letter word exactly once," *T.H.-Report 75-WSK-06*, Technological University Eindhoven, 1975.

44 J. D. Baena, J. Bonache, F. Martín, R. Marqués, F. Falcone, T. Lopetegi, M. A. G. Laso, J. García, I. Gil, M. Flores-Portillo, and M. Sorolla, "Equivalent circuit models for split ring resonators and complementary split rings resonators coupled to planar transmission lines," *IEEE Trans. Microw. Theory Tech.*, vol. 53, pp. 1451–1461, Apr. 2005.

45 F. Paredes, C. Herrojo, J. Mata-Contreras, and F. Martín, "Near-field chipless-RFID sensing and identification system with switching reading," *Sensors*, vol. 18, p. 1148, 2018.

46 F. Paredes, C. Herrojo, A. Moya, M. Berenguel-Alonso, D. Gonzalez, J. Bruguera, C. Delgado-Simao, and F. Martín, "Electromagnetic encoders screen-printed on rubber elevator belts and application to the measurement of cabin position and velocity," *Sensors*, vol. 22, paper 2044, 2022.

47 R. A. Foster, "A reactance theorem," *Bell Syst. Tech. J.*, vol. 3, pp. 259–267, Apr. 1924.

48 P. Vélez, J. Muñoz-Enano, A. Ebrahimi, J. Scott, K. Ghorbani, and F. Martín, "Step impedance resonator (SIR) loaded with complementary split ring resonator (CSRR): modeling, analysis and applications," *2020 IEEE-MTT-S Int. Microw. Symp. (IMS'20)*, Los Angeles, CA, USA, Jun. 2020.

49 J. Naqui, M. Durán-Sindreu, J. Bonache, and F. Martín, "Implementation of shunt connected series resonators through stepped-impedance shunt stubs: analysis and limitations," *IET Microw. Ant. Propag.*, vol. 5, pp. 1336–1342, Aug. 2011.

50 P. Vélez, J. Muñoz-Enano, A. Ebrahimi, C. Herrojo, F. Paredes, J. Scott, K. Ghorbani, and F. Martín, "Single-frequency amplitude-modulation sensor for dielectric characterization of solids and microfluidics," *IEEE Sensors J.*, vol. 21, no. 10, pp. 12189–12201, May 2021.

5

Frequency-Splitting Sensors

This chapter is focused on the theory, design, practical realizations, and applications of frequency-splitting sensors. Similar to most coupling-modulation sensors, the working principle of frequency-splitting sensors is symmetry disruption. Frequency-splitting sensors are implemented by means of two identical (not necessarily symmetric) sensing resonant elements, symmetrically loading a transmission line, or a transmission-line-based structure, and sensing is based on the splitting in the resonance frequency that results by symmetry perturbation.[1] The most canonical, but not exclusive, procedure for symmetry truncation in pairs of resonant elements is by placing asymmetric dielectric loads on top of such resonators. Consequently, most reported frequency-splitting sensors have been applied to the dielectric characterization of solids and liquids (or to the measurement of other variables related to it, including material composition). Such sensors are able to detect differences between the so-called reference (REF) sample, placed on top of one of the resonant elements of the pair, and the sample (or material) under test (MUT), situated on top of the other resonator. Nevertheless, it is also possible to exploit frequency splitting for the implementation of displacement sensors, in this case by using movable, or partially movable, resonant elements.

This chapter begins by presenting the general working principle that governs the functionality of frequency-splitting sensors. Then, a section is devoted to the analysis of transmission lines symmetrically loaded with pairs of resonant elements, as potential (and the most simple and canonical) candidates for the implementation of frequency-splitting sensors. In the considered preliminary structures, inter-resonator coupling is unavoidable due to the proximity between the resonators forming the pair. It is shown that electromagnetic coupling between the sensing resonators degrades the sensitivity, specifically in the limit of small perturbations, thereby affecting adversely sensor resolution. Thus, strategies to overcome this limitation, including the cascade configuration, the splitter/combiner approach, and other specific stratagems for coupling cancellation in closely spaced resonators, are then presented and applied to the design of specific sensors for the dielectric characterization of solids and liquids. The chapter also reports a two-dimensional displacement and alignment sensor based on frequency splitting. Finally, the advantages and drawbacks of frequency-splitting sensors, as compared with other microwave sensors, are highlighted.

---

1 Most microwave sensors based on frequency splitting are implemented by means of topologies exhibiting axial symmetry. An exception is the so-called cascade configuration, to be discussed later, where the two (identical) sensing resonators are placed at different planes in the host transmission line. Nevertheless, the working principle is also the frequency splitting, generated by an asymmetry in resonator's loading or by any other procedure able to modify the characteristics of one resonant element with regard to the other.

Planar Microwave Sensors, First Edition. Ferran Martín, Paris Vélez, Jonathan Muñoz-Enano, and Lijuan Su.
© 2023 John Wiley & Sons, Inc. Published 2023 by John Wiley & Sons, Inc.

5.1 Working Principle of Frequency-Splitting Sensors

The general working principle of frequency-splitting sensor is symmetry disruption of a pair of resonant elements, in most cases symmetrically loading a transmission line or a transmission-line-based structure. Figure 5.1 illustrates such principle by considering a coplanar waveguide (CPW) transmission line symmetrically loaded with a pair of split-ring resonators (SRRs). If the SRRs are unloaded, or if they are loaded by means of identical dielectric loads, symmetry is preserved, and a single notch (resonance) in the vicinity of the fundamental resonance frequency of the isolated (either loaded or unloaded) resonator arises.[2] As it will be shown, the frequency deviation of such single notch, with regard to the resonance of the isolated resonator, under perfect symmetry depends on the electromagnetic coupling between both resonators. However, asymmetrically loading the SRRs truncates the axial symmetry, and the original (single) notch splits into two resonances. Obviously, the separation between such resonances depends on the level of asymmetry. Therefore, these structures can be used for sensing, especially for measuring the differences (e.g. the dielectric constant) between a REF material and the MUT, each one placed on top of a resonant element of the pair. Thus, these sensors operate similar to differential-mode sensors, the canonical output variable being the difference in the resonance frequencies of the asymmetrically loaded resonators. However, there are applications where the measurement of two variables is needed (e.g. the real and the imaginary part of the complex permittivity of materials). In such cases, a second output variable, dependent at least on the additional input variable, is required.[3] In frequency-splitting sensors, the difference in the magnitude of the two notches is the second output variable that can be used for sensing, as it will be shown later in this chapter. Let us mention that these sensors are able to measure any variable involving symmetry perturbation. Therefore, they can also be applied to the measurement of displacements and velocities (as far as movable resonant elements are considered for sensing). An example of a two-dimensional displacement and alignment sensor

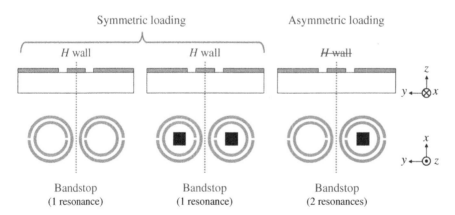

Figure 5.1 Sketch illustrating the working principle of frequency-splitting sensors, based on symmetry disruption. The figure considers the specific case of a CPW transmission line loaded with a pair of SRRs, and symmetry disruption is achieved by an asymmetric dielectric loading in the resonant elements (square inclusions). The CPW and the SRRs (etched in the back substrate side) are depicted in cross-sectional and bottom views, respectively, for better understanding.

---

2 Nevertheless, further notches are expected in the vicinity of the higher-order (harmonic) resonances of the SRRs.
3 In many sensors involving the measurement of several variables, the output variables exhibit cross sensitivities to the different input variables.

based on symmetry disruption in a transmission line loaded with a pair of broadside-coupled SRRs (BC-SRRs) will be later reported.

In frequency-splitting sensors, electromagnetic coupling between resonant elements (unavoidable if the resonators are closely spaced) degrades the sensitivity, since such inter-resonator coupling broadens the separation between the split resonances (when perfect symmetry is not fulfilled). Let us next analyze in detail the effects of inter-resonator's coupling in transmission lines symmetrically loaded with pairs of coupled resonators and discuss how such coupling affects the response of the structure and degrades the sensitivity.

5.2 Transmission Lines Loaded with Pairs of Coupled Resonators

In this section, three different structures consisting of transmission lines symmetrically loaded with a pair of coupled resonators are studied (such structures, especially the considered resonant elements, are representative and potential candidates for the implementation of frequency-splitting microwave sensors). The objective of the analysis is to obtain the resonance frequencies in terms of the resonator's parameters and inter-resonator coupling.

5.2.1 CPW Transmission Lines Loaded with a Pair of Coupled SRRs

The typical topology of a CPW transmission line symmetrically loaded with a pair of SRRs is depicted in Fig. 5.2(a) [1]. It can be appreciated that the SRRs, etched in the back substrate side, are closely spaced. This is unavoidable in practice, since the main mechanism of SRR excitation

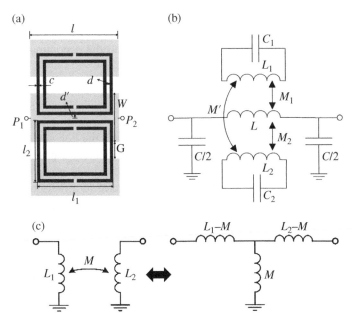

Figure 5.2 Typical topology of a CPW loaded with a pair of SRRs (a), lumped-element equivalent circuit model, considering magnetic coupling between SRRs (b), and equivalent T-circuit model of a two-port network consisting of two coupled inductors (c). The SRRs (in black) are etched in the back substrate side of the CPW (depicted in gray). The model considers the general case of different SRRs, or identical SRRs with different dielectric loading.

is the magnetic field generated by the line [2–4], which does not extend transversally beyond the slots of the CPW. Since for SRR excitation by means of the magnetic field, a significant portion of the magnetic flow generated by the line should penetrate the SRR loops, it follows that the separation between SRRs is limited. Therefore, for an accurate description of the structure, the mutual coupling between both SRRs cannot be neglected. Note also that in the structure of Fig. 5.2(a), the symmetry plane of the SRR is oriented transversely to the CPW line axis. By this means, SRR excitation by the electric field generated by the line is prevented, as discussed in [5, 6], and the model of the structure should only include a mutual inductance in order to describe the coupling mechanism between the line and the SRR pair.

Figure 5.2(b) depicts the lumped-element equivalent circuit model of the structure of Fig. 5.2(a). For the purpose of the present analysis, let us consider the most general case of an asymmetric structure. Line-to-resonator coupling is described by means of the mutual inductances M_1 and M_2, whereas inter-resonator coupling is accounted for through M' (i.e. the model assumes that magnetic coupling dominates the interaction between the SRRs, a reasonable hypothesis, since the face-to-face sides of the SRRs are expected to exhibit high current and hence a high magnetic field [7, 8]). The SRRs are modeled by the resonant tanks L_1–C_1 and L_2–C_2, and, finally, L and C are the inductance and capacitance, respectively, of the line. Note that the model of Fig. 5.2 (b) is a π-circuit. Since the parallel branches are purely capacitive, the poles of the series reactance provide the notch frequencies (or resonance frequencies of the coupled resonators). After some calculation, with the help of the equivalence indicated in Fig. 5.2(c), the reactance of the series branch is found to be

$$\chi_s(\omega) = \omega L + \omega^3 \cdot \left\{ \frac{C_1 M_1^2\left(1 - \dfrac{\omega^2}{\omega_2^2}\right) + C_2 M_2^2\left(1 - \dfrac{\omega^2}{\omega_1^2}\right) + 2\omega^2 M_1 M_2 M' C_1 C_2}{\left(1 - \dfrac{\omega^2}{\omega_1^2}\right)\left(1 - \dfrac{\omega^2}{\omega_2^2}\right) - \omega^4 M'^2 C_1 C_2} \right\} \tag{5.1}$$

where $\omega_1 = (L_1 C_1)^{-1/2}$ and $\omega_2 = (L_2 C_2)^{-1/2}$ are the angular resonance frequencies of the isolated resonators. The poles, or resonances, are given by those frequencies that null the denominator of the last term in (5.1), i.e.

$$\omega^2_\pm = \frac{\omega_1^2 + \omega_2^2 \pm \sqrt{(\omega_1^2 - \omega_2^2)^2 + 4M'^2\omega_1^4\omega_2^4 C_1 C_2}}{2(1 - M'^2\omega_1^2\omega_2^2 C_1 C_2)} \tag{5.2}$$

As it can be seen, the solutions do not depend on the mutual coupling between the line and the SRRs (M_1 and M_2). Thus, the single coupling mechanism with influence on the relative position between the resonance frequencies ω_+ and ω_- is inter-resonator coupling, described by means of M'. In situations where the SRRs are separated enough, M' can be neglected to a first-order approximation, and the solutions of (5.2) are simply ω_1 and ω_2. Nevertheless, as it has been pointed out, neglecting M' is not possible in most practical situations.

If the SRR-loaded line exhibits perfect symmetry, i.e. $M_1 = M_2 = M$, $\omega_1 = \omega_2 = \omega_0$, $L_1 = L_2 = L_r$, and $C_1 = C_2 = C_r$, expression (5.2) simplifies to

$$\omega_\pm = \frac{\omega_0}{\sqrt{1 \mp \dfrac{M'}{L_r}}} = \frac{\omega_0}{\sqrt{1 \mp k_M}} \tag{5.3}$$

where $k_M = M'/L_r$ is the magnetic coupling coefficient. However, the solution with the $(-)$ sign in the radicand of (5.3), ω_+, is not actually a transmission zero (notch) frequency. At this frequency, the numerator of the last term in (5.1) is also null, and by applying the L'Hôpital's rule, it follows that the series reactance is finite at ω_+ (therefore, there is not a pole at this frequency). Thus, it can be concluded that only one transmission zero (resonance) appears in the transmission coefficient of a CPW transmission line symmetrically loaded with two magnetically coupled SRRs. Moreover, such resonance appears to the left of ω_0, since $M' > 0$ (and therefore $\omega_- < \omega_0$).[4] Thus, this explains that in a perfectly symmetric structure, a single notch in the frequency response appears, despite the fact that the SRRs are coupled. In general, identical coupled resonators exhibit two split resonances, one being even and the other one being odd [7, 8]. However, since the excitation of the coupled SRRs is carried out by means of the fundamental (even) mode of the CPW transmission line, it follows that the odd resonance cannot be driven.

When symmetry is disrupted, e.g. as consequence of a change in the capacitance of one of the SRRs (due to dielectric loading), in general two notches appear (frequency splitting). However, a specific asymmetric case deserves special attention, i.e. identical resonance frequencies of the isolated resonators ($\omega_1 = \omega_2 = \omega_0$) with different inductances and capacitances ($L_1 \neq L_2$ and $C_1 = 1/L_1\omega_0^2 \neq C_2 = 1/L_2\omega_0^2$). In this case, expression (5.2) gives

$$\omega_\pm = \frac{\omega_0}{\sqrt{1 \mp \dfrac{M'}{\sqrt{L_1 L_2}}}} = \frac{\omega_0}{\sqrt{1 \mp k_M}} \tag{5.4}$$

Note that (5.4) is formally identical to (5.3). The difference is that the magnetic coupling coefficient is now given by $k_M = M'/\sqrt{L_1 L_2}$. The two solutions of (5.4) null the denominator of the last term in (5.1), and, in general, do not cancel the numerator of that term. However, if the following condition

$$C_1 M_1^2 + C_2 M_2^2 = 2 M_1 M_2 \sqrt{C_1 C_2} \tag{5.5}$$

equivalent to

$$\frac{M_1}{M_2} = \sqrt{\frac{L_1}{L_2}} = \sqrt{\frac{C_2}{C_1}} \tag{5.6}$$

is satisfied, then the numerator of the last term in (5.1) is also null at ω_+, and the impedance of the series branch is finite at this frequency (as the application of L'Hôpital's rule reveals). Thus, if the circuit elements satisfy (5.6), then only one transmission zero (at ω_-) is expected. Note that (5.6) represents a balance that forces the structure to behave similarly to the symmetric case. If (5.6) is not fulfilled, two notches at the frequencies given by (5.4), i.e. one above and the other one below ω_0, are expected.

For arbitrary SRR pairs, i.e. with different SRR frequencies ($L_1 \neq L_2$, $C_1 \neq C_2$, $\omega_1 \neq \omega_2$), the transmission zeros are given by the two solutions of (5.2), and the mutual coupling between the SRRs (M') enhances the distance between the transmission zeros, i.e.

$$\omega_+^2 - \omega_-^2 = \frac{\sqrt{(\omega_1^2 - \omega_2^2)^2 + 4M'^2\omega_1^4\omega_2^4 C_1 C_2}}{1 - M'^2\omega_1^2\omega_2^2 C_1 C_2} = \frac{\sqrt{(\omega_1^2 - \omega_2^2)^2 + 4k_M^2\omega_1^2\omega_2^2}}{1 - k_M^2} > \omega_1^2 - \omega_2^2 \tag{5.7}$$

where it has been assumed that $\omega_1 > \omega_2$. According to this result, in the limit of small perturbations, or small imbalances between the SRRs (i.e. for $\omega_1 \to \omega_2$), the difference between the notch frequencies, $\omega_+ - \omega_-$, the typical output variable, does not tend to zero. For the functionality of the structure as a sensor, this represents a degradation of the sensitivity in the limit of small perturbations, and therefore a limitation in the capability of the sensor to detect small changes between the SRRs (caused, e.g., by different dielectric loadings), representing a degradation in sensor resolution, as well.

---

4 The single notch frequency (at ω_-) for the symmetric situation can also be inferred by applying the magnetic wall concept, as discussed and demonstrated in [1].

(a) 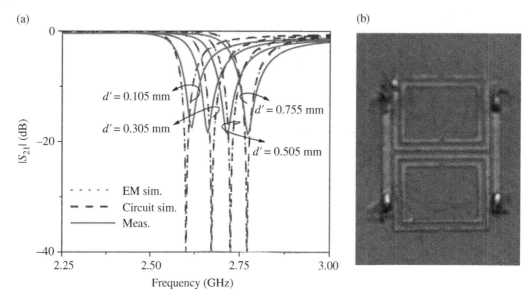 (b)

Figure 5.3 Frequency response of the symmetric SRR-loaded CPW specified in the text for different values of d' (a), and detail of one of the fabricated samples (b). Though not necessary under perfect symmetry, the ground planes of the CPW are connected through vias and backside strips, in order to prevent the appearance of the parasitic slot mode of the CPW, in case the SRRs are asymmetrically loaded. *Source:* Reprinted with permission from [1]; copyright 2015 IEEE.

For validation of the previous analysis, several SRR-loaded CPW structures were designed and fabricated in [1]. The shift in the (single) notch frequency for the symmetric case, as consequence of a variation of the distance between SRRs, was verified by fabricating four identical symmetric SRR-loaded CPWs, except in the SRR separation. SRR and CPW dimensions (see Fig. 5.2) were set to $l_1 = 4.8\,\text{mm}$, $l_2 = 3.8\,\text{mm}$, $c = d = 0.2\,\text{mm}$, $l = 5.6\,\text{mm}$, $W = 3\,\text{mm}$, and $G = 1.01\,\text{mm}$. Figure 5.3 depicts the transmission coefficient inferred by lossless electromagnetic simulations using *Agilent Momentum*, corresponding to the four structures with different values of inter-resonator distance, $2d'$ (the figure includes also one of the fabricated prototypes). The considered substrate was the *Rogers RO3010* with dielectric constant $\varepsilon_r = 11.2$ and thickness $h = 1.27\,\text{mm}$. The circuit parameters for the four considered cases (extracted from the method reported in [4]) are shown in Table 5.1. As d' increases, the mutual coupling M' decreases, and the resonance frequency increases. Note that the other circuit parameters do not significantly vary in the different implementations, and the agreement between circuit and electromagnetic simulations in the region of interest is very good, pointing out the validity of the model. The measured responses (also included in Fig. 5.3) are in good agreement with the simulations (slight discrepancies are due to

Table 5.1 Extracted circuit parameters (symmetric case) for different values of d'.

d' (mm)	L (nH)	C (pF)	C_r (pF)	L_r (nH)	M (nH)	M' (nH)
0.105	1.82	1.58	0.44	6.85	0.82	1.74
0.305	1.86	1.58	0.44	6.85	0.82	1.29
0.505	1.84	1.57	0.43	6.85	0.81	1.02
0.755	1.85	1.55	0.43	6.85	0.80	0.80

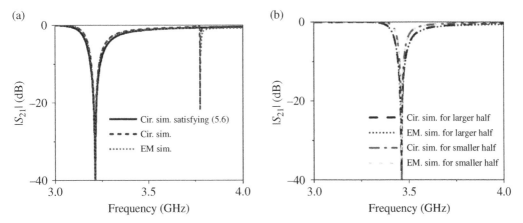

Figure 5.4 (a) Frequency response for the asymmetric SRR-loaded CPW with identical SRR resonance frequencies ($f_1 = f_2 = f_0 = \omega_0/2\pi = 3.47$ GHz), and circuit simulation corresponding to $L_1 = 5$ nH, $L_2 = 7.2$ nH, $M_2 = 0.774$ nH, $M_1 = 0.645$ nH, i.e. satisfying (5.6); (b) responses of the CPW loaded only with the larger or smaller SRR, to show that the isolated SRRs are tuned to identical frequency. Note that when both SRRs are present (and inter-resonator coupling is thereby activated), the first, f_-, and second, f_+, resonance frequencies appear to the left and right, respectively of $f_1 = f_2 = f_0$, the frequency of the isolated resonators. *Source:* Reprinted with permission from [1]; copyright 2015 IEEE.

the effects of losses, not included in the electromagnetic and circuit simulations, and to fabrication related tolerances).

Let us now consider an asymmetric SRR-loaded line with identical resonance frequency for both (isolated) SRRs ($\omega_1 = \omega_2 = \omega_0$) [1]. The considered geometrical parameters were set in this case to: for the CPW, $l = 5.6$ mm, $W = 3$ mm, $G = 1.01$ mm; for the smaller SRR, $l_1 = 4$ mm, $l_2 = 3$ mm, $c = 0.2$ mm, $d = 0.1$ mm, $d' = 0.305$ mm; for the larger SRR, $l_1 = 4.5$ mm, $l_2 = 4.1$ mm, $c = 0.2$ mm, $d = 0.725$ mm, $d' = 0.155$ mm.[5] The extracted parameters were found to be: $L = 1.77$ nH, $C = 1.6$ pF, $L_1 = 5.0$ nH, $C_1 = 0.42$ pF, $L_2 = 7.2$ nH, $C_2 = 0.29$ pF, $M_1 = 0.513$ nH, $M_2 = 0.774$ nH, $M' = 0.96$ nH. Figure 5.4 depicts the simulated response of this structure (the same substrate as the symmetric structure of Fig. 5.3 was considered). As far as condition (5.6) is not fulfilled, two notches (one above and the other below ω_0) are present in the response. The agreement between the circuit and electromagnetic simulation is also very good in this case. Nevertheless, note that the second notch is very soft. On the other hand, generating a structure layout with extracted parameters exactly satisfying (5.6) is difficult in practice. Nevertheless, by slightly modifying M_1 (see caption of Fig. 5.4), such condition is fulfilled (the other circuit parameters take the values indicated above). The circuit response, also included in Fig. 5.4, confirms that the second notch disappears in this case.

Finally, an asymmetric structure with different SRR resonance frequencies is considered [1]. The geometry is as follows: $l = 5.6$ mm, $W = 3$ mm, $G = 1.01$ mm; larger SRR: $l_1 = 4.8$ mm, $l_2 = 4.6$ mm, $c = d = 0.2$ mm; lower SRRs: $l_1 = 4.8$ mm, $l_2 = 3.8$ mm, $c = d = 0.2$ mm. The authors of [1] obtained the frequency response and the circuit parameters for four different values of d' (not shown), by considering the substrate parameters of the previous cases. The pair of notches (at $f_- = \omega_-/2\pi$ and $f_+ = \omega_+/2\pi$), depicted in Fig. 5.5, verifies that their distance increases as d' decreases, and $f_- < f_2$ and $f_+ > f_1$, in agreement with the theory.

---

5 Note that d' is actually defined in Fig. 5.2 as the distance between the SRR and the axial plane of the CPW. Thus in symmetric structures, the distance between SRRs is $2d'$. In nonsymmetric structures, d' may be different for both resonators.

(a) (b)

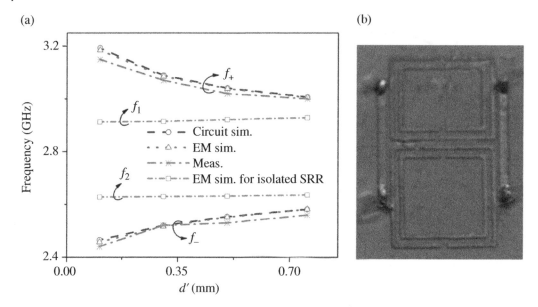

Figure 5.5 Variation of the notch frequencies as a function of d' for the asymmetric SRR-loaded CPW indicated in the text (a), and detail of the fabricated sample (bottom face) for $d' = 0.105$ mm (b). Reprinted with permission from [1]; copyright 2015 IEEE.

To end this section, let us mention that the considered model of Fig. 5.2, which describes a CPW loaded with a pair of coupled SRRs, suffices for the purpose of studying the effects of asymmetry and inter-resonator coupling on the frequency separation between the generated notches, the canonical output variable in frequency-splitting sensors.[6] Through the analysis of this model, and validation by means of electromagnetic simulation and experiment, it has been concluded that the coupling between the SRRs degrades the sensitivity at small perturbations. Alternative strategies to mitigate such coupling effects will be later presented in this chapter.

5.2.2 Microstrip Transmission Lines Loaded with a Pair of Coupled CSRRs

Figure 5.6(a) depicts a typical topology of a microstrip line loaded with a pair of coupled (i.e. closely spaced) complementary split-ring resonators (CSRRs). According to the circuit model, depicted in Fig. 5.6(b), the microstrip line is modeled by the inductance L and the capacitance C, the capacitances C_{c1} and C_{c2} account for the electric coupling between the line and the CSRRs, described by the resonant tanks L_1–C_1 and L_2–C_2, and C_M accounts for their mutual electric coupling. The model, valid as long as the CSRRs are electrically small, considers the general case of a microstrip line loaded with different CSRRs (asymmetric structure). Contrary to most CSRR-loaded lines, based

---

6 Apart from the model of Fig. 5.2(b) [1], and those reported in [2–5], other models of SRR-loaded CPW transmission lines have been presented. For example, in [9], CPWs loaded with multiple pairs of SRRs, with coupling between the SRR pairs of adjacent cells, are modeled. Paper [10] presents a circuit model of SRR-loaded lines, and lines loaded with other magnetic inclusions, that includes the effects of losses in the resonant elements. In [11, 12], wideband models of SRR-loaded CPWs are presented. Finally, let us mention that there are available models linking the circuit elements of SRRs with their geometrical parameters and parameters of the substrate [10, 13, 14]. However, such models are valid under restrictive conditions, not fulfilled in most practical situations where the SRR is the loading element of a transmission line. For this main reason, the synthesis of SRR-loaded lines from the circuit elements on the basis of parameter extraction [4, 6] is the usual procedure.

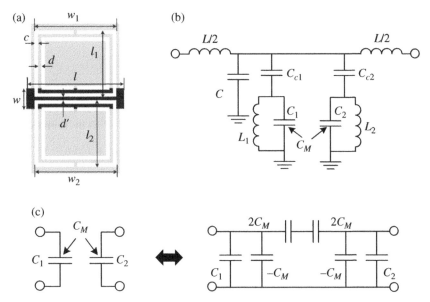

Figure 5.6 Typical topology of a microstrip line loaded with a pair of coupled CSRRs (a), lumped element circuit model (b), and equivalent π-circuit model of a two-port network consisting of two coupled capacitors (c). The ground plane, where the CSRRs are etched, is depicted in gray color. The conductor strip (upper metal) is depicted in black.

on a single CSRR per unit cell [15–23], the structure of Fig. 5.6 consists of a microstrip line loaded with two (coupled) CSRRs, since the intention is to use this structure for sensing, based on symmetry disruption and frequency splitting. There is a significant difference between the model of Fig. 5.6(b) and the model of the microstrip line loaded with a single CSRR reported in [13]. In [13], the capacitance of the line was considered the coupling capacitance between the line and the CSRR. The reason is that if a single CSRR is etched in the ground plane with its center aligned with the line axis, the conductor strip is located above the inner metallic region of the CSRR. Under these conditions, the electric field lines generated by the microstrip line entirely penetrate (at least to a first order approximation) the inner metallic region of the CSRRs (this is valid provided the conductor strip does not extend significantly beyond the extremes of the CSRR). However, in the structure of Fig. 5.6(a) not all the electric field lines penetrate the inner metallic regions of the CSRRs, because the bisecting plane separating both CSRRs coincides with the axis of the line. Hence, it is necessary to include a capacitor, C, between the conductor strip and ground in the model. Strictly speaking, this is not the line capacitance, but the capacitance between the inter-resonator metallic region and the conductor strip (note that this capacitance should increase by increasing the distance between CSRRs, at the expense of a decrease in C_{c1} and C_{c2}). It is also worth mentioning that the relative orientation between the CSRRs and the line axis prevents from the excitation of the CSRRs through the magnetic field generated by the line (i.e. mixed coupled is not allowed, and the single driving mechanism of the CSRR by the line is electric coupling [5]).

The transmission zeros of the circuit of Fig. 5.6(b) are given by those frequencies that null the reactance of the shunt branch (note that C does not have any influence on these frequencies). By forcing such reactance to be zero, the following bi-quadratic equation results [24]

$$A\omega^4 + B\omega^2 + D = 0 \tag{5.8}$$

where ω is the angular frequency and the coefficients of the previous equation are

$$A = L_1L_2\{C_M(C_{c1} + C_{c2})(C_{c1} + C_1)(C_{c2} + C_2) - C_{c1}C_{c2}C_M^2\}$$
$$+ L_1L_2\{C_{c1}C_{c2}(C_{c1} + C_1)(C_{c2} + C_2) - (C_{c1} + C_{c2})C_M^3\} \tag{5.9a}$$
$$= D \cdot L_1L_2\{(C_{c1} + C_1)(C_{c2} + C_2) - C_M^2\}$$

$$B = -D \cdot \{L_1(C_{c1} + C_1) + L_2(C_{c2} + C_2)\} \tag{5.9b}$$

$$D = C_{c1}C_M + C_{c2}C_M + C_{c1}C_{c2} \tag{5.9c}$$

Since D appears in the three coefficients of the biquadratic equation, dividing (5.8) by D gives

$$L_1L_2\{(C_{c1} + C_1)(C_{c2} + C_2) - C_M^2\}\omega^4 - \{L_1(C_{c1} + C_1) + L_2(C_{c2} + C_2)\}\omega^2 + 1 = 0 \tag{5.10}$$

providing,

$$\omega_\pm^2 = \frac{\omega_1^2 + \omega_2^2 \pm \sqrt{(\omega_1^2 - \omega_2^2)^2 + 4C_M^2\omega_1^4\omega_2^4 L_1L_2}}{2(1 - C_M^2\omega_1^2\omega_2^2 L_1L_2)} \tag{5.11}$$

where

$$\omega_i = \frac{1}{\sqrt{L_i(C_i + C_{ci})}} \tag{5.12}$$

with $i = 1, 2$ are the transmission zero frequencies of the isolated resonators (i.e. with their mutual coupling canceled). According to (5.11), if inter-resonator coupling is zero ($C_M = 0$), the two solutions of (5.8) are simply ω_1 and ω_2, as expected, since at these frequencies one of the two (uncoupled) parallel branches of the circuit of Fig. 5.6(b) is shorted to ground.

Let us now consider that $C_M \neq 0$, and that the structure exhibits perfect symmetry (i.e. $C_{c1} = C_{c2} = C_c$, $L_1 = L_2 = L_r$, and $C_1 = C_2 = C_r$, providing $\omega_1 = \omega_2 = \omega_0$). Under these conditions, the two solutions of (5.11) are:

$$\omega_\pm = \frac{\omega_0}{\sqrt{1 \mp \dfrac{C_M}{C_r + C_c}}} = \frac{\omega_0}{\sqrt{1 \mp k_C}} \tag{5.13}$$

where $k_C = C_M/(C_r + C_c)$ is the electric coupling coefficient. However, ω_- is not actually a physical solution. The reason is that both the numerator and the denominator of the reactance of the shunt branch are zero at that frequency, resulting in a finite reactance. Indeed, the single transmission zero frequency (ω_+), that arises when the CSRR-loaded line exhibits perfect symmetry, can be easily inferred by direct inspection of the circuit of Fig. 5.6(b) and the equivalence shown in Fig. 5.6(c) [7]. The coupling between CSRRs for this symmetric case has the effect of increasing the transmission zero frequency, as compared with the one of the isolated resonator (contrary to the situation of the previous section, where for perfect symmetry, the notch frequency of the SRR-loaded CPW shifts down). From a physical viewpoint, the odd resonance (ω_- for the microstrip line loaded with a pair of CSRRs) cannot be excited, since the propagating (quasi-microstrip) mode of the line is even (by contrast, in a pair of magnetically coupled SRRs, the odd resonance is the upper one).

Let us continue the parallelism with the case of the SRR-loaded CPW, by considering now an asymmetric CSRR-loaded line with identical CSRR resonance frequencies, i.e. $L_1 \neq L_2$, $C_1 \neq C_2$, but satisfying the condition $\omega_1 = \omega_2 = \omega_0$. In this case, expression (5.11) gives

$$\omega_\pm = \frac{\omega_0}{\sqrt{1 \mp \dfrac{C_M}{\sqrt{(C_{c1} + C_1)(C_{c2} + C_2)}}}} = \frac{\omega_0}{\sqrt{1 \mp k_C}} \tag{5.14}$$

In general, the two mathematical solutions given by (5.14) are both physical solutions (i.e. none of these frequencies provide a finite shunt reactance). The transmission zeros, or notches, are located to the left (ω_-) and right (ω_+) of ω_0, as expected. However, if the following balance condition is satisfied

$$\frac{C_{c1}}{C_{c2}} = \sqrt{\frac{C_1 + C_{c1}}{C_2 + C_{c2}}} = \sqrt{\frac{L_2}{L_1}} \qquad (5.15)$$

then only one transmission zero (at ω_+) arises, since the reactance is finite at ω_-.

For the general case, i.e. an asymmetric structure with arbitrary resonator frequencies ($L_1 \neq L_2$, $C_1 \neq C_2$) and $\omega_1 \neq \omega_2$, the transmission zeros are given by the two solutions of (5.11), and the coupling capacitance C_M increases the distance between the transmission zeros, i.e.

$$\omega_+^2 - \omega_-^2 = \frac{\sqrt{(\omega_1^2 - \omega_2^2)^2 + 4C_M^2\omega_1^4\omega_2^4 L_1 L_2}}{1 - C_M^2\omega_1^2\omega_2^2 L_1 L_2} = \frac{\sqrt{(\omega_1^2 - \omega_2^2)^2 + 4k_c^2\omega_1^2\omega_2^2}}{1 - k_C^2} > \omega_1^2 - \omega_2^2 \quad (5.16)$$

Moreover, $\omega_+ > \omega_1$ and $\omega_- < \omega_2$, where it is assumed that $\omega_1 > \omega_2$.

By comparing the analysis of this section with the one of Section 5.2.1, it follows that the behavior of microstrip lines loaded with pairs of electrically coupled CSRRs and CPWs loaded with pairs of magnetically coupled SRRs is very similar. The main difference is the relative position of the single notch frequency for the symmetric cases, or for the asymmetric balanced cases [i.e. those satisfying expressions (5.6) and (5.15)], to the right and left of ω_0 for the CSRR-loaded and for the SRR-loaded line, respectively. This similarity comes from the fact that the structures of Figs. 5.2 and 5.6 are dual structures. Indeed, using the following mapping between the elements of the circuits of Figs. 5.6(b) and 5.2(b),

$$C_M \leftrightarrow M' \qquad (5.17a)$$

$$L_{1,2} \leftrightarrow C_{1,2} \qquad (5.17b)$$

$$C_{1,2} + C_{c1,2} \leftrightarrow L_{1,2} \qquad (5.17c)$$

where the left- and right-hand sides refer to the elements of the circuit model of the CSRR-loaded microstrip line and SRR-loaded CPW, respectively, it follows that the expressions providing the resonance frequencies in all the considered (symmetric and asymmetric) cases, including the balance condition, are identical for both structures. It should be mentioned that, according to (5.17c), apparently the SRR- and the CSRR-based structures considered in the previous and in the present sections are not purely dual. The reason is that in the circuit of Fig. 5.6(b), the capacitances C_1 and C_2 are not actually the capacitances of the isolated resonators, but the capacitances of the CSRRs when they are coupled with the line. Contrarily, the inductances L_1 and L_2 of the circuit of Fig. 5.2(b) correspond to those of the isolated SRRs.[7] Actually, the capacitances of the isolated CSRRs are $C_1 + C_{c1}$ and $C_2 + C_{c2}$ (this justifies the definition of the coupling coefficient, k_C, as shown in

---

7 It should be mentioned that the (electric) coupling between the CSRR and the microstrip line has been modeled in the literature by directly connecting the mutual line-to-CSRR capacitance [C_{c1} or C_{c2} in Fig. 5.6(b)] between the resonant tank describing the CSRR [L_1–C_1 or L_2–C_2 in Fig. 5.6(b)] and the node where the whole shunt branch is connected. With this procedure, the capacitance of the CSRR resonant tank (C_1 or C_2) includes the effects of coupling with the line, and it cannot be considered to be the intrinsic capacitance of the isolated CSRR. By contrast, in the modeling of SRR-loaded lines, see Fig. 5.2(b), the mutual coupling between the line and the SRRs [M_1 or M_2 in Fig 5.2(b)] is explicitly considered, and therefore the corresponding inductances of the SRRs (the circuit duals of the capacitances of the CSRRs) are those corresponding to the isolated SRRs. This different treatment of the SRR- and the CSRR-based structures, in regard to line-to-resonator coupling, has been adopted in this book, in coherence with the related literature and, in particular, with references [1, 24].

5.13 and 5.14). Therefore, it is clear that both structures are purely dual and are described by dual circuits [25].

Three different CSRR-loaded microstrip structures were designed and fabricated in [24] for validation purposes. Two of such structures are symmetric, and the difference concerns the lengths $l_1 = l_2$ [see Fig. 5.6(a), which are set to $l_1 = l_2 = 3.8$ mm in one prototype (A) and to $l_1 = l_2 = 4.6$ mm in the other symmetric prototype (B). The asymmetric structure (prototype C) is implemented by considering the two previous CSRRs (i.e. $l_1 = 4.6$ mm and $l_2 = 3.8$ mm). The other parameters, in reference to Fig. 5.6(a), are $W = 1.18$ mm, $c = d = 0.2$ mm, $w_1 = w_2 = 4.8$ mm, and $d' = 0.2$ mm. The considered substrate is the *Rogers RO3010* with thickness $h = 1.27$ mm and dielectric constant $\varepsilon_r = 10.2$. The simulated frequency responses of these structures by excluding losses, inferred from the *Keysight Momentum* commercial software, are depicted in Fig. 5.7. Parameter extraction was carried out in [24] following the method reported in [26], with some modifications (as described in [24]), which are out of the scope of this book (see [24] for further details). Table 5.2 shows the extracted circuit parameters corresponding to the three fabricated prototypes (included in Fig. 5.7). The circuit simulations for the three considered prototypes are also depicted in Fig. 5.7. The good agreement with the lossless electromagnetic simulations validates the model of the CSRR-loaded lines. The measured responses are also in good agreement with the simulations (the slight

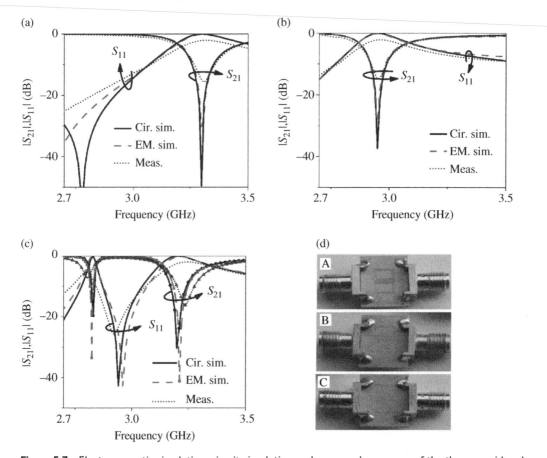

Figure 5.7 Electromagnetic simulation, circuit simulation and measured responses of the three considered CSRR-loaded microstrip structures described in the text. (a) Symmetric prototype A; (b) symmetric prototype B; (c) asymmetric prototype C; (d) photographs of prototypes A, B, and C. *Source:* Reprinted with permission from [24]; copyright 2016 IEEE.

Table 5.2 Extracted circuit parameters for the prototypes of Fig. 5.7

	L (nH)	C (pF)	C_1–C_2 (pF)	L_1–L_2 (nH)	$C_{c1,2}$ (pF)	C_M (pF)
A	3.47	0.086	1.96	1.10	0.336	0.177
B	3.53	0.092	2.27	1.18	0.360	0.177
C	3.50	0.085	1.96–2.27	1.10–1.18	0.347	0.177

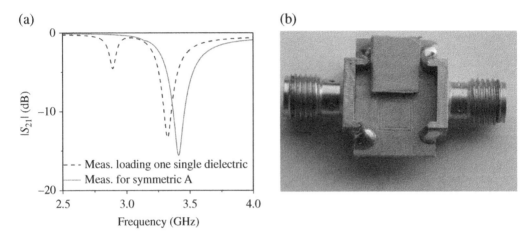

(a)   (b)

Figure 5.8 (a) Measured frequency response of the symmetric CSRR-loaded microstrip line corresponding to prototype A of Fig. 5.7 loaded with a dielectric slab placed on top of one of the CSRRs, compared with the response of the bare structure; (b) photograph of prototype A loaded with the dielectric slab in one of the CSRRs. *Source:* Reprinted with permission from [24]; copyright 2016 IEEE.

discrepancies can be attributed to fabrication-related tolerances and to the fact that losses have been excluded in the simulations).

CSRRs, like SRRs, are very sensitive resonators to the presence of a dielectric load on top of them. Therefore, let us consider an illustrative example of a preliminary prototype sensor/comparator (proof-of-concept) based on a microstrip line loaded with a pair of CSRRs, and operating as frequency-splitting sensor. The specific considered structure is, indeed, the prototype A of Fig. 5.7. The measured response of the bare device and the response when one of the CSRRs is loaded with a dielectric slab of permittivity identical to that of the substrate (10.2), thickness 1.27 mm, and dimensions 7.5 mm × 6.6 mm are depicted in Fig. 5.8 (the photograph of the sensor with the loaded sample is also included in the figure). As expected, frequency splitting can be observed in the asymmetrically loaded structure. However, inter-resonator coupling has two negative effects: (i) the resonance frequency of the unperturbed resonator is modified, and (ii) the first (lower frequency) notch is soft, as compared with the second one (conversely, in asymmetric SRR-loaded CPWs, the softer notch is the one located at the higher frequency). Thus, these results further confirm that inter-resonator coupling limits (degrades) the performance of these frequency-splitting sensors, as anticipated before. However, before discussing the strategies to mitigate such limitations, let us next consider a further structure (a microstrip line loaded with two step-impedance resonators – SIRs) as potential candidate for frequency-splitting sensor implementation.

(a)

(b)

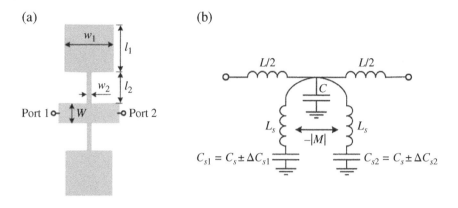

(c)

(d)

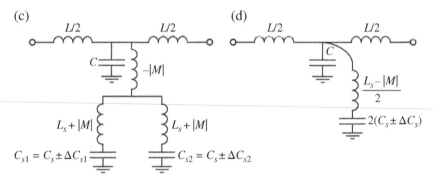

Figure 5.9 (a) Microstrip line loaded with a pair of shunt-connected SIRs; (b) equivalent circuit model with capacitive perturbations; (c) transformed equivalent circuit model; (d) transformed equivalent circuit model with symmetric perturbations.

5.2.3 Microstrip Transmission Lines Loaded with a Pair of Coupled SIRs

Step-impedance resonators (SIRs) are electrically small resonators that have been exhaustively used for filter design [27–30]. However, SIRs can also be of interest for sensing based on frequency splitting [31, 32]. The most canonical topology and circuit model of a frequency-splitting sensor based on a pair of SIRs and implemented in microstrip technology is depicted in Fig. 5.9 [31]. In this model, L and C are the line inductance and capacitance, respectively. The SIRs are modeled by means of shunt-connected series resonators, i.e. the narrow strip exhibits an inductive behavior, whereas the wide strip is a patch capacitor connected to ground [33]. Note that, in the model, the narrow (inductive) strips of the SIR are described by identical inductances L_s. Although this represents a certain loss of generality, for sensing applications the sensitive element is typically the patch capacitor (at least for dielectric characterization). Therefore, it is reasonable to consider that the inductance of both SIRs is identical.[8] By contrast, the SIR patches are modeled by different capacitors, C_{s1} and C_{s2}, where it is assumed that the specific values are due to perturbations with regard to the nominal capacitance, C_s (e.g. caused by a dielectric loading). Finally, due to the proximity between the SIRs, both symmetrically located at the same transverse plane of the microstrip line, inter-resonator coupling must be included in the model [31] for an accurate description of the structure of Fig. 5.9. In [31], it was assumed that SIRs' coupling is purely magnetic, reasonable

---

8 The model of Fig. 5.9 is adopted from [31], where identical SIR inductances are considered (as justified by the reasons explained in the text).

taking into account the significant separation between the capacitive patches. Moreover, it was argued in [31] that the magnetic coupling is negative due to the opposite currents flowing in the mirrored SIRs for the even resonance (this aspect will be later corroborated and discussed). Thus, the mutual inductance, $-|M|$, accounts for the electromagnetic interaction between both SIRs.

Using the equivalence depicted in Fig. 5.2(c), the circuit model of the SIR-loaded line of Fig. 5.9(b) can be transformed to the one shown in Fig. 5.9(c). The transmission zeros of the structure are given by those frequencies that null the reactance of the shunt branch. After some simple calculation, the following result (as given in [31]) is obtained:

$$\omega_\pm^2 = \frac{L_s(C_{s1} + C_{s2}) \pm \sqrt{[L_s(C_{s1} - C_{s2})]^2 + 4M^2 C_{s1} C_{s2}}}{2C_{s1}C_{s2}\left(L_s^2 - M^2\right)} \tag{5.18}$$

Let $\omega_1 = 1/\sqrt{L_s C_{s1}}$ and $\omega_2 = 1/\sqrt{L_s C_{s2}}$ be the resonance frequencies of the isolated SIRs and $k_M = -|M|/L_s$ the magnetic coupling coefficient. With these variables, (5.18) can be expressed as

$$\omega_\pm^2 = \frac{\omega_1^2 + \omega_2^2 \pm \sqrt{(\omega_1^2 - \omega_2^2)^2 + 4M^2\omega_1^4\omega_2^4 C_{s1} C_{s2}}}{2(1 - M^2\omega_1^2\omega_2^2 C_{s1} C_{s2})} \tag{5.19}$$

i.e. an expression formally identical to (5.2), and the difference of the squared solutions

$$\omega_+^2 - \omega_-^2 = \frac{\sqrt{(\omega_1^2 - \omega_2^2)^2 + 4M^2\omega_1^4\omega_2^4 C_{s1} C_{s2}}}{1 - M^2\omega_1^2\omega_2^2 C_{s1} C_{s2}} = \frac{\sqrt{(\omega_1^2 - \omega_2^2)^2 + 4k_M^2\omega_1^2\omega_2^2}}{1 - k_M^2} > \omega_1^2 - \omega_2^2 \tag{5.20}$$

is identical to (5.7), when it is written in terms of the resonance frequencies of the isolated resonators and coupling coefficient (right-hand-side term). There is, however, a fundamental difference between the circuit of Fig. 5.9 and that of Fig. 5.2. In the SRR-loaded line of Fig. 5.2, the single notch that appears under perfect balance is located at a frequency, ω_-, below the resonance frequency of the isolated SRRs, ω_0. However, in the SIR-loaded line of Fig. 5.9, perfect symmetry provides a notch situated above ω_0. To gain insight on this aspect, let us obtain the equivalent circuit to the one of Fig. 5.9(c) for perfect symmetry. Such circuit, depicted in Fig. 5.9(d), reveals that the notch frequency is

$$\omega_+ = \frac{\omega_0}{\sqrt{1 - \frac{|M|}{L_s}}} = \frac{\omega_0}{\sqrt{1 - k_M}} \tag{5.21}$$

which is certainly located above the resonance frequency of the isolated SIR. The negative magnetic coupling explains that for the even resonance of the SIR pair (the one that can be driven under perfect symmetry), the transmission zero shifts upward as compared with ω_0. Contrarily, in the line symmetrically loaded with two identical SRRs, the unique resonance shifts downward as compared with ω_0, because the magnetic coupling between the resonant elements is positive.[9]

---

9 Indeed, the relative position between the resonance frequency of an isolated resonator and the even-mode and odd-mode resonances that result when such resonator is magnetically coupled with an identical resonator determines the sign of the coupling coefficient. It is positive/negative if the even resonance appears at a lower/higher frequency than the one of the isolated resonator. For electric coupling between the resonant elements, the sign of the coupling coefficient shifts the even-mode and odd-mode resonances in the opposite direction, as compared with magnetic coupling.

For validation purposes, several symmetric and asymmetric SIR-based microstrip structures were simulated and characterized in [31, 32]. The fabricated structures are depicted in Fig. 5.10(a) and consist of three microstrip lines loaded with (i) a symmetric pair of SIRs, (ii) an asymmetric pair by increasing l_1 in one of the SIRs ($+\Delta C_{s2}$), and (iii) another asymmetric pair by decreasing l_1 ($-\Delta C_{s2}$). The procedure to extract the parameters was as follows. As a first step, the circuit parameters in Fig. 5.9(b) were extracted from electromagnetic simulation of the S-parameters by considering a single SIR without any perturbation. The parameter extraction procedure was similar to the one reported in [26], conveniently adapted to the considered circuit. Then the parameter extraction was repeated for the perturbed single SIR-loaded line. Finally, the determination of M was carried out by curve fitting the circuit simulation of the model in Fig. 5.9(b) to the electromagnetic simulation of the line loaded with symmetric SIRs [alternatively, expression (5.18) can also be used]. The extracted values are indicated in the caption of Fig. 5.10. As it can be observed, the agreement between the circuit and electromagnetic simulations is excellent for the symmetric structure (Fig. 5.10b), as well as for the microstrip line loaded with two unequal SIRs (Figs. 5.10c and 5.10d). The discrepancies in measurement are mainly due to fabrication-related tolerances and to some (unavoidable) uncertainty in the dielectric constant of the substrate. Figure 5.10 reveals that for the symmetric SIR-loaded line (without any perturbation in C_{s2}), the single notch (at 4.10 GHz) is located to the right of the corresponding notch in the single SIR-loaded line (at 3.83 GHz).[10]. Note that, as Figs. 5.10(b) and 5.10(c) reveal, the first notch is very soft, as compared with the second one, in the asymmetric structures.

To gain further insight on sensitivity degradation related to the effects of inter-resonator coupling in symmetrically loaded lines, let us depict the resonance frequencies of the circuit of Fig. 5.9(c). The results are plotted in Fig. 5.11(a), where unbalanced perturbations have been introduced by a capacitive variation in one of the SIRs ($\Delta C_{s1} = 0$, $\Delta C_{s2} \neq 0$) and different coupling coefficients $k_M = -|M|/L_s$ are considered [31]. Although the case with $k_M = 0$ cannot be implemented with a pair of shunt-connected SIRs sharing the same junction plane with the line, this situation illustrates the hypothetical case where uncoupled SIRs are used (nevertheless, pairs of uncoupled SIRs can be implemented by cascading them, rather than by parallel connecting them, as will be later pointed out in detail). As can be seen in Fig 5.11(a), when $k_M \neq 0$, the lower frequency f_- is more sensitive to $+\Delta C_{s2}$ than f_+, and, complementarily, the upper frequency f_+ is more sensitive to $-\Delta C_{s2}$. These curves tend to approach the resonance frequency of the isolated resonators as the perturbation increases. In addition, as k_M increases the frequency splitting $|f_+ - f_-|$ strengthens and degrades the sensitivity. Indeed, the maximum sensitivity corresponds to the case of uncoupled SIRs, and a significant drop in the sensitivity occurs for small perturbations. The sensitivity, shown in Fig. 5.11 (b), clearly depicts these effects.

Although the frequency-splitting phenomenon is interesting for sensing, it is clear that reducing inter-resonator coupling is convenient for sensitivity optimization. The next two sections discuss two different strategies for that purpose, i.e. the cascaded configuration and the splitter/combiner topology, and include several prototype sensors based on such approaches. Then, Section 5.5 presents two examples of frequency-splitting sensors where inter-resonator coupling is circumvented despite the fact that the resonant elements are located at the same axial position in the line. Nevertheless, it is important to mention that sensitivity degradation caused by inter-resonator coupling in frequency-splitting sensors is not a general property of planar sensors based on resonant

---

10 Note that in [31, 32], from which Figs. 5.10 and 5.11 are reprinted, the nomenclature used for the designation of the frequencies is different from the one in the present section (the caption of Fig. 5.10 indicates such differences).

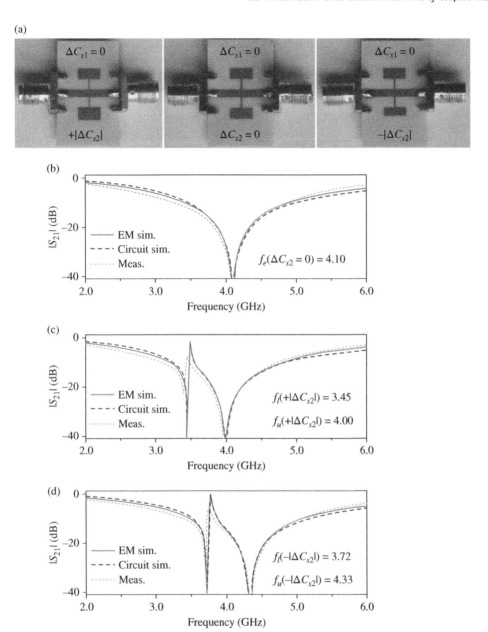

Figure 5.10 (a) Photograph of the fabricated SIR-loaded transmission lines, and transmission coefficient corresponding to the lossless electromagnetic simulation, circuit simulation, and measurement for a microstrip line loaded (b) with a pair of identical SIRs (symmetric case), and (c), (d) with a pair of unequal SIRs [these responses correspond to the fabricated structures shown in (a)]. The dimensions are $W = 1.83$ mm, $l_1 = l_2 = 2.6$ mm, $\Delta l_1 = \pm 0.5$ mm, $w_1 = 5.5$ mm, $w_2 = 250$ μm, and the length of the microstrip line is $l = 15.9$ mm. The substrate is *Rogers RO4003C* with considered dielectric constant $\varepsilon_r = 3.38$, thickness $h = 812.8$ μm, and loss tangent $\tan\delta = 0.0021$. The circuit values are $L = 1.81$ nH, $C = 0.57$ pF, $L_s = 2.45$ nH, $C_s = 0.70$ pF, $\Delta C_{s2} = 0.15C_s = \pm 0.11$ pF, and $|M| = 0.31$ nH ($k_M = -0.13$). Note that the frequencies $f_+ = \omega_+/2\pi$ and $f_- = \omega_-/2\pi$ were designated as f_2 and f_1 in [31], and as f_u and f_l in [32], respectively. Thus, the frequencies f_u and f_l indicated in (c) and (d) (the nomenclature of the original source [32] has been used) should be actually identified with f_+ and f_-, respectively. Let us also indicate that, in this section, f_1 and f_2 are actually the resonance frequencies of the isolated resonators. Moreover, the frequency designated as f_e in (b), corresponding to the (single) resonance frequency for the symmetric case, is actually f_+, as given by (5.21). The units of the indicated frequencies in the figures are GHz in all the cases. *Source:* Reprinted with permission from [32]; copyright 2016 EEE.

(a)

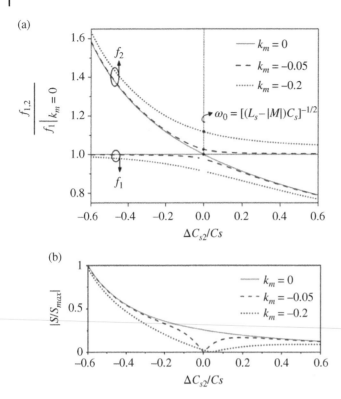

(b)

Figure 5.11 Normalized (a) resonance frequencies and (b) sensitivity magnitude in symmetric parallel-connected pair of SIRs loaded with the capacitive perturbations $\Delta C_{s1} = 0$ and $-0.6 \leq \Delta C_{s2}/C_s \leq 0.6$ for different values of k_M. The resonance frequencies are normalized to the constant frequency of the non-perturbed uncoupled SIR. The sensitivity is normalized to its maximum value S_{max}. *Source:* Reprinted with permission from [31]; copyright 2014 IEEE.

elements. In fact, there are examples of sensors, based on other principles, where inter-resonator coupling boosts up the sensitivity [34]. It has also been shown that coupled-line-based sensors are useful for sensitivity enhancement [35].

5.3 Frequency-Splitting Sensors Based on Cascaded Resonators

An obvious approach to eradicate inter-resonator coupling in frequency-splitting sensors is to consider a single sensing resonator and to perform two consecutive measurements, one with the reference (REF) sample, and the other one with the sample (or material) under test (MUT). However, this strategy prevents from the realization of real-time measurements and cannot be categorized within the class of frequency-splitting sensors; that is the inherent advantages of real-time frequency splitting (similar to the advantages of differential sensing) are lost with this approach.[11]

An alternative strategy for canceling inter-resonator coupling in frequency-splitting sensors based on two resonant elements is the cascade configuration [32]. The typical topology, by

---

11 Indeed, a sensor based on a single resonant element that exploits frequency shift and two independent measurements to compare a MUT with a REF sample is actually a frequency-variation sensor (these sensors are the subject of Chapter 2).

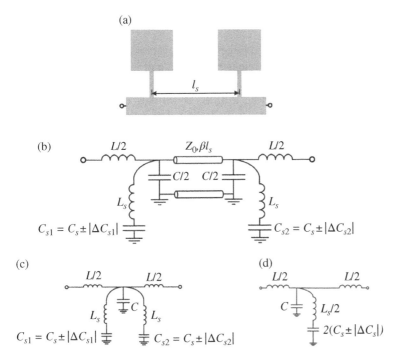

Figure 5.12 (a) Microstrip line loaded with a pair of identical SIRs in cascade connection. (b) Equivalent circuit model including arbitrary capacitive perturbations. (c) Simplified equivalent circuit model when the resonators are spaced half-wavelength apart, i.e. the in-between transmission line section has an electrical length of $\beta l_s = 180°$. (d) Simplified equivalent circuit model for $\beta l_s = 180°$ in the case of balanced perturbations. *Source:* Reprinted with permission from [32]; copyright 2016 IEEE.

considering SIR resonators and microstrip technology, is illustrated in Fig. 5.12(a). The sensor consists of a transmission line loaded with a cascade connection of two identical SIRs that are spaced apart by a transmission line section of length l_s. If the SIRs are loaded on the same side of the line, as is considered, the resonators may be coupled not only magnetically but also electrically (provided the SIR capacitive patches are closely spaced). However, the resonators are considered to be placed at sufficient distance, so that it can be assumed that the total coupling is negligible.

Figure 5.12(b) depicts the proposed circuit model of such cascaded configuration. Regardless of the length of the transmission line section between resonators, l_s, the transmission zero frequencies are given by

$$\omega_- = min\left\{\frac{1}{\sqrt{L_s(C_s \pm |\Delta C_{s1}|)}}, \frac{1}{\sqrt{L_s(C_s \pm |\Delta C_{s2}|)}}\right\} \tag{5.22a}$$

$$\omega_+ = max\left\{\frac{1}{\sqrt{L_s(C_s \pm |\Delta C_{s1}|)}}, \frac{1}{\sqrt{L_s(C_s \pm |\Delta C_{s2}|)}}\right\} \tag{5.22b}$$

where ω_- and ω_+ denote, again, the lower and upper resonance frequencies, respectively. According to the previous expressions, the two resonances can be shifted independently like using single SIRs, even for unbalanced perturbations. Clearly, the resonance frequency-splitting phenomenon, which emerges from unbalanced perturbations, is of different nature from the one of the parallel configuration. In cascaded SIRs, splitting occurs as a mere result of frequency shifting, whereas in parallel SIRs there is a combination of frequency shifting and inter-resonator coupling.

Let us now consider the particular case where the in-between transmission line section is half-wavelength long ($l_s = \lambda/2$, where λ is the guided wavelength). Since a half-wavelength line has the effect of translating the terminating impedance of one port to the other port, the circuit in Fig. 5.12 (b) is equivalent to that shown in Fig. 5.12(c). Therefore, the SIRs are virtually connected at the same junction, so that the structure behaves like if the SIRs were physically located at the same transverse plane (similar to the parallel configuration, but preventing from the presence of inter-resonator's coupling). Nevertheless, note that in cascaded SIRs intended to be separated a half-wavelength (or a multiple of a half-wavelength), such distance is satisfied at one frequency only (which should be the resonance frequency of the unperturbed resonators). However, the resonance frequencies, given by (5.22), do not depend on the inter-resonator distance (as long as inter-resonator coupling can be ignored). Therefore, the resonance frequencies of the model in Fig. 5.12(c) always coincide with those of the model in Fig. 5.12(b). If the perturbations are identical ($\omega_- = \omega_+ = \omega_0$), and the SIRs are separated a half-wavelength, the circuit in Fig. 5.12(c) can be approximated to the one in Fig. 5.12(d), which is formally identical to that with parallel SIRs depicted in Fig. 5.9(d).

To demonstrate the potential of the cascaded approach to solve the aforementioned coupling-related sensitivity degradation of the parallel configuration, the same structures considered in Fig. 5.10(a), but implemented by cascading the SIRs, were fabricated [see Fig. 5.13(a)] [32]. Accordingly, the same circuit elements were used to validate the equivalent circuit models.[12] As can be seen in Fig. 5.13(b), the agreement between the circuit simulations, the electromagnetic simulations, and measurements is good. The notch frequencies are the same as those that result by using single SIRs. It is apparent that the bandwidth of both notches is not narrow by nature, in contrast to what occurs at the lower resonance frequency employing parallel SIRs. The explicit comparison between the transmission zero frequencies resulting in the parallel and cascade configuration is depicted in Fig. 5.14. In view of this figure, it is clear that the two concerning topologies (parallel and cascaded) manifest a strongly different behavior caused by the presence or absence of inter-resonator coupling. Obviously, in the limit of small imbalances between both SIRs, the resonance frequencies tend to progressively approach, as far as the SIRs are cascaded. In such configuration, only the resonance frequency of the perturbed resonator shifts. Contrarily, frequency splitting in parallel SIRs is characterized by a shift in the two resonances (despite the fact that only one SIR is perturbed). According to all these results, the major capacity of the cascaded configuration to discriminate small perturbations between the SIRs (resolution) and to provide higher sensitivity, especially in the limit of small imbalances, is clear.

In the results presented so far concerning frequency-splitting structures based on SIRs (implemented by either the parallel or the cascade configuration), capacitive perturbations in the SIRs have been achieved by modifying the patch geometry. In a real scenario, the interest is typically the measurement of material properties related to changes in the capacitance of the SIRs, e.g. the dielectric constant of the MUT (sensor functionality), or the detection of differences (if they are present) between two samples (comparator functionality). In both cases, the operation principle

---

12 Let us clarify that the circuit parameters corresponding to the SIRs of Figs. 5.10 and 5.13 (both published in [32]) are identical, as far as their dimensions (see caption of Fig. 5.10) and considered substrate dielectric constant ($\varepsilon r = 3.38$) are also identical. Nevertheless, in [31], the considered substrate dielectric constant was $\varepsilon r = 3.10$, and, for this reason, the responses of the structures published in that paper are a bit different (despite the fact the dimensions of the SIRs are the same). Let us also mention that Fig. 4.11 has been reprinted from [31], with the extracted parameters indicated in that paper, which slightly vary as compared to those indicated in the caption of Fig. 5.10. However, the differences are so small that the circuit parameters of Fig. 4.11 have not been included in that figure (nevertheless, the interested readers can find such values in [31]).

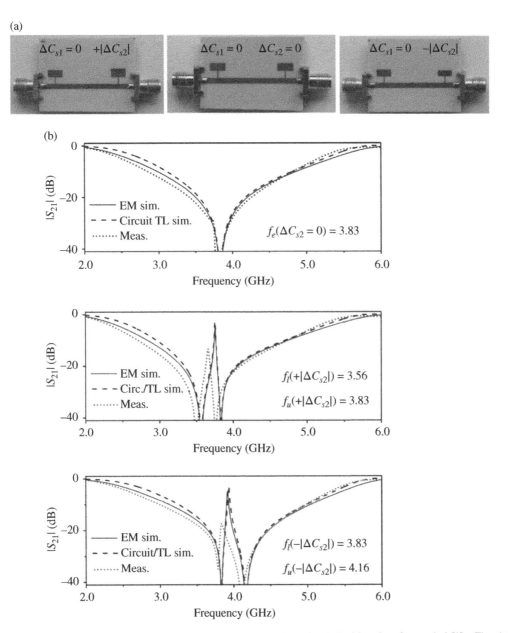

Figure 5.13 (a) Photograph of the considered microstrip lines loaded with pairs of cascaded SIRs. The right SIR is perturbed by $\pm|\Delta C_{s2}|$, whereas the left SIR is unperturbed ($\Delta C_{s1} = 0$). The dimensions, substrate, and circuit parameters are those of Fig. 5.10 (and $l_s = 23.9$ mm). (b) Transmission coefficient magnitude obtained from lossless electromagnetic and circuit simulations, and measurements. The indicated resonance frequencies correspond to simulations (in the figure, f_l and f_u are the designations adopted in [32], corresponding to the lower, f_-, and higher, f_+, notch frequencies, according to the definitions of this chapter). *Source:* Reprinted with permission from [32]; copyright 2016 IEEE.

is based on the variations of the capacitances of the SIRs (with identical and invariable dimensions) caused by the dielectric constant of the MUT and REF samples. Nevertheless, the REF material is typically invariant for a certain measurement. Therefore, for a sensor based on the cascade topology, the variations in the output variable, the difference between the notch frequencies, are only

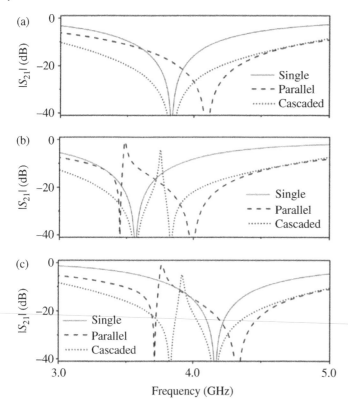

Figure 5.14 Transmission coefficient magnitude (lossless electromagnetic simulation) of the considered microstrip lines loaded with parallel or cascaded SIRs for $\Delta C_{s1} = 0$, and (a) $\Delta C_{s2} = 0$, (b) $+|\Delta C_{s2}|$, and (c) $-|\Delta C_{s2}|$. For comparison purposes, the response using a single SIR applying to it ΔC_{s2} is also plotted. *Source:* Reprinted with permission from [32]; copyright 2016 IEEE.

caused by variations in the resonance frequency of the SIR devoted to the MUT. The output variable can be expressed as $f_d = f_2 - f_1$, where f_2 and f_1 are the resonance frequencies of SIR-2 (MUT) and SIR-1 (REF), respectively. Since the resonance frequencies of the SIRs in cascade configuration are not affected by the loading of the other SIR, and f_2 may be higher or lower than f_1 (depending on the perturbation on the MUT), it is convenient to designate the notch frequencies as f_2 and f_1, rather than f_+ and f_-. By this means, the resonance frequencies referred to SIR-2 and SIR-1 are clearly indicated and distinguished. The input variable is the differential dielectric constant between the MUT and REF samples, i.e. $\varepsilon_{rd} = \varepsilon_{r2} - \varepsilon_{r1} \equiv \varepsilon_{MUT} - \varepsilon_{REF}$. Thus, the sensitivity can be expressed as

$$S = \frac{df_d}{d\varepsilon_{rd}} = \frac{d(f_2 - f_1)}{d(\varepsilon_{r2} - \varepsilon_{r1})} = \frac{df_2}{d\varepsilon_{r2}} \tag{5.23}$$

where the last equality is valid, provided ε_{r1} and f_1 remain invariable. In turn, (5.23) can be written as

$$S = \frac{df_2}{d\varepsilon_{r2}} = \frac{df_2}{dC_{s2}} \cdot \frac{dC_{s2}}{d\varepsilon_{r2}} \tag{5.24}$$

and the sensitivity depends on the specific placement of the MUT with regard to the SIR, since this determines the second term in the right-hand-side member of (5.24).

The most canonical and simple placement of the MUT is on top of the capacitive patch of the SIR. However, in microstrip technology, where the electric field lines are mostly confined within the

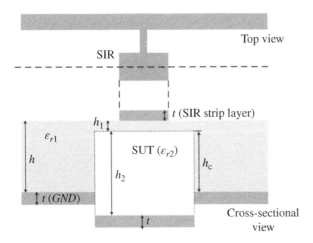

Figure 5.15 Sketch (not drawn to scale) of the top and cross-sectional view of the SIR and cavity loaded with the MUT (or sample under test, SUT).

substrate, it is expected that the MUT (with the indicated placement) causes soft variations in the capacitance, C_{s2}, of the SIR. In [32], the sensitivity was calculated by considering that the MUT is placed between the SIR patch and the ground plane of the microstrip line. This is the most favorable strategy for sensitivity optimization, but it is difficult to implement in practice, since it is equivalent to replace a portion of the substrate with the MUT. Nevertheless, an approach to this strategy was applied in [32]. Namely, a cavity was drilled below the SIR patch, in order to accommodate the MUT (a dielectric slab with a metallic adhesive tape, acting as ground plane), see Fig. 5.15.[13]

By considering that the MUT is circumscribed between the SIR patch and the ground plane, the capacitance C_{s2} is roughly the one of a parallel plate capacitor, i.e. proportional to the dielectric constant of the MUT, $\varepsilon_{r2} = \varepsilon_{MUT}$. Thus, the sensitivity can be approximated by

$$S = -\frac{f_2}{2\varepsilon_{r2}} \tag{5.25}$$

and the magnitude of the sensitivity decreases as the dielectric constant of the MUT increases (note also that f_2 decreases by increasing ε_{r2}). Figure 5.16(a) depicts the resonance frequencies, inferred by electromagnetic simulation, that result by introducing a MUT below SIR-2 (the substrate and geometry are those of the unperturbed SIRs in Figs. 5.10 and 5.13). The simulations have been carried out by considering different values of the MUT relative permittivity, ε_{r2}. For comparison purposes, the resonances corresponding to the parallel configuration are also included in the figure. The resulting sensitivities derived from the results in Fig. 5.16(a) are plotted in Fig. 5.16(b). Clearly, for small perturbations, the sensitivity in cascaded SIRs is superior to that obtained using parallel SIRs, as expected.

To clearly point out the higher capability of the cascaded configuration, over the parallel configuration, to discriminate tiny differences between the REF and the MUT samples, an important aspect in comparators, it was considered in [32] that two −3-dB notches suffice for reasonably discriminating a doubly notched response. The corresponding differential dielectric constant provides the resolution of the sensor/comparator. For the considered topologies, these values (in relative

---

13 In [31], a frequency-splitting sensor implemented by a parallel connection of two SIRs in CPW technology was considered. In that case, since the ground plane is coplanar to the SIR patches, the sensitivity that results by placing the MUT on top of the SIR patch is reasonable (however, in the limit of small perturbations, it is limited by the coupling between the resonant elements, as discussed).

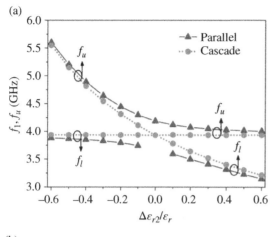

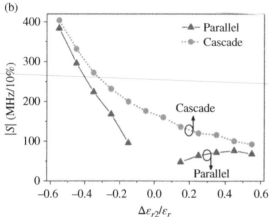

Figure 5.16 (a) Resonance frequencies for parallel and cascaded SIRs, obtained by electromagnetic simulation, versus a perturbation in the relative permittivity of an MUT loaded to one of the SIRs. The relative permittivity is perturbed by steps of $\Delta\varepsilon_{r2}/\varepsilon_r = \pm0.1$ ($\pm10\%$) so that $1.352 \leq \varepsilon_{r2} \leq 5.408$ ($-0.6 \leq \Delta\varepsilon_{r2}/\varepsilon_r \leq 0.6$) where $\varepsilon_r = 3.38$. (b) Magnitude of the sensitivity. The sensitivity in the jump discontinuity of the parallel configuration is not calculated. Losses are considered in the simulations. *Source:* Reprinted with permission from [32]; copyright 2016 IEEE.

terms) are roughly ±6 and $\pm2\%$ in the parallel and cascade configurations, respectively. The transmission coefficients for these inputs are plotted in Fig. 5.17. Clearly, the cascade configuration exhibits higher discrimination than the parallel one. It is important to mention that wideband notches are required. Otherwise, losses may mask the notches, as occurs in the lower notch for parallel SIRs when the input differential permittivity is close to zero. In summary, the discrimination is better with the cascaded configuration, where both notches are wide. However, under large perturbations, the parallel and cascade configuration have similar resonance frequencies and, therefore, comparable sensitivities (as Fig. 5.16 demonstrates).

In [32], measurements with permittivity perturbations were carried out by considering the reference structures of Figs. 5.10 and 5.13. In order to apply both positive and negative differential permittivity perturbations, the substrate in the vicinity of one of the SIRs was removed, as indicated before, by means of a drilling machine, thereby creating a cavity. Next, the cavity was either unfilled ($-|\Delta\varepsilon_{r2}|$) or filled with a *Rogers RO3010* substrate $\varepsilon_r = 11.2$ ($+|\Delta\varepsilon_{r2}|$). Figure 5.18 shows the photographs of the sensors with different loads, in order to illustrate the cavities, and their filling and

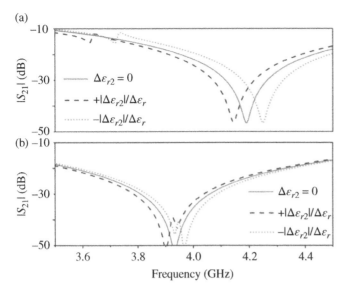

Figure 5.17 Transmission coefficient obtained by electromagnetic simulations for small differential permittivity perturbations in (a) parallel and (b) cascaded SIRs. The perturbations are those necessary to obtain two −3-dB notches: $\Delta\varepsilon_{r2}/\varepsilon_r = \pm0.06$ and ±0.02 (6 and 2%), respectively. *Source:* Reprinted with permission from [32]; copyright 2016 IEEE.

Figure 5.18 Photograph of the reference structures composed of microstrip lines loaded with pairs of (a) parallel and (b) cascaded SIRs. From left to right: (i) cavity in one of the SIRs; (ii) filled cavity with *Rogers RO3010* with $\varepsilon_{r2} = 11.2$; (iii) cavity covered with a metallic tape.

(a)

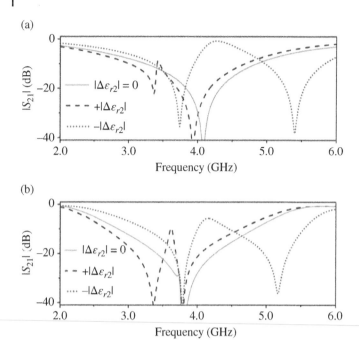

(b)

Figure 5.19 Measured transmission coefficient magnitude for the (a) parallel and (b) cascaded topologies under dielectric loading. Three scenarios are considered: (i) without cavity ($\Delta\varepsilon_{r2} = 0$); (ii) unfilled cavity so that $\varepsilon_{r2} = 1$ ($-|\Delta\varepsilon_{r2}|$); (iii) filled cavity with *Rogers RO3010* with $\varepsilon_{r2} = 11.2$ ($+|\Delta\varepsilon_{r2}|$). No perturbation is applied to the other SIR ($\Delta\varepsilon_{r1} = 0$). *Source:* Reprinted with permission from [32]; copyright 2016 IEEE.

covering. The measured results of the positive/negative perturbations, plotted in Fig. 5.19 together with those with no perturbation, are in accordance with theory. Even though the cavity dimensions cannot be controlled very accurately, at least with the in-house drilling system used in [32], these experiments validate the sensing principle under permittivity perturbation. Nevertheless, inspection of Fig. 5.19 reveals that the difference in notch frequencies is somehow smaller than the results of Fig. 5.16 for the considered dielectric constant values (1 for the unfilled cavity, and 11.2 for the filled cavity with the considered *Rogers* material). The reason is that the cavity was implemented by milling, and it was not possible to completely remove all the substrate material, since the SIR needs some material for mechanical stability (see Fig. 5.15). Therefore, the unfilled cavity, including the remaining substrate layer, has an effective dielectric constant larger than 1, and the cavity filled with the *Rogers* substrate (dielectric constant 11.2) has actually an effective dielectric constant smaller than 11.2. In other words, the measurement provides the effective dielectric constant of the structure below the SIR, including not only the MUT but also the presence of a narrow dielectric layer of relative permittivity 3.38, one of the considered substrate. Moreover, the thickness of the whole structure, layer on top the cavity plus MUT, is not necessarily the same as the thickness of the substrate. In the simulations that were carried out to obtain the results of Fig. 5.16, this substrate layer between the SIR metal level and the MUT was not considered. In practice, it is very difficult to precisely control the thickness of the remaining substrate between the SIR patch and the cavity. For this main reason, such layer was not considered in the simulations. Note, however, that with a more sophisticated fabrication technology (e.g. micromachining), such control would not be a problem.

Despite the presence of the dielectric layer (with thickness h_1, see Fig. 5.15) between the cavity and the SIR patch, a method to determine the dielectric constant of a piece of unknown material,

ε_{r2}, was reported in [32]. For that purpose, such thickness, h_1, must be first estimated (the reference permittivity, i.e. the one of the substrate, ε_{r1}, and the MUT thickness, h_2, necessary to determine ε_{r2}, are known). The method is based on the fact that the effective dielectric constant of the composite formed by the substrate layer on top of the cavity plus the MUT, ε_{eff}, is related to the respective dielectric constants by

$$\frac{\varepsilon_{eff}}{h} = \frac{\dfrac{\varepsilon_{r1}}{h_1} \cdot \dfrac{\varepsilon_{r2}}{h_2}}{\dfrac{\varepsilon_{r1}}{h_1} + \dfrac{\varepsilon_{r2}}{h_2}} \tag{5.26}$$

From the previous expression, the dielectric constant of the MUT can be isolated:

$$\varepsilon_{r2} = \frac{\varepsilon_{eff}\varepsilon_{r1}h_2}{\varepsilon_{r1}h - \varepsilon_{eff}h_1} \tag{5.27}$$

Nevertheless, as indicated, the thickness h_1 must first be obtained. To determine such thickness, a reference MUT with well-known dielectric constant and thickness is considered. The resulting ε_{eff} can be inferred from the split in frequency of the reference MUT and the curve of Fig. 5.16 corresponding to the cascade connection. Thus, if the dielectric constant and thickness of the reference MUT are known, h_1 can be obtained from (5.27). For the reference MUT corresponding to the *Rogers* substrate with $\varepsilon_{r2} = 11.2$ and $h_2 = 635\,\mu\text{m}$, and taking into account that the notch frequencies of Fig. 5.19(b) provide $\varepsilon_{eff} = 4.9$, according to Fig. 5.16, the resulting thickness of the layer on top of the cavity is found to be $h_1 = 369\,\mu\text{m}$. To verify the validity of this result, it was considered the curve of Fig. 5.19(b) corresponding to the unfilled cavity ($\varepsilon_{r2} = 1$). By introducing the corresponding effective dielectric constant $\varepsilon_{eff} = 1.69$ in (5.27), and $h_1 = 369\,\mu\text{m}$, the air thickness was found to be $h_2 = 371\,\mu\text{m}$, which is in reasonable agreement with the thickness of the cavity (note however that the metallic tape is somehow flexible and hence may reduce the effective value of the cavity thickness).

Once h_1 is known, the estimation of the dielectric constant of other MUT samples is possible. To validate the method, an *Arlon* slab with $\varepsilon_{r2} = 2.43$ and $h_2 = 490\,\mu\text{m}$ was the considered MUT in [32]. The measured transmission coefficient for the cascaded configuration is depicted in Fig. 5.20. The effective dielectric constant that results from Fig. 5.16 is $\varepsilon_{eff} = 2.70$, and using (5.27), the dielectric constant is found to be $\varepsilon_{r2} = 2.50$, very close to the nominal value of the dielectric constant of that material, 2.43.

The sensitivity of the sensor implemented by the cascade configuration of the SIRs (of the order of 0.6 GHz for small perturbations) is good by virtue of the broadside capacitance of the resonant

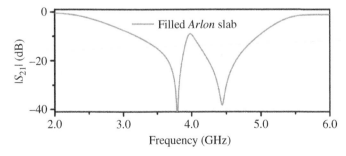

Figure 5.20 Measured transmission coefficient magnitude for the cascaded topology with the cavity filled by an *Arlon* substrate with the characteristics indicated in the text. *Source:* Reprinted with permission from [32]; copyright 2016 IEEE.

elements and the specific MUT placement, between the patch capacitance of the SIR and ground. Nevertheless, the realization of measurements and sample preparation with this approach is not simple. By contrast, other sensing planar resonant elements, such as the SRR or the CSRR, exhibit edge capacitances, not as sensitive to the effects of the dielectric constant of the MUT as the broadside capacitance of the SIR. However, with such resonators, the measurement procedure is intrinsically simpler (i.e. placing the MUT sample on top of the resonant element suffices for sensing,[14] and the MUT, in general, neither requires a specific thickness nor needs backside metallization). Thus, in the next section, where a different scheme for minimizing the effects of inter-resonator coupling is reported (the splitter/combiner configuration), the considered resonant elements of the two reported realizations (in microstrip technology) are SRRs and CSRRs. Let us also mention that it is possible to achieve reasonable sensitivities in SIR-based sensors implemented in planar technology, where the MUT is placed on top of the SIR. To optimize the dependence of the SIR capacitance with the dielectric constant of the MUT in such conditions, the idea is to surround the SIR patch by a metallization connected to the ground plane by means of vias [36] (however, this approach represents further fabrication complexity and cost, due to the presence of the vias).

To end this section, let us mention that a method to estimate the loss tangent of the MUT was reported in [32]. However, such method does not use frequency splitting, but a single resonator, and is similar to the method reported in Chapter 2, based on the measurement of the notch depth in resonator-loaded lines.

5.4 Frequency-Splitting Sensors Based on the Splitter/Combiner Configuration

The splitter/combiner configuration is an alternative to the cascade configuration (reviewed in the previous section) for moving the two sensing resonators away, thereby minimizing the effects of coupling. Figure 5.21 depicts three possible splitter/combiner sensor topologies that use the previously considered resonators (SRRs [37], CSRRs [38], or SIRs [39]) as sensing elements. If the structures are perfectly symmetric, the transmission coefficient exhibits a single notch (transmission zero) at the resonance frequency of the resonant element. For the SIR- and CSRR-based splitter/combiner structures of Fig. 5.21(a) and (b), such single notch appears because the parallel branches of the splitter/combiner are both grounded at the (unique) resonance frequency. By contrast, in the SRR-based sensor, the branches open at the intrinsic resonance frequency of the resonators, preventing from signal transmission at this frequency. However, if symmetry is truncated, e.g. as consequence of an asymmetric dielectric loading in the resonant elements, two transmission zeros are expected. However, in general, the transmission zeros are not given by the resonance frequencies of the individual (unequal) resonators, because at these frequencies one branch precludes signal transmission, but the injected power can be partially transmitted through the opposite branch of the splitter/combiner structure. In other words, the notches are, in general, consequence of signal interference.[15] Thus, if the splitter/combiner-based sensor is not properly designed, a

---

14 However, it should be mentioned that in order to minimize the effects of the air gap present between the planar resonant element and the MUT sample, it is necessary, in general, to pressure the MUT against the resonator's substrate (for that purpose, Teflon screws can be used, as pointed out in many sensors reported in this book).

15 The transmission zeros are not only dictated by the intrinsic resonance frequency of the resonators but also by the length of the lines of the splitter/combiner, and such transmission zeros occur, in general, at those frequencies where the signals at the end of the parallel branches exactly cancel.

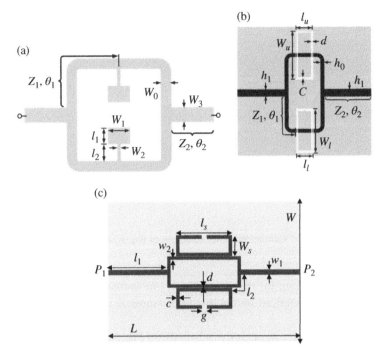

Figure 5.21 Typical topologies of the splitter/combiner frequency-splitting sensors based on SIR (a), CSRR (b), and SRR (c) sensing resonators.

phenomenology similar to the one related to inter-resonators' coupling (concerning the impossibility to discriminate small perturbations due to sensitivity degradation) arises. Without loss of generality, let us next consider in detail the analysis of the CSRR-based splitter/combiner sensor [38], in order to determine the design strategy to circumvent the above-cited sensitivity limitation (a prototype sensor will also be presented for validation purposes). The analysis of the SIR-based splitter/combiner sensor, reported in [39], is indeed simpler and it is not included in this book (moreover, SIRs exhibit an intrinsically lower sensitivity, as compared with CSRRs or SRRs, if the MUT is placed on top of them, the convenient approach in order to facilitate the measurement procedure, as discussed). Then, an example of a SRR-based splitter/combiner sensor combined with microfluidics and devoted to the dielectric characterization of liquid samples will be reported.

5.4.1 CSRR-Based Splitter/Combiner Sensor: Analysis and Application to Dielectric Characterization of Solids

The typical topology of the CSRR-based splitter/combiner microstrip sensor, depicted in Fig. 5.21 (b), consists of two transmission line branches, each one loaded with a CSRR etched in the ground plane [38]. In [38], such parallel lines were designed with a characteristic impedance of 50 Ω, the reference impedance of the ports. Thus, in order to match the structure with the 50-Ω ports, impedance inverters implemented by means of 35.35-Ω quarter-wavelength transmission line sections were cascaded between the ports and the T-junctions. Figure 5.22 depicts the circuit schematic of this structure. The circuit is not symmetric, but such imbalance refers only to the CSRR circuit elements in order to consider potential variations in such elements caused, e.g. by an asymmetric dielectric loading of the CSRRs. In the circuit, L_u (L_l) and C_u (C_l) model the inductance and capacitance of the microstrip line, respectively, above the CSRR in the upper (lower) parallel branch, and the resonators (CSRRs) are accounted for by the tanks L_{Cu}–C_{Cu} (upper CSRR) and L_{Cl}–C_{Cl} (lower

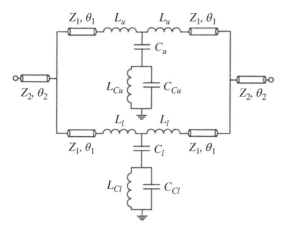

Figure 5.22 Circuit schematic of the CSRR-based splitter/combiner structure of Fig. 5.21(b).

CSRR) [13][16] The distributed elements account for the transmission line sections which are not located on top of the CSRRs and are characterized by the line impedances Z_i and the electrical lengths θ_i (with $i = 1, 2$).

To predict the transmission zero frequencies through the schematic of Fig. 5.22, the input and output transmission line sections (with characteristic impedance and electrical length Z_2 and θ_2, respectively) can be ignored, since such sections do not have any influence on the position of the notches. The resulting two-port network thus contains two parallel branches directly connected to the ports. The transmission zeros are given by those frequencies satisfying $S_{21} = S_{12} = 0$. However, since the structure under study consists of two parallel branches, it is convenient to deal with the admittance parameters. The transmission coefficient expressed in terms of the admittance parameters is [40]

$$S_{21} = \frac{-2Z_0 Y_{12}}{(1 + Z_0 Y_{11}) \cdot (1 + Z_0 Y_{22}) - Z_0^2 Y_{12} Y_{21}} \tag{5.28}$$

where Y_{11} and Y_{22} are the diagonal elements of the admittance matrix, and Y_{12} and Y_{21} are the anti-diagonal elements of such matrix. Since the considered structure is symmetric with regard to the bisecting plane between the input and the output port, it follows that $Y_{11} = Y_{22}$ and $Y_{12} = Y_{21}$ (and, obviously, $S_{12} = S_{21}$). According to (5.28), $Y_{12} = Y_{21} = 0$ is a sufficient condition to obtain $S_{21} = S_{12} = 0$, i.e. a transmission zero. By neglecting the input and the output transmission line sections in the network of Fig. 5.22, Y_{12} (or Y_{21}) can be expressed as

$$Y_{12} = Y_{21} = Y_{12,u} + Y_{12,l} \tag{5.29}$$

where $Y_{12,u}$ and $Y_{12,l}$ are the anti-diagonal elements of the admittance matrices of the upper and lower branches, respectively, of the network of Fig. 5.22.

The admittance matrix elements $Y_{12,u}$ and $Y_{12,l}$ can be determined by first obtaining the ABCD matrix of each branch. These matrices are given by the matrix product of the matrices corresponding to the three cascaded two-port networks for each branch, that is the pair of transmission line sections with characteristic impedance Z_1 and phase θ_1, and the sandwiched lumped two-port network. From the ABCD matrices for each branch, the elements of the right-hand-side member in

---

16 Although an asymmetric dielectric loading is expected to modify only the capacitance of the CSRRs, the more general case of an asymmetry in all the lumped elements of the circuit model is considered for analysis purposes.

(5.29) are given by $Y_{12,u} = -1/B_u$ and $Y_{12,l} = -1/B_l$, where B_u and B_l are the B elements of the ABCD matrix for the upper and lower branches, respectively [40]. Thus, the transmission zeros related to $Y_{21} = Y_{12} = 0$ are given by

$$\frac{1}{B_u} + \frac{1}{B_l} = 0 \tag{5.30}$$

with

$$B_u = j \left\{ (Z_1 \sin 2\theta_1 + 2\omega L_u \cos^2 \theta_1) + \frac{(Z_1 \sin \theta_1 + \omega L_u \cos \theta_1)^2}{\dfrac{\omega L_{Cu}}{1 - \dfrac{\omega^2}{\omega_{Cu}^2}} - \dfrac{1}{\omega C_u}} \right\} \tag{5.31a}$$

$$B_l = j \left\{ (Z_1 \sin 2\theta_1 + 2\omega L_l \cos^2 \theta_1) + \frac{(Z_1 \sin \theta_1 + \omega L_l \cos \theta_1)^2}{\dfrac{\omega L_{Cl}}{1 - \dfrac{\omega^2}{\omega_{Cl}^2}} - \dfrac{1}{\omega C_l}} \right\} \tag{5.31b}$$

and $\omega_{Cu}^2 = 1/L_{Cu} C_{Cu}$ and $\omega_{Cl}^2 = 1/L_{Cl} C_{Cl}$.

The analytical solution of (5.30) for the general case of an asymmetric structure is not simple (nevertheless, it is possible to obtain the pair of transmission zeros numerically). By contrast, if two identical CSRRs are considered, it follows that $B_u = B_l$, and expression (5.30) has a unique solution (transmission zero) with angular frequency given by

$$\omega_z = \frac{1}{\sqrt{L_C(C_C + C)}} \tag{5.32}$$

where $L_{Cu} = L_{Cl} = L_C$, $C_{Cu} = C_{Cl} = C_C$, and $C_u = C_l = C$. Note that (5.32) is the frequency that nulls the reactance of the (identical) shunt branches of the lumped two-port T-networks of Fig. 5.22, as expected. Except for the symmetric case, where the single transmission zero is simply given by the characteristics of the (identical) resonators (L_C and C_C) and their coupling to the line (through C), the two transmission zeros of the general (asymmetric) case are consequence of an interfering phenomenon between the parallel CSRR-loaded line sections. That is, the notches appear at those frequencies that provide out-of-phase signals at the output ports of the individual parallel branches.

Let us now consider an alternative situation providing also a transmission zero, that is $Y_{11} = Y_{22} = \infty$ with $Y_{21} = Y_{12} \neq \infty$. It is apparent, according to (5.28), that such combination of admittance matrix parameters gives $S_{21} = S_{12} = 0$. Nevertheless, it does not mean that if $Y_{21} = Y_{12} = \infty$, a transmission zero does not appear. In other words, $Y_{21} = Y_{12} \neq \infty$, with $Y_{11} = Y_{22} = \infty$, is a sufficient condition, but it may not be necessary. Let us analyze this aspect in detail. Note that $Y_{11,u} = \infty$ and/or $Y_{11,l} = \infty$ suffices to guarantee that $Y_{11} = Y_{11,u} + Y_{11,l} = \infty$ ($Y_{11,u}$ and $Y_{11,l}$ are the diagonal elements of the admittance matrices of the upper and lower branches, respectively, of the network of Fig. 5.22). Therefore, let us calculate, e.g., $Y_{11,u}$. This parameter can be inferred from the elements of the ABCD matrix as $Y_{11,u} = D_u/B_u$ [40]. Even though B_u has been calculated before, see (5.31a), this element can be simplified by designating by Y_u the admittance of the shunt branch (formed by C_u, L_{Cu}, C_{Cu}). Using the same procedure, the element D_u can be expressed in terms of Y_u. Once B_u and D_u are inferred, $Y_{11,u}$ can be expressed as a function of Y_u, i.e.

$$Y_{11,u} = \frac{\cos 2\theta_1 - \frac{\omega L_u}{Z_1} \sin 2\theta_1 + jY_u\left(\omega L_u \cos 2\theta_1 + Z_1 \frac{\sin 2\theta_1}{2} - \omega^2 L_u^2 \frac{\sin 2\theta_1}{2Z_1}\right)}{j(Z_1 \sin 2\theta_1 + 2\omega L_u \cos^2\theta_1) - Y_u(Z_1 \sin\theta_1 + \omega L_u \cos\theta_1)^2} \quad (5.33)$$

Inspection of (5.33) reveals that $Y_u = \infty$ (corresponding to a short circuit) provides $Y_{11,u} = \infty$, as far as the following condition is satisfied

$$Z_1 \sin\theta_1 + \omega_0 L_u \cos\theta_1 = 0 \quad (5.34)$$

where ω_0 is the frequency giving $Y_u = \infty$. From (5.34), it follows that the phase of the lines of the parallel branches, as defined in Fig. 5.21(b), must be

$$\theta_1 = \arctan\left(-\frac{\omega_0 L_u}{Z_1}\right) = \pi - \arctan\left(\frac{\omega_0 L_u}{Z_1}\right) \equiv \theta_{1,\infty} \quad (5.35)$$

That is, (5.35) gives the electrical length at ω_0 that is necessary to obtain $Y_{11,u} = \infty$ and hence $Y_{11} = \infty$. However, this condition is not enough to guarantee that a transmission zero at ω_0 arises. As indicated before, $Y_{21} = Y_{12} \neq \infty$ ensures that $S_{21} = 0$ at ω_0, but note that if $Y_{11} = \infty$ but D is finite (and consequently $B = 0$), then necessarily $Y_{12} = \infty$.

To gain insight on this last aspect, let us express the denominator and the numerator of (5.33), B_u and D_u, respectively, as follows:

$$B_u = j2\cos\theta_1(Z_1 \sin\theta_1 + \omega_0 L_u \cos\theta_1) - Y_u(Z_1 \sin\theta_1 + \omega_0 L_u \cos\theta_1)^2 \quad (5.36)$$

$$D_u = 1 - \frac{2\sin\theta_1}{Z_1}(Z_1 \sin\theta_1 + \omega_0 L_u \cos\theta_1) + jY_u(Z_1 \sin\theta_1 + \omega_0 L_u \cos\theta_1)\left(\cos\theta_1 + \frac{\omega_0 L_u \sin\theta_1}{Z_1}\right) \quad (5.37)$$

The first term in the right-hand side of (5.36) is null if (5.34) is fulfilled. As for the second term, $Y_u = \infty$ in the limit when $\omega \rightarrow \omega_0$, but this term is multiplied by the square of (5.34), null at ω_0. This indeterminacy is solved by applying the L'Hôpital rule to $(Z_1 \sin\theta_1 + \omega_0 L_u \cos\theta_1)^2/Y_u^{-1}$. The first derivative of the numerator with frequency is null at ω_0. However, the derivative of any reactance with frequency should be finite at the resonance frequencies (where the impedance nulls, or $Y_u^{-1} = 0$ in our case). Consequently, $B_u = 0$ at ω_0, with θ_1 set to the value given by (5.35), and therefore $Y_{12,u} = \infty$ and $Y_{12} = \infty$. For which concerns D_u, if (5.34) is satisfied, the second term is null. The third term is neither null nor infinite, but finite at ω_0 (where $Y_u = \infty$). The reason is that, in this case, application of the L'Hôpital rule to $(Z_1 \sin\theta_1 + \omega_0 L_u \cos\theta_1)/Y_u^{-1}$ provides a finite value of both the derivative of the numerator and the derivative of the denominator. According to these words, it follows that D_u is a finite number satisfying $D_u \neq 1$ (a relevant results as it is shown below). It is clear, then, that at ω_0, with θ_1 set to the value given by (5.35), $Y_{11} = \infty$ and $Y_{12} = \infty$.

To determine if a notch is present at ω_0 with θ_1 satisfying (5.35), resulting in $Y_{11} = \infty$ and $Y_{12} = \infty$, it is necessary to express the transmission coefficient (5.28) as a function of the involved ABCD parameters for each parallel branch of the structure, namely

$$S_{21} = \frac{2Z_0\left(\frac{1}{B_u} + \frac{1}{B_l}\right)}{\left(1 + Z_0\left(\frac{D_u}{B_u} + \frac{D_l}{B_l}\right)\right)^2 - Z_0^2\left(\frac{1}{B_u} + \frac{1}{B_l}\right)^2} \quad (5.38)$$

where D_l and B_l are finite at ω_0, provided we are now analyzing an asymmetric structure, and therefore the lower parallel branch does not exhibit any singularity at ω_0. In the limit $B_u \rightarrow 0$, (5.38) can thus be expressed as

$$S_{21} = \frac{2Z_0 B_u}{Z_0^2 D_u^2 - Z_0^2} \tag{5.39}$$

and $S_{21} = 0$, provided $D_u \neq 1$, a condition that is satisfied as demonstrated in the preceding paragraph. Thus, if the phase of the parallel lines, as defined in Figs. 5.21(b) or 5.22, is set to the value given by (5.35) at ω_0 (the angular resonance frequency of one of the CSRRs, e.g. the one of the upper branch), a transmission zero at that frequency arises. Such notch does not depend on the resonance frequency of the other CSRR, and it is not due to signal interference, but to the fact that the virtual ground at the central position of the branch at ω_0 is translated to the input and output T-junctions of the splitter/combiner, as consequence of the specific electrical length of the line, $\theta_{1,\infty}$.[17] Such virtual ground at these T-junctions, generated by one of the parallel branches, suffices to prevent from signal transmission between the input and the output ports of the whole structure. However, if the structure is asymmetric, the other notch depends on the resonance frequencies of both CSRRs, and it is due to signal interference (i.e. it is related to the destructive interference of the two branches).

In order to validate the previous analysis, as well as the circuit model of Fig. 5.22, lossless electromagnetic and circuit simulations of different structures were compared in [38]. The circuit elements describing the transmission line sections loaded with CSRRs were extracted following the procedure described in [26]. For that purpose, the considered CSRR-loaded microstrip line sections were independently simulated in [38]. First, the symmetric CSRR-loaded splitter/combiner section as depicted in Fig. 5.21(b) was considered. The frequency response (magnitude of the transmission coefficient) inferred from electromagnetic simulation is depicted in Fig. 5.23(a). This figure also depicts the response resulting from circuit simulation by using the extracted element values, indicated in the caption. Then, two asymmetric structures were considered. To achieve the asymmetry, the dimensions of the lower CSRR were modified, specifically W_l, by increasing or decreasing ΔW_l, leaving the upper CSRR unaltered. In one case, this dimension was increased and in the other it was decreased. The responses (electromagnetic and circuit simulations) are depicted in Fig. 5.23(b) and (c), where the corresponding sets of extracted parameters are indicated (see caption). In all the cases, there is very good agreement between the electromagnetic and circuit simulations, pointing out the validity of the model. Figure 5.23 also includes the measured responses, inferred from the *Agilent N5221A* vector network analyzer. Further electromagnetic simulations with different values of W_l were carried out in [38]. Figure 5.24(a) depicts the pairs of transmission zeros as a function of $\Delta W_l / W_l$. As ΔW_l approaches zero, corresponding to the symmetric structure, the separation between the transmission zeros decreases. However, it can be appreciated in Fig. 5.24(a) that both notches do not converge (a sudden jump occurs when the structure is symmetric, with only one transmission zero, as anticipated before).

In the structure of Fig. 5.21(b), providing the responses of Fig. 5.23 and the pairs of transmission zeros of Fig. 5.24(a) for different values of W_l, the electrical length of the transmission line sections between the T-junctions and the position of the CSRRs, θ_1, does not satisfy (5.35), i.e. $\theta_1 < \theta_{1,\infty}$. To gain insight on the effects of θ_1, the electromagnetic simulations of the structures considered in Fig. 5.24(a), but considering $\theta_1 > \theta_{1,\infty}$, were carried out in [38]. The pairs of transmission zeros that result by varying W_l are depicted in Fig. 5.24(b). The behavior is very similar to the one observed in Fig. 5.24(a). However, the single transmission zero for the symmetric structure belongs in this case

---

17 In view of (5.35), if $L_u = 0$, the electrical length is $\theta_{1,\infty} = \pi$. This is an expected result, since it is well known that a half-wavelength transmission line translates the termination impedance present at one port to the other port. Indeed, if the considered resonators are SIRs, the lines with impedance and phase Z_1 and θ_1, respectively, are directly connected to the shunt LC series resonator modeling the SIR (i.e., $L_u = 0$), and $\theta_{1,\infty} = \pi$, as it is demonstrated in [39].

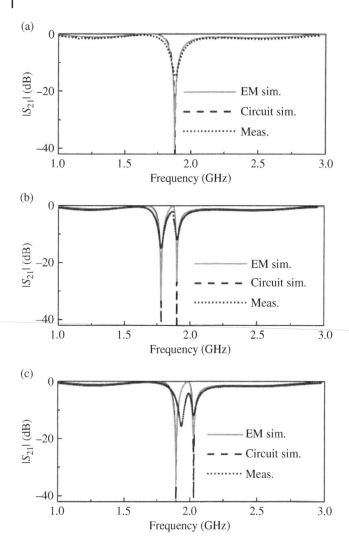

Figure 5.23 (a) Magnitude of the transmission coefficient corresponding to the symmetric structure of Fig 5.21(b) by considering the *Rogers RO3010* substrate with thickness $h = 1.27$ mm and dielectric constant $\varepsilon_r = 10.2$, and with geometrical parameters as follows: $h_0 = 1.15$ mm, $h_1 = 2.22$ mm, $c = 0.2$ mm, $d = 0.2$ mm, $W_u = W_l = 7.86$ mm, $l_u = l_l = 4.8$ mm; (b) magnitude of the transmission coefficient obtained by increasing the length of the lower CSRR with $\Delta W_l = 0.1 W_u = 0.786$ mm; (c) magnitude of the transmission coefficient obtained by decreasing the length of the lower CSRR with $\Delta W_l = -0.1 W_u = -0.786$ mm. The extracted parameters are: (a) $L_u = L_l = 2.18$ nH, $C_u = C_l = 0.82$ pF, $L_{Cu} = L_{Cl} = 1.91$ nH, and $C_{Cu} = C_{Cl} = 2.94$ pF; (b) $L_l = 2.30$ nH, $C_l = 0.82$ pF, $L_{Cl} = 2.08$ nH, and $C_{Cl} = 3.09$ pF; (c) $L_l = 2.07$ nH, $C_l = 0.82$ pF, $L_{Cl} = 1.70$ nH, and $C_{Cl} = 2.84$ pF. For both (b) and (c), the rest of extracted element values are the same as (a). *Source:* Reprinted with permission from [38]; copyright 2016 IEEE.

to the opposite curve. Finally, the phase satisfying $\theta_1 = \theta_{1,\infty}$, see Fig. 5.24(c), was considered. In this case, the pair of transmission zeros merge when the structure is symmetric, and the two curves cross. This is an expected result since it was demonstrated before that when condition (5.35) is satisfied, one of the transmission zeros is given by the frequency that nulls the reactance of the upper shunt branch, regardless of the dimensions of the CSRR present at the other (lower) branch. Concerning the frequency response, a typical characteristic when $\theta_1 = \theta_{1,\infty}$ is the similarity between the two notches (depth and width) for asymmetric structures, as it can be appreciated in Fig. 5.25.

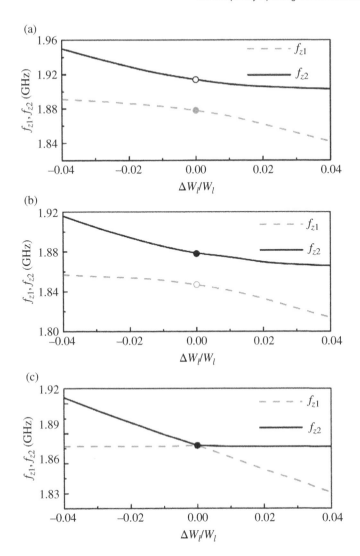

Figure 5.24 Dependence of the transmission zero frequencies with the relative increment/decrement of the width of one of the CSRRs ($\Delta W_l/W_l$) for different electrical lengths of the transmission lines: (a) $\theta_1 = 0.672\pi < \theta_{1,\infty}$; (b) $\theta_1 = 1.008\pi > \theta_{1,\infty}$; (c) $\theta_1 = 0.84\pi = \theta_{1,\infty}$. *Source:* Reprinted with permission from [38]; copyright 2016 IEEE.

Note the similarity between the curves of Fig. 5.24 and those depicted in Figs. 5.11(a) and 5.16(a), in reference to the pair of notches in lines loaded with a pair of SIRs (either parallel-connected or cascaded to the line). Obviously, such similarity extends also to the sensitivity. In the use of the CSRR-loaded splitter/combiner structures as sensors or comparators based on frequency splitting, the sensitivity is defined as the variation of the frequency difference between the two transmission zeros (f_{z1} and f_{z2}) with the variable that generates the asymmetry (typically a difference in the dielectric constant between two samples). Nevertheless, for analysis purposes, we can adopt as input variable the variation in the dimension of the CSRR of the lower branch, ΔW_l. Thus, without loss of generality, the sensitivity is

$$S = \frac{d\Delta f_z}{d\Delta W_l} \tag{5.40}$$

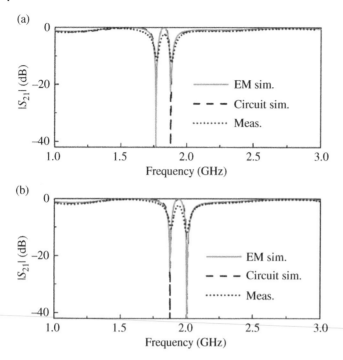

Figure 5.25 Response of the CSRR-loaded splitter/combiner structure with $\theta_1 = \theta_{1,\infty}$ for two asymmetric structures. (a) W_l = 8.65 mm (achieved by increasing the nominal value in $\Delta W_l = 0.1 W_u = 0.786$ mm); (b) W_l = 7.07 mm (achieved by decreasing the nominal value in $-\Delta W_l = -0.1 W_u = -0.786$ mm). *Source:* Reprinted with permission from [38]; copyright 2016 IEEE.

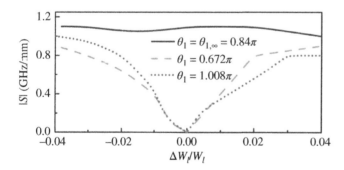

Figure 5.26 Sensitivity as a function of $\Delta W_l / W_l$ for different values of the electrical length of the parallel transmission lines of the splitter/combiner structure. *Source:* Reprinted with permission from [38]; copyright 2016 IEEE.

where $\Delta f_z = f_{z1} - f_{z2}$. The optimum structure in terms of sensitivity, see Fig. 5.26, is the one satisfying (5.35), i.e. the one providing the transmission zeros of Fig. 5.24(c). The sensitivity for small unbalanced perturbations is clearly superior when $\theta_1 = \theta_{1,\infty}$ (0.84π in the structure under study), but for large perturbations the curves tend to merge, similarly to Figs. 5.11(b) and 5.16(b). In Fig. 5.24, the asymmetries are caused by varying the dimensions of one of the CSRRs, as indicated, but this behavior (the dependence of the transmission zero curves with θ_1) does not depend on the cause of the asymmetry, and it is general for unbalanced loads. In summary, for sensitivity optimization, the electrical length of the parallel lines must be forced to satisfy (5.35) at the frequency corresponding to the resonance frequency of the reference (REF) CSRR. This situation is very similar to the one of uncoupled resonators in simple lines loaded with a pair of resonant elements (achieved by the cascade configuration).

To demonstrate the potential of the structure satisfying $\theta_1 = \theta_{1,\infty}$ as sensor, the lower CSRR was loaded with small dielectric slabs with different dielectric constant in [38] (the upper CSRR was kept unloaded). Specifically, the MUT samples were square-shaped pieces of un-metalized commercial microwave substrates with dielectric constants of 10.2 (*Rogers RO3010*), 3.55 (*Rogers RO4003C*), and 2.43 (*Arlon CuClad 250*). The measured responses are depicted in Fig. 5.27(a), whereas the variation of frequency splitting, Δf_z, with the dielectric constant, exhibiting roughly a linear variation, is shown in Fig. 5.27(b). This curve is useful to determine the dielectric constant of unknown substrates/samples from the measurement of the resulting frequency splitting (it is

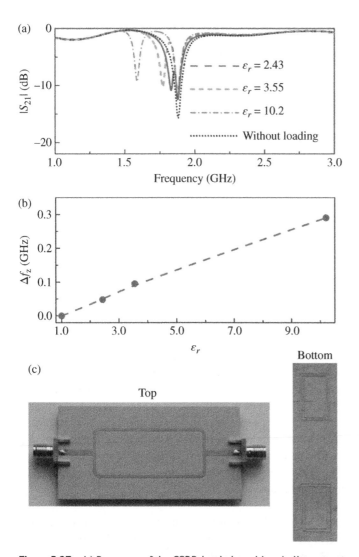

Figure 5.27 (a) Response of the CSRR-loaded combiner/splitter structure with $\theta_1 = \theta_{1,\infty}$ to different dielectric loads, (b) variation of Δf_z with the dielectric constant of the considered load, and (c) photograph (top view) of the sensor (the detail of the CSRRs, etched in the ground plane is also included). *Source:* Reprinted with permission from [38]; copyright 2016 IEEE.

assumed that the effects of losses can be neglected, provided losses in the MUT samples are reasonably small). In principle, the dielectric constant of the MUT sample can be arbitrarily small. Nevertheless, for a sufficiently small differential dielectric constant, it is expected that the two notches merge in a single one by the effect of losses of the device (metallic and dielectric). Therefore, the discrimination is limited by this effect. The curve of Fig. 5.27(b) exhibits a significant variation of Δf_z with the dielectric constant, i.e. good sensitivity. As a test example, the device was loaded in [38] with a square slab of un-cladded *FR4* substrate (with nominal dielectric constant 4.5). The resulting frequency splitting was found to be $\Delta f_z = 0.112$ GHz, providing a dielectric constant of 4.56, according to the curve of Fig. 5.27(b), i.e. in close agreement to the nominal value. Obviously, the MUTs should extend beyond the region occupied by the CSRR, in order to avoid any dependence of the transmission zero frequencies with the relative position between the MUT and the sensing CSRR. Moreover, if the samples are not semi-infinite in the vertical direction, their thickness should be comparable in order to use the calibration curve of Fig. 5.27(b) for measuring the dielectric constant of unknown samples. For semi-infinite samples, the sensitivity can be analytically inferred, but this analysis is out of the scope of this chapter, since it is very similar to previous sensitivity analysis where CSRR sensing elements, or other similar resonators, are considered.

To end this section, let us mention that multi-section CSRR-based splitter/combiner sensors, able to provide multiple measurements simultaneously, have been reported in [41] (the SIR-based counterparts of the splitter/combiner sensors reported in this section, including also multi-section devices, are analyzed in [39], where several prototype devices are included). The estimation of the loss tangent of the MUT samples by means of the CSRR-based splitter/combiner sensor was not reported in [38]. The next section is devoted to the implementation and validation of a microfluidic SRR-based splitter/combiner sensor devoted to the dielectric characterization of liquids (solutions of ethanol in deionized – DI – water). As it will be shown, a method for measuring both the real and the imaginary part of the dielectric constant of the considered liquid samples is reported.

5.4.2 Microfluidic SRR-Based Splitter/Combiner Frequency-Splitting Sensor

The splitter/combiner frequency-splitting sensing approach is a strategy to uncouple the resonant elements by spacing them away. However, such extra separation between the sensing resonators is also useful for the analysis of samples with moderate dimensions (larger than those of the resonators), or for the implementation of sensors requiring additional elements for sensing, e.g. fluidic channels. This section presents a microfluidic splitter/combiner frequency-splitting sensor implemented in microstrip technology and based on a pair of SRRs [37]. The sensor is applied to the characterization of the solute content in mixtures of ethanol in DI water, as well as to the determination of the complex dielectric constant of such liquid samples. As it will be shown, two output variables will be needed for the determination of the real and the imaginary part of the complex permittivity, i.e. the differential notch frequency, and the differential notch depth. The former is mainly related to the dielectric constant (as discussed in the previous sensor implementations). The differential notch depth is intimately related to the loss tangent of the MUT.

The device is composed of a microstrip SRR-loaded splitter/combiner configuration etched on a low-loss microwave substrate (the *Rogers RO3010* with thickness $h = 1.27$ mm, dielectric constant $\varepsilon_r = 10.2$, and loss tangent $\tan\delta = 0.0023$), and two microfluidic channels (the channels are placed on top of the gap region of the SRRs, where the electromagnetic energy is concentrated). The layout of the microwave (microstrip) circuitry is depicted in Fig. 5.21(c). The impedance of the parallel

(a)

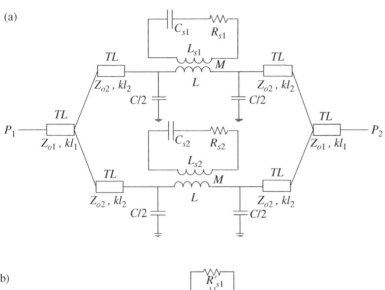

(b)

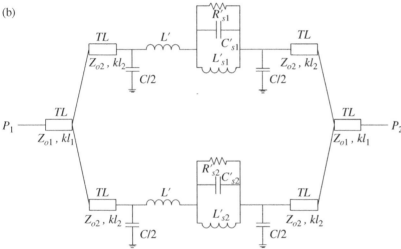

Figure 5.28 Circuit model corresponding to the topology of Fig. 5.21(c), including lumped and distributed components (a), and transformed model (b).

transmission lines is $Z_{02} = 50\,\Omega$. Thus, in order to optimize the matching with the reference imped-ance (50 Ω) of the ports, impedance inverters implemented by means of quarter-wavelength trans-mission lines, with impedance $Z_{01} = 35.35\,\Omega$, are cascaded between the T-junctions and the ports.[18] The SRRs were designed in order to exhibit their fundamental resonance in the vicinity of $f_0 = 1$ GHz. The proposed equivalent circuit model of the SRR-loaded splitter/combiner is depicted in Fig. 5.28(a) [2–4]. The resonant tanks L_s–C_s account for the SRRs, magnetically coupled with the parallel microstrip line sections through mutual inductances M. In this model, intrinsic SRR losses (i.e. without the presence of the MUT liquid of top them) are taken into account through the resistance R_s (line losses are neglected). Indeed, losses should be considered in this model, since

18 Note that the designation of the line impedances is different from that of the line sections in reference to the CSRR-based splitter/combiner structure of Figs. 5.21(b) or 5.22 (i.e. the designation of the original sources [37, 38] has been adopted).

liquids are lossy samples. Thus, for an accurate description of the structure, an additional resistor modeling MUT losses will be necessary, as it will be shown later. The line sections in proximity to the SRRs are accounted for by the inductance L and the capacitance C. Finally, the impedance inverters and the transmission line sections between the T-junctions and the SRRs are modeled by distributed components with the indicated characteristic impedance and electrical length.

The circuit of Fig. 5.28(a) can be transformed to the circuit of Fig. 5.28(b) through formulas given in [2–4, 6]. Parameter extraction of the lumped elements in Fig. 5.28(b) was carried out in [37] through the procedure reported in [4, 6] (applied only to each SRR and transmission line section coupled with it). Typically, parameter extraction is carried out by excluding losses in both the electromagnetic and circuit simulation. Then, losses are introduced in the electromagnetic simulation, and R'_s is inferred by curve fitting. As indicated, resonator losses are included in the model since the notch depth, intimately related to SRR losses, is a relevant parameter in the proposed sensor. It should be clarified that the models of Fig. 5.28 do not include the effects of the fluidic channels plus the liquids inside. To account for the effects of the channels and liquids, additional elements will be introduced in the model. Nevertheless, let us first validate the model corresponding to the microwave circuitry of the sensor by comparing electromagnetic and circuit simulations (the latter inferred from the extracted parameters).

Figure 5.29(a) and (b) depict the frequency response of the symmetric circuit of Fig. 5.21(c) inferred through electromagnetic simulation by means of the *Ansys HFSS* electromagnetic solver (the geometrical parameters, in reference to Fig. 5.21(c), are indicated in the caption). The response exhibits a single notch, as expected, at $f_0 = 1.040$ GHz, where the resonant tank $L'_s - C'_s$ is an open circuit. For the symmetric case (identical resonators), SRR reactive elements satisfy $L'_{s1} = L'_{s2} = L'_s$ and $C'_{s1} = C'_{s2} = C'_s$. The circuit simulation (inferred by means of *Keysight ADS*) with the extracted parameters is also depicted in Fig. 5.29(a) and (b). As can be seen, there is very good agreement in both the magnitude and phase responses. The structure that is obtained by varying the gap dimensions of the lower SRR (with $g = 3.6$ mm for such SRR), i.e. providing axial asymmetry, was also simulated in [37]. In this case, a pair of transmission zeros are visible, see Fig. 5.29(c) and (d). The transmission coefficient (magnitude and phase) obtained from circuit simulation is also depicted in the figures, and there is also very good agreement between the simulations in both domains (circuit and electromagnetic). Note that in the circuit simulations of Fig. 5.29(c) and (d), only those parameters relative to the lower SRR have been modified. With these results, the circuit model of the microwave part of the sensor (i.e. by excluding the effects of the channels and liquids) is validated.

Let us now focus the attention on the mechanical and fluidic parts of the sensor [37]. The channels are located on top of the SRR gap region, where the electromagnetic energy is mainly concentrated. Figure 5.30 shows the top and lateral side views of the mechanical and fluidic parts of the sensor, including relevant dimensions. Channel dimensions are designated by h_{ch}, w_{ch}, and l_{ch}. The mechanical parts consist of a polyether ether ketone (PEEK) structure, designed in order to accommodate the fluidic connectors for liquid injection in a controllable way through a syringe, the reference (REF) liquid in channel 1 and the MUT liquid in channel 2. Polydimethylsiloxane (PDMS) was used to fabricate the fluidic channels due to their biocompatibility and easy fabrication process.

The photograph of the fabricated sensor is depicted in Fig. 5.31 (details of the fabrication process are given in [37]). It is important to mention that due to substrate absorption (modifying substrate permittivity and hence inducing systematic errors in the sensor measurements), a thin film (0.12 mm) of glass ($\varepsilon_r = 5.5$) was placed between the microwave substrate and the channels. By this means, absorption is completely circumvented, although sensitivity is somehow degraded by the

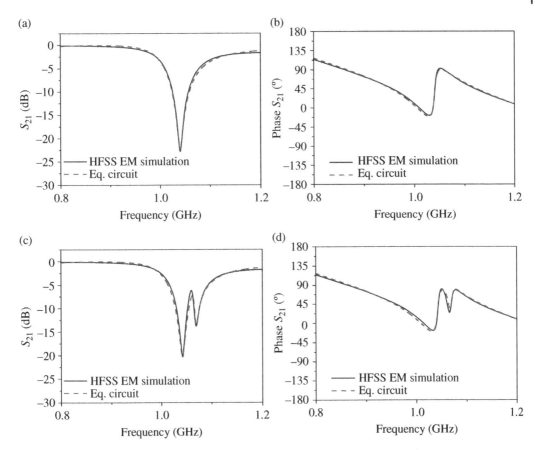

Figure 5.29 Magnitude and phase of the transmission coefficient in symmetric (a, b) and asymmetric cases (c, d) for the topology of Fig. 5.21(c). The element values [in reference to Fig. 5.28(b) considering the symmetry case] are $L' = 6.405$ nH, $C = 6.744$ pF, $L'_s = 0.585$ nH, $C'_s = 40$ pF, $C'_s = 510$ Ω, $Z_{01} = 35.35$ Ω, $Z_{02} = 50$ Ω, and $kl_1 = 90°$, and $kl_2 = 38°$. The element values [in reference to Fig. 5.28(b) considering the asymmetry case] are $C'_{s2} = 38.7$ pF and $R'_{s2} = 676$ Ω (the other elements remain unchanged). The dimensions of the symmetric structure, in reference to Fig. 5.21(c) are $L = 86$ mm, $W = 62$ mm; inverter dimensions are $l_1 = 27$ mm and $w_1 = 2.22$ mm; SRR dimensions are $l_s = 25$ mm, $W_s = 9$ mm, $c = 1.4$ mm, $g = 2.4$ mm; the slot separation between the lines and the SRRs is $d = 0.2$ mm; the dimensions of the transmission line sections between the T-junctions and the SRRs are $l_2 = 9.21$ mm, $w_2 = 1.34$ mm. *Source:* Reprinted with permission from [37]; copyright 2017 IEEE.

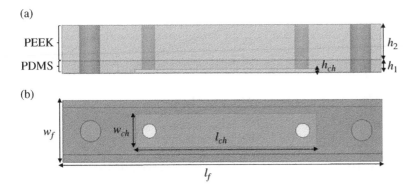

Figure 5.30 Lateral (a) and top (b) views of the mechanical and fluidic parts of the microwave sensor and relevant dimensions. $h_{ch} = 1.5$ mm, $l_{ch} = 26$ mm, $w_{ch} = 4.6$ mm, $l_f = 46$ mm, $w_f = 12.6$ mm, $h_1 = 3$ mm, and $h_2 = 9$ mm.

Figure 5.31 Photograph of the SRR-based splitter/combiner microfluidic sensor.

presence of the glass layer. Nevertheless, such separation layer is necessary, at least in most cases, to prevent direct contact between the liquids and the substrate. The presence of the mechanical and fluidic parts on top of the microwave substrate slightly modifies the element values of the circuit of Fig. 5.28(b). The response of the whole sensor (with air in the channels), including the measurement, electromagnetic simulation (inferred with the *Ansys HFSS* simulator), and circuit simulation is depicted in Fig. 5.32. The good agreement validates the circuit model, as well as the electrical definition of the different materials in the *Ansys HFSS* electromagnetic simulator (see further details in [37]).

Let us now consider the model of the whole sensor, i.e. including the liquids inside the channels. For that purpose, a capacitor, $C_{ch,i}$, and a resistor, $R_{ch,i}$ (with $i = 1, 2$), should be added to the model of Fig. 5.28, as indicated in the model of Fig. 5.33. The real part of the permittivity of the liquid, related to the density of electromagnetic energy stored inside the liquid, is modeled by the capacitor, $C_{ch,i}$, whereas $R_{ch,i}$ accounts for the imaginary part of the permittivity (which gives a measure of how dissipative the liquid is). To validate the complete model of the sensor, the measured responses of various combinations of air and liquids in the channels were compared with the circuit simulations with extracted parameters [37], see Fig. 5.34. The circuit parameters, excluding those corresponding to the liquid model, are those given in the caption of Fig. 5.32. For coherence, such parameters must be the same regardless of the channel content. Thus, only the parameters modeling the channel

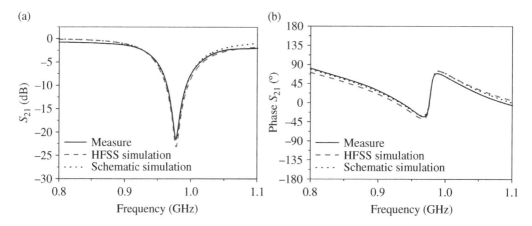

Figure 5.32 Measurement, electromagnetic simulation, and circuit simulation of the transmission coefficient magnitude (a) and phase (b) of the microfluidic sensor of Fig. 5.31 with air inside the channels (i.e. corresponding to a symmetric structure). The extracted element values are: $L' = 7.405$ nH, $C = 6.744$ pF, $L'_s = 0.697$ nH, $C'_s = 38$ pF, $R'_s = 518.8$ Ω, $Z_{01} = 35.35$ Ω, $Z_{02} = 50$ Ω, and $kl_1 = 90°$, and $kl_2 = 38°$. *Source:* Reprinted with permission from [37]; copyright 2017 IEEE.

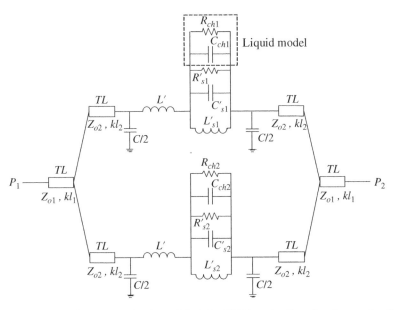

Figure 5.33 Equivalent circuit model of the SRR-based splitter/combiner microfluidic sensor that includes the effects of the presence of liquids inside the channels.

content have been adjusted in order to obtain a good match between the measured responses and the circuit simulations (these parameters are given in Table 5.3). It is worth mentioning that for the air/DI water combination, a finite capacitance and resistance for the air model were obtained, but the resulting capacitance is very small and the resulting resistance is very high, so that these values are within the tolerance limits of the parameter extraction method. Note that the parameters modeling the DI water are identical for all considered combinations involving such component.

Despite the fact that the model of Fig. 5.33 provides a good description of the sensor, the determination of the complex permittivity of the MUT liquid, or the solute content, from the measurement of the differential notch depth and position using analytical expressions is not simple. Therefore, calibration curves constitute a reasonable alternative. Figure 5.35(a) depicts the measured responses of the sensor for variations in liquid mixtures (i.e. different volume fractions of ethanol in DI water) injected to the MUT channel (the REF liquid being DI water in all cases). The recorded differential notch depth and position, as a function of the volume fraction of ethanol, are depicted in Fig. 5.35(b). Note the sudden jump that appears in both the differential notch depth, ΔS_{21}, and position, Δf_z, between pure DI water (symmetric case) and a solution with 10% of ethanol concentration. This jump is because the dimensions of the sensor designed and fabricated in [37], the one reported in this section, were not optimized. As indicated in the previous section, there is an optimum length of the parallel lines for sensitivity optimization in the limit of small perturbations. Particularly, such length must be set to the value that generates a virtual ground in the input and output T-junctions, when the frequency of the input signal is the one of the REF resonator. In the sensor under study, the optimum electrical length of these lines must satisfy [37][19]

---

19 This is the equivalent condition to (5.35) in reference to CSRR-based splitter/combiner frequency-splitting sensors. Note, however, that, following the nomenclature of the original papers, the electrical lengths of the different line sections (parallel lines and inverters), as well as the characteristic impedances of such sections, are designated with different variables. At SRR resonance, there is an open in the symmetry plane between the input and the output ports. This means that the impedance seen from the T-junctions, looking at the SRRs, is the one corresponding to a half parallel line (with impedance Z_{o2} and phase kl_2) terminated with the capacitor $C/2$, see the model of Fig. 5.28 or 5.33. Such impedance is null, provided C and kl_2 satisfy (5.41). In the hypothetical limit when $C = 0$ (a condition not achievable in practice), the termination impedance at resonance is simply an open circuit, and the length of the parallel line must be a quarter wavelength or an odd multiple, or $kl_2 = (2n + 1) \cdot \pi/2$, as derived from (5.41).

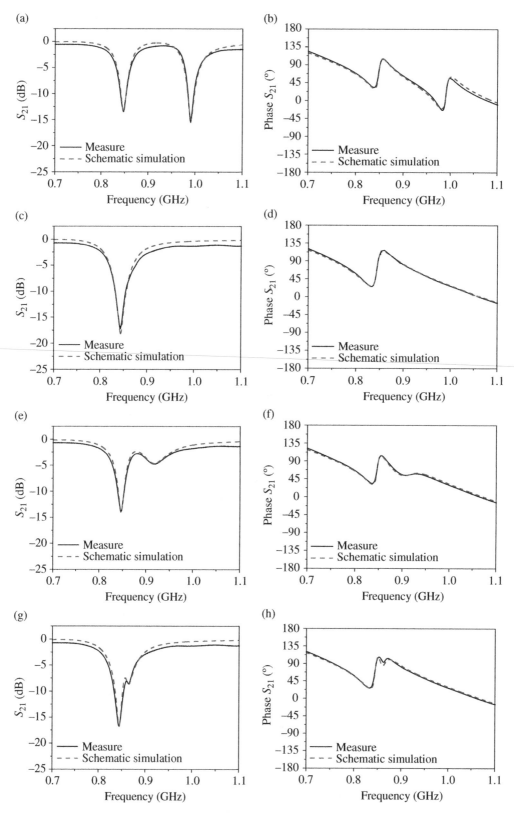

Figure 5.34 Magnitude (left) and phase (right) of the transmission coefficient for various combinations of channel content. (a, b) Air/DI water, (c, d) DI water/DI water, (e, f) DI water/ethanol, and (g, h) DI water/DI water + 10% ethanol. *Source:* Reprinted with permission from [37]; copyright 2017 IEEE.

Table 5.3 Extracted element parameters of the fluid model for channels 1 and 2.

Channel 1	$C_{ch,1}$ (pF)	$R_{ch,1}$ (Ω)	Channel 2	$C_{ch,2}$ (pF)	$R_{ch,2}$ (Ω)
DI water	13	1000	Air	0.1	2930
DI water	13	1000	DI water	13.0	1000
DI water	13	1000	Ethanol	6.7	80
DI water	13	1000	DI water + 10% ethanol	12.2	300

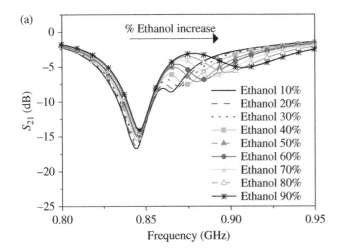

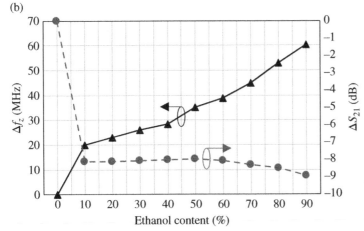

Figure 5.35 Measured transmission coefficient (magnitude) for different mixtures of DI water and ethanol (MUT liquid), considering DI water as REF liquid (a), and representation of Δf_z and $\Delta|S_{21}|$ as a function of the ethanol concentration (b). *Source:* Reprinted with permission from [37]; copyright 2017 IEEE.

$$kl_2 = \arctan\left(\frac{2}{\omega_0 C Z_{02}}\right) \tag{5.41}$$

As mentioned, forcing the fulfilment of (5.41) optimizes the sensitivity. However, since the physical length of the lines is fixed, in particular l_2, it is obvious that a change in the REF liquid modifies ω_0, and, consequently, the optimum length is no longer satisfied. In other words, it is not possible to optimize the sensitivity in the limit of small perturbations regardless of the considered REF liquid. Due to this main reason, the intention in [37] was to implement a sensor with application to the analysis and characterization of a diversity of liquid samples. Thus, rather than sensitivity optimization, the main aim was to reduce sensor dimensions by considering a short (arbitrary) length of the parallel lines. Nevertheless, in view of the calibration curves of Fig. 5.35(b), sensor functionality for the determination of volume fraction of ethanol in DI water is feasible with the proposed implementation [37]. The sensitivity, and therefore the accuracy, is better for high volume fractions (i.e. for large perturbations), rather than for small volume fractions, since the length l_2 was not optimized, an aspect discussed previously in this chapter (see Section 5.4.1).

For the estimation of the real and the imaginary part of the complex dielectric constant of the MUT liquid, the usual approach [42] of assuming a linear dependence between the output and the input variables was adopted in [37]. Thus, the variation of the real, $\Delta\varepsilon'$, and imaginary, $\Delta\varepsilon''$, part of the complex permittivity can be expressed as [42]

$$\Delta\varepsilon' = k_{11}\,\Delta f_z + k_{12}\,\Delta\,|S_{21}| \tag{5.42a}$$

$$\Delta\varepsilon'' = k_{21}\,\Delta f_z + k_{22}\,\Delta\,|S_{21}| \tag{5.42b}$$

where $\Delta\varepsilon' = \varepsilon'_{MUT} - \varepsilon'_{REF}$, $\Delta\varepsilon'' = \varepsilon''_{MUT} - \varepsilon''_{REF}$, $\Delta f_z = f_{z,MUT} - f_{z,REF}$, and $\Delta|S_{21}| = |S_{21}|_{MUT} - |S_{21}|_{REF}$. In order to obtain the four coefficients in (5.42), it is necessary to at least use the input and output variables corresponding to two volume fractions of ethanol in DI water (calibration). The output variables can be directly inferred from Fig. 5.35(b). As for the input variables, the complex permittivity of pure ethanol and DI water are well known [43]. Thus, $\Delta\varepsilon'$ and $\Delta\varepsilon''$ for 100% ethanol are known (the REF liquid is DI water). The other pair of input variables, $\Delta\varepsilon'$ and $\Delta\varepsilon''$, is necessary to determine the four coefficients, may be that corresponding to a certain volume fraction of ethanol, e.g. 10%. The complex permittivity of the liquid mixture can be inferred from the complex permittivity of the constitutive components, ethanol and DI water in this case, using well-known models, e.g. the Weiner model [44]. Using this procedure, the coefficients in (5.42) were found to be $k_{11} = -0.944$ MHz^{-1}, $k_{12} = -0.545$ dB^{-1}, $k_{21} = 0.127$ MHz^{-1}, and $k_{22} = 0.260$ dB^{-1}. Note that in (5.42), the left-hand-side members actually correspond to the differential real and imaginary part of the complex dielectric constant, or relative complex permittivity, and are therefore dimensionless.

Once the coefficients of (5.42) are known, the determination of the complex permittivity of the MUT sample is as simple as obtaining the frequency response and, from it, inferring Δf_z and $\Delta|S_{21}|$. Obviously, as proof-of-concept, the data points of Fig. 5.35(b) can be used to determine the complex permittivity of the mixtures corresponding to the indicated volume fractions of ethanol in DI water. The results, plotted in Fig. 5.36, lie between the static upper (WU) and lower (WL) limits of the real and imaginary parts of the complex dielectric constant, as given by the Weiner model. These results validate the functionality of the device as a sensor for the determination of the complex dielectric constant of liquids.

5.5 Other Approaches for Coupling Cancelation in Frequency-Splitting Sensors

This section presents two examples of frequency-splitting sensors based on a transmission line loaded with a pair of resonant elements, where the mutual coupling between the resonators is canceled despite the fact that the sensing resonators are located in the same axial position of the line.

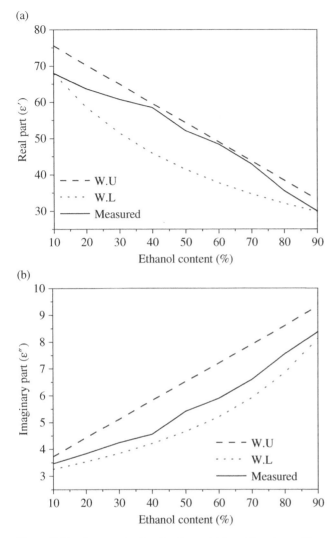

Figure 5.36 Extracted value for the real (a) and imaginary (b) parts of the complex dielectric constant in mixtures of DI water/ethanol. The static Weiner model (upper −WU− and lower −WL− limits) is also included for comparison purposes. *Source:* Reprinted with permission from [37]; copyright 2017 IEEE.

In one case, an electric wall between the otherwise electrically coupled resonators is added to the structure (specifically, the magnetic-LC −MLC− resonator is considered in the reported example [45]). In the second case, coupling is prevented, or roughly prevented, by spacing the resonators (SRRs) away; a canonical solution provided the SRRs are excited by means of wide microstrip lines [46].

5.5.1 MLC-Based Frequency-Splitting Sensor

The MLC resonator [47] is the dual counterpart of the electric-LC (ELC) resonator [48] (see Section 2.2 for further details). Since the MLC is a slot resonator, it can be etched in the ground plane of a microstrip line and used for sensing. Particularly, MLC resonators can be applied to the dielectric characterization of solids and liquids (similar to CSRRs, SRRs or SIRs). The MLC, like the ELC resonator, exhibits two resonant modes, the even and the odd mode, and behaves

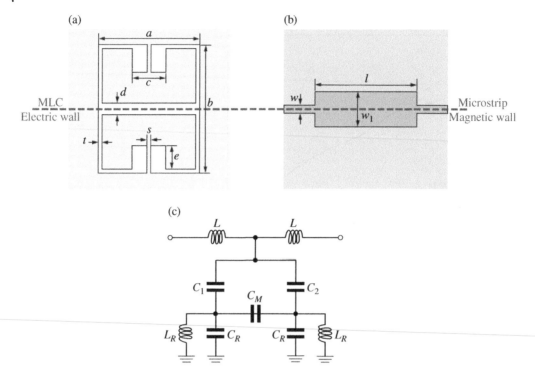

Figure 5.37 Bottom (a) and top (b) views of the topology of a microstrip transmission line loaded with a MLC resonator, and equivalent circuit model (c). *Source:* Reprinted with permission from [45]; copyright 2020 MDPI.

similarly to a pair of electrically coupled resonators. The typical topology and circuit model of a microstrip line symmetrically loaded with an MLC resonator is depicted in Fig. 5.37. The circuit model is roughly identical to the one of the microstrip line loaded with a pair of electrically coupled CSRRs, see Fig. 5.6(b). The difference is the capacitance C of Fig. 5.6(b), not present in the model of Fig. 5.37. However, such capacitance does not determine the position of the resonance frequencies (as it was discussed in Section 5.2.2), and it is therefore irrelevant from the point of view of sensor functionality. Thus, the analysis carried out in Section 5.2.2 applies to the structure of Fig. 5.37. In particular, if the structure exhibits perfect symmetry, the MLC odd-mode resonance (the fundamental, or lowest, one) cannot be excited, and a single notch in the frequency response is visible, as reported in [45]. Naturally, by truncating symmetry, e.g. by asymmetrically loading the MLC resonator, two notches are expected. However, due to the coupling capacitance, designated as C_M in the circuit of Fig. 5.37, it is expected that the sensitivity of the differential notch frequencies (the canonical output variable in frequency-splitting sensors) with the differential dielectric constant (or any other input variable causing symmetry imbalance) is degraded, especially for small perturbations.

A possible stratagem to circumvent the effects of C_M on sensitivity degradation is to add an electric wall in the axial symmetry plane of the structure, as proposed in [45]. By doing this, if the structure is symmetric, a single transmission zero is also expected. However, the new transmission zero will be located at the odd resonance of the MLC, due to the presence of such electric wall. If symmetry is truncated, the electric wall prevents from any interaction between the two parts of the MLC resonator, and the structure behaves as a transmission line loaded with a pair of uncoupled resonators. In other words, the notch frequencies are determined by each MLC half and do not depend on the characteristics of the other half. The circuit model of the structure that includes the electric wall, for arbitrary loads in the MLC halves (accounted for by means of different MLC capacitances, C_{R1} and C_{R2}), is depicted in Fig. 5.38. The resonance frequencies are

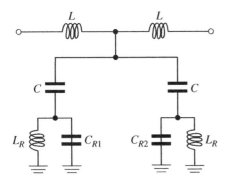

Figure 5.38 Circuit model of the MLC-loaded line with arbitrary loads in the MLC halves, and with an electric wall in the axial symmetry plane. *Source:* Reprinted with permission from [45]; copyright 2020 MDPI.

$$\omega_1 = \frac{1}{\sqrt{L_R(C + C_{R1})}} \tag{5.43a}$$

$$\omega_2 = \frac{1}{\sqrt{L_R(C + C_{R2})}} \tag{5.43b}$$

where $C_{R1} = C_R + C_M/2 + \Delta C_{R1}$ and $C_{R2} = C_R + C_M/2 + \Delta C_{R2}$ (C_R and C_M are the capacitance of each unperturbed MLC half and the coupling capacitance, respectively, as depicted in the circuit of Fig. 5.37(c), corresponding to the symmetric structure without the presence of the electric wall in the axial symmetry plane).

Figure 5.39 depicts a prototype of a frequency-splitting sensor based on a MLC resonator, where the electric wall was implemented by means of a metallic frame (device dimensions and substrate characteristics are indicated in the caption) [45]. The figure includes the measured frequency responses obtained by leaving the REF half of the MLC resonator unaltered and the other half (devoted to the MUT) covered by dielectric slabs of different dielectric constant. The considered slabs are uncoated microwave substrates, specifically, 1.5-mm thick *Rogers RT5880* with $\varepsilon_r = 2.2$ and $\tan\delta = 0.0004$, 1.524-mm thick *Rogers RO4003* with $\varepsilon_r = 3.55$ and $\tan\delta = 0.0027$, 1.5-mm thick *FR4* with $\varepsilon_r = 4.7$ and $\tan\delta = 0.025$, 1.28-mm thick *Rogers RO3006* with $\varepsilon_r = 6.5$ and $\tan\delta = 0.002$, and 1.27-mm thick *Rogers RO6010* with $\varepsilon_r = 10.7$ and $\tan\delta = 0.0023$. The samples were cut in fragments of 15×20 mm^2, covering completely the sensing area. As it can be appreciated, the notch at roughly 2.2 GHz, the one generated by the bare half of the MLC resonator, does not vary when the other MLC half is loaded with different MUT samples.

The results of Fig. 5.39 demonstrate the potential of the approach in effectively canceling the effects of mutual coupling between the MLC halves, thereby allowing for the implementation of frequency-splitting sensors exhibiting high sensitivity in the limit of small perturbations, see Fig. 5.39(d). However, sensor implementation requires a metallic frame, emulating an electric wall, and this increases fabrication costs and sensor dimensions (in the vertical direction). Note also that this strategy requires wide (and hence mismatched) microstrip lines, but this does not prevent from detecting the notches, necessary for sensing.

5.5.2 SRR-Based Frequency-Splitting Sensor Implemented in Microstrip Technology

Section 5.2.1 was focused on the analysis of CPW transmission lines loaded with a pair of magnetically coupled SRRs. It is well known that CPWs are suitable lines for SRR excitation. The reason is that by etching the SRRs in the back substrate side beneath the CPW slots, significant inductive

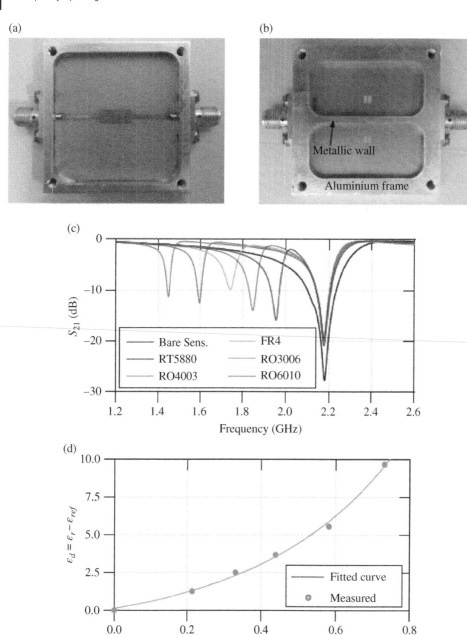

Figure 5.39 Photograph of the top (a) and bottom (b) sides of a fabricated frequency-splitting sensor based on a microstrip line loaded with a MLC resonator, frequency responses (transmission coefficient) for different dielectric loads in the half of the MLC resonator devoted to the MUT (c), and dependence between the differential dielectric constant, ε_d, and the differential notch frequency, f_d (d). The electric wall in the axial symmetry plane is implemented by means of a metallic wall of finite width. The geometrical dimensions, in reference to Fig. 5.37, are a = 10.2 mm, b = 15.4 mm, c = 2.6, d = 2.2 mm, e = 2.6 mm, s = t = 0.2 mm, w = 1.2 mm, and w_1 = 6.2 mm. The sensor was fabricated on the *Rogers RT6002* substrate with dielectric constant ε_r = 2.93, thickness h = 0.508 mm, and loss tangent tan δ = 0.0037. *Source:* Reprinted with permission from [45]; copyright 2020 MDPI.

coupling between the line and the SRRs results. This is a consequence of the fact that the magnetic field generated by the line produces an important magnetic flux passing through the SRRs.[20] Nevertheless, it does not mean that other transmission line types cannot be able to magnetically drive such resonant elements. In particular, SRR-loaded microstrip lines with the SRRs etched in the upper side of the substrate, in proximity to the line strip, have been reported [49]. To a first-order approximation, the dominant coupling mechanism between the line and the SRRs in these structures is also magnetic. However, since the SRRs are separated at least the distance corresponding to the strip width (provided the structure is symmetrically loaded with a pair of SRRs), the mutual coupling between SRRs is naturally minimized, and such coupling can be neglected for sufficiently wide lines. In [46], this strategy was used to avoid inter-resonator coupling in frequency-splitting sensors based on SRR-loaded microstrip lines.

The topology of the sensor designed and fabricated in [46] is depicted in Fig. 5.40(a). The circuit model is identical to the one shown in Fig. 5.2, except by the fact that the mutual inductance M' (modeling inter-resonators' coupling) is not present (thus, the model is not depicted). Figure 5.40(b) shows the photograph of the fabricated device (dimensions and substrate are indicated in the caption), whereas Fig. 5.40(c) and (d) depict the measured frequency responses for various SRR loadings and the dependence of the differential notch frequency with the dielectric constant of the MUT, respectively. The considered dielectric loads are un-metallized microwave substrates with dielectric constants of 2.2 (*Rogers RT5880*), 3.66 (*Rogers RO4350*), 4.7 (*FR4*), 6.5 (*Rogers RO3006*), and 10.7 (*Rogers RO6010*). The thickness of these MUT samples is larger than 1 mm and all the samples were cut in order to exhibit an area of $14 \times 14 \text{ mm}^2$, covering the whole SRR area. It is apparent, in view of Fig. 5.40(c), the presence of a notch at 2.1 GHz, regardless of the dielectric load present in the SRR devoted to the MUT sample. Such notch frequency is the one of the bare SRR, and it is clear that changes in the other SRR do not alter it, which means that inter-resonators' coupling is negligible.

5.6 Other Frequency-Splitting Sensors

Let us briefly review in this section two alternative approaches for the implementation of frequency-splitting sensors. In one case [50], the difference, as compared with the sensors reported so far, concerns sensor functionality, i.e. as bandpass filter, but the working principle, sensing mechanism, and applications are similar to those of the splitter/combiner sensors reported in Section 5.4. In the second case [51], the main originality concerns sensor application, i.e. the measurement of linear displacements in two dimensions. Moreover, in this sensor, the necessary symmetry perturbations for sensing are achieved by means of a movable slab with a metallic pattern. Due to these significant differences as compared with the frequency-splitting sensors studied so far, mainly concerning the sensing mechanism and sensor application, this sensor is analyzed in more detail.

5.6.1 Frequency-Splitting Sensors Operating in Bandpass Configuration

In the frequency-splitting sensors reported so far, the structures operate as stopband (or, more specifically, notch) filters. However, it is possible to implement sensors based on the same principle, i.e. frequency splitting, by considering passband topologies. For example, in [50], a frequency-splitting

---

20 This is true if the substrate is sufficiently thin. As the substrate thickness broadens, the separation between the SRRs and the CPW metal layer increases, and the magnetic coupling decreases.

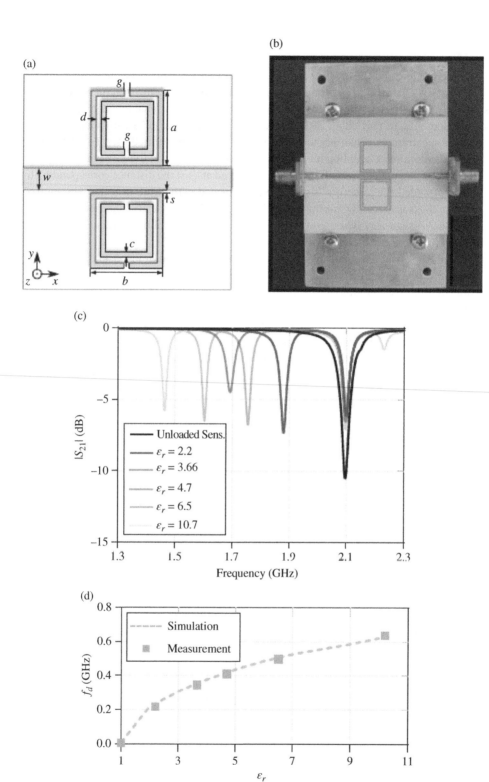

Figure 5.40 Topology (a) and photograph (b) of the SRR-based microstrip frequency-splitting sensor, frequency responses (transmission coefficient) for different dielectric loads in the SRR devoted to the MUT (c), and dependence between the differential notch frequency, f_d, and the dielectric constant of the MUT (d). Dimensions are $a = b = 10.8$ mm, $c = 0.4$ mm, $d = 0.4$ mm, $g = 0.2$ mm, $s = 0.2$ mm, and $w = 1.6$ mm. The sensor was implemented on a 0.762-mm thick *Rogers RO4350* substrate with a relative permittivity of 3.66 and a loss tangent of 0.0037. *Source:* Reprinted with permission from [46]; copyright 2018 IEEE.

sensor based on the splitter/combiner configuration but exhibiting a pair of peaks (rather than notches) under imbalanced conditions was reported. The detailed analysis of this sensor structure is out of the scope of this chapter. Nevertheless, let us mention that switching from the stopband to the bandpass response was achieved by merely etching a gap in the parallel lines of the splitter/combiner, in the position where the resonant elements (SRRs in [50]) are located. By adding such gap, each branch behaves as a coupled-resonator bandpass filter, the sensor exhibits a single peak for balanced loads, and it splits into two peaks when symmetry is truncated. The functionality of these sensors is very similar to the one of their stopband counterparts, the output variable being the difference between the peak frequencies and, eventually, the difference in the insertion loss of both peaks.

5.6.2 Frequency-Splitting Sensors for Two-Dimensional Alignment and Displacement Measurements

The application of the frequency-splitting sensing approach to the implementation of two-dimensional alignment and displacement sensors was investigated in [51]. The sensing elements in these sensors are variants of the so-called BC-SRR [14, 52], a very appropriate resonant element for displacement sensing, as it was discussed in Chapter 4. Figure 5.41 depicts the side and top views of the sensor proposed in [51]. A right-angle bended microstrip line was considered for sensing, since the device was intended to measure linear displacements in the two dimensions (x and y) of space. The pairs of resonant elements necessary for sensing based on frequency splitting were located symmetrically with regard to the axis of each line section, as depicted in Fig. 5.41. Moreover, each pair of resonators was designed in order to resonate at different frequencies. In Chapter 4,

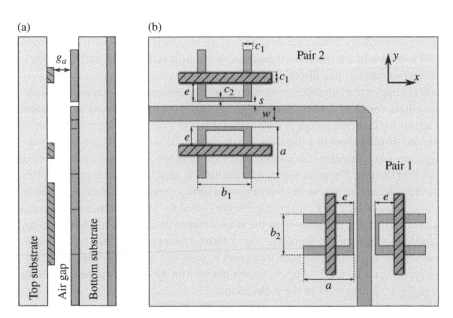

Figure 5.41 (a) Side view and (b) top view of the two-dimensional frequency-splitting displacement sensor based on pairs of modified BC-SRRs. The *Rogers RO4003* substrate with relative permittivity $\varepsilon_r = 3.38$ and thickness $h = 0.81$ mm was used for the bottom and top substrates, which are separated with an air gap $g_a = 0.25$ mm. The width of the microstrip line is $w = 1.85$ mm, corresponding to a 50-Ω line. The dimensions of the resonators are $a = 12.2$ mm, $b_1 = 15.5$ mm, $b_2 = 10.5$ mm, $c_1 = 2.5$ mm, $c_2 = 1.2$ mm, $s = 0.2$ mm, and at initial position $e = 4.8$ mm. *Source:* Reprinted with permission from [51]; copyright 2014 Elsevier.

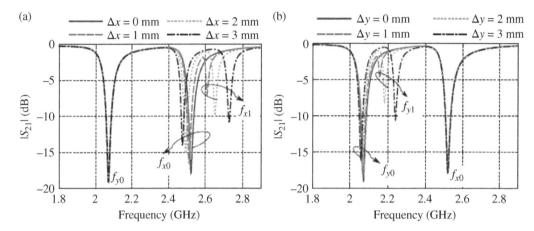

Figure 5.42 Simulated transmission coefficients of the frequency-splitting two-dimensional displacement sensor for different values of displacement in the *x* direction from 0 to 3 mm in steps of 1 mm, while $\Delta y = 0$ mm (a), and for different values of displacement in the *y* direction from 0 to 3 mm in steps of 1 mm, while $\Delta x = 0$ mm (b). *Source:* Reprinted with permission from [51]; copyright 2014 Elsevier.

Section 4.3.2.2, it was shown that the resonance frequency of a BC-SRR depends on the relative position between the upper and lower metallic ring, or loop, of the particle. Thus, by etching one of the BC-SRR loops in a dielectric slab in relative motion to a static part, where the other loop is etched, the measurement of the relative displacement between both loops is possible. However, the typical topology of the BC-SRR (see Section 4.3.2.2) was modified in [51], so that the resonant frequency is altered by the displacement in one direction only.[21] For that purpose, the upper (movable) loops were replaced with straight strips, as shown in Fig. 5.41. With this configuration, provided the strips are long enough, the resonance frequency of the modified BC-SRR varies only when the strips experience a motion in the direction transverse to the strip axis. That is, each BC-SRR is only sensitive to a displacement in one direction.

Figure 5.42(a) and (b) depict the simulated response of the two-dimensional sensor to displacements in *x* and *y* directions, respectively. Specifically, Fig. 5.42(a) shows the simulated transmission coefficients of the sensor for different values of displacement in the *x* direction, from 0 to 3 mm in steps of 1 mm, when no displacement in *y* direction is applied, i.e. $\Delta y = 0$ mm. At the initial position, corresponding to perfect symmetry ($\Delta x = 0$ mm), only two notches at f_{x0} and f_{y0} appear in the transmission spectrum of the line. The resonances at frequencies f_{x0} and f_{y0} are associated with Pair 1 and Pair 2, respectively. However, as shown in the figure, a displacement in *x* direction results in the splitting of the resonance of the Pair 1 into two notches at f_{x0} and f_{x1}. The difference between the two frequencies, i.e. $\Delta f_x = f_{x1} - f_{x0}$, increases with the displacement in *x* direction. Therefore, Δf_x can be used for sensing the value of Δx. Similarly, Fig. 5.42(b) shows the simulated transmission coefficients of the sensor for variation of Δy from 0 to 3 mm in steps of 1 mm, while the Pair 1 exhibits perfect symmetry, i.e. $\Delta x = 0$ mm. The figure clearly shows that $\Delta f_y = f_{y1} - f_{y0}$ increases with Δy, allowing for sensing a displacement in the *y* direction.

The photograph of the fabricated prototype sensor is depicted in Fig. 5.43. The experimental setup used for measurements included micrometer actuators for adjusting the air gap space between the two substrates, as well as two pairs of micrometer actuators for accurate displacement in the *x* and *y* directions, as reported in [51]. Figure 5.44(a) depicts the measured transmission coefficients of the

---

21 This is necessary in order to distinguish the motion in the *x* and *y* directions.

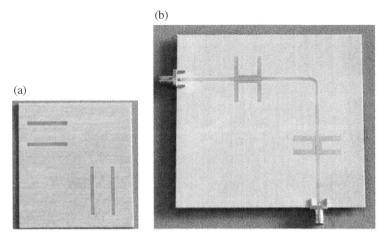

Figure 5.43 Photographs of the fabricated frequency-splitting two-dimensional displacement sensor: (a) top and (b) bottom substrates. *Source:* Reprinted with permission from [51]; copyright 2014 Elsevier.

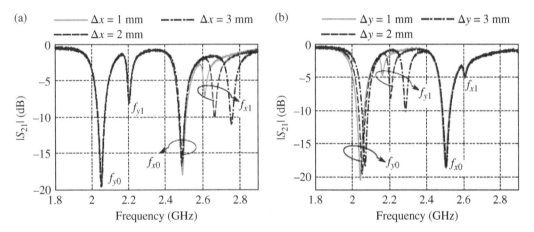

Figure 5.44 Measured transmission coefficients for the sensor of Fig. 5.43 for three different values of the displacement in the *x* direction, while Δ*y* = 2 mm (a), and for three different values of displacement in the *y* direction, while Δ*x* = 0.5 mm (b). *Source:* Reprinted with permission from [51]; copyright 2014 Elsevier.

sensor for different values of $\Delta x = 1$, 2, and 3 mm, while the sensor has a fixed displacement of 2 mm in *y* direction. The figure shows that while the frequency difference Δf_x increases from 105 to 267 MHz, Δf_y does not change. Similarly, Fig. 5.44(b) shows the measured results for a fixed displacement of 0.5 mm in *x* direction, resulting in fixed notches at $f_{x0} = 2.5$ GHz and $f_{x1} = 2.6$ GHz, while Δy changes from 1 to 3 mm in steps of 1 mm, which can be sensed from the increase in Δf_y from 95 to 210 MHz. The experiment shows that each pair of the modified BC-SRRs are exclusively responsive to a displacement either in *x* or *y* direction. Thus, the structure can be efficiently used as a two-dimensional alignment and displacement sensor.

In Fig. 5.45(a) and (b), the measured Δf_x and Δf_y versus displacement in *x* and *y* directions, respectively, are compared with those of the simulation results. The sensor exhibits reasonable good linearity. Note that the air gap distance can modify the resonance of the notch, but if there is perfect alignment, only one notch for each resonator pair is expected. Thus, for alignment purposes the sensor benefits from robustness against environmental changes. Nevertheless, tolerances in the

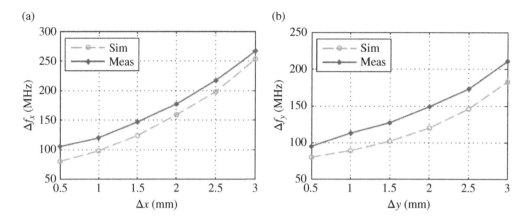

Figure 5.45 Comparison between the measured and simulated differential notch frequencies versus displacement: (a) Δf_x versus Δx and (b) Δf_y versus Δy. *Source:* Reprinted with permission from [51]; copyright 2014 Elsevier.

gap distance may also influence the results. This difficulty in controlling the air gap (along with the fabrication and measurement setup tolerances) is the reason for the discrepancy between measurements and simulations in the results of Fig. 5.45, as indicated in [51], and the main drawback of this approach. In addition, similar to the coupling-modulation displacement sensors of Section 4.3.1, the sensors reported in this section exhibit a limited input dynamic range, intimately related to the dimensions of the considered sensing resonators.

5.7 Advantages and Drawbacks of Frequency-Splitting Sensors

Like other sensors based on symmetry disruption, frequency-splitting sensors are robust against cross sensitivities related to potential changes in ambient factors (temperature, atmospheric pressure, humidity, etc.). The reason, as pointed out before in this book, is the invariability of the environmental parameters at the typical scales of the sensors (typically, few centimeters, or even less). Thus, environmental changes are seen as common-mode stimuli in frequency-splitting sensors, thereby having small or negligible effect on the sensor's response. By contrast, in sensors that are not based on electromagnetic symmetry properties, e.g. frequency-variation sensors, such changes may be the cause of undesired drifts in the output variable(s).

Frequency-splitting sensors cannot be considered true differential sensors, since two independent sensors do not constitute them. However, frequency-splitting sensors, like differential sensors, provide a differential output variable(s) in response to a differential input variable(s) and require two sensing elements, typically, a pair of resonators. Therefore, frequency-splitting sensors are generally larger than other sensors based on a single sensing resonant element (e.g. frequency-variation sensors). For measurement purposes, a frequency-sweeping signal covering at least the output dynamic range is needed (similar to frequency-variation sensors). This is a clear disadvantage as compared with phase-variation and coupling-modulation sensors, where a single-tone (harmonic) signal suffices for sensing, representing a reduction in the cost of the associated electronics. Although in most prototype sensors reported in the literature, and particularly in frequency-splitting sensors, vector network analyzers (VNAs) are commonly used for sensing, voltage-controlled oscillators (VCOs) in a real scenario should replace this high-cost equipment. However, although not compared with the cost of VNAs, broadband VCOs able to cover the required output dynamic range may represent a significant portion of the overall cost of the sensor (especially, if the output dynamic range forces the use of various VCOs, each one covering a portion of the required spectrum). Thus, from the point of view of sensor costs, the sensors studied in this chapter cannot compete against single-frequency sensors.

It should also be mentioned that in the design of frequency-splitting sensors, caution concerning the effects of inter-resonators' coupling on sensitivity degradation must be taken, as it has been exhaustively discussed in the chapter. Several strategies to circumvent the electromagnetic interaction between the sensing resonators, illustrated by means of prototype examples, have been discussed in this chapter. However, such approaches are not exempt of certain design or fabrication complexity.

In reference to sensor performance, frequency-splitting sensors are comparable to frequency-variation sensors. For example, the sensitivity of the differential notch frequency, the usual output variable, is comparable with the sensitivity of the frequency-variation sensor counterparts (provided inter-resonators' coupling is prevented), an aspect exhaustively discussed in Chapter 2. For dielectric constant measurements, relatively large variations of the input signal can also be measured with frequency-splitting sensors. Thus, such sensors exhibit significant input dynamic ranges, but at the expense, as mentioned before, of a relatively high cost of the electronics needed for sensing.

Frequency-splitting sensors, like differential-mode sensors, are of especial interest as comparators, able to detect tiny differences between a certain MUT sample and the REF sample in real time. Combined with fluidic channels, these sensors can be applied to the characterization of liquid samples, e.g. to the determination of their composition and to the detection of impurities. In particular, monitoring changes in liquid composition with regard to a well-known reference liquid in real time is a canonical application of frequency-splitting sensors, of interest in real scenarios (e.g. industrial processes, food industry, and biosensing). In such applications as comparators, sensor resolution, intimately related to the sensitivity at small perturbations, is a key sensor parameter. Reasonable resolutions in microfluidic sensors based on frequency splitting, as well as on frequency variation, have been reported. However, in general, the resolution achievable with other sensors, such as phase-variation sensors (Chapter 3) and differential-mode sensors, to be discussed in the next chapter, is superior.

To end this chapter, let us dedicate a few words to a brief comparative analysis between the two-dimensional displacement and alignment prototype sensor reported in the previous section, and the displacement sensors studied in Chapters 3 and 4. With regard to the input dynamic range, the sensor of Section 5.6.2 (Fig. 5.43) is comparable to other sensors implemented by means of transmission lines loaded with a resonant element, typically based on coupling modulation (e.g. the sensors of Figs. 4.5, 4.7, and 4.9). In all these sensors, including the one of Fig. 5.43, the input dynamic range is intimately related to the size of the sensing resonant elements. By contrast, in the displacement and velocity sensors based on electromagnetic encoders, e.g. those of Figs. 4.47 and 4.52, the input dynamic range is, theoretically, unlimited, since the encoder can be arbitrarily large. Displacement sensors based on phase variation and operating in reflection have also been reported (see Fig. 3.29). In such sensors, the input dynamic range is determined by the length of the sensing line, a half-wavelength (or multiple) line, as dictated by sensitivity optimization requirements (see Chapter 3). In general, these sensors and electromagnetic linear encoders are useful for moderate and large-range linear measurements. By contrast, the sensor reported in Fig. 5.43 and the coupling-modulation displacement sensors of Figs. 4.5, 4.7, and 4.9 are more appropriate for short-range displacement measurements, or when the interest is to detect potential lack of alignment between two surfaces. Comparing the sensitivity between these displacement sensors is not easy due to the different output variables involved. Nevertheless, the sensor reported in Section 5.6.2 exhibits a reasonable sensitivity, and a relatively good linearity within the considered input dynamic range.

References

1 L. Su, J. Naqui, J. Mata-Contreras, and F. Martín, "Modeling metamaterial transmission lines loaded with pairs of coupled split-ring resonators," *IEEE Ant. Wireless Propag. Lett.*, vol. 14, pp. 68–71, 2015.

2 F. Martín, F. Falcone, J. Bonache, R. Marqués, and M. Sorolla, "Split ring resonator based left handed coplanar waveguide," *Appl. Phys. Lett.*, vol. 83, pp. 4652–4654, Dec. 2003.

3 F. Aznar, J. Bonache, and F. Martín, "Improved circuit model for left handed lines loaded with split ring resonators," *Appl. Phys. Lett.*, vol. 92, paper 043512, Feb. 2008.

4 F. Aznar, M. Gil, J. Bonache, J. D. Baena, L. Jelinek, R. Marqués, and F. Martín, "Characterization of miniaturized metamaterial resonators coupled to planar transmission lines through parameter extraction," *J. Appl. Phys.*, vol. 104, paper 114501-1-8, Dec. 2008.

5 J. Naqui, M. Durán-Sindreu, and F. Martín, "Modeling split ring resonator (SRR) and complementary split ring resonator (CSRR) loaded transmission lines exhibiting cross polarization effects," *IEEE Ant. Wireless Propag. Lett.*, vol. 12, pp. 178–181, 2013.

6 F. Martin, *Artificial Transmission Lines for RF and Microwave Applications*, John Wiley, Hoboken, NJ, USA, 2015.

7 J. S. Hong and M. J. Lancaster, *Microstrip Filters for RF/Microwave Applications*, John Wiley, Hoboken, NJ, USA, 2001.

8 J. S. Hong and M. J. Lancaster, "Couplings of microstrip square open-loop resonators for cross-coupled planar microwave filters," *IEEE Trans. Microwave Theory Tech.*, vol. 44, pp. 2099–2109, Dec. 1996.

9 J. Naqui, A. Fernández-Prieto, F. Mesa, F. Medina, and F. Martín, "Effects of inter-resonator coupling in split ring resonator (SRR) loaded metamaterial transmission lines," *J. Appl. Phys.*, vol. 115, no. 19, paper 194903, 2014.

10 F. Bilotti, A. Toscano, L. Vegni, K. Aydin, K. Alici, and E. Ozbay, "Equivalent-circuit models for the design of metamaterials based on artificial magnetic inclusions," *IEEE Trans. Microwave Theory Tech.*, vol. 55, no. 12, pp. 2865–2873, Dec. 2007.

11 V. Sanz, A. Belenguer, A. L. Borja, J. Cascon, H. Esteban, and V. E. Boria, "Broadband equivalent circuit model for a coplanar waveguide line loaded with split ring resonators," *Int. J. Ant. Propag.*, vol. 2012, pp. 1–6, Oct. 2012.

12 M. A. G. Elsheikh and A. M. E. Safwat, "Wideband modeling of SRR-loaded coplanar waveguide," *IEEE Trans. Microwave Theory Tech.*, vol. 67, no. 3, pp. 851–860, Mar. 2019.

13 J. D. Baena, J. Bonache, F. Martín, R. Marqués, F. Falcone, T. Lopetegi, M. A. G. Laso, J. García, I Gil, M. Flores-Portillo, and M. Sorolla, "Equivalent circuit models for split ring resonators and complementary split rings resonators coupled to planar transmission lines," *IEEE Trans. Microwave Theory Tech.*, vol. 53, pp. 1451–1461, Apr. 2005.

14 R. Marqués, F. Martín, and M. Sorolla, *Metamaterials with Negative Parameters: Theory, Design and Microwave Applications*, John Wiley, Hoboken, NJ, USA, 2007.

15 F. Falcone, T. Lopetegi, J. D. Baena, R. Marqués, F. Martín and M. Sorolla, "Effective negative-ε stopband microstrip lines based on complementary split ring resonators," *IEEE Microwave Wireless Compon. Lett.*, vol. 14, pp. 280–282, Jun. 2004.

16 F. Falcone, T. Lopetegi, M. A. G. Laso, J. D. Baena, J. Bonache, R. Marqués, F. Martín, and M. Sorolla, "Babinet principle applied to the design of metasurfaces and metamaterialsm," *Phys. Rev. Lett.*, vol. 93, paper 197401, Nov. 2004.

17 J. Bonache, F. Martín, I. Gil, J. García-García, R. Marqués, and M. Sorolla, "Microstrip bandpass filters with wide bandwidth and compact dimensions," *Microw. Opt. Technol. Lett.*, vol. 46, pp. 343–346, Aug. 2005.

18 J. Bonache, F. Martín, J. García-García, I. Gil, R. Marqués, and M. Sorolla, "Ultra wide band pass filtres (UWBPF) based on complementary split rings resonators," *Microw. Opt. Technol. Lett.*, vol. 46, pp. 283–286, Aug. 2005.

19 P. Mondal, M. K. Mandal, A. Chaktabarty, and S. Sanyal, "Compact bandpass filters with wide controllable fractional bandwidth," *IEEE Microwave Wireless Compon. Lett.*, vol. 16, pp, 540–542, Oct. 2006.

20 J. Bonache, I. Gil, J. García-García, and F. Martín, "Novel microstrip band pass filters based on complementary split rings resonators," *IEEE Trans. Microwave Theory Tech.*, vol. 54, pp. 265–271, Jan. 2006.

21 J. Bonache, G. Sisó, M. Gil, A. Iniesta, J. García-Rincón, and F. Martín, "Application of composite right/left handed (CRLH) transmission lines based on complementary split ring resonators (CSRRs) to the design of dual band microwave components," *IEEE Microwave Wireless Compon. Lett.*, vol. 18, pp. 524–526, Aug. 2008.

22 G. Sisó, J. Bonache, M. Gil, and F. Martín, "Application of resonant-type metamaterial transmission lines to the design of enhanced bandwidth components with compact dimensions," *Microw. Opt. Technol. Lett.*, vol. 50, pp. 127–134, Jan. 2008.

23 S. Eggermont and I. Huynen, "Leaky wave radiation phenomena in metamaterial transmission lines based on complementary split ring resonators," *Microw. Opt. Technol. Lett.*, vol. 53, pp. 2025–2029, Sep. 2011.

24 L. Su, J. Naqui, J. Mata-Contreras, and F. Martín, "Modeling and applications of metamaterial transmission lines loaded with pairs of coupled complementary split ring resonators (CSRRs)," *IEEE Ant. Wireless Propag. Lett.*, vol. 15, pp. 154–157, 2016.

25 F. Aznar, M. Gil, J. Bonache, and F. Martín, "SRR- and CSRR-loaded metamaterial transmission lines: a comparison to the light of duality," *2nd Int. Congr. Adv. Electromagn. Mater. Microwaves Opt. (Metamaterials'08)*, Pamplona, Spain, Sep. 2008.

26 J. Bonache, M. Gil, I. Gil, J. Garcia-García, and F. Martín, "On the electrical characteristics of complementary metamaterial resonators," *IEEE Microwave Wireless Compon. Lett.*, vol. 16, pp. 543–545, Oct. 2006.

27 M. Makimoto and S. Yamashita, "Compact bandpass filters using stepped impedance resonators," *Proc. IEEE*, vol. 67, no. 1, pp. 16–19, Jan. 1979.

28 M. Makimoto and S. Yamashita, "Bandpass filters using parallel coupled stripline stepped impedance resonators," *IEEE Trans. Microwave Theory Tech.*, vol. 28, no. 12, pp. 1413–1417, Dec. 1980.

29 J. T. Kuo and E. Shih, "Microstrip stepped impedance resonator bandpass filter with an extended optimal rejection bandwidth," *IEEE Trans. Microwave Theory Tech.*, vol. 51, (5), pp. 1554–1559, 2003.

30 J. K. Lee, D. H. Lee, and Y. S. Kim, "A compact low-pass filter with double-step impedance shunt stub and defected ground structure for wideband rejection," *Microw. Opt. Technol. Lett.*, vol. 52, no. 1, pp. 132–134, 2010.

31 J. Naqui, C. Damm, A. Wiens, R. Jakoby, L. Su, and F. Martín, "Transmission lines loaded with pairs of magnetically coupled stepped impedance resonators (SIRs): modeling and application to microwave sensors," *2014 IEEE MTT-S Int. Microw. Symp. (IMS2014)*, Tampa, FL, USA, Jun. 2014.

32 J. Naqui, C. Damm, A. Wiens, R. Jakoby, L. Su, J. Mata-Contreras, and F. Martín, "Transmission lines loaded with pairs of stepped impedance resonators: modeling and application to differential permittivity measurements," *IEEE Trans. Microwave Theory Tech.*, vol. 64, no. 11, pp. 3864–3877, Nov. 2016.

33 J. Naqui, M. Durán-Sindreu, J. Bonache, and F. Martín, "Implementation of shunt connected series resonators through stepped-impedance shunt stubs: analysis and limitations," *IET Microwave Ant. Propag.*, vol. 5, pp. 1336–1342, Aug. 2011.

34 C. G. Juan, B. Potelon, C. Quendo, E. Bronchalo, and J. M. Sabater-Navarro, "Highly-sensitive glucose concentration sensor exploiting inter-resonators couplings," *2019 49th Europ. Microw. Conf. (EuMC2019)*, Paris, France, Oct. 2019, pp. 662–665.

35 I. Piekarz, J. Sorocki, K. Wincza, and S. Gruszczynski, "Microwave sensors for dielectric sample measurement based on coupled-line section," *IEEE Trans. Microwave Theory Tech.*, vol. 65, no. 5, pp. 1615–1631, May 2017.

36 A. Ebrahimi, J. Scott, and K. Ghorbani, "Ultrahigh-sensitivity microwave sensor for microfluidic complex permittivity measurement," *IEEE Trans. Microwave Theory Tech.*, vol. 67, no. 10, pp. 4269–4277, Oct. 2019.

37 P. Vélez, L. Su, K. Grenier, J. Mata-Contreras, D. Dubuc, and F. Martín, "Microwave microfluidic sensor based on a microstrip splitter/combiner configuration and split ring resonators (SRR) for dielectric characterization of liquids," *IEEE Sensors J.*, vol. 17, pp. 6589–6598, Oct. 2017.

38 L. Su, J. Mata-Contreras, J. Naqui, and F. Martín, "Splitter/combiner microstrip sections loaded with pairs of complementary split ring resonators (CSRRs): modeling and optimization for differential sensing applications," *IEEE Trans. Microwave Theory Tech.*, vol. 64, no. 12, pp. 4362–4370, Dec. 2016.

39 L. Su, J. Mata-Contreras, and F. Martín, "Configurations of splitter/combiner microstrip sections loaded with stepped impedance resonators (SIRs) for sensing applications," *Sensors*, vol. 16, no. 12, paper 2195, 2016.

40 D. M. Pozar, *Microwave Engineering*, 3rd ed. John Wiley, Hoboken, NJ, USA, 2005.

41 L. Su, J. Naqui, J. Mata-Contreras, and F. Martín, "Cascaded splitter/combiner microstrip sections loaded with complementary split ring resonators (CSRRs): modeling, analysis and applications," *2016 IEEE MTT-S Int. Microw. Symp. (IMS16)*, San Francisco, CA, USA, May 2016.

42 W. Withayachumnankul, K. Jaruwongrungsee, A. Tuantranont, C. Fumeaux, and D. Abbott, "Metamaterial-based microfluidic sensor for dielectric characterization," *Sensor Actuators A Phys.*, vol. 189, pp. 233–237, Jan. 2013.

43 B. L. Hayes, *Microwave Synthesis: Chemistry at the speed of Light*, CEM Publishing, Matthews, NY, USA, 2006.

44 O. Weiner, "Die theorie des Mischkorpers fur das Feld der statonare Stromung i. die mittelwertsatze fur kraft, polarisation und energie," *Der Abhandlungen der Mathematisch-Physischen Klasse der Koniglvol.* 32, Sachsischen Gesellschaft der Wissenschaften, 1912, pp. 509–604.

45 A. Ebrahimi, G. Beziuk, J. Scott, and K. Ghorbani, "Microwave differential frequency splitting sensor using magnetic-LC resonators," *Sensors*, vol. 20, p. 1066, 2020.

46 A. Ebrahimi, J. Scott, and K. Ghorbani, "Differential sensors using microstrip lines loaded with two split-ring resonators," *IEEE Sensors J.*, vol. 18, pp. 5786–5793, 2018.

47 J. Naqui, M. Durán-Sindreu, and F. Martín, "Differential and single-ended microstrip lines loaded with slotted magnetic-LC (MLC) resonators," *Int. J. Ant. Propag.*, vol. 2013, Article ID 640514, 8 pages, 2013.

48 D. Schurig, J. J. Mock, and D. R. Smith, "Electric-field-coupled resonators for negative permittivity metamaterials," *Appl. Phys. Lett.*, vol. 88, no. 4, Article ID 041109, pp. 1–3, 2006.

49 J. García-García, F. Martín, F. Falcone, J. Bonache, I. Gil, T. Lopetegi, M. A. G. Laso, M. Sorolla, and R. Marqués, "Spurious passband suppression in microstrip coupled line band pass filters by means of split ring resonators," *IEEE Microwave Wireless Compon. Lett.*, vol. 14, pp. 416–418, Sep. 2004.

50 M. H. Zarifi, S. Farsinezhad, B. D. Wiltshire, M. Abdorrazaghi, N. Mahdi, P. Kar, M. Daneshmand, and K. Shankar, "Effect of phosphonate monolayer adsorbate on the microwave photoresponse of TiO_2 nanotube membranes mounted on a planar double ring resonator," *Nanotechnology*, vol. 27, paper 375201, Sep. 2016.

51 A. K. Horestani, J. Naqui, Z. Shaterian, D. Abbott, C. Fumeaux, and F. Martín, "Two-dimensional alignment and displacement sensor based on movable broadside-coupled split ring resonators," *Sens. Act. A.*, vol. 210, pp. 18–24, Apr. 2014.

52 R. Marqués, F. Medina, and R. Rafii-El-Idrissi, "Role of bianisotropy in negative permeability and left handed metamaterials," *Phys. Rev. B*, vol. 65, paper 144441, 2002.

6

Differential-Mode Sensors

Prototype examples of differential-mode planar microwave sensors have been reported in the previous chapters. Indeed, rather than a working principle, differential-mode operation denotes a specific sensing approach, where the response of two independent single-ended sensors is compared. By this means, a differential sensor provides a differential output variable in response to a differential input variable, and the device is robust against cross-sensitivities caused, e.g. by external factors, such as temperature, humidity, and pressure, as far as these variables are seen as common-mode stimuli by the sensor. Nevertheless, the differential-mode concept can be applied to different types of sensors, operating under different principles, such as frequency-variation or phase-variation sensors. For example, differential-mode phase-variation sensors devoted to dielectric characterization are reported in Sections 3.2.1.1, 3.2.2.1, 3.2.2.4, and 3.4.2 of Chapter 3.

According to the previous words, differential-mode sensors do not constitute a category of sensors at the same level as those corresponding to the principles discussed in Chapters 2–5. A sensor can be designated as differential as far as it is based on two independent single-ended sensors, but the working principle can be of any type (frequency variation or phase variation, typically). Coupling-modulation sensors based on symmetry properties, studied in Chapter 4, are also robust against the effects of external variables unable to disrupt symmetry, but such sensors are not based on two independent sensing elements. By contrast, frequency-splitting sensors, discussed in Chapter 5, consist of a transmission line (or a transmission-line-based structure) symmetrically loaded with two sensing resonant elements.[1] The output variable in such sensors is the difference between the split resonance frequencies caused by symmetry disruption (in turn generated by a differential input variable, e.g. a differential dielectric constant). However, frequency-splitting sensors are not true differential-mode sensors since they are not based on two independent single-ended sensors.[2]

Despite the fact that some differential-mode sensors have been already analyzed in this book (mainly in Chapter 3),[3] there is a set of scientific contributions focused on differential sensors implemented by means of a pair of uncoupled resonator-loaded [8–12] or resonator-terminated [13, 14] transmission lines. This chapter is dedicated to the study and applications of these sensors, mainly devoted to material characterization and composition. In transmission-mode sensors, a difference in the dielectric loads of the sensitive elements (the resonators) generates a difference

---

1 The exception are the frequency-splitting sensors based on the cascade configuration.
2 Nevertheless, due to the similarity with differential-mode sensors, frequency-splitting sensors have been considered to be differential-mode structures by some researchers.
3 These differential sensors and other similar sensors are based on phase variation and are also reported in references [1–7].

Planar Microwave Sensors, First Edition. Ferran Martín, Paris Vélez, Jonathan Muñoz-Enano, and Lijuan Su.
© 2023 John Wiley & Sons, Inc. Published 2023 by John Wiley & Sons, Inc.

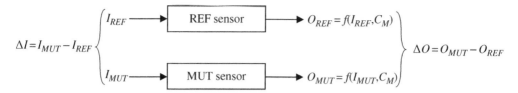

Figure 6.1 Sketch showing the differential-mode sensor concept.

in the transmission coefficient of the (independent) lines, proportional to the cross-mode transmission coefficient, the considered output variable. In reflective-mode sensors, the output variable is the cross-mode reflection coefficient, proportional to the difference between the reflection coefficients of the individual lines. Several prototype examples of differential sensors based on different resonant elements are reported in the chapter, including a biosensor useful for the determination of the total electrolyte concentration in urine samples.

6.1 The Differential-Mode Sensor Concept

Figure 6.1 depicts the sketch of the differential-mode sensor concept. A differential-mode sensor consists of a pair of identical, independent, single-ended sensors, sensitive to the variable of interest. In a differential type measurement, the variable to be measured (measurand) is a differential variable, defined as the difference between the measurand in the reference (REF) sensor, I_{REF}, and the measurand in the other sensor, usually designated as test sensor, or MUT sensor,[4] I_{MUT}. Hence, the differential input variable is $\Delta I = I_{MUT} - I_{REF}$, whereas the differential output variable is simply the difference between the output variables of each independent sensor, in response to the respective input variable, that is $\Delta O = O_{MUT} - O_{REF}$. Naturally, if the measurands in both sensing elements are identical, the differential input and output variables are both (ideally) null.

As mentioned, differential sensors are scarcely sensitive to the effects of external variables, or ambient factors, such as temperature, humidity, and atmospheric pressure, which do not vary at the typical scales of the sensors (of the order of centimeters, or even less). The reason is that such variables are seen as common-mode signals, and thereby their effects are canceled by the sensor, as far as it operates differentially. Nevertheless, this aspect requires further analysis, since absolute elimination of cross-sensitivities to common-mode variables is not possible, in general. Thus, let us consider that the output variable of the REF and MUT sensors is also sensitive to an external common-mode variable, designated as C_M. Let us define the transfer function of the sensors as

$$\Delta O = O_{MUT} - O_{REF} = f(I_{MUT}, C_M) - f(I_{REF}, C_M) \tag{6.1}$$

---

4 Note that this nomenclature applies to differential sensors devoted to dielectric characterization or material composition determination, where the test sensor is covered by the material under test (MUT). In such sensors, a typical differential input variable is the differential dielectric constant between the MUT and the REF material. Nevertheless, without loss of generality, this nomenclature can be adopted to designate the input (I_{MUT}) and output (O_{MUT}) variables of the test sensor, regardless of the type of measurand. In the sketch of Fig. 6.1, the independent sensing elements are designated by the REF and the MUT labels.

Obviously, if the input signals at both independent sensors are identical, $I_{MUT} = I_{REF}$ (or $\Delta I = 0$), the differential output variable is also null ($\Delta O = O_{MUT} - O_{REF} = 0$), indicating that the measurands are identical in both sensors. However, if $I_{MUT} \neq I_{REF}$, in general

$$\Delta O = f(I_{MUT}, C_M) - f(I_{REF}, C_M) \neq f(I_{MUT}, C_M = 0) - f(I_{REF}, C_M = 0) \tag{6.2}$$

where the right-hand side member is the expected sensor response without the presence of the external common-mode stimulus. Expression (6.2) indicates that, in general, the effects of cross-sensitivities to common-mode signals cannot be completely suppressed. However, it is reasonable to consider that such common-mode signals generate a small variation in the output variables of both sensors, as compared with the effects generated by the input variable of interest. Under such small-signal conditions, the response of each independent sensor can be approximated by

$$O_{MUT} = f(I_{MUT}, C_M) = f(I_{MUT}, C_M = 0) + \left.\frac{df(I_{MUT}, C_M)}{dC_M}\right|_{C_M = 0} \cdot C_M \tag{6.3a}$$

$$O_{REF} = f(I_{REF}, C_M) = f(I_{REF}, C_M = 0) + \left.\frac{df(I_{REF}, C_M)}{dC_M}\right|_{C_M = 0} \cdot C_M \tag{6.3b}$$

and the differential output variable is found to be

$$\Delta O = O_{MUT} - O_{REF} = f(I_{MUT}, C_M) - f(I_{REF}, C_M)$$

$$= f(I_{MUT}, C_M = 0) - f(I_{REF}, C_M = 0) + \left(\left.\frac{df(I_{MUT}, C_M)}{dC_M}\right|_{C_M = 0} - \left.\frac{df(I_{REF}, C_M)}{dC_M}\right|_{C_M = 0}\right) \cdot C_M \tag{6.4}$$

The derivatives in the right-hand-side member of (6.4) are not, in general, identical, unless $I_{MUT} = I_{REF}$. However, it is expected that the difference between these derivative terms is small (especially if the MUT and REF measurands do not differ so much). Therefore, the differential output variable can be approximated by

$$\Delta O = f(I_{MUT}, C_M) - f(I_{REF}, C_M) \approx f(I_{MUT}, C_M = 0) - f(I_{REF}, C_M = 0) \tag{6.5}$$

and it can be concluded that, to a first order approximation, the effects of cross-sensitivities to common-mode external signals can be neglected in differential-mode sensors. By contrast, in the single-ended counterparts, the presence of the external signal C_M generates a difference in the output variable O_{MUT} (with regard to the expected value) given by the last term in (6.3a).

Differential measurements can be performed by means of single-ended sensors in a two-step process, where first the response to the REF variable is recorded, and then the sensor is stimulated with the MUT variable. If this pair of measurements is carried out in a time interval where changes in external factors are highly improbable, such external variables can be considered to be common-mode stimuli, and the advantages of real-time differential measurements with regard to robustness against cross-sensitivities are preserved. Obviously, the advantage of performing differential measurements with a single-ended sensor in a two-step process is the lower cost of the sensing structure, consisting of a single sensor. Nevertheless, there are many applications involving real-time monitoring of a certain variable, or a continuous comparison between two variables. In such circumstances, real-time differential measurements by means of true differential sensors are necessary.

6.2   Differential Sensors Based on the Measurement of the Cross-Mode Transmission Coefficient

This section is devoted to the analysis, design, validation, and applications of differential-mode sensors operating in transmission and implemented by means of pairs of resonator-loaded transmission lines. As it will be shown, such sensors are highly sensitive to differences in the material properties of the REF and MUT samples and are therefore of interest for the analysis of solids and liquids, including not only the measurement of dielectric properties (e.g. the complex permittivity) but also the determination of material composition (for example the solute content in liquid mixtures), or the real-time monitoring of changes in a certain substance, as compared with a reference, among others. One of the proposed devices was found to be applicable to monitoring the total electrolyte concentration in urine samples, of potential interest as a pre-screening system in medical diagnosis or to detect potential pathologies.

6.2.1   Working Principle

The general working principle of differential sensors based on pairs of transmission lines is mode conversion, caused by imbalances between the sensing elements. Examples of differential permittivity sensors based on ordinary or artificial lines have been presented in Chapter 3 [1–7]. In such sensors, the difference in the permittivity between the MUT and REF samples generates a difference in the phase and characteristic impedance of the MUT and REF sensing lines and consequently a difference in the transmission coefficients (transmission-mode sensors) or reflection coefficients (reflective-mode sensors). Such different transmission or reflection coefficients, obviously, generate mode conversion from the common to the differential mode, and vice versa. The canonical differential output variable in transmission-line-based differential sensors is the difference in the phase of the transmission or reflection coefficients (see the example reported in Section 3.2.1.1). However, the output variable in transmission-mode and reflective-mode differential sensors can also be the differential transmission coefficient, $S_{21,MUT} - S_{21,REF}$, and the differential reflection coefficient, $\rho_{MUT} - \rho_{REF}$, respectively, both indicative of mode conversion. Note that the differential transmission and reflection coefficients are complex numbers. Therefore, it is potentially possible to measure two independent input variables by considering such differential output variables, e.g. the real and the imaginary part of the complex permittivity of the MUT sample. Nevertheless, in many reported differential sensors, the modulus of the differential transmission or reflection coefficient is the (single) considered output variable.[5]

In transmission-mode differential sensors, the differential transmission coefficient can be obtained by adding a pair of rat-race hybrid couplers, as depicted in Fig. 3.10 (see a specific implementation in Fig. 3.13, where the independent sensors are two identical meander lines) [6]. By this means, the four-port differential sensor is transformed to a two-port device with the transmission coefficient proportional to the differential transmission coefficient of the sensor pair, provided the isolated ports are adequately terminated. Thus, for isolated ports terminated with matched loads ($\rho = \rho' = 0$, see Fig. 3.10), the transmission coefficient is identical (except the sign) to the cross-mode transmission coefficient of the four-port differential sensor, S_{21}^{cd}, see expression (3.56), which is in turn proportional to the differential transmission coefficient, i.e.

$$S_{21}^{cd} = \frac{1}{2}\left(S_{21,MUT} - S_{21,REF}\right) \tag{6.6}$$

---

5 As it will be shown later, in sensors devoted to the measurement of the complex permittivity of the MUT sample (real and imaginary parts), the maximum value of the magnitude of the differential transmission coefficient and the frequency position of such maximum are the two required output variables.

Actually, the previous expression is valid as far as the differential sensor is based on a pair of independent and uncoupled sensing elements.[6] However, such sensing elements can be of any type, not necessarily ordinary (including meander) lines, the case of Fig. 3.10, corresponding to the implementation of Fig. 3.13. It should be mentioned that in the device reported in Fig. 3.13, based on a pair of ordinary lines, the isolated ports were actually terminated by means of open circuits, since, by properly choosing the electrical length of the sensing lines (set to $n \cdot \pi$), the transmission coefficient (the measurable quantity) is identical to the differential transmission coefficient. By this means, the factor 1/2 that appears in expression (6.6) is circumvented, and the sensitivity is enhanced. However, this is valid in the limit of small perturbations, as discussed in Chapter 3 [6].

In reflective-mode differential sensors, the differential reflection coefficient can be inferred from the transmission coefficient of the two-port structure that results by adding a rat-race hybrid coupler to the differential sensor, as depicted in Fig. 3.32 [7]. Such transmission coefficient, given by expression (3.123), is half the differential reflection coefficient of the reflective-mode differential sensing structure, and this result is valid regardless of the sensing elements (in the sensor of Fig. 3.33, step-impedance open-ended sensing lines were considered).

According to the preceding paragraphs, the differential output variable in either transmission-mode or reflective-mode differential sensors can be inferred by means of a simple two-port measurement by simply adding rat-race couplers to the original differential sensing structure. Alternatively, the differential transmission or reflection coefficients can be inferred by means of four-port vector network analyzers. Regardless of the specific approach to obtain the differential output variable, in the differential sensors reported in Chapter 3, the sensing principle is the phase variation experienced by the sensing lines when such lines are covered by the MUT and REF samples.

In this chapter, the main objective is to highlight the potential of differential sensors based on resonator-loaded or resonator-terminated lines for sensitivity and resolution optimization. In such sensors, the sensing elements are the loading resonators, sensitive to the dielectric properties of the surrounding medium. These sensors are especially suited for dielectric characterization and material composition analysis, and their high sensitivity and good resolution are related to the significant variation experienced by the frequency response of the sensing lines, when the sensing resonant elements loading the lines are covered by the REF and MUT samples. Indeed, in most cases, the single-ended counterparts of the differential sensors belong to the category of frequency-variation sensors, studied in Chapter 2. Thus, the differential-mode sensors to be discussed next could have been coherently included in Chapter 2. Nevertheless, it has been preferred by the authors to study these sensors in a separate chapter in order to avoid an excessively long Chapter 2 (the analysis, design, validation, and applications of the reported implementations cannot be treated superfluously). Let us also mention that the canonical output variable in the differential sensors based on resonator-loaded lines is the magnitude of the cross-mode transmission coefficient (either at the resonance frequency of the REF resonator or the maximum value), very sensitive to differences in the dielectric load of the sensing resonators. Eventually, when two output variables are necessary for sensing, the frequency where the cross-mode transmission coefficient is a maximum is the additional output variable, as it will be shown later.

6.2.2 Examples and Applications

Let us next review in detail three reported transmission-mode differential sensors, one of them based on open complementary split-ring resonator (OCSRR)-loaded lines [8], a second one implemented by means of transmission lines loaded with SRRs [9], and a third prototype based on

---

6 Moreover, under these conditions, $S_{21}^{cd} = S_{21}^{dc}$.

transmission lines loaded with dumbbell-shaped defect ground structure (DB-DGS) resonators [11]. In all these prototypes, the sensors are equipped with microfluidic channels, and, consequently, are devoted to the characterization of liquid samples. The use of one of such prototypes as a biosensor devoted to the measurement of electrolyte content in urine samples is also included in this section.

6.2.2.1 Microfluidic Sensor Based on Open Complementary Split-Ring Resonators (OCSRRs) and Application to Complex Permittivity and Electrolyte Concentration Measurements in Liquids

It was shown in Chapter 2 that OCSRRs are electrically small planar resonators very sensitive to the dielectric properties of the material in contact with it [15]. In [8], a differential-mode microfluidic sensor based on a pair of OCSRR-loaded lines was proposed. The output variable is the cross-mode transmission coefficient, and the sensor was applied to the determination of electrolyte content in diluted solutions of deionized (DI) water, as well as to the measurement of the complex dielectric constant of liquids.

Figure 6.2 depicts the layout of the pair of OCSRR-loaded lines, a zoom view of the resonant element, and the equivalent circuit model. The equivalent circuit model takes into account the presence of the reference (REF) liquid and the liquid under test (LUT) on top of the OCSRRs of the REF and MUT (or LUT) lines, respectively. For this reason, the circuit model is asymmetric. In the model, the lines are described by its characteristic impedance, Z_c, and electrical length, kl (k and l being the phase constant and the physical length). The OCSRRs are modeled by lossy resonant tanks, connected to the respective host lines through inductances, L_z. Such inductances and the inductances of the resonant tanks, L_p, are not influenced by the presence of liquids in both channels, contrary to the capacitance and conductance of the OCSRRs, which depend on the liquid on top of them. Thus, C_p is the OCSRR capacitance corresponding to the empty channel, whereas the effects of the liquid (REF or LUT) on top of the OCSRR are taken into account by parallel connecting a capacitance, C_{REF} and C_{LUT}, respectively, in each channel, to C_p.[7] Similarly, the conductance of the empty channels, G, is connected parallel to G_{REF} and G_{LUT} in order to include the effects of the REF and LUT liquids, respectively, in the conductance of the OCSRRs. It should be clarified that the inductance L_z is not always present in the circuit models describing an OCSRR-loaded line [16]. This inductance is not an intrinsic reactive element of the OCSRR. However, the distance between the host lines and the OCSRRs in the microfluidic sensors reported in [8], with the layout depicted in Fig. 6.2, is significant, and thereby such inductance must be included in the model for an accurate description of the structure. This substantial separation between the host lines and the OCSRRs is a requirement forced by the need to accommodate the fluidic channels on top of the OCSRRs, the sensing elements, at sufficient distance from the host lines (in order to avoid any influence of the liquids and channels on such lines).

To validate the model, as well as sensor functionality, a differential OCSRR-based microfluidic sensor with resonator dimensions indicated in the caption of Fig. 6.2, and lines with a width and separation of 1.33 and 32 mm, respectively, was considered in [8]. Such line width provides a

---

7 Note that C_{REF} and C_{LUT} in the model of Fig. 6.2(c) are the capacitances that should be added to the capacitance of the OCSRRs without liquid in the channels, C_p, in order to describe the total capacitance of the particle (for the REF and LUT channels, respectively) when it is covered by the corresponding liquid (REF or LUT). In other words, C_{REF} and C_{LUT} do not correspond to the contribution of the OCSRR capacitance of the channel half-space where the REF and LUT liquids are present. By contrast, in Chapter 2 (Fig. 2.5), the designation C_{MUT} was attributed to the contribution to the total capacitance of the sensing resonator (loaded with the MUT) corresponding to the half-space where the MUT is present. In that chapter, see, e.g., Fig. 2.20, the incremental capacitance needed to describe the total capacitance of the sensing resonator loaded with the MUT is called $C_{e,MUT}$, equivalent to C_{LUT} in the model of Fig. 6.2(c).

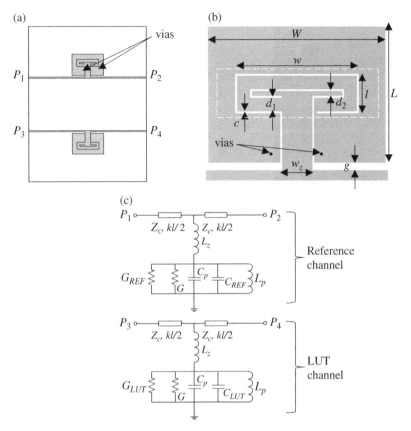

Figure 6.2 Layout of the OCSRR-based differential-mode microfluidic sensor (a), zoom view of the OCSRR (b), and circuit model (c). OCSRR dimensions (in mm) are $W = 20$, $L = 15.1$, $w = 14$, $l = 4.35$, $c = 0.2$, $d_1 = 1.5$, $d_2 = 0.55$, $g = 0.79$, $w_z = 3.4$ and $l_z = 6.49$. The radius of the vias is 0.1 mm. The channel limits are indicated by the dashed lines. *Source:* Reprinted with permission from [8]; copyright 2018 IEEE.

characteristic impedance of $Z_c = 77 \, \Omega$ in the considered substrate, i.e. *FR4* with thickness $h = 1.6$ mm, dielectric constant $\varepsilon_r = 4.4$, and loss tangent $\tan\delta = 0.02$. The fluidic channels, made of poly-dimethylsiloxane (PDMS), are identical to those of the sensor of Fig. 5.31. The fabricated sensor is shown in Fig. 6.3, where the capillaries for liquid injection can be appreciated. Similar to the sensor of Fig. 5.31, a dry film of clear polyester (with estimated dielectric constant and thickness of 3.5 and 55 μm, respectively) was placed between the substrate and the fluidic channels in order to avoid liquid absorption by the substrate.

Several scenarios were considered in [8] for model validation. First, the responses of the individual lines with empty channels (i.e. loaded with air), inferred from electromagnetic simulation (using *CST Microwave Studio*) and experimentally (by means of the *Agilent PNA N5221A* vector network analyzer) were obtained, see Fig. 6.4 (the reference impedance of the ports is $Z_0 = 50 \, \Omega$). Using a parameter extraction procedure similar to other methods reported previously in this book (and detailed, e.g., in [17, 18]), the circuit elements of the equivalent circuit model of Fig. 6.2 were obtained (see Table 6.1). The simulation of the circuit schematic with the extracted element values is also depicted in Fig. 6.4. The good agreement with the measured responses points out the validity of the circuit model for the description of the sensing structure, including the (empty) fluidic channels.

Figure 6.3 Photograph of the OCSRR-based differential-mode microfluidic sensor, including the fluidic channels and connectors.

(a)

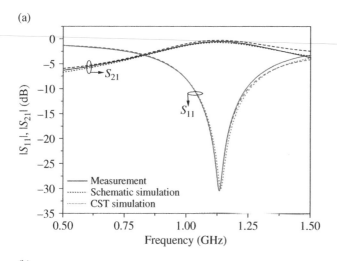

(b)

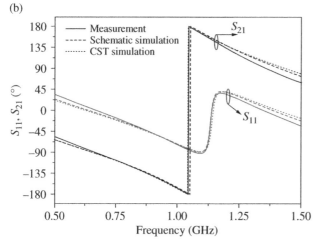

Figure 6.4 Insertion and return loss of the individual OCSRR-loaded line loaded with empty channel. (a) Magnitude response; (b) phase response. *Source:* Reprinted with permission from [8]; copyright 2018 IEEE.

In a second stage, deionized (DI) water was injected in both channels. The corresponding responses (identical for both OCSRR-loaded lines) are depicted in Fig. 6.5. The extracted parameters, also indicated in Table 6.1, were used to obtain the response of the circuit schematic, which is in good agreement with the measured response. These results confirm that the presence of liquid

Table 6.1 Extracted element parameters of the equivalent circuit model of the OCSRR-based differential microfluidic sensor.

Channel 1/Channel 2	L_z (nH)	L_p (nH)	C_p (pF)	C_{ref} (pF)	C_{LUT} (pF)	G (mS)	G_{ref} (mS)	G_{LUT} (mS)
Air/Air	4.87	7.79	2.04	0	0	0.59	0	0
DI water/DI water	4.87	7.79	2.04	2.4	2.4	0.59	0.28	0.28
DI water/Na solution	4.87	7.79	2.04	2.4	2.52	0.59	0.28	3.03

(a)

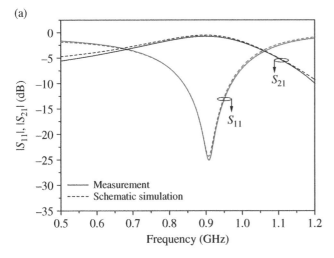

(b)

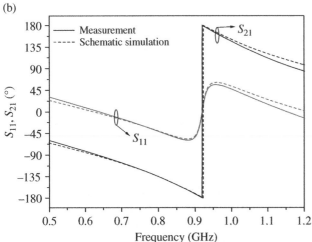

Figure 6.5 Insertion and return loss of the individual OCSRR-loaded line loaded with DI water in both channels. (a) Magnitude response; (b) phase response. *Source:* Reprinted with permission from [8]; copyright 2018 IEEE.

(a)

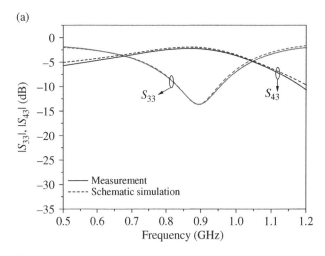

(b)

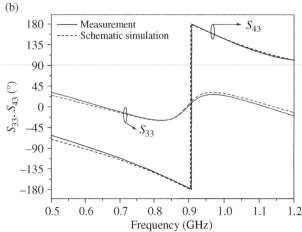

Figure 6.6 Insertion and return loss of the individual OCSRR-loaded line loaded with a solution of sodium in DI water in the channel. (a) Magnitude response; (b) phase response. *Source:* Reprinted with permission from [8]; copyright 2018 IEEE.

in the channel modifies the capacitance and conductance of the corresponding OCSRR, but not the OCSRR inductance and strip inductances, L_p and L_z, respectively.

The third validation experiment was done by injecting DI water (REF liquid) in one channel and a solution of sodium in DI water (LUT), with a solute concentration of 20 g/L, in the other channel. The frequency responses of the OCSRR-loaded line with sodium solution in the channel are depicted in Fig. 6.6. The extracted circuit parameters are also included in Table 6.1. The circuit simulation, also included in Fig. 6.6, is in good agreement with the measured response. The presence of sodium does not modify significantly the capacitance of the OCSRR (as compared with the one of the OCSRR with DI water in the channel), as revealed by the fact the resonance frequency does not vary significantly. However, the conductance is substantially altered (as revealed by the value of G_{LUT}). This means that the presence of this small quantity of sodium in DI water has small effect on the dielectric constant of the resulting solution, but significant effect on its loss tangent (or on the imaginary part of the complex permittivity), revealing a significant level of dielectric loss in DI water caused by the presence of sodium ions.

(a)

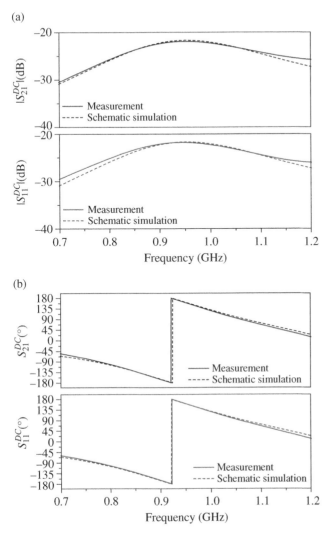

(b)

Figure 6.7 Cross-mode insertion and return loss of the sensor with DI water in the reference channel and a sodium solution in DI water in the LUT channel. (a) Magnitude response; (b) phase response. *Source:* Reprinted with permission from [8]; copyright 2018 IEEE.

Finally, Fig. 6.7 depicts the cross-mode transmission, S_{21}^{cd}, and reflection, S_{11}^{cd}, coefficient measured in the last asymmetric loading case (DI water as reference liquid and the sodium solution as LUT). The same response inferred from the circuit model with extracted parameters (included in Table 6.1 for both lines, i.e. the one loaded with DI water and the one loaded with sodium solution) is also included in Fig. 6.7. Again, both responses are in good agreement. With this set of experiments, the proposed circuit model is validated. The presence of liquid in the channel modifies the capacitance and conductance of the OCSRRs, leaving unaltered the inductive elements of the circuit model.

Let us next analyze in detail the circuit model of Fig. 6.2(c) in order to infer the influence of the involved parameters on the cross-mode transmission coefficient, the output variable. The general analysis, by considering an arbitrary line impedance, Z_c, is complex. However, the analysis simplifies if it is assumed that $Z_c = Z_0$, the reference impedance of the ports. Under these conditions, the line sections of length $l/2$ only introduce a phase shift in the response of both lines, and the effects of such line sections can be neglected, provided the output variable is the magnitude of the cross-mode transmission coefficient. If the lines are uncoupled, the considered situation, the cross-mode transmission coefficient is given by (6.6), or

$$S_{21}^{cd} = \frac{1}{2}(S_{21} - S_{43}) \tag{6.7}$$

according to the port definition of Fig. 6.2(c). By neglecting the effects of the host lines (as justified above), the transmission coefficients of the individual lines are given by

$$S_{21} = \frac{1}{1 + \dfrac{Y_{REF} \cdot Z_0}{2}} \tag{6.8a}$$

$$S_{43} = \frac{1}{1 + \dfrac{Y_{LUT} \cdot Z_0}{2}} \tag{6.8b}$$

Y_{REF} and Y_{LUT} being the admittances of the shunt branches for the REF and LUT channels, respectively. Such admittances can be expressed as:

$$Y_{REF} = \frac{\dfrac{1}{L_p L_z}\left(\dfrac{1}{\omega_{REF}^2} - \dfrac{1}{\omega^2}\right) - j\dfrac{G'_{REF}}{L_z\omega}}{G'_{REF} + j\left\{\dfrac{\omega}{L_p}\left(\dfrac{1}{\omega_{REF}^2} - \dfrac{1}{\omega^2}\right) - \dfrac{1}{L_z\omega}\right\}} \tag{6.9a}$$

$$Y_{REF} = \frac{\dfrac{1}{L_p L_z}\left(\dfrac{1}{\omega_{LUT}^2} - \dfrac{1}{\omega^2}\right) - j\dfrac{G'_{LUT}}{L_z\omega}}{G'_{LUT} + j\left\{\dfrac{\omega}{L_p}\left(\dfrac{1}{\omega_{LUT}^2} - \dfrac{1}{\omega^2}\right) - \dfrac{1}{L_z\omega}\right\}} \tag{6.9b}$$

where ω is the angular frequency, $G'_{REF} = G + G_{REF}$, $G'_{LUT} = G + G_{LUT}$, and ω_{REF} and ω_{LUT} are the angular frequencies of the parallel resonant tanks of the reference and LUT channels, respectively, i.e.

$$\omega_{REF} = \frac{1}{\sqrt{L_p(C_p + C_{REF})}} \tag{6.10a}$$

$$\omega_{LUT} = \frac{1}{\sqrt{L_p(C_p + C_{LUT})}} \tag{6.10b}$$

Evaluating (6.9a) and (6.9b) at ω_{REF}, and introducing the results in 6.8(a) and 6.8(b), respectively, the transmission coefficients of the REF channel line and LUT channel line evaluated at ω_{REF} are found to be

$$S_{21,\omega_{REF}} = \frac{1}{1 - \dfrac{\dfrac{G'_{REF}Z_0}{L_z\omega_{REF}}\left(jG'_{REF} - \dfrac{1}{L_z\omega_{REF}}\right)}{2\left(G'^2_{REF} + \dfrac{1}{L_z^2\omega_{REF}^2}\right)}} \tag{6.11a}$$

$$S_{43,\omega_{REF}} = \cfrac{1}{1 - \cfrac{\cfrac{G'_{LUT}Z_0}{L_z\omega_{REF}}\left[\cfrac{\omega_{REF}}{L_p\omega_L^2} - \cfrac{1}{L_z\omega_{REF}}\right] - \cfrac{G'_{LUT}Z_0}{L_pL_z\omega_L^2} + j\left\{\cfrac{Z_0}{L_pL_z\omega_L^2}\left[\cfrac{\omega_{REF}}{L_p\omega_L^2} - \cfrac{1}{L_z\omega_{REF}}\right] + \cfrac{Z_0G'^2_{LUT}}{L_z\omega_{REF}}\right\}}{2\left\{G'^2_{LUT} + \left[\cfrac{\omega_{REF}}{L_p\omega_L^2} - \cfrac{1}{L_z\omega_{REF}}\right]^2\right\}}}$$

$$(6.11b)$$

where $\omega_L^{-2} = L_p(C_{LUT} - C_{REF})$.

Despite the fact that liquids exhibit non-negligible values of the loss tangent (as compared with the typical values of low-loss solid dielectrics), it is reasonable to assume that losses are small, i.e. G'_{REF} and G'_{LUT} small. This hypothesis is reasonable even for liquid solutions containing ions, provided their concentration is small.[8] Let us now consider that the REF liquid and the LUT exhibit similar dielectric constants, a reasonable hypothesis in situations where the LUT is a diluted solution of the REF liquid. Under these approximations (low-losses and small perturbations), the right-hand side of the denominator of (6.11a) and (6.11b) is small. Consequently, signal transmission at ω_{REF} is roughly total in both channels. Thus, using the approximation $1/(1 - x) = 1 + x$, valid for small values of x [as occurs in expressions (6.11a) and (6.11b)], and taking into account that for small perturbations ($C_{LUT} \approx C_{REF}$) and low-losses (G'_{REF} and G'_{LUT} small), the following approximations hold

$$\frac{\omega_{REF}^2 L_z}{\omega_L^2 L_p} \ll 1 \tag{6.12}$$

$$G'^2_{REF}L_z^2\omega_{REF}^2 \ll 1; G'^2_{LUT}L_z^2\omega_{REF}^2 \ll 1, \tag{6.13}$$

the real and the imaginary parts of the cross-mode transmission coefficient can be expressed as:

$$Re\left\{S_{21,\omega_{REF}}^{cd}\right\} = \frac{Z_0}{4}\left[G_{LUT} - G_{REF} + G'_{REF}L_z(C_{LUT} - C_{REF})\omega_{REF}^2\right] \tag{6.14a}$$

$$Im\left\{S_{21,\omega_{REF}}^{cd}\right\} = \frac{Z_0\omega_{REF}}{4}\left[C_{LUT} - C_{REF} + L_z\left(G'^2_{REF} - G'^2_{LUT}\right)\right] \tag{6.14b}$$

and the modulus of the cross-mode transmission coefficient is

$$\left|S_{21,\omega_{REF}}^{cd}\right| = \sqrt{\left(Re\left\{S_{21,\omega_{REF}}^{cd}\right\}\right)^2 + \left(Im\left\{S_{21,\omega_{REF}}^{cd}\right\}\right)^2} \tag{6.15}$$

If the terms dependent on L_z in expressions (6.14) cannot be neglected, the sensitivity of the cross-mode transmission coefficient to variations in the complex dielectric constant of the LUT (as compared with the reference liquid and manifested as variations in the capacitance and conductance of the resonant tanks) does not exhibit a simple dependence on the circuit parameters. However, if L_z is sufficiently small and the corresponding terms can be neglected, the modulus of the cross-mode transmission coefficient is given by:

$$\left|S_{21,\omega_{REF}}^{cd}\right| = \frac{Z_0}{4}\sqrt{(G_{LUT} - G_{REF})^2 + \omega_{REF}^2(C_{LUT} - C_{REF})^2} \tag{6.16}$$

and it exhibits a relatively simple dependence with the circuit parameters describing the effects of the liquids in both channels. Indeed, under the considered approximations, it has been found that

---

8 The OCSRR-based differential sensor will be applied to the measurement of ion concentration in very diluted solutions of DI water.

the modulus of the cross-mode transmission coefficient at ω_{REF} is expressed in terms of the differential conductance, $G_{LUT} - G_{REF}$, and the differential capacitance, $C_{LUT} - C_{REF}$ (see 6.16). In situations where the LUT and the REF liquid exhibit similar conductance ($G_{REF} \approx G_{LUT}$), it follows that $\left|S_{21,\omega_{REF}}^{cd}\right|$ is proportional to the differential capacitance. If the capacitances of both channels are similar ($C_{REF} \approx C_{LUT}$), a situation reasonable when a REF liquid is compared with a solution of ions (e.g. electrolytes) in it, then the output variable is proportional to the differential conductance.

Let us now try to express (6.16) in terms of the dielectric constant and loss tangent of the LUT, the typical input parameters in sensors devoted to dielectric characterization. According to the definition of C_{LUT} and C_{REF} in Fig. 6.2(c), these capacitances are simply the difference between the total capacitance of the OCSRR covered with the corresponding liquid (LUT or REF), and the capacitance of the OCSRR when it is surrounded by air, C_p. Thus, C_{LUT} and C_{REF} can be expressed as

$$C_{LUT} = \frac{C_p(\varepsilon_{LUT} - 1)}{\varepsilon_r + 1} \tag{6.17a}$$

$$C_{REF} = \frac{C_p(\varepsilon_{REF} - 1)}{\varepsilon_r + 1} \tag{6.17b}$$

and the differential capacitance is thus

$$C_{LUT} - C_{REF} = \frac{C_p(\varepsilon_{LUT} - \varepsilon_{REF})}{\varepsilon_r + 1} \tag{6.18}$$

It has been assumed, as usual in this book, that the LUT and the REF liquids are semi-infinite in the vertical direction. The loss tangents of the LUT and REF liquids can be expressed in terms of the conductances as [19]

$$\tan \delta_{LUT} = \frac{G_{LUT}(\varepsilon_r + 1)}{C_p \omega_{REF} \varepsilon_{LUT}} \tag{6.19a}$$

$$\tan \delta_{REF} = \frac{G_{REF}(\varepsilon_r + 1)}{C_p \omega_{REF} \varepsilon_{REF}} \tag{6.19b}$$

Thus, the differential conductance is

$$G_{LUT} - G_{REF} = \frac{C_p \omega_{REF}}{\varepsilon_r + 1} \left(\varepsilon_{LUT} \cdot \tan \delta_{LUT} - \varepsilon_{REF} \cdot \tan \delta_{REF}\right) \tag{6.20}$$

and it cannot be expressed in terms of the differential loss tangent.

Taking into account that the resonance frequency of the OCSRR of the REF line is

$$\omega_{REF} = \frac{1}{\sqrt{L_p C_p \dfrac{\varepsilon_r + \varepsilon_{REF}}{\varepsilon_r + 1}}} \tag{6.21}$$

expression(6.16) can be written as

$$\left|S_{21,\omega_{REF}}^{cd}\right| = \frac{Z_0}{4} \cdot \sqrt{\frac{C_p}{L_p}} \cdot \frac{\sqrt{\left(\varepsilon_{LUT} \cdot \tan \delta_{LUT} - \varepsilon_{REF} \cdot \tan \delta_{REF}\right)^2 + \left(\varepsilon_{LUT} - \varepsilon_{REF}\right)^2}}{\sqrt{(\varepsilon_r + \varepsilon_{REF})(\varepsilon_r + 1)}} \tag{6.22}$$

where (6.18) and (6.20) have been used. Expression (6.22) reveals that either an imbalance in the dielectric constant or a difference in the loss tangent of the LUT and REF liquids (or both simultaneously) generate a finite (i.e. not null) cross-mode transmission coefficient. Nevertheless, unless the dielectric constants are roughly identical, the cross-mode transmission coefficient is dominated by the effects of the imbalance in the dielectric constant (the reason is that, in general, $\varepsilon_{LUT} - \varepsilon_{REF}$ $\gg \varepsilon_{LUT} \cdot \tan \delta_{LUT} - \varepsilon_{REF} \cdot \tan \delta_{REF}$, since, typically, $\tan \delta_{LUT} \ll 1$ and $\tan \delta_{REF} \ll 1$). According to (6.22), increasing the reference impedance of the ports, Z_0, boosts up the sensitivity. Nevertheless, Z_0 (typically set to $Z_0 = 50\,\Omega$) is not a design parameter. It is also apparent from (6.22) that designing the OCSRR with a high capacitance and low inductance contributes to sensitivity optimization. However, it is difficult in practice to implement OCSRRs with high C_p/L_p ratio. Finally, choosing a substrate with a small dielectric constant favors sensitivity, since ε_r appears in the denominator of (6.22).

As indicated, to simplify the previous analysis, it has been considered that the impedance of the host lines is $Z_c = Z_0$, the impedance of the ports. However, it has been demonstrated by electromagnetic simulation that increasing Z_c (i.e. $Z_c > Z_0$ with $Z_0 = 50\,\Omega$) has the effect of enhancing the sensitivity [8]. For this main reason, the impedance of the host line was set to $Z_c = 77\,\Omega$ in the prototype of Fig. 6.3. The analysis in this case is not so simple, but for relatively small variations of Z_c with regard to Z_0, it is expected that the effects of Z_c on $\left|S_{21,\omega_{REF}}^{cd}\right|$ are similar to those of Z_0 when the line impedance and the impedance of the ports are identical.

In [8], the sensor of Fig. 6.3 was focused on the measurement of the solute concentration of diluted solutions of sodium in DI water, as well as on the characterization of the complex dielectric constant of liquid mixtures (particularly, solutions of ethanol and methanol in DI water). Thus, the reference liquid is DI water (the solvent). Let us next focus on the first application. Figure 6.8 depicts the cross-mode transmission coefficient corresponding to different levels of sodium concentration, between 0.25 g/L (the minimum level that can be reasonably resolved) and 80 g/L (the solutions were prepared by carefully weighting the sodium content by means of a precision weighting machine, model *Rs-Pro*, with 1 mg resolution). The cross-mode transmission coefficient experiences significant variations in the overall magnitude when the sodium concentration increases. Figure 6.8(b) depicts the dependence of the maximum value of $\left|S_{21}^{cd}\right|$ for each curve as a function of the sodium concentration. For small sodium concentrations, $\left|S_{21}^{cd}\right|_{max}$ varies significantly with the sodium content, and such variation progressively decreases. Thus, the sensitivity is maximum for small concentrations and then it progressively decreases (Fig. 6.8c). It is remarkable that the variation experienced by $\left|S_{21}^{cd}\right|_{max}$ is approximately linear (and hence the sensitivity is roughly constant) up to sodium concentrations of 2.5 g/L, as it can be appreciated in the zoom view shown in the inset of Fig. 6.8(b). The maximum value of the sensitivity is $S = 0.0092\ (g/L)^{-1}$.

From the results of Fig. 6.8(b), a calibration curve, consisting of an order-3 polynomial providing a correlation coefficient of $R^2 = 0.9983$, was inferred in [8], i.e.

$$[Na](g/L) = 36807 \cdot \left|S_{21}^{cd}\right|_{max}^3 - 3480.1 \cdot \left|S_{21}^{cd}\right|_{max}^2 + 322.68 \cdot \left|S_{21}^{cd}\right|_{max} - 4.76 \tag{6.23}$$

This calibration curve does not exactly predict a null concentration of sodium when $\left|S_{21}^{cd}\right|_{max} = 0$. The reason is that when the LUT is pure DI water, i.e. without sodium content, the measured cross-mode transmission coefficient has a finite value (Fig. 6.8a). Nevertheless, by measuring the cross-mode transmission coefficient, the sodium concentrations of unknown samples can be inferred with high accuracy using the calibration curve. This aspect is confirmed from the small differences between the nominal values of sodium concentration and the values provided by the calibration curve inferred from the measured values of $\left|S_{21}^{cd}\right|_{max}$, also included in Fig. 6.8(b).

The sensor of Fig. 6.3 is able to resolve concentration levels as small as 0.25 g/L, smaller than the typical levels of sodium concentration in blood plasma or urine. This good resolution, intimately related to the sensitivity at small concentrations, is due to the fact that a small sodium density suffices to substantially modify the conductivity of DI water. Therefore, a small sodium content is

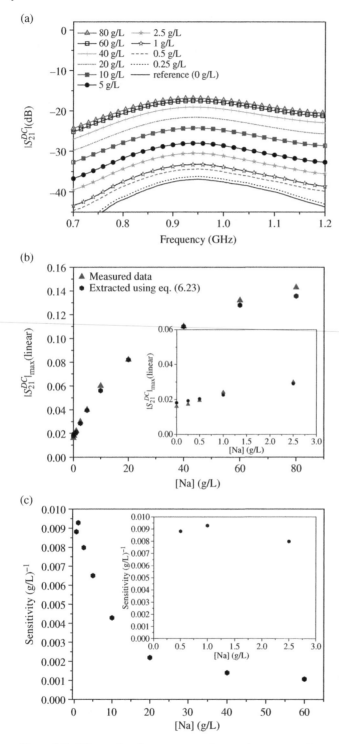

Figure 6.8 Effects of sodium concentration on the cross-mode transmission coefficient for the sensor of Fig. 6.3. (a) Dependence of the cross-mode transmission coefficient on frequency; (b) relation between sodium concentration and $\left|S_{21}^{cd}\right|_{max}$ in linear form; (c) sensitivity. The inset in (b) depicts a zoom view for small concentrations. *Source:* Reprinted with permission from [8]; copyright 2018 IEEE.

expected to significantly alter the loss tangent, or the imaginary part of the dielectric constant, of DI water, resulting in significant mode conversion. However, from the comparison between Figs. 6.5 (a) and 6.6(a), it can be concluded that the presence of sodium does not significantly alter the real part of the dielectric constant, as results from the small variation of the resonance frequencies in those figures.

The measured responses of Fig. 6.8(a) are referred to $50\,\Omega$ ports, that is, $Z_0 = 50\,\Omega$. In order to validate the approximations in (6.15) and (6.16), from the results of Fig. 6.8(a), $\left|S_{21}^{cd}\right|$ referred to ports with impedance $Z_0 = Z_c = 77\,\Omega$ (using *Keysight ADS*) was obtained [8]. The results of $\left|S_{21}^{cd}\right|$ at ω_{REF}, i.e. $\left.\left|S_{21}^{cd}\right|\right|_{\omega_{REF}}$ (see Fig. 6.9), are compared with the prediction given by (6.15) and (6.16), and also with the exact analytical solution. The agreement between the measurement, the exact analytical solution, and the approximate solution given by (6.15) is very good in the considered range of sodium concentrations. However, the rough approximation (6.16) is only valid for very small perturbations. The reason is that L_z is not very small in the considered sensing structure, and the influence of the terms that depend on L_z increases with the sodium concentration. The results of Fig. 6.9 validate the previous analysis and point out the effects of circuit parameters on the cross-mode transmission coefficient at ω_{REF}. Note, however, that in Fig. 6.8(b), the considered output variable is $\left.\left|S_{21}^{cd}\right|\right|_{max}$, rather than $\left.\left|S_{21}^{cd}\right|\right|_{\omega_{REF}}$ (and the reference impedance of the ports is $Z_0 = 50\,\Omega$). Nevertheless, the responses of Fig. 6.8(a) exhibit a very soft dependence on frequency in the vicinity of ω_{REF}. Therefore, the conclusions derived from the previous analysis, relative to the effects of model parameters on the cross-mode transmission coefficient at ω_{REF}, can be extended to the output variable, $\left.\left|S_{21}^{cd}\right|\right|_{max}$, even by considering the usual reference impedance of the ports, $Z_0 = 50\,\Omega$.

Let us next focus on the dielectric characterization of liquid mixtures. In [8], calibration curves inferred from mixtures of DI water and ethanol were obtained. Since pure DI water and pure ethanol exhibit complex dielectric constants of $\varepsilon_{water} = \varepsilon' - j\varepsilon'' = 80.86 - j3.04$ and $\varepsilon_{ethanol} = \varepsilon' - j\varepsilon'' = 27.86 - j9.79$, respectively (at ω_{REF}), it follows that the method is especially suitable for dielectric characterization of liquids with complex dielectric constant within the previous ranges. The measured S-parameters, including the cross-mode transmission coefficient, that are obtained by introducing pure DI water in the REF channel and different mixtures of DI water and ethanol, with varying ethanol concentration, in the LUT channel, are depicted in Fig. 6.10.

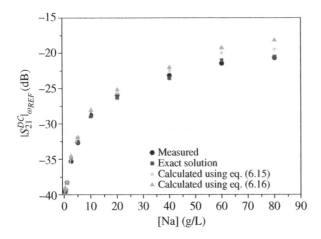

Figure 6.9 Variation of the cross-mode transmission coefficient at ω_{REF} as a function of sodium concentration. The considered port impedances are $Z_0 = 77\,\Omega$, as indicated in the text. *Source:* Reprinted with permission from [8]; copyright 2018 IEEE.

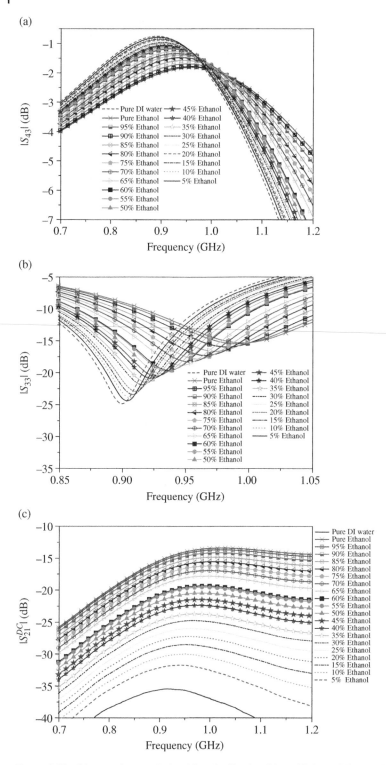

Figure 6.10 Measured transmission (a) and reflection (b) coefficient of the single-ended LUT line loaded with different mixtures of DI water and ethanol, and cross-mode transmission coefficient (c). *Source:* Reprinted with permission from [8]; copyright 2018 IEEE.

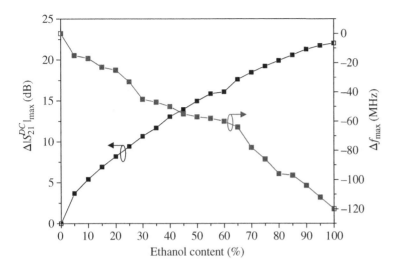

Figure 6.11 Dependence of $\Delta|S_{21}^{cd}|_{max}$ (in dB) and Δf_{max} with the ethanol concentration. *Source:* Reprinted with permission from [8]; copyright 2018 IEEE.

As the ethanol content increases, absorption, due to losses, increases. This is an expected result provided the imaginary part of the dielectric constant in ethanol is larger than in DI water. However, the real part of the complex dielectric constant is smaller in ethanol, hence decreasing the capacitance contribution of the LUT when the ethanol content increases, and consequently shifting up the resonance frequency, see Fig. 6.10(b). In order to determine both the real and the imaginary parts of the complex dielectric constant of the LUT, it is necessary to deal with two output variables. In [8], the considered output variables were (i) the variation experienced by the maximum value of the cross-mode transmission coefficient by loading the LUT channel with the LUT liquid, as compared with the REF liquid, $\Delta|S_{21}^{cd}|_{max} = |S_{21}^{cd}|_{max,REF} - |S_{21}^{cd}|_{max,LUT}$, and (ii) the variation of the frequency position of the maximum value of the transmission coefficient, $\Delta f_{max} = f_{max,REF} - f_{max,LUT}$. Let us insist that the sub-indexes *max,REF* and *max,LUT* denote the maximum value of the cross-mode transmission coefficient and the corresponding frequency when the LUT channel is loaded with the REF liquid (symmetric case) and LUT, respectively. The output variables are depicted in Fig. 6.11 as a function of the volume fraction of ethanol. From these data, and from the knowledge of the complex dielectric constant of the different mixtures (given by the Weiner model [20], see Fig. 6.12), the calibration curves for both the incremental real, $\Delta\varepsilon'$, and imaginary, $\Delta\varepsilon''$, parts of the dielectric constant were obtained (using multiple linear regression). Such curves are

$$\Delta\varepsilon' = k_{11}\Delta|S_{21}^{cd}|_{max} + k_{12}\Delta f_{max} \tag{6.24a}$$

$$\Delta\varepsilon'' = k_{21}\Delta|S_{21}^{cd}|_{max} + k_{22}\Delta f_{max} \tag{6.24b}$$

with coefficients $k_{11} = -1.805\,\text{dB}^{-1}$, $k_{12} = 0.107\,\text{MHz}^{-1}$, $k_{21} = -0.092\,\text{dB}^{-1}$, and $k_{22} = -0.063$ MHz^{-1}, and a correlation coefficient of $R^2 = 0.994$.

Using equations 6.24(a) and 6.24(b) and the data of Fig. 6.11, the real and imaginary parts of the complex dielectric constant for the different mixtures of ethanol in DI water were obtained (Fig. 6.12) [8]. The results are in good agreement with the prediction given by the Weiner model (the upper and the lower limits of that model are also indicated in the figure). Indeed, such good

(a)

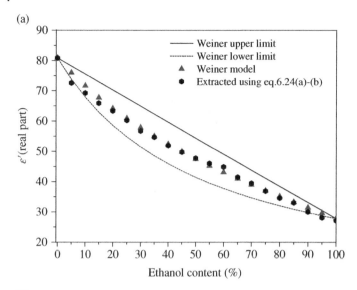

(b)

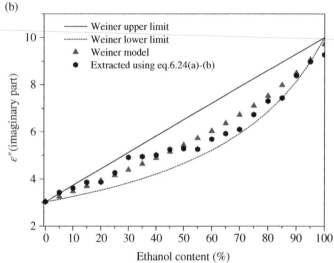

Figure 6.12 Extracted value for the real (a) and imaginary (b) parts of the complex dielectric constant in mixtures of DI water/ethanol. The static Weiner model is also included for comparison purposes. *Source:* Reprinted with permission from [8]; copyright 2018 IEEE.

agreement is expected provided the calibration curves were obtained from the Weiner model applied to ethanol/DI water mixtures.

Nevertheless, to properly validate the functionality of the sensor, the calibration curves were used to determine the complex dielectric constant in several mixtures of methanol in DI water. For that purpose, pure DI water was injected to the REF channel, whereas different mixtures of DI water and methanol were subsequently injected to the LUT channel by varying the methanol concentration (from 0 to 100% in steps of 5%). The cross-mode transmission coefficient for each measurement is plotted in Fig. 6.13. From the values of $\Delta\left|S_{21}^{cd}\right|_{max}$ and Δf_{max} (depicted in Fig. 6.14), the variation of the real and imaginary parts of the complex dielectric constant was obtained [using expressions 6.24(a) and (b)] for all the mixtures (with reference to the values of DI water, the REF liquid).

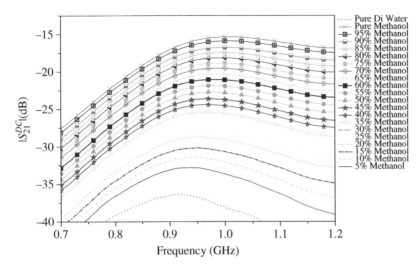

Figure 6.13 Measured cross-mode transmission coefficient corresponding to different mixtures of DI water and methanol. *Source:* Reprinted with permission from [8]; copyright 2018 IEEE.

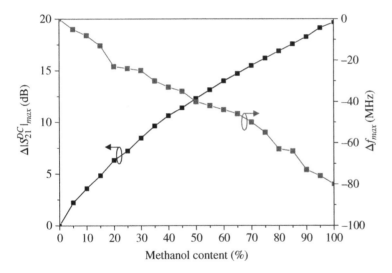

Figure 6.14 Dependence of $\Delta|S_{21}^{cd}|_{max}$ (in dB) and Δf_{max} with the methanol concentration. *Source:* Reprinted with permission from [8]; copyright 2018 IEEE.

The results, depicted in Fig. 6.15, are compared with the predictions of the Weiner model. The good agreement is indicative of the validity of the proposed sensor to determine the complex dielectric constant of liquids. In particular, the nominal complex dielectric constant of methanol at the frequency of sensor operation, i.e. $30.87 - j6.48$, is reasonably predicted.

6.2.2.2 Microfluidic Sensor Based on SRRs and Application to Electrolyte Concentration Measurements in Aqueous Solutions

In this section, a differential microfluidic sensor similar to the one presented in the previous section, but based on a pair of microstrip lines loaded with SRRs, is succinctly reviewed [9]. In the

(a)

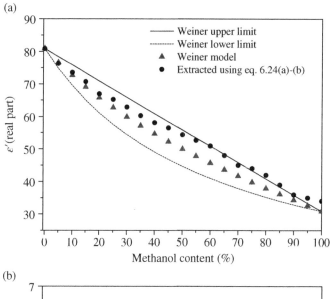

(b)

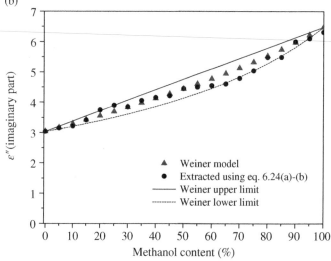

Figure 6.15 Extracted value for the real (a) and imaginary (b) parts of the complex dielectric constant in mixtures of DI water/methanol. The static Weiner model is also included for comparison purposes. *Source:* Reprinted with permission from [8]; copyright 2018 IEEE.

sensor of Fig. 6.3, vias were required in order to connect the OCSRRs to ground, see Fig. 6.2(a). By contrast, in the SRR-based differential sensor subject of this section, vias are not necessary, an interesting aspect to provide more robustness to the sensor against potential imbalances related to fabrication-related tolerances. It should be mentioned that sensor resolution is related to sensitivity at small perturbations, but it is also influenced by the level of symmetry of the bare sensor. Namely, imperfections in sensor fabrication may generate asymmetries between the sensor lines, which may obscure the detection of small differences between the REF liquid and the LUT liquid.

The layout of the differential-mode SRR-based sensor is depicted in Fig. 6.16, where dimensions are indicated. The sensor was implemented on the *FR4* substrate with dielectric constant $\varepsilon_r = 4.4$ and thickness $h = 1.6$ mm. The fluidic channels and mechanical parts are identical to those used in the sensor of the previous section. Also, to avoid substrate absorption, a dry film of clear polyester,

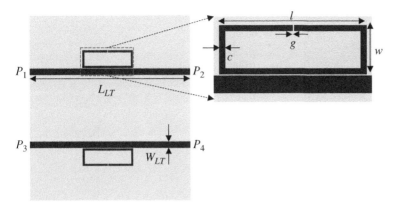

Figure 6.16 Topology of the microwave structure of the SRR-based differential sensor. Dimensions (in mm) are L_{LT} = 76.5, W_{LT} = 2.79, l = 24, c = 1, w = 8.17, g = 0.2. The ground plane is depicted in gray. *Source:* Reprinted with permission from [9]; copyright 2019 IEEE.

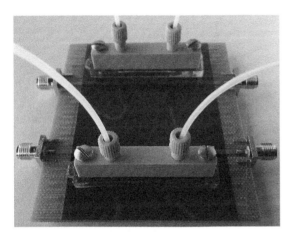

Figure 6.17 Photograph of the SRR-based differential sensor including the microwave structure, the fluidic channels (plus mechanical accessories) and connectors.

with an estimated thickness of 50 μm and dielectric constant of 3.5, was deposited on top of the SRRs. The photograph of the whole sensing structure is shown in Fig. 6.17. The fluidic channels are located in such a way that the LUT and REF liquids lie on top of the gaps of the SRRs, the more sensitive regions of the resonators, due to the high fringing fields in those regions.

The fabricated sensor of Fig. 6.17 was validated in [9] by considering DI water solutions with different types of electrolytes, particularly, NaCl, KCl and CaCl$_2$. Nevertheless, in this book, only the first set of experiments, carried out by injecting DI water solutions with different concentrations of NaCl in the LUT channel, are reported. The REF liquid is, obviously, pure DI water, injected in the REF channel. The measured cross-mode transmission coefficient for the different NaCl concentrations is depicted in Fig. 6.18(a). The cross-mode transmission coefficient resulting when the LUT channel is filled with pure DI water, the REF sample (corresponding to a balanced loading), is also included. The maximum value of the modulus of the cross-mode transmission coefficient for this case is -38.95 dB. This is a small value, indicative of a good balance between both channels. Thanks to such small value, concentrations of NaCl as small as 0.25 g/L can be resolved. By increasing the electrolyte concentration, the cross-mode transmission coefficient also increases, as expected. Figure 6.18(b) depicts the dependence of the maximum value of the cross-mode transmission

(a)

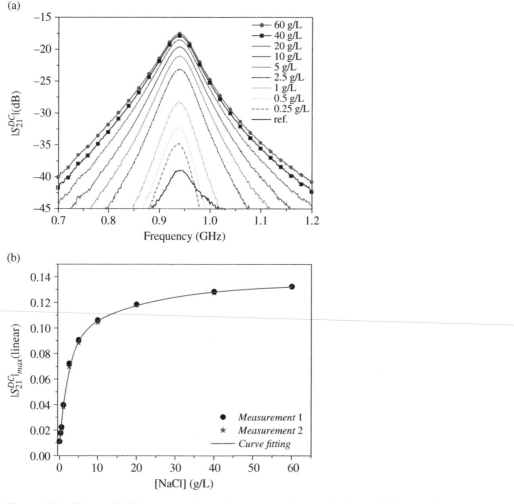

(b)

Figure 6.18 Effects of NaCl concentration on the cross-mode transmission coefficient for the sensor of Fig. 6.17. (a) Dependence of the cross-mode transmission coefficient on frequency; (b) relation between sodium concentration and $\left|S_{21}^{cd}\right|_{max}$ in linear form. *Source:* Reprinted with permission from [9]; copyright 2019 IEEE.

coefficient, $\left|S_{21}^{cd}\right|_{max}$, with the concentration of NaCl, [NaCl]. Actually, a pair of measurements was carried out in [9], in order to ensure that the results are repetitive (such measurements are distinguished by the labels 1 and 2 in Fig. 6.18b). For small concentrations, the sensitivity is very high, and it progressively decreases as [NaCl] increases. The maximum value of the sensitivity is found to be 0.033 (g/L)$^{-1}$, i.e. substantially higher than the maximum sensitivity of the OCSRR-based differential sensor reported in the previous section, also validated by considering solutions of NaCl in DI water. From the results of Fig. 6.18(b), the following calibration curve, with a correlation coefficient of $R^2 = 0.99966$, is obtained

$$[\text{NaCl}]\,(\text{g/L}) = -0.43 + 0.45 \cdot 10^{\left(\frac{\left|S_{21}^{cd}\right|_{max}}{0.03649}\right)} + 2 \cdot 10^{\left(\frac{\left|S_{21}^{cd}\right|_{max} - 5}{0.0091}\right)} \tag{6.25}$$

Using such curve, the concentration of NaCl can be inferred by measuring $\left|S_{21}^{cd}\right|_{max}$.

6.2.2.3 Microfluidic Sensor Based on DB-DGS Resonators and Application to Electrolyte Concentration Measurements in Aqueous Solutions

The third prototype example is a differential microfluidic sensor consisting of a pair of microstrip lines loaded with a DB-DGS resonator, and the corresponding fluidic channels, plus mechanical accessories. In [11], where such sensor was reported, a detailed sensitivity analysis, similar to the one of Section 6.2.2.1 in reference to the OCSRR-based differential sensors, was carried out. Moreover, sensor validation was performed by measuring the electrolyte content, particularly NaCl, in solutions of DI water and by determining the complex permittivity of solutions of isopropanol, also in DI water, as well as the volume fraction of the solute. This subsection is focused on such analysis, developed on the basis of the equivalent circuit model of the sensor, as well as on the cited applications. Interestingly, identical conclusions to those derived in Section 6.2.2.1 in regard to the requirements of the reactive element values of the sensing resonators (OCSRRs) for sensitivity optimization are inferred from the analysis of the DB-DGS-based differential sensor. The reason is that, as it will be shown, the modulus of the cross-mode transmission coefficient at the resonance frequency of the REF resonator, the considered output variable in the analysis, is given by an expression formally identical to (6.22).

The layout of the DB-DGS-based differential-mode microfluidic sensor and its equivalent circuit model are depicted in Fig. 6.19 [11], whereas the fabricated device is shown in Fig. 6.20 (the device was implemented in the *Rogers RO3010* substrate with dielectric constant $\varepsilon_r = 10.2$, thickness $h = 1.27$ mm and loss tangent $\tan\delta = 0.0035$). The fluidic channels, identical to those of the previous sensors, are placed in the ground plane, on top of the capacitive regions of the DB-DGSs (the narrow slots), which are the parts sensitive to a variation in the dielectric properties of the materials (liquids) in contact with the resonant elements. Also, similar to the microfluidic sensors of the previous sections, a dry film is placed between the channels and the DB-DGS resonators, in order to avoid substrate absorption.

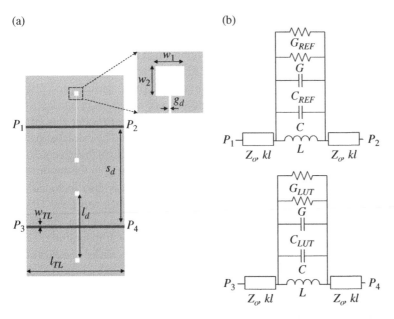

Figure 6.19 Layout of the DB-DGS-based differential sensor (a) and equivalent circuit model (b). Dimensions (in mm) are $w_1 = w_2 = 2$, $w_{TL} = 1.14$, $l_{LT} = 50$, $l_d = 28$, $g_d = 0.2$, $S_d = 44$. In (a), the ground plane is depicted in light gray. *Source:* Reprinted with permission from [11]; copyright 2019 MDPI.

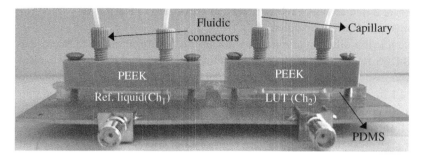

Figure 6.20 Photograph of the fabricated DB-DGS-based differential microfluidic sensor.

The DB-DGSs are modeled as series-connected parallel resonant tanks [21]. Using the nomenclature of the original source [11], the reactive elements L and C are the inductance and capacitance, respectively, of the DB-DGSs without liquid in the channels, whereas G accounts for the conductance, mainly related to substrate losses (such conductance does not take into account the effects of the liquid in the channel). By introducing liquid into the channels, the elements that are expected to experience a variation are the capacitance and conductance of the DB-DGSs. Thus, C_{REF} and G_{REF} account for the effects of the liquid in the REF channel (if it is present), whereas C_{LUT} and G_{LUT} are the corresponding element values for the LUT channel.[9] Finally, the transmission lines are described by their characteristic impedance, Z_0, identical to the reference impedance of the ports in the present analysis, and by the electrical length kl, where l is the physical length and k is the phase constant.

Model validation was carried out in [11] by comparing electromagnetic simulations (inferred by means of *ANSYS HFSS*) with circuit simulations (obtained by means of the schematic simulator included in *Keysight ADS*). Since the REF and LUT lines are uncoupled, it suffices to consider the simulation (circuit and electromagnetic) of any of the lines, for example the REF line. It was first obtained the transmission coefficient of the REF line with an empty channel (i.e. with air) by means of electromagnetic simulation, see Fig. 6.21(a). The figure also includes the circuit simulation, with the extracted parameters indicated in the caption (note that the presence of the channel and dry film was taken into account in the electromagnetic simulations, from which the circuit parameters were extracted). The excellent agreement in both the magnitude and phase responses is indicative of the validity of the model. The REF line was also simulated by considering the channel full of DI water. For that purpose, the complex dielectric constant of water was introduced in the *ANSYS HFSS* simulator. The response is depicted in Fig. 6.21(b). The extracted parameters with unloaded channel, i.e. $C, L,$ and G were maintained, whereas to include the effects of DI water in the channel, C_{REF} and G_{REF} were adjusted (the values are also indicated in the caption of Fig. 6.21). The circuit response corresponding to these circuit parameters is also included in Fig. 6.21, and again the agreement with the full wave electromagnetic simulation is very good. Obviously, the simulations reveal that considering the presence of DI water in the channel has the effect of decreasing the resonance frequency of the DB-DGS resonator and the notch magnitude. This result is consequence of the high dielectric constant of DI water (which increases significantly

---

9 Note that the circuit model of Fig. 6.19 is the differential counterpart of the circuit model of Fig. 2.20. However, the incremental capacitance of the resonant element necessary to take into account the presence of either the REF liquid or the LUT (or MUT) liquid in the corresponding channel is designated as C_{REF} and C_{LUT}, respectively, in the model of Fig. 6.19, whereas in the model of Fig. 2.20, the incremental capacitance associated with the presence of the MUT was designated as $C_{e,MUT}$.

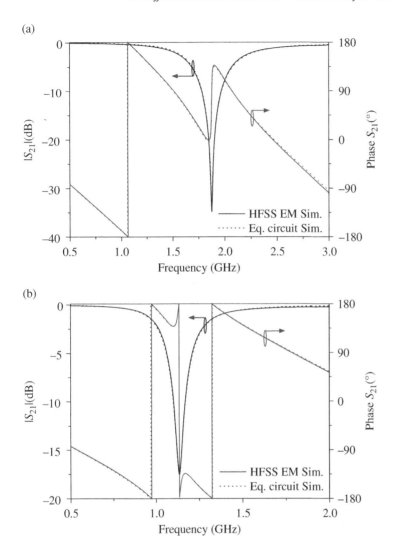

Figure 6.21 Transmission coefficient (magnitude and phase) of the REF channel without liquid (a), and with DI water (b), inferred from electromagnetic and circuit simulation. The element values of the equivalent circuit model are (in reference to Fig. 6.19b): $Z_0 = 50\ \Omega$, $kl = 78.51°$, $L = 2.40$ nH, $C = 2.51$ pF, $C_{REF} = 5.74$ pF, $G = 0.18$ mS, and $G_{REF} = 1.31$ mS. Note that in the circuit simulations of (a), $C_{REF} = 0$ pF and $G_{REF} = 0$ mS, as corresponds to the absence of any liquid in the channel. *Source:* Reprinted with permission from [11]; copyright 2019 MDPI.

the overall capacitance of the DB-DGS resonator) and the high dissipation factor of DI water, corresponding to a high value of the imaginary part of the complex dielectric constant, or loss tangent (which enhances the losses of the resonant element).

Let us now focus on the analysis of the sensor on the basis of the equivalent circuit model of Fig. 6.19(b).The cross-mode transmission coefficient of the structure depends on the length and characteristic impedance of the host lines. Nevertheless, it is considered that the impedance of the lines coincides with the reference impedance of the ports, Z_0. Under these conditions, the magnitude of the cross-mode transmission coefficient depends only on the impedance of the DB-DGSs, and it is given by the modulus of the cross-mode transmission coefficient by considering that the ports are located at the planes of the resonant element. The cross-mode transmission coefficient is given by (6.7).[10] By neglecting the host lines, the transmission coefficients of the individual DB-DGS resonators are

$$S_{21} = \frac{1}{1 + \dfrac{1}{2Z_0} Z_{REF}} \tag{6.26a}$$

$$S_{43} = \frac{1}{1 + \dfrac{1}{2Z_0} Z_{LUT}} \tag{6.26b}$$

where the impedance of the DB-DGSs, Z_{REF} and Z_{LUT}, for the REF and LUT channels, respectively, are given by

$$Z_{REF} = \frac{j\omega L}{j\omega L G'_{REF} - \dfrac{\omega^2}{\omega_{REF}^2} + 1} \tag{6.27a}$$

$$Z_{LUT} = \frac{j\omega L}{j\omega L G'_{LUT} - \dfrac{\omega^2}{\omega_{LUT}^2} + 1} \tag{6.27b}$$

In (6.27a) and (6.27b), the following variables have been used

$$C'_{REF} = C + C_{REF} \tag{6.28a}$$

$$G'_{REF} = G + G_{REF} \tag{6.28b}$$

$$C'_{LUT} = C + C_{LUT} \tag{6.28c}$$

$$G'_{LUT} = G + G_{LUT} \tag{6.28d}$$

$$\omega_{REF}^{-2} = L C'_{REF} \tag{6.28e}$$

$$\omega_{LUT}^{-2} = L C'_{LUT} \tag{6.28f}$$

Similar to the analysis of Section 6.2.2.1, let us consider that the output variable is the cross-mode transmission coefficient evaluated at the resonance frequency of the reference channel, ω_{REF}. For small perturbations and low losses, ω_{REF} and ω_{LUT} are not very different, and both G'_{REF} and G'_{LUT} are small as compared with $Y_0 \equiv 1/Z_0$. With these approximations, S_{21} and S_{43} at ω_{REF} can be expressed as

$$S_{21,\omega_{REF}} = \frac{1}{1 + \dfrac{1}{2Z_0 G'_{REF}}} \cong 2Z_0 G'_{REF} \tag{6.29a}$$

$$S_{43,\omega_{REF}} = \frac{1}{1 + \dfrac{1}{2Z_0\left\{ G'_{LUT} - j\dfrac{\omega_{REF}}{L}\left(\dfrac{1}{\omega_{REF}^2} - \dfrac{1}{\omega_{LUT}^2}\right)\right\}}} \cong 2Z_0\left\{ G'_{LUT} - j\dfrac{\omega_{REF}}{L}\left(\dfrac{1}{\omega_{REF}^2} - \dfrac{1}{\omega_{LUT}^2}\right)\right\} \tag{6.29b}$$

and the modulus of the cross-mode transmission coefficient is simply

$$\left| S_{21,\omega_{REF}}^{cd} \right| = Z_0 \sqrt{(G_{LUT} - G_{REF})^2 + \omega_{REF}^2 (C_{LUT} - C_{REF})^2} \tag{6.30}$$

---

10 It is assumed that the REF and LUT DB-DGS-loaded lines are uncoupled.

i.e. an expression formally identical to (6.16). Thus, it is obvious that the dependence of the magnitude of the cross-mode transmission coefficient at ω_{REF} with the relevant variables, namely the loss tangent and the dielectric constant of the LUT and REF liquids, can be expressed as

$$\left|S_{21,\omega_{REF}}^{cd}\right| = Z_0 \cdot \sqrt{\frac{C}{L}} \cdot \frac{\sqrt{(\varepsilon_{LUT} \cdot \tan \delta_{LUT} - \varepsilon_{REF} \cdot \tan \delta_{REF})^2 + (\varepsilon_{LUT} - \varepsilon_{REF})^2}}{\sqrt{(\varepsilon_r + \varepsilon_{REF})(\varepsilon_r + 1)}} \tag{6.31}$$

The previous result is inferred from an analysis identical to that of Section 6.2.2.1, see expressions (6.17)–(6.22). As expected, (6.31) and (6.22) are identical, except by the proportionality factor and the different variables used to designate the reactive elements of the bare resonators.

Since for sensitivity optimization, the ratio C/L must be high, as it is inferred from (6.31), the DB-DGS resonators were designed in [11] with the specific shape of Fig. 6.19(a). The narrow and elongated slots increase the capacitance, whereas the small square apertures at the extremes of the slots contribute to limit the inductance value. For the validation of the previous analysis and the indicated approximations, let us consider that the REF liquid is DI water (with complex dielectric constant $80.66 - j4.92$ at ω_{REF}), and that the LUT channel contains an hypothetical liquid exhibiting a complex dielectric constant variation of 3% (smaller) as compared with the one of the REF liquid (i.e. $78.24 - j4.77$). The responses of both lines, as well as the cross-mode transmission coefficient, inferred from electromagnetic and circuit simulation are depicted in Fig. 6.22. Figure 6.22(b) includes the results derived from the approximate analytical expressions (6.30) and (6.31). These analytical results are undistinguishable and are in reasonable agreement with the results inferred from the electromagnetic and circuit simulations at ω_{REF}. Thus, the analytical (and approximate) formulas predict the value of the cross-mode transmission coefficient at ω_{REF} to a good approximation. Such good agreement further validates the equivalent circuit model of Fig. 6.19(b), including the parameters describing the liquid properties (C_{REF}, G_{REF}, C_{LUT}, and G_{LUT}) and confirms the validity of the low-loss and small perturbation assumptions (of application in the considered case).

Sensor performance was first evaluated in [11] by measuring the cross-mode transmission coefficient for different solutions of NaCl in DI water injected in the LUT channel (pure DI water was the considered REF liquid). The measured cross-mode transmission coefficient for the different mixtures of DI water and NaCl is plotted in Fig. 6.23(a). The maximum value of the magnitude of the cross-mode transmission coefficient when both channels are filled with the REF liquid is -31.94 dB. This is a small value, indicating that the structure is quite balanced when identical liquids are present in both channels. As expected, as the concentration of NaCl in DI water increases, the maximum value of the cross-mode transmission coefficient also increases, due to an increasing symmetry imbalance between both channels. The results indicate that the frequency corresponding to the maximum value of magnitude of the cross-mode transmission coefficient (for each NaCl concentration level) does not experience a significant change, which is coherent with the fact that the main effect of the presence of electrolytes in DI water is the variation of the dissipation factor (loss tangent or imaginary part of the complex permittivity).

Figure 6.23(b) depicts the variation of the maximum value of the cross-mode transmission coefficient, $\left|S_{21}^{cd}\right|_{max}$, with the concentration of NaCl (in g/L). The following calibration curve with correlation coefficient $R^2 = 0.99992$ was obtained

$$[\text{NaCl}](\text{g/L}) = 3.425 \cdot 10^{\left(\frac{\left|S_{21}^{cd}\right|_{max}}{0.085}\right)} + 3.295 \cdot 10^{\left(\frac{\left|S_{21}^{cd}\right|_{max} - 4}{0.019}\right)} - 4.98 \tag{6.32}$$

The sensitivity increases as the NaCl concentration decreases (see the zoom inset in Fig. 6.23b), and the highest value of the sensitivity was found to be 0.035 $(\text{g/L})^{-1}$, with a sensor resolution

(a)

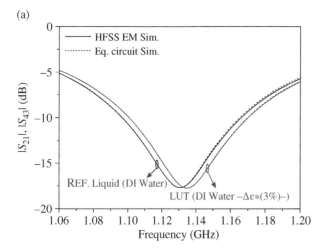

(b)

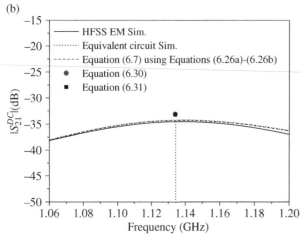

Figure 6.22 (a) Simulated transmission coefficient (magnitude) of the REF and LUT lines by considering DI water in the REF channel, and an hypothetical liquid with complex permittivity 3% smaller (real and imaginary part) than the one of water in the LUT channel; (b) magnitude of the cross-mode transmission coefficient inferred from the equivalent circuit model, the *ANSYS HFSS* simulator, and the indicated analytical expressions. *Source:* Reprinted with permission from [11]; copyright 2019 MDPI.

of 0.25 g/L. To evaluate the repetitiveness of the sensor response, a set of mixtures of DI water and NaCl (corresponding to the concentrations of the first campaign) was injected in random order in the LUT channel. The results, designated as *Measurement 2* in Fig. 6.23(b), indicate that the results are repetitive.

The functionality of the device of Fig. 6.20 as permittivity sensor, able to determine the complex dielectric constant of liquid samples, was also demonstrated in [11]. In this case, the considered output variables are identical to those considered for the sensor of Fig. 6.3, i.e. (i) the variation experienced by the maximum value of the cross-mode transmission coefficient by loading the LUT channel with the LUT liquid, as compared with the REF liquid, $\Delta |S_{21}^{cd}|_{max} = |S_{21}^{cd}|_{max,REF} - |S_{21}^{cd}|_{max,LUT}$, and (ii) the variation of the frequency position of the maximum value of the transmission coefficient, $\Delta f_{max} = f_{max, REF} - f_{max, LUT}$. Two output variables are needed because the complex dielectric constant is composed of the real and the imaginary parts. Figure 6.24(a) depicts the cross-mode

(a)

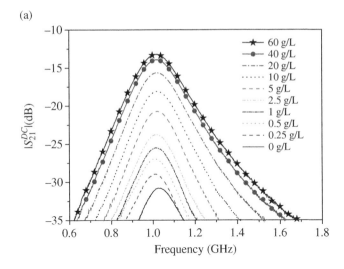

(b)

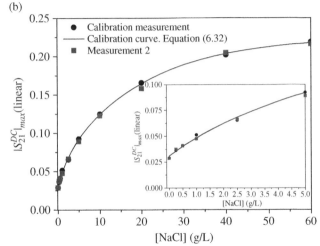

Figure 6.23 Cross-mode transmission coefficient for different concentrations of NaCl in DI water (a) and variation of $|S_{21}^{cd}|_{max}$ with NaCl content (b). *Source:* Reprinted with permission from [11]; copyright 2019 MDPI.

transmission coefficient for different mixtures of isopropanol in DI water. From this figure, the output variables for each volume fraction of isopropanol in DI water are obtained, see Fig. 6.24(b). Using a procedure similar to the one reported in Section 6.2.2.1, it was found that the variation of the real and imaginary parts of the complex permittivity can be expressed according to (6.24), with coefficients $k_{11} = 3.789\,\text{dB}^{-1}$, $k_{12} = 0.441\,\text{MHz}^{-1}$, $k_{21} = 0.009\,\text{dB}^{-1}$, and $k_{22} = 0.010\,\text{MHz}^{-1}$. From equation (6.24) and the data of Fig. 6.24(b), the real and the imaginary parts of the complex dielectric constant, as a function of isopropanol content in DI water, were obtained (Fig. 6.25). In the same figure, the Weiner model [20] was used to establish the upper and lower limits of the real and imaginary parts of the complex dielectric constant for mixtures of both liquids. The calculated values for the complex dielectric constant of mixtures of DI water and isopropanol are between the limits predicted by the Weiner model. These results validate the functionality of the proposed sensor for dielectric characterization of liquids. It should also be pointed out that the sensor is useful to monitor the concentration of isopropanol in DI water, with a resolution of at least 5% of volume fraction, as derived from the results of Fig. 6.24(a).

(a)

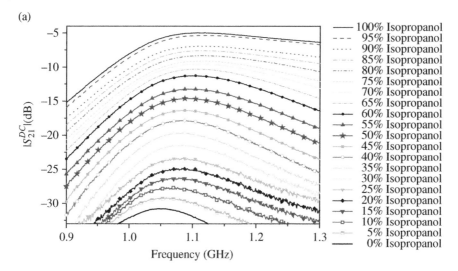

(b)

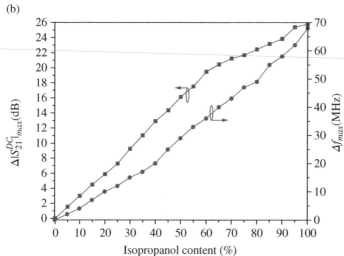

Figure 6.24 Cross-mode transmission coefficient for different mixtures of isopropanol and DI water (a) and dependence of $\Delta\left|S_{21}^{cd}\right|_{max}$ and Δf_{max} with isopropanol content (b). *Source:* Reprinted with permission from [11]; copyright 2019 MDPI.

6.2.2.4 Prototype for Measuring Electrolyte Content in Urine Samples

The sensors of Figs. 6.17 and 6.20 can be used as biosensors devoted to the characterization of the electrolyte concentration in urine samples [9, 12]. Let us report in this section the results presented in [9]. Nevertheless, let us first mention that the real-time monitoring of the total electrolyte concentration in urine may be important, since changes in the level of electrolytes may be indicative of certain pathologies. Thus, sensors able to detect such changes during disease treatment in hospital environments can be used as low-cost pre-screening systems. Note that the sensors presented in this section are especially suited for that purpose, since, by nature, these sensors act as comparators able to detect differences between a REF sample and the LUT sample.

Several horse urine samples, with different levels of electrolyte concentrations (Na$^+$, K$^+$ and Cl$^-$), were characterized in [9]. The specific concentration of each ion was measured by means of an

(a)

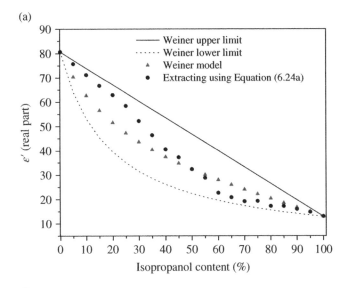

(b)

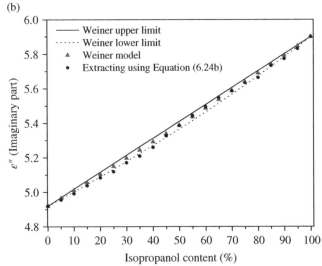

Figure 6.25 Extracted value for the real (a) and imaginary (b) parts of the complex dielectric constant in mixtures of DI water/isopropanol. The static Weiner model (upper and lower limits) is also included for comparison purposes. *Source:* Reprinted with permission from [11]; copyright 2019 MDPI.

ion-selective electrode (ISE) electrochemical system by the Biochemistry and Molecular Biology Department of Universitat Autònoma de Barcelona. From the measured values of the individual electrolyte concentrations (i.e. Na^+, K^+ and Cl^-) for each sample, expressed in mEq/L, the total concentrations were obtained by a simple addition.

The total ion concentration of the urine samples provided to the authors in [9] are indicated in the right-hand-side column of Fig. 6.26(a). By considering as REF liquid the urine sample with the highest level of ion concentration (615.8 mEq/L), the cross-mode transmission coefficient that results by injecting the different samples to the LUT channel are those depicted in Fig. 6.26(a). It should be mentioned that for monitoring changes in total ion concentration with time in a real

(a)

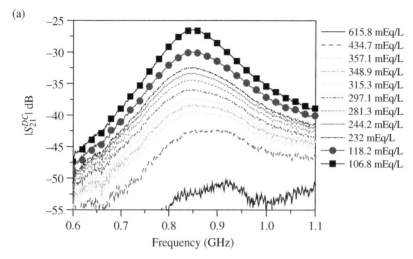

(b)

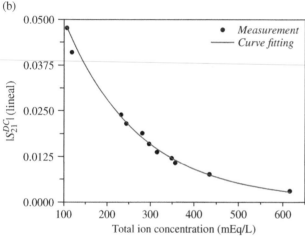

Figure 6.26 Cross-mode transmission coefficient for different values of total ion concentration in urine samples (a), and variation of $\left|S_{21}^{cd}\right|_{max}$ with total ion concentration in urine (b). *Source:* Reprinted with permission from [9]; copyright 2019 IEEE.

scenario, the reference sample should be, obviously, the urine at the beginning of the test campaign. In the measurements of Fig. 6.26(a), it has been (arbitrarily) considered that the reference sample is the one with 615.8 mEq/L ion concentration, as mentioned above. The variation of $\left|S_{21}^{cd}\right|_{max}$ with the total ion concentration is depicted in Fig. 6.26(b). The calibration curve, with $R^2 = 0.994\,22$, is

$$\frac{[*]\mathrm{mEq}}{\mathrm{L}} = -46.62 + 517.26 \cdot 10^{\left(\frac{-\left|S_{21}^{cd}\right|_{max}}{0.037}\right)} + 428.68 \cdot 10^{\left(\frac{-\left|S_{21}^{cd}\right|_{max}}{0.0036}\right)} \tag{6.33}$$

where the total ion concentration has been designated as [*]. The maximum sensitivity is -0.00058 $(\mathrm{mEq/L})^{-1}$. These results point out the potential of these differential microfluidic sensors to monitor electrolyte concentration variations in biological samples, such as urine (or even blood), in a real scenario.

6.3 Reflective-Mode Differential Sensors Based on the Measurement of the Cross-Mode Reflection Coefficient

Differential sensors operating in reflection were presented in Chapter 3 (Section 3.2.2.4). In those sensors, the working principle is the variation of the phase of the reflection coefficient of the pair of open-ended step-impedance transmission lines.[11] Indeed, the difference in the phase of the reflection coefficient between the REF and the LUT (or MUT) lines, assuming that there is a line imbalance, generates a finite cross-mode reflection coefficient. Such cross-mode reflection coefficient can be inferred by connecting the open-ended lines to the isolated ports of a rat-race hybrid coupler, as indicated in Chapter 3. The transmission coefficient of the resulting two-port structure gives the cross-mode reflection coefficient of the reflective-mode differential sensor. Alternatively, the cross-mode reflection coefficient can be inferred by direct subtraction of the reflection coefficients of the individual lines.

Reflective-mode differential sensors based on a pair of lines terminated with a resonator, the sensing element, have also been reported [13, 14]. In [14], the sensing resonant element is a grounded open split-ring resonator (OSRR) [22], which can be modeled as a series resonator connected to ground. In [13], the lines are terminated by means of step-impedance resonators (SIRs) [23], also described by means of grounded series resonant tanks. There is, however, a fundamental difference between the sensors reported in [13, 14]. In the SIR-based differential sensor of [13], 50-Ω resistors are series connected to the resonators. This means that the lines are terminated by a matched load at the resonance frequency, thereby generating a notch in the individual reflection coefficients at that frequency. Line imbalance, e.g. due to an asymmetric dielectric load, not only generates a difference in the phase of the reflection coefficient but also a difference in the magnitude of the reflection coefficients. This contributes to enhance the sensitivity of the cross-mode reflection coefficient with the differential input variable, typically the differential permittivity. Contrarily, in the OSRR-based reflective-mode differential sensor reported in [14], where resistors are not included, line imbalance mainly generates a relative variation in the phase of the reflection coefficients of the lines. Nevertheless, if there are significant differences in the loss factor of the REF and LUT (or MUT) samples, differences in the modulus of the reflection coefficient may also arise. In any case, line imbalance generates a perceptible cross-mode reflection coefficient, the output measurable variable.

Let us next show the performance of these sensors. Figure 6.27 depicts the differential sensor based on the SIR-terminated lines [13]. This sensor was equipped with PDMS fluidic channels for liquid characterization. The REF liquid was pure DI water, whereas mixtures of glycerol-water were the considered LUT samples. Figure 6.28(a) shows the reflection coefficients of the LUT line measured for different mixtures of glycerol and DI water injected in the LUT channel. The corresponding cross-mode reflection coefficients are depicted in Fig. 6.28(b), whereas the magnitude of the cross-mode reflection coefficient for each LUT sample at the frequency of the REF resonator (SIR loaded with DI water) is depicted in Fig. 6.28(c). The average sensitivity is found to be $S = 0.446$ dB/%. The linearity of the sensor over the whole range of volume fractions of glycerol is very good.

The reflective-mode differential microfluidic sensor based on OSRR-terminated lines is depicted in Fig. 6.29 [14]. Note that the OSRRs are grounded by means of quarter-wavelength open-ended lines, acting as impedance inverters, and thereby virtually connecting the OSRRs to ground. The transmission coefficient of the two-port device, proportional to the cross-mode reflection

---

11 For this reason, these sensors were included in Chapter 3, devoted to phase-variation sensors.

(a)

(b)

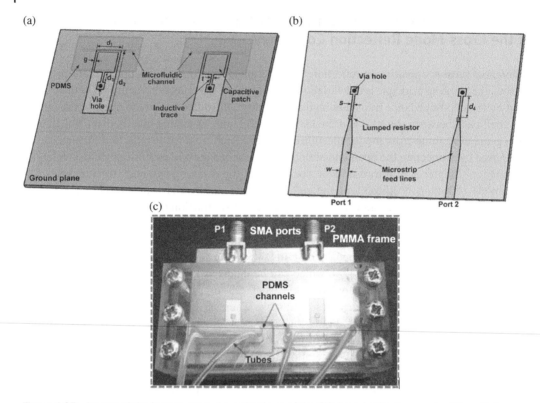

(c)

Figure 6.27 Layout of the bottom (a) and top (b) views of the SIR-based reflective-mode differential microfluidic sensor, and fabricated device (c). Dimensions (in mm) are: d_1 = 4.7 mm, d_2 = 12.3 mm, d_3 = 2 mm, d_4 = 4 mm, g = 0.45 mm, s = 0.4 mm, and w = 1.65. The sensor was implemented on the *Rogers RO4350* substrate with thickness h = 0.762 mm, dielectric constant ε_r = 3.66 and loss tangent tanδ = 0.0037. For details of the fluidic channels, see [13]. *Source:* Reprinted with permission from [13]; copyright 2020 Elsevier.

coefficient, is depicted in Fig. 6.30(a) for different solutions of isopropanol in DI water (the REF liquid is DI water). The measured transmission coefficient measured at the operating frequency of the rat-race coupler (or the frequency of the REF resonator loaded with DI water) is depicted in Fig. 6.30(b). The maximum sensitivity is found to be in this case 1.6 dB/%.

6.4 Other Differential Sensors

Many other differential microwave sensors, devoted to various applications, have been reported in the literature [24–30]. It is not the intention of this section to review all these implementations in detail, but to comment some relevant aspects that distinguish such sensors from the differential sensing structures discussed previously. The sensors reported in [24, 28] are reflective-mode differential structures based on resonant elements, where the differential output variable is the difference in the resonance frequency (and eventually the difference in the magnitude of the individual resonances) of both sensing elements. Thus, these sensors are similar to those of the previous section, the main difference being the considered output variable.

In the sensors of [29, 30], the main differential aspect concerns the fact that the sensing elements (resonators) are different, contrary to the differential sensors of the previous sections. The main relevant advantage of differential sensing is the robustness against environmental factors, which

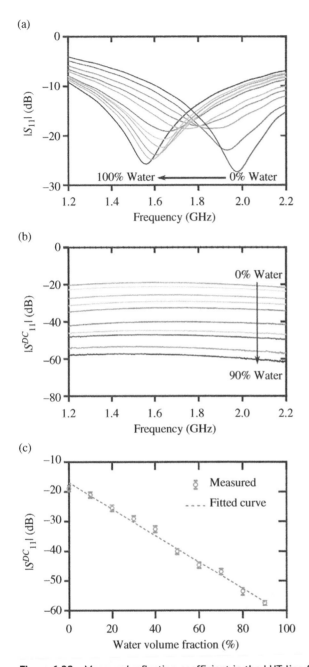

(a)

(b)

(c)

Figure 6.28 Measured reflection coefficient in the LUT line for various volume fractions of glycerol–water solutions (a), cross-mode reflection coefficient (b), and cross-mode reflection coefficient measured at 1.56 GHz, the resonance frequency of the SIR loaded with DI water (c). *Source:* Reprinted with permission from [13]; copyright 2020 Elsevier.

are seen as common mode signals, and their effects are therefore minimized by the common-mode cancelation capability of differential sensors, as discussed previously in this chapter. However, it is argued in [30] that potential alterations in the environmental parameters are monitored by the REF resonator, while the sensing resonator, devoted to the MUT (or LUT), is used to characterize the dielectric properties of the MUT. Indeed, with this mode of operation, these sensors do not actually

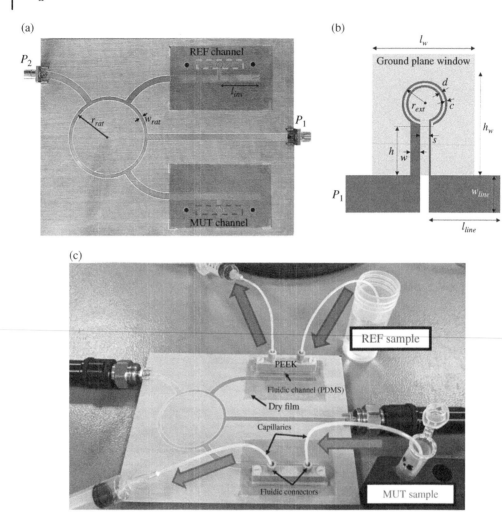

Figure 6.29 Top view of the OSRR-based reflective-mode differential microfluidic sensor excluding the fluidic parts (a), detail of the OSRR layout (b), and perspective view of the whole fabricated sensor (c). Dimensions (in mm) are: r_{ext} = 2.1, d = c = 0.2, s = 0.8, w = 0.9, h = 4.4, l_{line} = 30 and w_{line} = 3.4, l_w = 8.2, h_w = 10.8, l_{inv} = 23.473, w_{rat} = 1.8474 and r_{rat} = 23.3. Channel dimensions (in mm) are h_{ch} = 1.5 mm (height), l_{ch} = 26 mm (length), and w_{ch} = 4.6 mm (width), identical to those of Fig. 5.30. The substrate is the *Rogers RO4003C* with thickness h = 1.524 mm, dielectric constant ε_r = 3.55, and loss tangent tanδ = 0.0022. *Source*: Reprinted with permission from [14]; copyright 2020 Cambridge University Press.

behave as true comparators.[12] It is worth mentioning that the device reported in [30] is based on a splitter/combiner configuration, but contrary to the frequency-splitting sensors reported in Chapter 5, the pair of (unequal) resonator-loaded lines exhibit a bandpass behavior. The correct functionality of the sensor, with the resonance peak of one resonator not affected by the loading of the other resonator, is ensured by implementing the splitter/combiner by means of a Wilkinson configuration [31].

---

12 Actually, if the resonant elements of the sensors in [29, 30] are not identical, they cannot be considered to be true differential sensors, according to the definition given in this book.

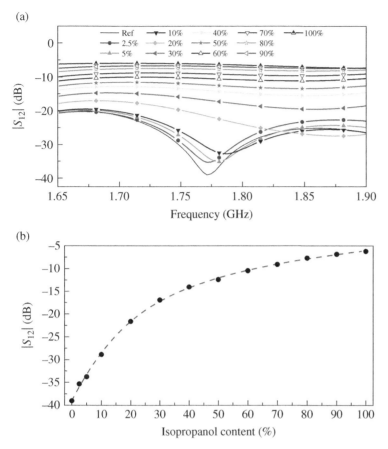

Figure 6.30 Frequency response of the differential sensor of Fig. 6.29 for different concentrations of isopropanol in DI water injected in the MUT channel, and pure DI water injected in the REF channel (a), and variation of the transmission coefficient measured at 1.77 GHz with the isopropanol content (b). *Source: Reprinted with permission from [14]; copyright 2020 Cambridge University Press.*

Special attention deserves the differential sensor reported in [26], since the working principle of this device is quite different from those working principles that govern the functionality of most differential-mode microwave sensors. In [26], differential sensing is based on quasi-microstrip mode to slot-mode conversion. Thus, the device consists of a pair of identical CSRR-loaded microstrip line segments connected through a T-junction to the input access line, see Fig. 6.31. The CSRRs are symmetrically etched with regard to the axial plane of a slotline, implemented in the ground plane of the microstrip line, as shown in Fig. 6.31. According to this configuration, if the CSRRs are symmetrically loaded, the symmetry plane is a magnetic wall, and the (odd) slot mode cannot be excited. Under these conditions, the injected input power is reflected back to the source and partially radiated, due to the effects of the CSRRs. By contrast, if symmetry is disrupted by an imbalance in the dielectric loads of the resonators, the generation of the slot mode in the slotline is expected. Obviously, mode conversion depends on the level of asymmetry. Hence, the structure of Fig. 6.31 is useful for differential sensing. To monitor the mode conversion level, a slotline/microstrip transition is added to the structure, as depicted in Fig. 6.31 [32]. Thus, the output variable in the resulting two-port differential sensor is simply the transmission coefficient between the input and the output ports. In [26], validation was carried out by comparing a REF sample with

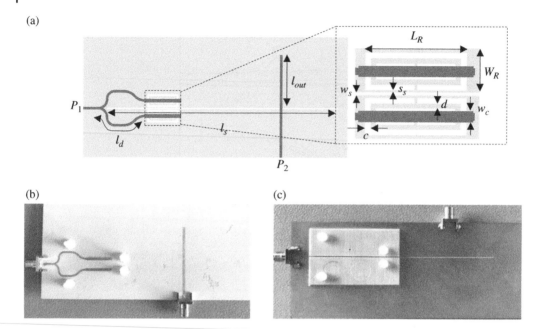

Figure 6.31 Topology of the CSRR differential sensor based on microstrip to slot-mode conversion (a), and photograph of the top (b) and bottom (c) views, in the latter case with the REF and MUT samples attached to the substrate. Dimensions (in mm) are: L_R = 14.64, W_R = 6.50, c = d = 0.80, w_c = 1.68, w_s = S_s = 0.40, l_{out} = 23.46, l_s = 107.60, l_d = 26.32. The ground plane where the CSRRs and the slotline resonator are etched is depicted in gray color in (a). The sensor substrate is the *Rogers RO3010* with dielectric constant ε_r = 10.2, thickness h = 1.27 mm, and loss tangent tanδ = 0.0027. The input and output access microstrip lines, as well as the pair of lines between the T-junction and the CSRRs, are 50-Ω lines (with width of 1.12 mm). The microstrip line sections on top of the CSRRs are wider in order to enhance the coupling with the CSRRs. *Source:* Reprinted with permission from [26]; copyright 2019 IEEE.

defected samples, identical to the REF sample, but with arrays of holes of different densities (it was shown that the device is able to detect the presence of tiny defects in the MUT samples). Moreover, the responses of the sensor by considering the REF resonator surrounded by air (i.e. $\varepsilon_{REF} = 1$) and the MUT resonator loaded with different samples were also obtained in [26], see Fig. 6.32. From the maximum value of the differential transmission coefficient magnitude, $\Delta|S_{21}|_{max} = |S_{21}|_{max,REF} - |S_{21}|_{max,MUT}$, depicted as a function of the differential dielectric constant, $\Delta\varepsilon_r = \varepsilon_{REF} - \varepsilon_{MUT}$, a calibration curve, useful for measuring the dielectric constant of unknown samples, was inferred (see Fig. 6.32b). The resulting maximum sensitivity (the one in the limit where $\varepsilon_{MUT} = \varepsilon_{REF} = 1$) is $S = 8.1$ dB, a competitive value. However, the modeling of this sensor to predict the sensitivity is not easy, since it involves mode conversion.

6.5 Advantages and Drawbacks of Differential-Mode Sensors

As discussed throughout the chapter, the main relevant advantage of differential sensors over their single-ended counterparts is their inherent capability to suppress the effects of common-mode stimuli caused, e.g. by environmental factors (such as temperature, humidity or pressure). Thus, differential sensors are robust against cross-sensitivities caused by variations in ambient conditions. Differential sensors are also of interest in certain applications where, rather than the measurement of a certain variable, the main interest is to compare a variable, or a sample, with a well-known

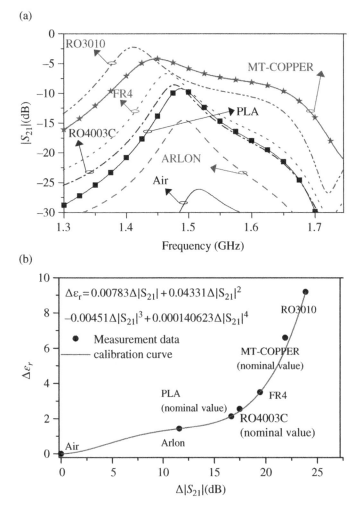

Figure 6.32 (a) Measured transmission coefficient of the sensor of Fig. 6.31 by considering air as REF material and different MUT samples, all of similar thickness (roughly 1.5 mm); (b) dependence of the differential dielectric constant with the differential transmission coefficient, and calibration curve. *Source:* Reprinted with permission from [26]; copyright 2019 IEEE.

reference. The functionality of differential sensors as comparators has been illustrated through several examples, including the detection of tiny defects in samples (see the phase-variation comparators reported in Chapter 3), the determination of volume fraction of solute in diluted solutions (based on comparison measurements), or the monitoring of electrolyte concentration in biosamples (horse urine). For the comparator functionality, sensor resolution (intimately related to sensitivity at small perturbations) is the key aspect. In general, the reported differential sensors of this chapter, as well as those reported in Chapter 3 (based on phase variation) exhibit good sensitivity and resolution, and are therefore useful as comparators.

The main drawback of differential sensors is the need of using two independent sensing elements, at least for real-time measurements, which increases sensor size and cost. Moreover, sensor size further increases if additional planar elements (such as couplers, and dividers), devoted to collect the output variable by means of two-port measurements, are included. On the other hand, it should be mentioned that symmetry imbalances caused by inaccuracies in the fabrication process tend to degrade sensor resolution, since such imbalances are seen as a differential stimulus by the sensor. Nevertheless, despite these disadvantages, differential sensors do not need, in general, a calibration process, since differential-mode operation inherently cancels potential drifts caused by ambient factors (the main cause of inaccuracies). Additionally, differential measurements with single-ended sensors are also possible. By this procedure, real-time monitoring of the variable of interest is not possible, but such an approach can be applied in many situations where external factors are controlled or do not vary during measurement. Note also that differential measurements in a two-step process by means of single-ended sensors inherently solve the above-mentioned issue concerning fabrication-related symmetry imbalances.

References

1 C. Damm, M. Schussler, M. Puentes, H. Maune, M. Maasch, and R. Jakoby, "Artificial transmission lines for high sensitive microwave sensors," *2009 IEEE Sensors,* Christchurch, New Zealand, Oct. 2009, pp.755–758.

2 F. J. Ferrández-Pastor, J. M. García-Chamizo, and M. Nieto-Hidalgo, "Electromagnetic differential measuring method: application in microstrip sensors developing," *Sensors*, vol. 17, no. 7, paper 1650, 2017.

3 J. Muñoz-Enano, P. Vélez, M. Gil, and F. Martín, "An analytical method to implement high sensitivity transmission line differential sensors for dielectric constant measurements," *IEEE Sensors J.*, vol. 20, pp. 178–184, Jan. 2020.

4 M. Gil, P. Vélez, F. Aznar, J. Muñoz-Enano, and F. Martín, "Differential sensor based on electro-inductive wave (EIW) transmission lines for dielectric constant measurements and defect detection," *IEEE Trans. Ant. Propag.*, vol. 68, pp. 1876–1886, Mar. 2020.

5 M. Gil, P. Vélez, F. Aznar, A. Mesegar, J. Muñoz-Enano, M. Duque, and F. Martín, "Electro-inductive wave transmission line based microfluidic microwave sensor," *2020 IEEE MTT-S International Microwave Biomedical Conference (IMBioC 2020)*, Toulouse, France, Dec. 2020.

6 J. Muñoz-Enano, P. Vélez, M. Gil, J. Mata-Contreras, and F. Martín, "Differential-mode to common-mode conversion detector based on rat-race couplers: analysis and application to microwave sensors and comparators," *IEEE Trans. Microw. Theory Techn.*, vol. 68, pp. 1312–1325, Apr. 2020.

7 C. Herrojo, P. Vélez, J. Muñoz-Enano, L. Su, P. Casacuberta, M. Gil, and F. Martín, "Highly sensitive defect detectors and comparators exploiting port imbalance in rat-race couplers loaded with step-impedance open-ended transmission lines," *IEEE Sensors J.*, vol. 21, pp. 26731–26745, Dec. 2021.

8 P. Vélez, K. Grenier, J. Mata-Contreras, D. Dubuc, and F. Martín, "Highly-sensitive microwave sensors based on open complementary split ring resonators (OCSRRs) for dielectric characterization and solute concentration measurements in liquids," *IEEE Access*, vol. 6, pp. 48324–48338, Dec. 2018.

9 P. Vélez, J. Muñoz-Enano, K. Grenier, J. Mata-Contreras, D. Dubuc, and F. Martín, "Split ring resonator (SRR) based microwave fluidic sensor for electrolyte concentration measurements," *IEEE Sensors J.*, vol. 19, no. 7, pp. 2562–2569, Apr. 2019.

10 P. Vélez, J. Muñoz-Enano, and F. Martín, "Electrolyte concentration measurements in DI water with 0.125 g/L resolution by means of CSRR-based structures," *2019 49th European Microwave Conference (EuMC 2019)*, Paris, France, Oct. 2019.

11 P. Vélez, J. Muñoz-Enano, M. Gil, J. Mata-Contreras, and F. Martín, "Differential microfluidic sensors based on dumbbell-shaped defect ground structures in microstrip technology: analysis, optimization, and applications," *Sensors*, vol. 19, no. 14, p. 3189, 2019.

12 J. Muñoz-Enano, P. Vélez, M. Gil, E. Jose-Cunilleras, A. Bassols, and F. Martín, "Characterization of electrolyte content in urine samples through a differential microfluidic sensor based on dumbbell-shaped defect ground structures," *Int. J. Microw. Wireless Technol.*, vol. 12, no. 9, pp. 817−824, 2020.

13 A. Ebrahimi, F. J. Tovar-López, J. Scott, and K. Ghorbani, "Differential microwave sensor for characterization of glycerol-water solutions," *Sensors Actuators B Chem.*, vol. 321, paper 128561, Oct. 2020.

14 J. Muñoz-Enano, P. Vélez, M. Gil, and F. Martín, "Microfluidic reflective-mode differential sensor based on open split ring resonators (OSRRs)," *Int. J. Microw. Wireless Technol.*, vol. 12, pp. 588−597, Sep. 2020.

15 A. Velez, F. Aznar, J. Bonache, M. C. Velázquez-Ahumada, J. Martel, and F. Martín, "Open complementary split ring resonators (OCSRRs) and their application to wideband CPW band pass filters", *IEEE Microw. Wireless Compon. Lett.*, vol. 19, pp. 197−199, Apr. 2009.

16 P. Vélez, J. Mata-Contreras, L. Su, D. Dubuc, K. Grenier, and F. Martín, "Modeling and analysis of pairs of open complementary split ring resonators (OCSRRs) for differential permittivity sensing," *2017 IEEE MTT-S International Microwave Workshop Series on Advanced Materials and Processes (IMWS-AMP 2017)*, Pavia, Italy, Sep. 2017.

17 J. Bonache, M. Gil, I. Gil, J. Garcia-García, and F. Martín, "On the electrical characteristics of complementary metamaterial resonators," *IEEE Microw. Wireless Compon. Lett.*, vol. 16, pp. 543−545, Oct. 2006.

18 F. Martín, *Artificial Transmission Lines for RF and Microwave Applications*, John Wiley, Hoboken, NJ, USA, 2015.

19 L. Su, J. Mata-Contreras, P. Vélez, A. Fernández-Prieto, and F. Martín, "Analytical method to estimate the complex permittivity of oil samples," *Sensors*, vol. 18, p. 984, 2018.

20 O. Weiner, "Die theorie des mischkorpers fur das feld der statonare stromung i. die mittelwertsatze fur kraft, polarisation und energie," *Trans. Math.-Phys. Class Roy. Saxon Soc. Sci.*, vol. 32, pp. 509−604, 1912.

21 D. Ahn, J. S. Park, C. S. Kim, Y. Qian, and T. Itoh, "A design of the low-pass filter using the novel microstrip defected ground structure," *IEEE Trans. Microw. Theory Tech.*, vol. 49, no. 1, pp. 86−93, Jan. 2001.

22 J. Martel, R. Marqués, F. Falcone, J. D. Baena, F. Medina, F. Martín, and M. Sorolla, "A new LC series element for compact band pass filter design," *IEEE Microw. Wireless Compon. Lett.*, vol. 14, pp. 210−212, May 2004.

23 J. Naqui, M. Durán-Sindreu, J. Bonache, and F. Martín, "Implementation of shunt connected series resonators through stepped-impedance shunt stubs: analysis and limitations," *IET Microw. Ant. Propag.*, vol. 5, pp. 1336−1342, Aug. 2011.

24 A. Ebrahimi, J. Scott, and K. Ghorbani, "Transmission lines terminated with LC resonators for differential permittivity sensing," *IEEE Microwave and Wireless Components Letters*, vol. 28, no. 12, pp. 1149−1151, Dec. 2018.

25 H. Hallil, P. Bahoumina, K. Pieper, J. L. Lachaud, D. Rebière, A. Abdelghani, K. Frigui, S. Bila, D. Baillargeat, Q. Zhang, P. Coquet, E. Pichonat, H. Happy, and C. Dejous, "Differential passive microwave planar resonator-based sensor for chemical particle detection in polluted environments," *IEEE Sensors Journal*, vol. 19, no. 4, pp. 1346−1353, Feb. 2019.

26 P. Vélez, J. Muñoz-Enano, and F. Martín, "Differential sensing based on quasi-microstrip-mode to slot-mode conversion," *IEEE Microw. Wireless Compon. Lett.*, vol. 29, pp. 690−692, Oct. 2019.

27 B. Camli, E. Altinagac, H. Kizil, H. Torun, G. Dundar, and A. D. Yalcinkaya, "Gold-on-glass microwave split-ring resonators with PDMS microchannels for differential measurement in microfluidic sensing," *Biomicrofluidics*, vol. 14, no. 5, paper 054102, Sep. 2020.

28 H. Y. Gan, W. S. Zhao, Q. Liu, D. W. Wang, L. Dong, G. Wang, and W. Y. Yin, "Differential microwave microfluidic sensor based on microstrip complementary split-ring resonator (MCSRR) structure," *IEEE Sensors J.*, vol. 20, no. 11, pp. 5876–5884, Jun. 2020.

29 M. C. Jain, A. V. Nadaraja, B. M. Vizcaino, D. J. Roberts, and M. H. Zarifi, "Differential microwave resonator sensor reveals glucose-dependent growth profile of *E. coli* on solid agar," *IEEE Microw. Wireless Compon. Lett.*, vol. 30, no. 5, pp. 531–534, May. 2020.

30 S. Mohammadi and M. H. Zarifi, "Differential microwave resonator sensor for real-time monitoring of volatile organic compounds," *IEEE Sensors J.*, vol. 21, no. 5, pp. 6105–6114, Mar. 2021.

31 D. M. Pozar, *Microwave Engineering*, 4th ed., John Wiley, Hoboken, NJ, USA, 2011.

32 X. Guo, L. Zhu, K. W. Tam, and W. Wu, "Wideband differential bandpass filters on multimode slotline resonator with intrinsic common-mode suppression," *IEEE Trans. Microw Theory Techn.*, vol. 63, pp. 1587–1594, May. 2015.

7

RFID Sensors for IoT Applications

The previous chapters have been mainly devoted to the study and analysis of several strategies for sensing using microwaves and planar structures. The main objectives have been to present the working principles, to discuss those aspects that are influential on sensor performance (sensitivity, resolution, dynamic range, linearity, accuracy, etc.), and to report several prototype examples. Special effort has been dedicated to the analysis of the reported sensors (at least in most cases) on the basis of the underlying physics and equivalent circuits, with the main goal of obtaining useful guidelines for sensor performance optimization (typically, although not exclusively, to boost up the sensitivity). In some cases, the associated sensor electronics needed for post-processing has also been discussed. However, the main effort in this book has been dedicated to the analysis and study of the sensing elements (e.g. electrically small planar resonators or transmission line sections, conveniently excited) and the effects of the measurand (e.g. a displacement or a change in the composition of a sample, among others) on them. Such sensing elements plus the structures needed for their excitation (at microwave frequencies in the considered devices) constitute the electromagnetic module of the sensor, the main subject of this book. Nevertheless, within today's paradigm of the Internet of Things (IoT) [1–4], hundreds (or even thousands) of sensors should provide information of the surrounding environment (or monitor the evolution of certain variables of interest) wirelessly and in real time. Thus, providing the capability of wireless data communication and, even further, wireless feeding (or wireless power transfer – WPT) to the sensors becomes a real need in many applications.

In traditional wireless sensor networks (WSNs), the sensor nodes consist of a sensing element (or a transducer, able to convert the parameter of interest in an electrical or electromagnetic variable), a microcontroller, a radiofrequency (RF)/microwave transceiver (for communication purposes), and a power source, typically a battery. The main drawback of such architecture is the presence of the battery. Since batteries have a limited lifetime, the implementation of WSNs with many sensor nodes does not seem to be realistic (moreover, batteries add an extra cost to the sensor nodes). Many research efforts are dedicated nowadays to the topic of energy scavenging, or harvesting (e.g. from ambient power sources) [5]. Potentially, by collecting natural energy by means of dedicated circuits, it is possible to self-power the wireless sensor nodes. Nevertheless, solutions with simultaneous wireless information and power transfer (SWIPT), such as the one offered by passive (i.e. battery less) radiofrequency identification (RFID) technology [6–9], are in general preferred over those based on energy harvesting. Thus, a class of wireless sensors, designated as RFID-enabled sensors [10–22], emerged in parallel to the deployment of the RFID technology. There are commercially available RFID sensors, or RFID sensing tags, with relatively high cost. This justifies and explains the significant efforts worldwide focused on the research and development of low-cost

Planar Microwave Sensors, First Edition. Ferran Martín, Paris Vélez, Jonathan Muñoz-Enano, and Lijuan Su.
© 2023 John Wiley & Sons, Inc. Published 2023 by John Wiley & Sons, Inc.

RFID-enabled sensors, specially, although not exclusively, sensors based on the so-called chipless-RFID concept [23–35]. Without the presence of the integrated circuit (IC), chipless-RFID sensing tags can be implemented in full-planar technology, compatible with additive processes [30], and "green" sensing nodes can be envisaged [13].[1] Nevertheless, it should be mentioned that the advantageous aspects related to the absence of batteries and ICs in chipless-RFID sensing tags (mainly, a reduced cost, a longer lifetime, as well as the possibility to implement eco-friendly devices) have a negative counterpart, i.e. a limited sensor performance (typically, less accuracy, data storage capability, and read distance).[2] Thus, there is still plenty of room for research activities in the field of RFID sensors. Indeed, many research efforts devoted to improving the performance of chipless-RFID sensors and labels are currently being done, with the aim of pushing this low-cost and potentially "green" technology toward the market.

In this chapter, a succinct review of RFID-based sensors, considering both chip-based and chipless-based solutions, is carried out. A classification scheme of the main sensor types is reported, and the sensor working principle for each case is discussed. Specific examples, representative of the potential of the RFID sensing approach, are included in the chapter. Moreover, the chapter reports the main applications of RFID sensors within the IoT world, points out some commercial solutions, and briefly discusses the processes and materials that are compatible with the implementation of environmentally friendly RFID sensing tags. Within this framework, a first section should be dedicated to briefly introduce the fundamentals of RFID.

7.1 Fundamentals of RFID

RFID is a wireless technology devoted to identification, tracking, and authentication of items, animals, persons, consumer products, etc. [6–9]. RFID belongs to the category of automatic identification and data capture (AIDC) systems, similarly to optical barcodes (or QR codes), voice recognition systems, biometric systems, smart cards, optical character recognition (OCR), etc. In an RFID system, the items of interest are equipped with a tag containing the identification (ID) code. In chip-based tags, those that are commercially available and profusely utilized, the information is stored in an IC, or chip. By contrast, in chipless-RFID tags the data is contained in a printed encoder, typically, although not exclusively, consisting of a set of planar resonant elements, functional or detuned, depending on the logic state associated with them.

The wireless communication between the tag and the reader proceeds either by far field or by means of near-field coupling. For chip-based RFID, in those systems operating at the HF (13.56 MHz) band (HF-RFID), the communication between the tag and the reader proceeds by means of inductive coupling and the tags consist of the chip plus a loop antenna (Fig. 7.1a). In these HF-RFID systems, the communication is contactless, but proximity (typically in the sub-cm scale) between the tag and the reader is required, similar to near-field communication (NFC) systems. By contrast, in RFID systems operating in the UHF band, between 840 and 960 MHz (the specific working frequency depends on the regulation of each world region [36]), the communication between the UHF-RFID tag (Fig. 7.1b) and the reader proceeds by far-field backscattering (Fig. 7.2).

Typically, HF- and UHF-RFID tags are powered by the interrogation signal (passive tags) [37]. However, in certain RFID systems, where very large read distances are required (i.e. above 10 m),

---

1 In the framework of RFID technology, strictly speaking, "green" sensing tags are battery-less and chipless-RFID sensors implemented on recyclable or biodegradable substrates with organic inks (e.g. graphene- or carbon-based inks) and organic sensing films, if they are present.
2 It is also well known that chipless-RFID tags merely devoted to identification, tracking, and authentication, i.e. without sensing capability, do not exhibit a performance comparable with the one of the chip-based counterparts.

(b)

(a)

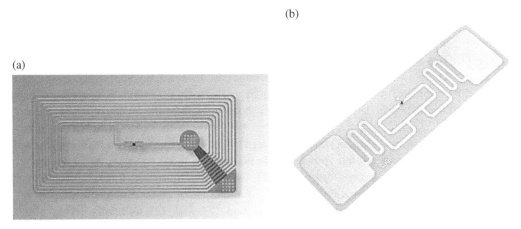

Figure 7.1 Photograph of a commercial near-field HF-RFID chipped tag (a) and far-field UHF-RFID chipped tag (b).

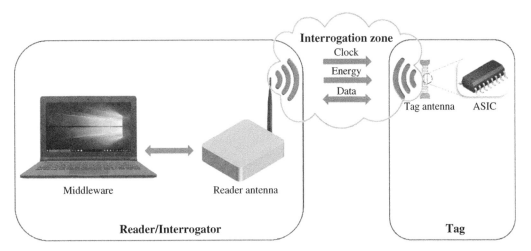

Figure 7.2 Schematic of a far-field backscattering chip-based UHF-RFID system. *Source:* Reprinted with permission from [21]; copyright 2020 Springer.

e.g. for toll collection or real-time location tracking, among others, active tags (i.e. equipped with batteries) are needed [38]. Such active tags typically operate at the 2.4 GHz microwave frequency band (nevertheless, although less common, there are also passive tags operating at this frequency, as well as active tags operating at the UHF band). Let us also mention that there are RFID tags classified as semi-passive (or battery-assisted passive – BAP) [39–42]. The difference with active tags is that in BAP tags, the battery is exclusively devoted to aid the communication with the reader by powering the chip, and the communication relies on a backscattering scheme, where the tag responds to the interrogation signal sent by the reader by reflecting and modulating these waves. Thus, the functionality of BAP RFID is very similar to the one of purely passive systems. By contrast, purely active tags contain not only the battery but also a transmitter, and these two components are used to broadcast the information to the interrogator. Indeed, in many active RFID systems, broadcasting does not depend on whether the interrogator is present or not (in this case, the tags are called beacon tags). Nevertheless, in certain active RFID systems, the tags act as transponders responding to an interrogation query.

Concerning chipless-RFID systems [23–25], this technology emerged as a low-cost alternative to chipped-RFID. The absence of the IC represents a very significant reduction of the cost of the tag, but at the expense of larger size, shorter read distances and, in general, lower data storage capability. Except surface acoustic wave (SAW) based tags [43–49], which can be considered to belong to the category of chipless-RFID tags, but with significant cost, there are no commercially available chipless-RFID systems. The research activity focused on improving the limitative aspects of such chipless-RFID technology is significant, in part due to the interest for the implementations of low-cost "green" systems, as mentioned before. Chipless-RFID tags neither contain a chip (the information is stored in a printed encoder) nor are powered by batteries. There are two main approaches for the implementation of printed chipless-RFID systems: (i) those operating in time domain [50–59] and (ii) frequency-domain-based tags [23, 24, 60–76]. Moreover, there are frequency-domain tags that exploit additional domains, such as amplitude or phase, among others, in order to enhance the data density of the tags (hybrid tags [77–90]). Let us briefly review these approaches next.

In most time-domain RFID systems based on printed chipless-RFID tags reported so far, encoding is achieved by means of a delay line loaded with reflectors at specific positions determined by the ID code. Thus, the tag consists of a delay line plus an antenna used for communication with the reader. The working principle of such systems is time-domain reflectometry (TDR) (see Fig. 7.3).[3]

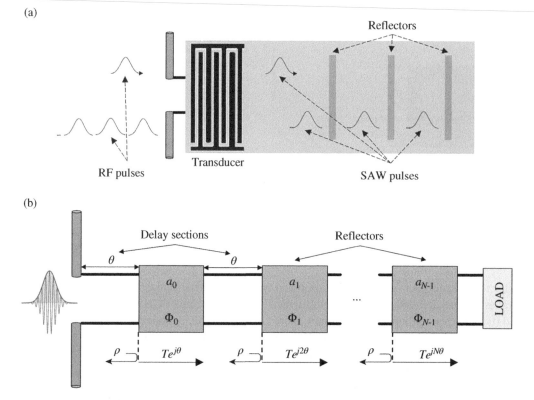

Figure 7.3 Working principle of a TDR-based chipless-RFID system. (a) Using SAW technology; (b) using delay lines. *Source:* Reprinted with permission from [21]; copyright 2020 Springer.

---

3 SAW-based chipless-RFID tags also exploit TDR.

Specifically, the ID code is contained in the echoes generated by the delay line, in response to a pulsed interrogation signal in time domain. TDR-based tags, as such tags are usually referred to, exhibit fast responses compared with frequency-domain tags, but their bit-encoding capability is limited, and large delay lines or very narrow pulses are needed to avoid overlapping of the reflected pulses. The exception are SAW-based tags, which operate under the same principle, but exhibit competitive performances (data storage capacity) by virtue of the electro-acoustic transducers. Nevertheless, such high data capacity is achieved at the expense of a higher cost, as previously indicated.

For the reasons explained in the precedent paragraph, most fully planar chipless-RFID tags are implemented on the basis of a frequency-domain (rather than time-domain) working principle. Frequency-domain-based tags, also known as spectral signature barcodes [23, 24], consist of a set of resonant elements, each tuned to a different frequency. Typically, each resonant element provides a bit of information, and the corresponding logic state, "1" or "0," depends on whether the resonant element is functional or detuned (inoperative) at the fundamental frequency. Tag reading proceeds by sending a multifrequency interrogation signal (covering the spectral bandwidth) to the tag, and the ID code is given by the presence or absence of abrupt spectral features in the amplitude, phase, and group-delay responses. Like TDR-based tags, spectral signature barcodes are read through far field, and there are two main types of tags, namely retransmission based [61, 62] and backscattered based [60, 67] (Fig. 7.4). In retransmission tags, a transmission line is loaded with the resonant elements constituting the printed encoder and terminated by two cross-polarized antennas for reception/transmission from/to the reader. In backscattered chipless-RFID tags, the resonant elements provide the spectral signature through the peaks in their radar cross section (RCS) response, and antennas are avoided, hence reducing the size of the tags. Despite the fact that a considerable number of bits (35 bits) in spectral-signature chipless-RFID tags has been reported [61] (see Fig. 7.5), the required spectral bandwidth for tag reading is very wide.[4] This represents a penalty in terms of the reader costs in a real scenario, where the sweeping interrogation signal must be generated by means of a voltage-controlled oscillator (VCO), rather than by using a vector network analyzer (VNA), the usual procedure in proof-of-concept demonstrators.

Strategies to increase the bit density per frequency (DPF) exploiting more than one domain simultaneously have been reported in the literature [77–90]. In hybrid tags, the designation given to these multi-domain chipless-RFID labels, the main aim is to increase the number of bits by providing more than two logic states per resonant element, using two independent parameters (or domains) simultaneously for coding. By that method, the DPF and the data density per surface (DPS), two figures of merit in chipless tags, can be effectively improved. Various proposals for hybrid chipless-RFID tags can be found in the literature, including tags where the frequency position is combined with the phase deviation [77], polarization diversity [78], and notch bandwidth [89], as well as tags where the frequency is combined with the peak [82, 83] and notch [84, 85] magnitude, among others [79–81, 86–88, 90]. Hybrid tags use frequency for coding; therefore, such tags can be considered to belong to the class of frequency-domain tags, as well.

Although the number of bits achievable with hybrid tags can potentially be improved (a 64-bit tag was reported in [79]), the read distances achievable with these far-field tags, including also purely frequency-domain retransmission and backscattering-based tags, are very small (typically few dozens of centimeters, in the best of the cases). Such read ranges cannot compete against those

---

4 Actually, such 35-bit chipless-RFID tag was implemented in a low-loss metallized microwave substrate by means of an etching process on the Cu layer. The tag performance, including the number of bits, achievable by implementing such tags with printing processes on low-cost flexible substrates by means of conducive or organic inks is expected to be significantly degraded.

(a)

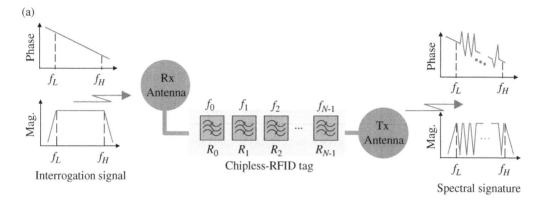

(b)

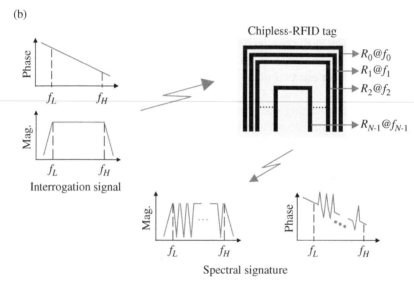

Figure 7.4 Working principle of frequency-domain retransmission-based (a) and backscattering-based (b) chipless-RFID systems. *Source:* Reprinted with permission from [21]; copyright 2020 Springer.

of chip-based UHF-RFID passive tags (up to various meters), with well-stablished communication protocols and standards (that make possible the interoperability of RFID products), as well as modulation and encoding schemes [6–9, 36]. Particularly, the EPC UHF Gen 2 or ISO/IEC 18000-63 is the considered global standard protocol for RFID. The identification of the ID code, generally called electronic product code (EPC) in UHF-RFID, is achieved by means of a sequence of commands sent between the reader and the tag. In passive UHF-RFID tags, the reader initiates a round of interrogation with a query command. This query command "wakes up" the tag, which answers with the information that it has stored. The query command that the reader sends is a message encoded in a wave, modulated by means of amplitude shift keying (ASK). This wave propagates through the air, and when it reaches the tag antenna, the RF power is converted into DC power through a rectifier circuit, so that the antenna reflection coefficient is modulated. The wave is modulated with the encoded information of the chip and backscattered from the tag to the interrogator. The encoding system to transmit the information is the pulse interval encoding, similar to a Morse code, i.e. it uses short and long pauses to represent the "0" and "1" logic states [91]. In order to avoid collision

(a)

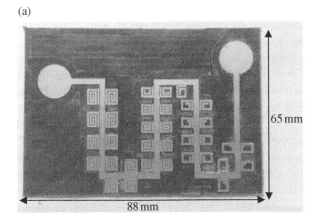

65 mm

88 mm

(b)

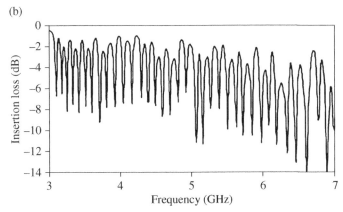

Figure 7.5 Photograph (a) and measured response (b) of a 35-bit retransmission-based chipless-RFID tag implemented by means of spiral resonators, where all bits are set to "1." *Source:* Reprinted with permission from [61]; copyright 2009 IEEE.

between tags that answer at the same time, the readers are provided with anti-collision algorithms. In practice, this means that quasi-simultaneous tag reading is possible, a capability that is not possible with chipless-RFID systems.

Thanks to the previously mentioned communication protocols, modulation and encoding schemes, and anti-collision algorithms, relatively long read distances can be achieved in chipped-UHF-RFID systems. By contrast, effects such as clutter, shadowing, environmental reflections, and multi-tag interference prevent from the achievement of long read ranges in backscattering- or retransmission-based chipless-RFID. In certain applications, however, where the read distance is not a critical issue, the low cost of chipless-RFID tags may justify the adoption of this technology. Particularly, chipless-RFID can substitute chipped-HF-RFID systems operating in the near field, thereby forced to situate the interrogator and the tag at short distances. However, a key aspect is to achieve a competitive number of bits, sufficient for the considered application. In this regard, let us mention that a near-field chipless-RFID approach with an unprecedented number of bits was presented in [92] (and subsequently further developed [21, 34, 93–104]). The system works identical to the position sensors based on electromagnetic encoders reported in Chapter 4. The difference is that for chipless-RFID applications, the encoders should be printed by means of conductive inks on a low-cost (typically) flexible substrate, including plastic or organic (e.g. paper) substrates. Thus, the tags consist of a chain of printed metallic inclusions, and tag reading proceeds by proximity between the tag and the reader, sequentially, i.e. bit to bit, in a time-division multiplexing scheme. In this time-domain chipless-RFID system, the working principle is completely different from that of TDR-based tags. Encoding is achieved by the presence or absence of functional metallic inclusions, typically resonant elements, at predefined position in the tag chain. For tag reading, an element able to discern the functional and inoperative inclusions is needed, and for that purpose, a transmission line loaded with resonant elements or with other elements sensitive to the presence of functional tag inclusions on top of it is typically used, as discussed in Chapter 4 (see Fig. 7.6). One relevant advantage of this chipless-RFID system is that the ID code is contained in the amplitude modulation (AM) signal generated by tag motion over the sensitive part of the reader in response to a harmonic interrogation signal injected to the transmission line of the reader. That is, a single-tone signal suffices for tag reading, contrary to other chipless systems, which require broadband signals for reading purposes. Since the ID code is contained in the time-domain envelope function, these chipless-RFID tags have been designated as time-domain signature barcodes, in parallelism with the spectral signature barcodes of frequency-domain systems [21].

The number of bits achievable with time-domain signature barcodes is only limited by tag size. Thus, by reducing the period of the tag chain to small values, a relatively high number of bits can be achieved with small-sized tags (Fig. 7.7). Nevertheless, tag reading requires mechanical motion between the tag and the reader. With these characteristics, time-domain signature barcodes are considered very useful in secure paper applications, or authentication of premium products, where tag reading by proximity (and alignment), such as these systems require, does not seem to represent an issue. For example, as it was demonstrated in [21], time-domain signature barcodes, similar to those of Fig. 7.7, printed in Spanish medical prescriptions (a paper of poor quality) are correctly read with the system developed in [97, 98], the one used to obtain the results of Fig. 7.7. Indeed, reading at short distance may be interesting in scenarios where confidence against spying or eavesdropping is a requirement. Due to the similarity between these chipless-RFID systems based on time-domain signature barcodes and the displacement sensors based on electromagnetic encoders reported in Chapter 4, it could be reasonable to accept that those sensors are indeed RFID sensors. However, we have not included the electromagnetic encoder-based displacement sensors in this chapter since the encoder (not necessarily printed) cannot be considered to be a true chipless-RFID tag (despite the fact that in the quasi-absolute electromagnetic encoders of Chapter 4, an ID code to determine the absolute position is necessary).

To end this section, let us briefly compare in the subsequent paragraphs chipped-RFID, chipless-RFID, and optical barcodes (or QR codes). As compared with optical barcodes or QR codes, chip-based RFID tags can store significantly much more information, enabling the identification of individual items. Chip-based RFID tags do not require a direct line of sight with the reader (except HF-RFID tags, operating in the near field), and reading distances of several meters are possible with far-field passive tags operating in the UHF band (UHF-RFID tags). Such distances can be further extended to dozens of meters with battery-equipped tags (active or BAP tags), although with higher cost and size. Despite the fact that passive UHF-RFID tags are relatively cheap (several US$ cents), their use is prohibitive in many applications involving low-cost items. The tag price is dictated by the presence and placement of the IC (by contrast, the tag size is determined by the antenna), and it is difficult to envision a future scenario (at least, at short term) where the price of chipped tags drops to US$0.01 or below.

As indicated previously, chipless RFID emerged as an alternative to chip-based RFID for partially alleviating the high cost of silicon chips. By replacing ICs with encoders, the tag cost can be dramatically reduced, and reasonable predictions forecast that the price of massively manufactured chipless-RFID tags will fall below US$0.01 [23]. The tag price is dictated by the (progressively

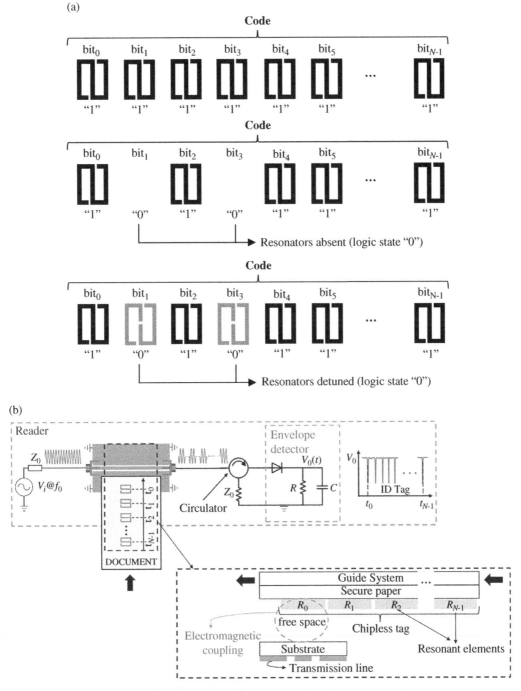

Figure 7.6 (a) Time-domain signature barcodes based on linear chains of resonant elements (S-shaped split-ring resonators – S-SRRs – in the considered example) with all resonant elements present and functional (upper chain), with some resonators absent (central chain), and with all resonant elements present but some of them detuned (lower chain). (b) Sketch showing the working principle of the near-field chipless-RFID system based on time-domain signature barcodes. In (a), the bits are indicated. Note that the same effect ("0") is achieved by the absence of resonant element or by resonator detuning (achieved by cutting it). *Source:* Reprinted with permission from [21]; copyright 2020 Springer.

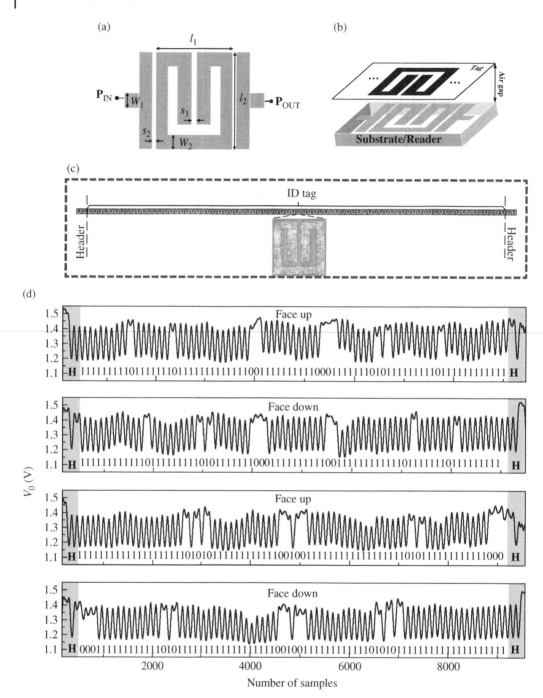

Figure 7.7 Chipless-RFID system based on time-domain signature barcodes implemented by means of SRRs. (a) Detail of the sensitive part of the reader, a SRR-loaded microstrip line in bandpass configuration; (b) scheme of the sensitive part of the reader with a functional tag inclusion (also a SRR resonator, but oppositely oriented) on top of it; (c) photograph of a 80-bit tag with header bits inkjet-printed in ordinary paper; (d) envelope functions corresponding to two different ID codes read with the tag faced up or down (with regard to the reader). Reader dimensions (in mm) are l_1 = 3.16, l_2 = 3.35, s_1 = 0.2, s_2 = 0.2, W_1 = 0.56, and W_2 = 0.5. Tag dimensions are length 267.2 mm and width 3.35 mm. The estimated air gap is 0.25 mm. The substrate of the reader line is the *Rogers RO3010* with thickness h = 0.635 mm and dielectric constant ε_r = 10.2. *Source:* Reprinted with permission from [97, 98]; copyright 2018 IEEE and MDPI.

Table 7.1 Comparison between chipless-RFID tags, chipped-RFID tags, and optical ID codes.

	Optical ID codes	Chipless RFID	Chipped RFID
Cost	Ultra low	Low	Medium
Size	Very small	Large	Medium
Read range	Very small	Small/moderate	High
Data storage	Medium	Medium	High
Simultaneous reading	No	No	Yes
Reprogrammable	No	Yes (with limits)	Yes
Single band operation	Yes	No	Yes
Power level (reader)	—	Low	Moderate
Harsh environment	No	Yes (with limits)	No
Easy to copy	Yes	No	No

decreasing) cost of conductive inks and the printing-fabrication processes (rotogravure, rotary and planar screen-printing, offset, and inkjet). The final cost is determined mainly by the amount of ink the encoder requires (which is, in turn, related to the data-storage capability or number of bits) and by the tag antenna, if it is present.

Chipless-RFID tags fall between chip-based tags and optical barcodes (or QR codes) in terms of performance, that is the read range,[5] and cost. Table 7.1 provides a comparison between optical and RFID (chip and chipless) technologies that includes performance, cost, size, and other aspects relevant in certain applications [34]. According to Table 7.1, chipless-RFID technology is inferior to the chip-based counterpart concerning tag storage capacity, size, and read range (exception to this are the time-domain signature barcodes, with potentially very high data capacity, but with extremely low read ranges, as discussed). Progress in any of these areas, combined with the low tag cost, low power needed by the reader, robustness against harsh environments (due to the absence of electronic circuits), and potential for the implementation of "green" tags, may push chipless-RFID technology toward a progressive penetration into the market.

An important aspect highlighted in Table 7.1, related to tag costs, concerns writing and erasing the tags, a feature not possible in optical barcodes and QR codes. By massively printing identical chipless tags and programming them in a later stage, tag-manufacturing expenses can be significantly reduced, since a single mask is needed for such massive (all-identical) tag printing. Tag programming (and erasing) is possible under certain circumstances, and it is believed to be a key factor for the potential success of chipless RFID in the future. For example, if the ID code depends on the shape of the elements forming the encoder, it does not seem reasonable to massively manufacture identical tags and write the ID code in a later stage. By contrast, using tags as those of Fig. 7.6, where a simple cut suffices to detune the resonant elements, tag programing (and erasing) is viable.[6]

---

5 The time-domain signature barcodes are an exception, since these chipless tags require proximity and alignment with the reader, as indicated in the text.
6 One possible scenario would be to massively print all the tags with their resonant elements cut (detuned) and then program (write) the tags by means of inkjet-printing, by adding conductive ink in the cuts of those resonators whose binary state should be switched. To erase the tags, the cuts should be recovered (a process that can be potentially achieved, e.g. by laser ablation).

Nevertheless, it is not realistic to write/erase the tags' ID code as many times as with chip-based tags. Taking into account these limitations, this aspect has been marked "Yes (with limits)" in Table 7.1.

Another important issue, which is very sensitive in applications devoted to security and authentication, is copying and plagiarism. Optical barcodes (or QR codes) can simply be photocopied, and the copies have the same functionality as the originals. Therefore, low-cost copying is possible in optical ID systems; by contrast, copying RFID tags (chipless and chipped) is possible, but more sophisticated (and, hence, more expensive) systems are necessary. In particular, chipless-RFID tags based on printed conductive inks can be reverse-engineered (unless the encoders are buried) and consequently can be reproduced. However, to that end, high-cost printers, conductive inks, and custom printing processes are required. On the other hand, although photocopied chipless-RFID tags contain the ID code, tag reading cannot be achieved by means of a specific (dedicated) chipless-RFID reader (able to read only the ink-based printed tags). Thus, the counterfeiting of items and goods is prevented unless high-cost systems are used for the copying.

Further differences between the considered ID technologies concern simultaneity of tag reading, the power required by the reader, the bandwidth of the interrogation signal, and robustness against hostile environmental conditions. Simultaneous tag reading is only possible in chip-based RFID systems, by virtue of the anti-collision algorithms, as mentioned. Concerning the reader's required power level, it can be considered to be moderate in chipped-RFID systems, as long as a minimum power level is needed to activate the tag chip. By contrast, in chipless-RFID systems, especially in those systems operating in near field that use time-domain signature barcodes, the power level required by the reader is small. Concerning the bandwidth of the interrogation signal, it is narrow in optical systems and chipped RFID, whereas in most chipless-RFID systems, wideband signals are needed for tag reading. The exception is the chipless-RFID system based on time-domain signature barcodes, where a single-tone (harmonic) signal suffices for reading. Finally, for tag operation in harsh environments, it should be noted that silicon ICs exhibit a limited robustness against extreme ambient factors (temperature, humidity, and radiation). Therefore, chipless-RFID tags can be considered (in general) superior in that aspect, although conductive-ink properties may degrade if the ink is subjected to extreme conditions. In certain environments, subjected to pollution, the functionality of optical barcodes and QR codes may be limited, contrary to the superior robustness of RF systems against dirtiness.

This section has been limited to present and compare the different RFID systems and to qualitatively explain their behavior. For a detailed analysis of the different RFID approaches, out of the scope of this manuscript, there are many sources, including books and journal papers, quoted in the preceding paragraphs. The main interest has been to briefly explain the RFID fundamentals in order to provide the necessary information to understand the behavior of the different RFID sensing strategies and systems, to be discussed in the next section.

7.2 Strategies for RFID Sensing

RFID sensors constitute a new paradigm for the IoT. RFID sensor tags exhibit unique features, including wireless sensing, wireless information and power transfer, lightweight and low profile, compatibility with additive manufacturing processes, non-line-of-sight transmission, negligible maintenance, and environmental sustainability,[7] which are essential for adopting this RFID-sensing technology in IoT applications (e.g. manufacturing, logistics, healthcare, agriculture and

---

7 Let us clarify that, in general, not all these features are simultaneously present in RFID sensors.

food, civil engineering, automotive industry, and smart cities). In this section, the state-of-the-art in RFID sensor technology is reviewed, with a focus on the system/architecture implementation perspective (other excellent reviews on this topic can be found in [19, 20, 22]). There are commercially available RFID sensors, with relatively high cost, based on chipped tags. However, there is a huge research activity on the topic aimed to address the current challenges, mainly relative to size and cost reduction, performance optimization, and compatibility with the "green" paradigm (i.e. the implementation of fully recyclable/biodegradable sensor tags).

RFID sensors can be classified according to various criteria, for example according to their operating frequency band or according to their application, among others. However, following the classification scheme proposed in the review paper [20], two main groups of RFID sensors can be distinguished, i.e. chip-based and chipless-based sensors. Nevertheless, before analyzing and discussing the main architectures and specific topologies of each class, let us briefly present the architecture and main characteristics of the sensing nodes in WSNs, as well as the main limitative aspects [105]. Such nodes consist of the sensing element, a microcontroller, a transceiver, and a power source (Fig. 7.8). Typically, the element representing the major contribution to energy-consumption is the transceiver. Therefore, it is very important to choose a low-power-consumption transceiver connecting the sensor node to the network. By this means, energy consumption is optimized, thereby extending the battery (and sensor) lifetime. Nevertheless, the need for batteries in WSNs is a handicap for two main reasons: (i) batteries are, in general, not cheap, and, most important, (ii) in WSNs with many nodes (hundreds or thousands), the continuous need of battery replacement, if possible, may jeopardize the viability of the network. For these reasons, sensor schemes that collect energy from the ambient or wirelessly from the communications network (RF power) seem to be the solution to this problem. In this regard, RFID technology with sensing capabilities (or RFID-enabled sensors) can represent the turn-around point for the massive deployment of the IoT [20].

Figure 7.9 depicts the classification of RFID sensors, as given by [20], where the two main groups, chip-based and chipless-based, are further divided into subgroups. In opinion of the authors of this book, such classification is very pertinent and convenient because it obeys to the working principle of the sensor, including the communication with the interrogator, and therefore it provides a natural path for analyzing the different sensor types, to be discussed next.

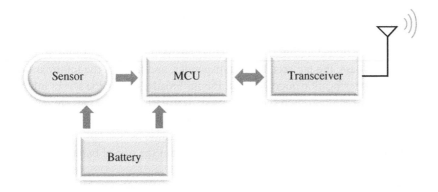

Figure 7.8 Sketch of the system architecture of a sensor node in a wireless sensor network.

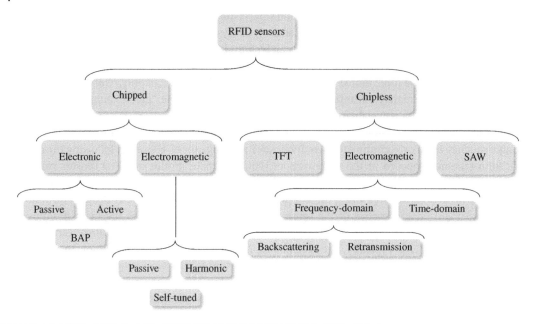

Figure 7.9 Classification of the RFID sensors.

7.2.1 Chip-Based RFID Sensors

According to Fig. 7.9, there are two types of chip-based RFID sensors, the so-called electronic sensors and the electromagnetic sensors. Both sensor types contain the same functional units (see Fig. 7.10), but sensing proceeds by means of radically different schemes. In electronic sensors, the sensing unit interacts with the IC, whereas in the electromagnetic configuration, the sensing operation is based on the change in the tag frequency response, achieved by modifying the characteristics of the antenna [20].

7.2.1.1 Electronic Sensors

The main relevant and distinctive characteristic of the so-called electronic RFID sensor tags is the separation between the sensing and communication functions. That is, a specific sensor module (an electronic sensor) able to monitor certain variables, such as temperature, humidity, or strain, to cite some of them, is either integrated inside the chip or interfaced through an external microcontroller. The sensed information is digitally encoded and transmitted to the reader together with the ID code. In order to achieve a significant read range, the energy efficiency of the tag is a key issue. Eventually, a battery might be included within the tag (battery-assisted tag) to boost up the read range, if it is necessary. In this case, the tag acts like a transponder, and it sends the information to the reader in response to an interrogation query. If a continuous broadcasting of the sensor information is required, then an active configuration should be used. Active RFID sensor tags are robust and efficient, exhibiting read ranges up to hundreds of meters in some cases, but they are expensive and the lifetime of the batteries is limited as far as such sensors operate as beacons, rather than transponders.[8] Obviously, the read range of active tags is higher than the one of battery-assisted

---

8 In active beacon RFID sensor tags, rather than waiting to "hear" the interrogator's signal, the tags "beacon," or send out, their specific information every certain periods of time. By contrast, RFID tags operating as transponders (passive tags, battery-assisted tags, and some active tags) send out their information upon receiving the reader's signal.

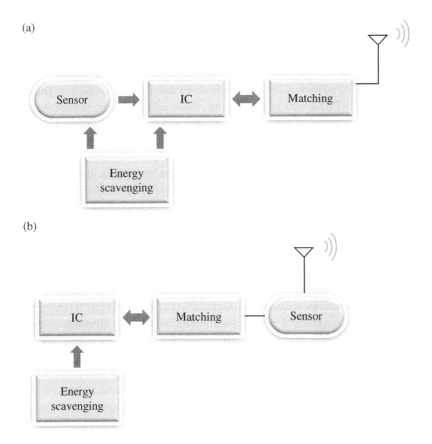

Figure 7.10 Sketch of the system architecture of an electronic (a) and electromagnetic (b) chip-based and battery-free RFID sensor operating in far field. Energy scavenging from ambient sources (e.g. light) is optional (alternatively, the IC and sensor module are powered by the interrogation signal). In (b), the sensor is an integral part of the antenna.

tags, in turn higher than the one of passive tags, but the cost follows the opposite trend. Thus, choosing a specific RFID electronic sensor (there are many sensors of this type in the market) is a trade-off between cost and read range. To give some indicative numbers, certain active RFID sensor tags with severe specifications and able to operate in harsh conditions, as encountered in many scenarios, may cost US$100, or even more, but their read distances are situated above 100 m in some cases. By contrast, commercial passive, or semi-passive, RFID sensor tags may cost between few US$ (generally tags operating in the HF band) and few dozens of US$ (UHF tags) [20]. Note that the higher cost of commercial RFID sensor tags as compared with the cost of ordinary (i.e. sensor free) RFID tags is due to the presence of the electronic sensing module in the former. Applications of electronic RFID sensor tags include temperature and moisture measurements, force and strain measurements, motion sensing, real-time location, and asset orientation monitoring (acceleration) [106–120] (some of these sensors operate in the near field).

7.2.1.2 Electromagnetic Sensors
Electromagnetic RFID sensors exploit the physics of the RFID response. Such sensors do not contain a dedicated electronic sensor module able to monitor specific parameters (for this reason, such sensors have been called self-sensors in [12]). Rather than that, the tag antenna acts as the sensing

element, able to monitor changes in the materials, or environment, surrounding the tag through the variation of the antenna impedance and gain. This variation is communicated to the reader typically by analog modulation of the backscattered power. However, in the so-called RFID sensor grids, with multiple chips included in a single tag or with several single-chipped tags in proximity, the information of interest is encoded by means of the ID code of the different chips [12, 20]. Thus, with such digital encoding scheme, RFID sensor grids are more robust against potential changes in the relative position and orientation between the reader and the tag.

Unless additional elements/materials are attached/deposited on top of the tag antenna, such radiator acts as a permittivity sensor, sensitive to the "effective" complex permittivity of the surrounding material, since such permittivity has direct influence on the antenna impedance and gain. The antenna impedance and gain are relevant parameters determining the power level of the backscattered signal of the tag (in particular, the impedance of the antenna affects the matching between the antenna and the chip[9]). Therefore, it follows that any variable affecting the complex permittivity is susceptible to be sensed. These RFID sensors, where the (unperturbed/unmodified) tag antenna is the sensing element, have been designated as sensor-less RFID tags in [12]. Nevertheless, the antenna impedance and gain can also be altered, or further altered, by adding elements (certain materials or specific antenna loads) able to perturb the reactive and/or resistive part of the antenna impedance by means of an external stimulus. It can be considered that in these "true" RFID sensors, a specific physical sensor is included as integral part of the antenna, altering its electromagnetic properties. Materials acting as transducers can be deposited or printed on top of the antenna in order to detect physical or chemical changes in the tagged object or tag environment. For example, carbon nanotube (CNT)-based films [121] or *Kapton* films, among others, are useful for measuring strain, temperature, gas/liquid composition, etc. (sensing materials, also designated as "smart" or functional materials, are the subject of Section 7.3). Alternatively to the use of functional materials, the antenna can be loaded by means of certain topologies, e.g. interdigital capacitors, exhibiting a high sensitivity to the dielectric constant of the surrounding material, and thereby altering the antenna reactance [122]. The impedance of the sensing element should preferably be reactive in order to avoid the introduction of further electric losses, which would reduce the read range of the tag.

One limitative aspect of electromagnetic RFID sensors based on a single chip/tag is the double functionality of the antenna (or antenna/sensor), acting simultaneously as sensing element and communicator. Both functions cannot be optimized simultaneously because sensing is based on antenna detuning, and this intrinsically degrades the communication link (namely the tag must suffer a certain level of mismatch in order to provide information relative to the intended measurement). One solution to this problem consists of using the so-called self-tuning chips [123, 124]. The working principle is based on autonomously adjusting the impedance of the chip, in response to a change in the antenna impedance (caused in a sensing operation), by means of a set of capacitors that can be selectively activated. Such capacitors form a switching network that automatically selects the optimum configuration to ensure perfect (or quasi-perfect) matching between the antenna and the chip, consequently maximizing the power transfer from the antenna to the chip. In these adaptive matching RFID sensor tags, the self-tunable chip sends to the reader the information written in its memory (ID code), plus the state of the switching capacitors (also sent digitally). As far as the latter information is related to the level of detuning of the antenna, it

---

9 Conjugate matching between the antenna and the chip is required in order to maximize the power level of the backscattered signal from the tag. Any variation in the antenna impedance alters such optimum matching (detuning), thereby affecting such power level.

provides an indirect measurement of the physical variable of interest. Obviously, sensor resolution is limited by the number of capacitors in the switching network and by their range of values [124].

Let us now succinctly comment the main relevant characteristics of the so-called harmonic RFID sensors (or harmonic tags). In these systems, the tag receives the interrogation at a certain frequency, and it scatters back at a harmonic frequency, typically at twice the frequency of the interrogation signal. The main motivation for such (more complex) approach is the major immunity to the effects of clutter, since it can be reasonably assumed that objects surrounding the sensor tag exhibit linear responses (therefore, not being able to produce interferences at any harmonic frequency of the interrogation signal). Frequency multipliers, e.g. implemented by means of a Schottky diode, are needed in order to achieve the harmonic tag function [125, 126]. Moreover, the matching between the antenna and the chip should be guaranteed at both the fundamental and harmonic frequencies. These aspects add design complexity to the harmonic tags, as compared with conventional (i.e. single frequency) RFID tags.

The level of maturity of chipped electromagnetic RFID sensors as commercial components is not as high as the one of electronic sensors, and electromagnetic RFID sensors are still a subject of investigation. Let us next briefly review some specific implementations as illustrative examples. In [122], a dual-tag capacitive haptic sensor (see Fig. 7.11a) was presented. The dual tag was implemented as a means to perform a calibration-free detection system able to circumvent the effects of the medium surrounding the tag. This device is an electromagnetic RFID sensor where the antenna of the sensing tag is connected to a meandered sensing element acting as a series resonator, with a capacitance affected by the sensing target. In [127], a tag for can detection in smart refrigerator system of beverages/drinks was reported (Fig. 7.11b). The main relevant characteristic of the proposed solution is the fact that the can acts as part of the radiator of the RFID sensing tag. In both implementations [122, 127], the tags were inkjet-printed using conductive inks, an aspect to be discussed later. Near-field electromagnetic sensors have also been reported. For example, in [128], an RFID sensor devoted to food quality and security assessment was presented. The sensor consists of an inductive loop antenna and the silicon chip (Fig. 7.11c). The loop acts as the sensing element, sensitive to the presence of substances, or changes, in the surrounding medium, manifested as alterations in the electromagnetic field generated by the loop antenna. This modifies the impedance of the antenna, which can be measured via mutual inductive coupling between the RFID sensor antenna and the pickup coil of the sensor reader. Other electromagnetic chipped-RFID sensors devoted to ambient monitoring and crack detection can be found in [129] and [130], respectively.

7.2.2 Chipless-RFID Sensors

Chipless-RFID sensors constitute a low-cost alternative to chip-based sensors since the tag does not include any microchip. Moreover, these sensors are also battery free and are therefore potential candidates for the implementation of "green" sensing [13], provided the materials used for the implementation of the sensing tags, including the inks and the tag substrate, are fully recyclable/biodegradable. Organic and graphene-based inks are commercially available, and there are many organic substrates, including paper or compostable materials, that can be useful for tag printing. Moreover, the absence of batteries and ICs provides a potentially unlimited lifetime to these sensors, as well as the possibility to operate under hostile conditions [131–137]. By contrast, due to the absence of communication protocols, reliable tag reading in chipless-RFID sensors can be guaranteed only under restrictive conditions.

The most extended approach in chipless-RFID sensor implementation exploits the changes generated in an antenna or resonant element by the physical variable to be sensed. Thus, in most reported chipless-RFID sensors, sensing is achieved through the frequency shift of the resonant

(a)

(b)

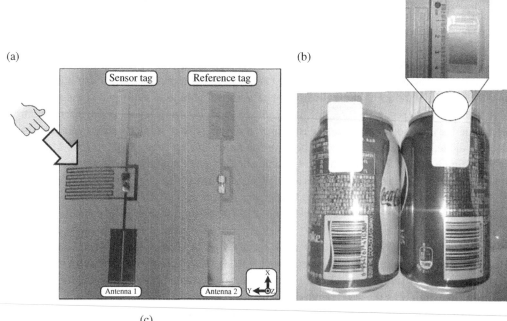

(c)

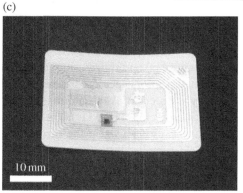

Figure 7.11 Examples of electromagnetic RFID sensors. (a) Dual-tag RFID haptic sensor; (b) RFID sensor for can detection; (c) RFID near-field electromagnetic sensor for food quality and safety assessment. *Source:* Reprinted with permission from [122] (a), [127] (b) and [128] (c). Copyright 2015 IEEE, 2019 IEEE and 2012 American Chemical Society.

peaks of the backscattered response that are generated by a change in the immediate environment surrounding the tag (e.g. a change in the permittivity, or in a variable related to it). This means that the sensed information in these devices, as well as the ID code, is inferred by obtaining the spectral response of the tag [131]. Concerning the transduction mechanisms, capacitance and surface resistance changes are the most common, but displacement can also be used for sensing [135, 137]. These sensors based on changes in the spectral signature are designated as frequency-domain electromagnetic chipless-RFID sensors. Nevertheless, as shown in Fig. 7.9, there are other chipless-RFID sensor types, including time-domain, SAW-based, and thin-film transistor (TFT) sensors. Let us succinctly review next the main characteristic and some implementations of electromagnetic chipless-RFID sensors, including frequency- and time-domain devices (SAW- and TFT-based sensors are out of the scope of this book, but the interest readers can find information in [138–146]).

7.2.2.1 Time-Domain Sensors

Concerning the time-coded chipless-RFID sensors, detailed information is available in [18], where the authors discuss the main strategies for tag and reader implementation based on ultra-wideband (UWB) time-domain technology, as well as processing techniques to mitigate the effects of signal-to-noise ratio (SNR), cross coupling (between the transmitting and receiving antenna), or clutter. See also [16] for an overview of time-domain chipless-RFID sensing techniques. Similar to TDR-based chipless-RFID systems, the time-coded sensor counterparts exploit the echoes of a pulsed signal in time domain caused by the tag, typically a UWB antenna loaded with a delay line. The sensing element is the delay line (eventually terminated with a sensing load), and the sensed information can be encoded either in the time delay [147], the most usual approach, or in the amplitude [148] of the pulses (backscattered signal) corresponding to the so-called tag mode. Besides such mode (the mode of interest), the receiver antenna captures the signal coming from the transmitter antenna (cross-coupling), the clutter from the environment, the structural mode of the tag antenna, as well as other potential sources of interference. Background subtraction[10] and time-gating processing techniques are implemented in these systems in order to remove the effects of clutter (reflections caused by nearby objects) and cross-coupling from the transmitter to the receiver. In order to overcome the detection problems related to a low SNR in the receiver, and thus enhance the read range, the so-called continuous wavelet transform (CWT) can be applied [149–151].

Let us now give some more details concerning the operation principle of these time-domain chipless-RFID sensor tags. When the pulse radiated by the transmitting antenna hits the tag, part of it is backscattered toward the receiver (structural mode), whereas a portion of the incident signal is transmitted to the delay line of the tag, reflected back, and finally re-transmitted to the reader (tag mode). Thus, these two scattering modes are reflected from the tag (and impinge the reader) at different instants of time. The structural mode mainly depends on the antenna shape, material, and size, whereas the delay line connected to the tag antenna mainly determines the tag mode. In tags where the sensed information is contained in the delay time of the tag mode (true time-coded tags), the delay line is terminated with an open circuit, since a maximum magnitude of the reflection coefficient seen from the antenna (and, hence, a maximum of the tag-mode amplitude) is obtained [147]. This time-coding approach is useful, for instance, for measuring variables able to modify the propagation velocity of the delay line, e.g. permittivity and material composition. Alternatively, terminating the delay line with a sensing load able to partially absorbing the energy impinging on it (as a function of the variable to be sensed) results in a modulation of the magnitude of the reflection coefficient of the sensing load, with direct impact on the amplitude of the tag mode [148]. An advantage of these amplitude-coding chipless-RFID sensors concerns the fact that the delay time of the tag mode (with regard to the structural mode) can be used to identify the tag (ID encoding) by controlling the length of the delay line, whereas the amplitude of such mode is used for sensing purposes. By contrast, when the information relative to the variable under measurement is contained in the delay time of the tag mode, providing an identification to the tag is also possible, for example tags can be distinguished by assigning them different delay line lengths, similar to amplitude-encoding tags. However, this approach adds detection/measurement difficulty since both the ID and the sensing information have influence on the same variable, the delay time. An alternative is to use reflectors located at certain positions of the delay line (the usual approach in pure chipless-RFID based on TDR), but the bit capacity or capability to differentiate tags is very limited due to pulse overlapping [16].

---

10 Assuming that clutter and cross-coupling are stationary, their effect is mitigated by subtracting the response of the environment (or background response), which includes the object where the sensor tag is attached (but not the tag), from the response with the presence of the tag.

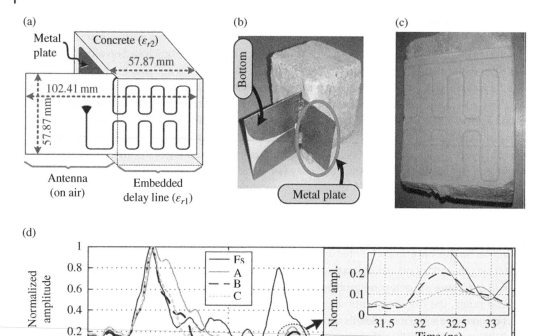

Figure 7.12 Wireless sensor for concrete composition measurements. (a) Schematic of the concrete block with embedded tag; (b) photograph of the embedded tag (back face); (c) cut of the concrete layer after fabrication; (d) measured time-domain response of the tag embedded in three concrete blocks and on air. *Source:* Reprinted with permission from [147]; copyright 2015 IEEE.

As illustrative examples of time-domain chipless-RFID sensor tags, Fig. 7.12 depicts a wireless sensor for measuring concrete mixture composition [147], whereas Fig. 7.13 shows a wireless temperature sensor [148]. The former system is based on tags implemented by means of meander delay lines and UWB Vivaldi antennas, whereas the reader is a *Time-Domain PulsON P400 monostatic radar module (MRM)* impulse radar. Sensing is based on the variation in the permittivity of concrete for the different mixtures, which modifies the electrical length of the delay line and, consequently, the time delay. In order to avoid the effects of surface roughness, the delay line of the tag is embedded in concrete, as shown in Fig. 7.12. Experimental validation in [147] was carried out by considering three concrete blocks (A, B, and C) of different relative concentrations of portland cement and sand. Figure 7.12 depicts also the measured time-domain response of the blocks, performed at a 40 cm tag-reader distance, after applying the background subtraction and CWT, as mentioned. A concrete block without any tag embedded was used in order to emulate the real-case background (portion of wall without tag). The responses of the different blocks can be distinguished from the different delays between the structural mode and the tag mode. Note that the tag mode amplitude is reduced when the tag is embedded in concrete, but the peaks and the time position of each peak (delay time) can be clearly discerned.

The wireless temperature sensor, reported in [148], is implemented by means of tags consisting of a meander delay line, plus a broadband eccentric annular monopole antenna, see Fig. 7.13. Two readers were considered in [148]. In one case, the reader uses the step-frequency (sweeping) technique, and it is based on a VNA to perform the frequency sweep; the other reader is based on a

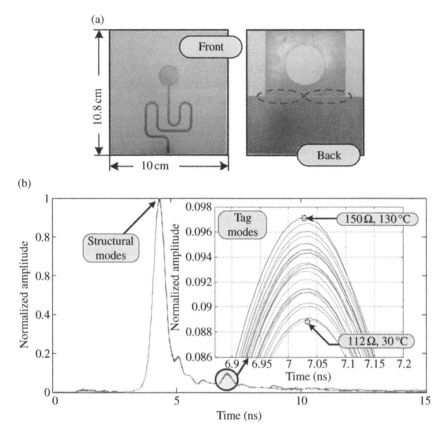

Figure 7.13 Wireless sensor for temperature measurements. (a) Front and back views of the tag, excluding the presence of the temperature transducer; (b) measured time-domain response of the tag for different temperatures, inferred from the step-frequency technique. *Source:* Reprinted with permission from [148]; copyright 2012 IEEE.

low-cost UWB radar, similar to the impulse radar used in regard to the concrete sensor of the preceding paragraph. Obviously, for the step-frequency reader, a sine wave is continuously swept over the frequency band of interest with the VNA, and using the inverse Fourier Transform of the transmission coefficient, the time-domain response is obtained. Figure 7.13 depicts the top and bottom views of the tag. Although not shown in the figure, for temperature sensing, a temperature-to-resistance transducer (model *Vishay PTS 100 Ω*) was soldered to the open end of the meander line of the tag.[11] By varying the temperature (achieved by means of a heat gun able to provide a maximum temperature of 130 °C in [148]), the sensor resistance changes. Thus, a temperature variation modulates the magnitude of the reflection coefficient seen from the end termination of the delay line, and an indirect temperature measurement can be inferred from the amplitude of the backscattered signal (tag mode). The time-domain response for different temperatures obtained by means of the step-frequency technique and a reader-tag distance of 100 cm is also depicted in Fig. 7.13 [148]. It can be appreciated that temperature measurements in the range between 30 and 130 °C can be performed. Further details concerning sensor calibration and other aspects, such as the comparison between the frequency-swept and radar impulse reader techniques, are given in [148]. Other time-domain chipless-RFID sensors can be found in [152–157].

---

11 As far as an external transducer for sensing is used, this sensor is not a true electromagnetic chipless-RFID sensor.

7.2.2.2 Frequency-Domain Sensors

There has been an intensive research activity in the last decade devoted to the design and implementation of frequency-domain chipless-RFID sensors [158–174]. In these sensors, a set of printed resonators, tuned to different frequencies, is used for identification and sensing. Typically, a single resonant element of the set is devoted to sensing purposes, but it is potentially possible to dedicate several resonators to perform a multivariable measurement. The additional resonators, if they are present, provide the ID code of the tag. However, in certain implementations, the tags are not provided with identification capability, i.e. they are exclusively composed of the sensing resonator/s [164]. Nevertheless, such sensors are also considered to belong to the category of frequency-domain chipless-RFID sensors, since their working principle is identical to the one of frequency-domain chipless-RFID identification tags. The sensing principle relies on the resonance changes (frequency position or magnitude, or both) caused by the variable under measurement, similar to the frequency-variation sensors discussed in Chapter 2. In some implementations, the sensing resonator/s is/are covered by materials (typically thin films) that act as transducers, able to modify their permittivity or conductivity with the variable under measurement, e.g. temperature, humidity, and gas composition.

Similar to the frequency-domain chipless-RFID identification tags, the sensor counterparts can be implemented as retransmission-based devices, or as backscattering sensing tags. The advantage of backscattering over retransmission chipless-RFID sensors is the absence of dedicated antennas in the former. The tag resonators simply reflect back the interrogation signal, when such signal impinges the tag, and the ID code, as well as the sensing information is contained in the spectral signature of the response. By eliminating the tag antenna, the area occupied by the sensing tag can be substantially reduced, as compared with the one of retransmission-based tags. In retransmission-based tags, the receiving and transmitting antennas of the tag, connected as port terminations of the resonator-loaded transmission line, are typically cross-polarized in order to prevent from cross-coupling between the receiving and transmitting antennas of the reader, also cross polarized. One limitation of backscattering sensing tags is the fact that it is difficult to distinguish the backscattered signal from the background response (indeed, it also applies to retransmission-based tags, but the effect is, in general, less severe), and background subtraction is, typically, necessary. Nevertheless, to alleviate this limitation of backscattering sensing tags, various techniques have been proposed, including cross-polarization [164, 175, 176] (similar to retransmission-based tags), circular polarization [177], and dual polarization [178], among others.

For example, sensing tags based on depolarizing resonators (or depolarizing tags) have been reported in [164]. The resonators are bended dipoles, as shown in Fig. 7.14, and have the ability to backscatter a cross-polarized signal when the impinging signal sent by the reader is vertically polarized. For such purpose, the resonators should work at their third resonance, where the current in the bended arms is comparable to the current in the central part of the resonant element. In the implementation of [164], four identical bended resonators, exclusively devoted to sensing, are used in order to increase the level of the reflected signal and facilitate sensor detection. Sensing is carried out by virtue of the fact that at the corners of the bended dipoles, a via connects the upper metal (the resonator) with the ground plane through circular slots acting as capacitive elements. The schematic of the RFID sensing system is depicted in Fig. 7.14. The reader, a VNA in [164], sends a vertically polarized wave (e.g. using a vertically oriented Vivaldi antenna) and then it sweeps the interrogation frequency in order to obtain the resonance frequency of the tag (which depends on the material in contact with it) through the measured cross-polar component of the response. For that purpose, a cross-polarized antenna (e.g. a horizontally oriented Vivaldi antenna) is used, as depicted in Fig. 7.14. By this means, the receiving antenna filters the main contribution to background reflection and receives the backscattered field generated by the tag. Moreover, with such

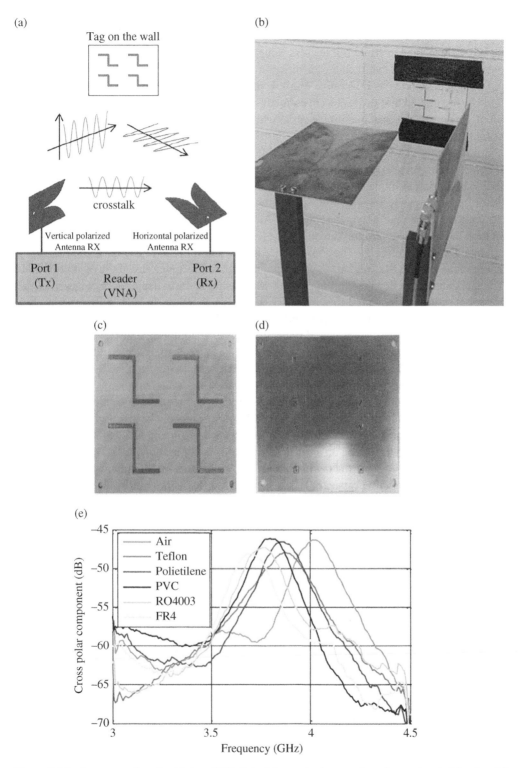

Figure 7.14 Frequency-domain chipless-RFID dielectric constant sensor based on a depolarizing tag. (a) Schematic illustrating the working principle; (b) experimental setup; (c) top view of the sensing tag; (d) bottom view of the sensing tag; (e) measured cross-polar response after background subtraction for different materials. The tag substrate is the *Rogers 4003C* with thickness h = 1.62 mm and dielectric constant ε_r = 3.43. Resonator's dimensions are width 2 mm, length of the external arms 20 mm, length of the central arm 22 mm. *Source:* Reprinted with permission from [164]; copyright 2018 IEEE.

depolarizing tags, the cross-polarized antennas of the reader minimize the coupling interference between both radiators. Such chipless-RFID sensors can be applied to dielectric constant measurements and were applied to structural health testing in [164]. Here, the measured cross-polar component, after background subtraction, for different materials (with different dielectric constants) behind the tag is reported as illustrative example of the potential of the approach (see Fig. 7.14). It should be mentioned that the tags were correctly detected in [164] by considering read distances up to 1 m. Obviously, the peak magnitude varies with the read distance, but not the peak frequency, the output variable in the considered sensor. Further details on this sensor and its application to structural health monitoring can be found in [164].

Let us next report three examples of frequency-domain chipless-RFID sensors with identification capability. In the first example, a humidity sensor, the tag contains three slot resonators, devoted to identification, and one electric-LC (ELC) resonator, used for sensing [158]. The ELC resonator is coated with a hygroscopic polymer (Polyvinyl Alcohol -PVA) able to absorb water molecules. A variation in the water vapor pressure alters the level of water absorption by the polymer, which in turn modifies its permittivity. The consequence is a variation in the capacitance and resonance frequency of the sensing ELC resonator, which provides an indirect measure of the relative humidity. The fabricated tag is depicted in Fig. 7.15(a), whereas Fig. 7.15(b) depicts a photograph of the experimental setup, implemented in a humidity-controlled chamber. In such configuration, two horn antennas are connected to a VNA and are oriented face-to-face, with the sensing tag in between. Fig. 7.15(c) depicts the measured transmission coefficient for different levels of humidity. The slot resonators generate three resonances, useful for tag identification, which do not depend on the relative humidity. By contrast, the resonance of the ELC resonator, near 6.5 GHz, experiences a detectable variation with the changes in humidity level.

In the previous proof-of-concept wireless sensing system, the antennas are oriented face-to-face, in relative proximity, sandwiching the tag. Such configuration might not be practical in many applications. Let us next present another example of a frequency-domain chipless-RFID sensor system with identification capability, based on a true backscattering configuration [174]. The sensor is also devoted to humidity sensing and the tag uses seven ELC resonators for identification and one resonator, based on an interdigital capacitor (IDC), for sensing. The working principle is very similar to the one of the previously reported sensor, namely the sensing resonator is coated with PVA, whereas the other resonant elements are left uncovered. The sensing tag, as well as the experimental setup, are shown in Fig. 7.16. A nonreflective temperature and humidity chamber was manufactured using styrofoam in [174], and the tag was placed in the central position of the chamber. For measuring the tag response, a VNA with two horn antennas disposed as depicted in Fig. 7.16 was utilized (the measurements were carried out in an anechoic chamber). Figure 7.16 includes also the measured bistatic RCS characteristics of the 7-bit chipless-RFID tag for relative humidity varying between 50 and 80%. As it can be seen, the resonant peaks corresponding to the ELC identifying resonators do not appreciably change with the relative humidity level, contrary to the resonance associated with the IDC-based resonator.

The last example of a frequency-domain chipless-RFID sensing system corresponds to a retransmission-mode temperature sensor [170]. The sensor has identification and sensing functionality, achieved by using six different resonant elements coupled with a microstrip transmission line. Four of these resonators provide two bits of information, that is, can be configured in order to generate two notches in the frequency response, to produce only one notch (at either frequency), or to be transparent. Thus, these four multistate resonators contribute with eight bits to the encoding capacity of the tag. These four resonators consist of a coupled line section with two arms of different length. The electrical length of these arms is a quarter-wavelength at the desired notch frequency. Thus, the four different states of each resonator (corresponding to two bits) are generated by selectively connecting, or not, the quarter-wavelength line sections (arms) to the host coupled line

(b)

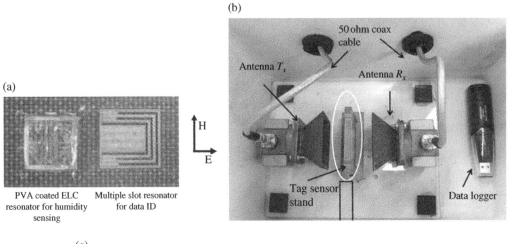

(a)

PVA coated ELC resonator for humidity sensing   Multiple slot resonator for data ID

(c)

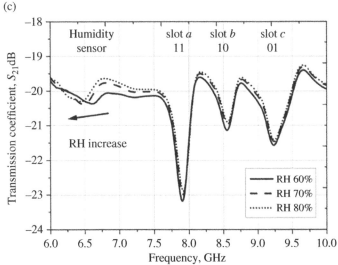

Figure 7.15 Frequency-domain chipless-RFID humidity sensor with (3-bit) identification capability. (a) Photograph of the tag; (b) experimental setup; (c) measured transmission coefficient (calibrated). The total area occupied by the tag is 15 mm × 6.8 mm. The tag was implemented on the *Taconic TLX_0* substrate with height $h = 0.5$ mm, dielectric constant $\varepsilon_r = 2.45$, and loss tangent $\tan\delta = 0.0019$. *Source:* Reprinted with permission from [158]; copyright 2014 IEEE.

element. Namely, if one arm is connected, the corresponding notch arises. The other two resonators are more complex, since they can be configured in order to generate eight different states (corresponding to three bits). Thus, the topology is similar to the one of the previous (four-state) resonators, but, in this case, the resonator consists of three quarter-wavelength arms, rather than two. Nevertheless, one of the arms in one of these two resonators, devoted to sensing, is connected to the coupled element, since the corresponding resonance should be present in the spectral response. A thin film made of a material exhibiting a temperature-dependent permittivity covers such arm, and the temperature can be indirectly measured from the frequency shift experienced by the notch frequency of the sensing arm. Note that a temperature change alters the electrical length of the sensing arm through the variation of the effective dielectric constant. In [170], a thin-film composite polyamide with a linear dependence of the dielectric constant with temperature was used. According to these words, the sensing resonator contributes with two bits to tag encoding. Concerning the other (3-bit) resonator, one arm, connected to the coupled line section, is used

(a)

(b)

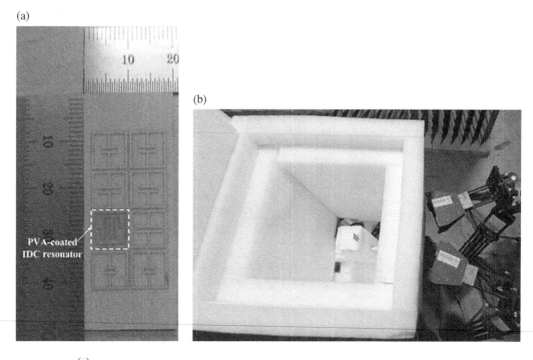

(c)

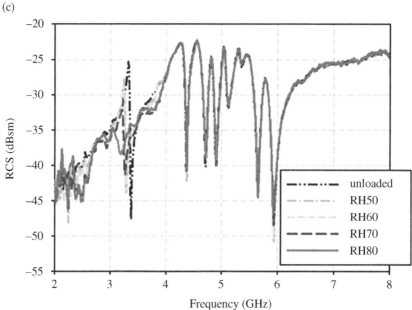

Figure 7.16 Frequency-domain chipless-RFID humidity sensor with (7-bit) identification capability. (a) Photograph of the tag; (b) experimental setup showing the temperature and humidity chamber with the tag inside it; (c) measured bistatic radar cross-section response. The tag is implemented on the *RF-301* substrate with relative permittivity ε_r = 2.97, thickness h = 0.8 mm, and loss tangent tanδ = 0.0012. *Source:* Reprinted with permission from [174]; copyright 2021 MDPI.

as a reference to improve the accuracy of the temperature measurements, and therefore this resonator provides also two bits of information. In summary, 14 notches are generated in the spectral response when all the arms are connected to the coupled line section, but two of them are devoted to sensing (those generated by the sensing and reference arms), and, consequently, the sensor is a 12-bit chipless-RFID tag.

Figure 7.17 depicts the layout of the tag (excluding the cross-polarized antennas necessary for communication purposes), as well as the simulated transmission coefficient, when all the arms are connected to the coupled lines of the respective resonators. Some tags with different ID codes were fabricated in [170], and their responses were also measured (see Fig. 7.18). In one of the tags, the sensing arm was coated with the polyamide film, thereby generating certain shift in the corresponding resonance, as can be appreciated in Fig. 7.19. Two-port temperature measurements carried out by directly connecting the ports of the resonator-loaded transmission line of the tag to a VNA were successfully demonstrated in [170] (Fig. 7.20 depicts the detail of the measured

(a)

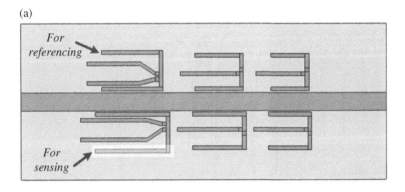

(b)

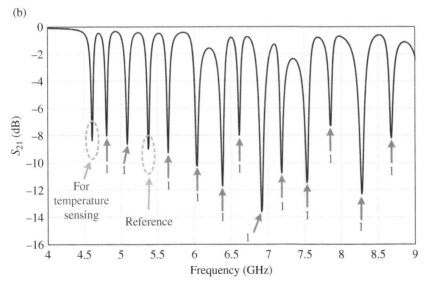

Figure 7.17 Layout (a) and frequency response (b) of the chipless-RFID temperature sensor based on multistate resonators. The small metallic squares in the junctions of the resonators may be present, or not, depending on the specific ID code. The considered substrate for the simulations is the *RT/Duroid* with dielectric constant $\varepsilon_r = 2.2$ and thickness $h = 0.78$ mm. The host line exhibits a characteristic impedance of 50 Ω. The total length of the tag, excluding the connectors and the access lines, is roughly 3 cm. *Source:* Reprinted with permission from [170]; copyright 2020 Elsevier.

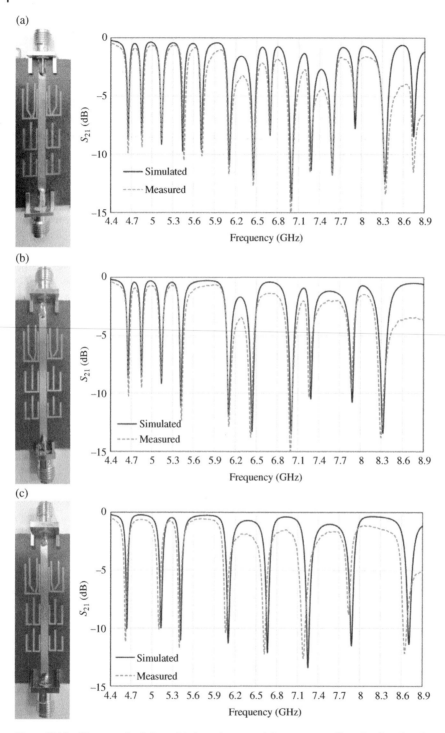

Figure 7.18 Photograph of three fabricated tags and the corresponding simulated and measured spectral responses. These tags neither have temperature sensing capability (the sensing arm is not coated with the temperature-dependent polyamide thin film), nor exhibit wireless connectivity with the reader (the measured responses have been inferred by directly connecting the VNA to the tag ports). Tag ID codes are (a) 11111111111111; (b) 11110110110110; (c) 10110101010101. *Source:* Reprinted with permission from [170]; copyright 2020 Elsevier.

(a)

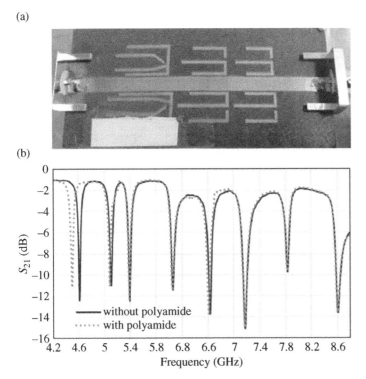

(b)

Figure 7.19 Photograph of the tag of Fig. 7.18(c) with the polyamide film covering the sensing arm (a) and measured spectral response compared with the one without the polyamide film (b). The effect of the polyamide film is a left displacement of the notch frequency associated with the sensing arm. *Source:* Reprinted with permission from [170]; copyright 2020 Elsevier.

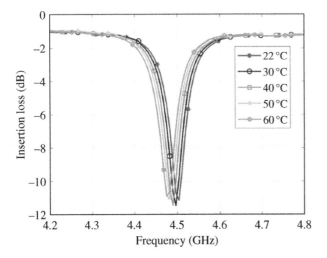

Figure 7.20 Detail of the frequency response of the temperature sensing tag around the notch frequency generated by the sensing arm for various values of the temperature. *Source:* Reprinted with permission from [170]; copyright 2020 Elsevier.

frequency responses in the vicinity of the notch frequency of the sensing arm for different temperatures). In order to demonstrate the viability of performing wireless measurements, the tags were equipped with planar cross-polarized UWB antennas. For such measurements, the VNA was connected to two UWB horn antennas, as well. Figure 7.21 shows the experimental setup, as well as the measured tag responses, compared with those inferred in the two-port measurements. As it can be seen, the system is able to read the tags. Nevertheless, it should be mentioned that the measurements were performed inside an anechoic chamber.

In the present section, the main design strategies, working principles, and several illustrative examples of chipless-RFID sensors have been reported. It has not been the objective of the authors to provide an exhaustive overview of these wireless, battery-free, and chipless sensors. For that purpose, further information can be found in several surveys on the topic and references therein [15, 16, 19, 20, 22, 179–181].

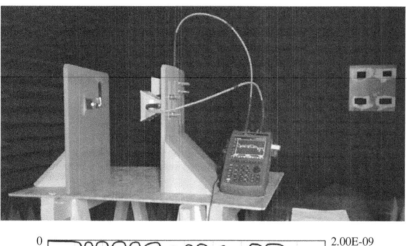

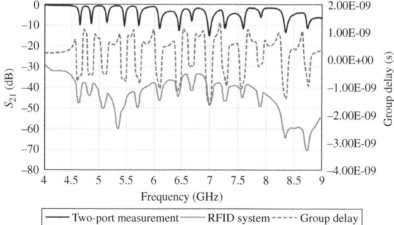

Figure 7.21 Photograph of the setup used for the validation of the wireless temperature sensing system based on chipless-RFID tags equipped with cross-polarized antennas, and measured spectral responses, including the group delay, of the tags of Fig. 7.18. *Source:* Reprinted with permission from [170]; copyright 2020 Elsevier.

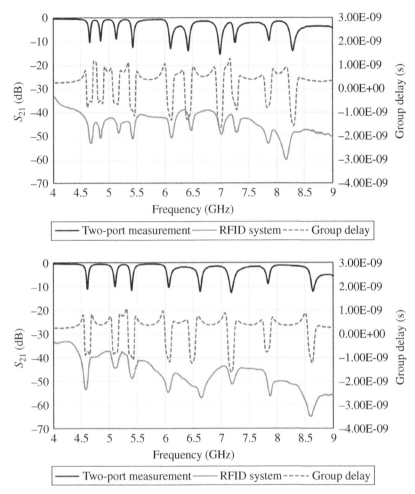

Figure 7.21 (Continued)

7.3 Materials and Fabrication Techniques

This section focuses on briefly describing the materials involved in the implementation of RFID sensors and the fabrication techniques. In the most general case, a RFID sensor tag consists of three layers: the substrate, the conductive layer (with the antenna and/or the printed encoder patterns), and the sensitive layer, typically made of a thin film of a material (able to transduce a physical variable to an electromagnetic variable) deposited or printed in specific regions of the tag.[12] Thus, specific materials for each structural layer are needed, and such materials determine the tag cost, performance, and sensing properties. Concerning tag fabrication, the processes involved in chipless-RFID sensors are, in general, simpler and cheaper than those needed for the implementation of chipped-RFID sensors. Due to the lack of the IC, as well as other electronic blocks (mainly, an

---

12 Obviously, electronic RFID sensors do not include sensing materials, as far as a dedicated module performs the sensing function, as discussed in Section 7.2.1.1. Nevertheless, in many chip-based electromagnetic and chipless-RFID sensors, the sensing functionality is achieved without the need to incorporate specific sensing films.

electronic sensor module and a microcontroller), there is no need for microelectronic processes in the chipless-RFID manufacturing chain. Moreover, wire-bonding or soldering are avoided, since it is not necessary to connect any electronic module (including the chip) to the tag antenna or substrate. Thus, chipless-RFID sensor tags are, in general, more robust than chip-based sensors against harsh conditions susceptible to cause damage in electronic components (extreme ambient conditions, radiation, etc.), and the processes needed for chipless-RFID tag fabrication are relatively cheap (in particular, low-cost massive fabrication processes based on additive techniques, i.e. printing, can be applied to chipless-RFID tag manufacturing [182–188]). Nevertheless, sensors based on specific "smart" materials (acting as transducers) might be subjected to further tag fabrication complexity, depending on the particularities of such materials.

In RFID sensors where the cost is not a critical issue, tag implementation in printed-circuit board (PCB) or low-loss microwave substrates might constitute an interesting option for performance optimization. Typically, such substrates are metallized with copper films of several tens of microns thick. The conductivity of copper as well as the loss factor of most PCB and low-loss microwave substrates typically suffice to achieve reasonable performance in chipless-RFID sensor tags. In chip-based RFID sensors, the electronic and electromagnetic quality of the substrate and conductive materials is not so critical. Thus, such sensors are preferably implemented on plastic/polymer substrates using conductive inks (or metallic laminates). This does not mean that chipless-RFID sensing tags cannot be implemented with such substrates and inks, but at the penalty of performance (number of bits and read range) degradation.

The canonical fabrication process for sensing tags based on PCB or cladded microwave substrates is photolithography (a well-known subtractive technique also called photoetching), at least for massive manufacturing. Nevertheless, at laboratory level, or for the fabrication of proof-of-concept prototypes on rigid substrates, generating the metallic patterns of the tag by means of milling or drilling is an interesting solution. Photoetching can also be applied to the fabrication of tags implemented on other substrates, such as plastics or fabric, which should be conveniently metallized or covered with a layer of conductive material. In some cases, commercial chip-based UHF-RFID tags are fabricated by lithography on flexible substrates coated with aluminum.

Additive processes, such as screen-printing, gravure printing, flexography, or offset, among others, are the most common industrial techniques for conductive material deposition on flexible substrates [182–191]. These techniques are especially suited for mass production, whereas inkjet printing is of special interest for low-cost rapid prototyping [53, 97, 98, 104, 122, 136, 192–194]. Among the flexible substrates, polymer-based substrates, such as polyethylene terephthalate (PET) or polyethylene naphthalate (PEN), fabric, and organic substrates, including paper or compostable materials, can be potentially used for tag implementation. Concerning the conductive inks, those that exhibit higher conductivities are based on silver nanoparticles, with values close to 10^7 S/m, i.e. roughly one order of magnitude lower than the conductivity of most common metals used in electronics industry, like copper, silver, gold, or aluminum [195, 196]. Despite such limited conductivity, silver-based inks suffice for many RFID applications. Other inks based on graphene [197–200] or organic conducting polymers [e.g. poly (3,4-ethylenedioxythiophene) polystyrene sulfonate – PEDOT:PSS] [104] are of interest since, combined with organic substrates (paper, for instance), can potentially be applied to the implementation of "green" sensing tags (i.e. recyclable, biodegradable, or eco-friendly) [13, 201]. Nevertheless, the typical conductivities of such inks are significantly smaller than those of silver-based inks. Therefore, the implementation of full "green" sensing tags, based not only on organic substrates but also on organic inks (as well as sensing materials if they are present), requires further research efforts. In this regard, let us mention that processes to improve the electromagnetic performance of graphene inks, based on compression techniques, have been carried out [198, 202, 203]. On the other hand, the functionality of near-field chipless-RFID tags based on time-domain signature barcodes (without sensing capability) implemented by means of PEDOT:PSS inkjet-printed on paper substrates has been demonstrated [104].

Concerning the sensing materials, it is not the objective of this section to provide an exhaustive overview of their physical/electromagnetic properties and involved transduction mechanisms (for that purpose, the authors recommend the sources [17, 134, 179, 180], and references therein). The main aim is to succinctly present some of the most common "smart" sensing materials useful for measuring environmental variables, to briefly indicate the stimuli to which they are sensitive (and how such stimuli affect the electromagnetic properties of the materials), and to report their target applications (e.g. humidity sensing, temperature sensing, and gas composition determination). Nevertheless, "smart" sensing materials, or films, are not always present in RFID sensors (as it can be appreciated in some of the prototypes reported so far). For example, electronic RFID sensors use dedicated electronic modules to perform the measurements and, consequently, they do not need additional materials for sensing. However, in many reported chip-based RFID electromagnetic sensors and chipless-RFID sensors, "smart" materials are not used, as well. This applies especially to sensors focused on dielectric characterization of materials and measurement of related variables [147, 164], or to displacement/position sensors [137, 204]. Typically, although not exclusively, sensing is based on the changes that the variable under test generate in the complex permittivity (or dielectric constant and loss tangent) of the material surrounding the sensitive elements (for example a resonant element in frequency-domain chipless-RFID sensors). Thus, for dielectric characterization, it suffices to contact the MUT to the sensing element. By this means, any change in the dielectric properties of the MUT is able to modify relevant parameters of the sensing elements, such as resonance frequency, amplitude level, quality factor, electrical length, or phase, the typical output variables (as it has been profusely discussed throughout this book). However, environmental variables, such as humidity, or temperature, among others, are not able to significantly alter, in general, the characteristics of the sensing element, unless it is coated with a material whose properties (typically the complex permittivity) are sensitive to such variables.

Probably, the most utilized "smart" materials in RFID sensing are polymers [205, 206]. For example, various works focused on temperature sensing use polyamide thin films [170, 207] (such films exhibit a temperature-dependent dielectric constant with enough sensitivity for appreciably modifying the equivalent capacitance or electrical length of the considered sensing resonators by temperature variation [208]). However, most polymers, especially those exhibiting hygroscopic properties, such as polyvinyl alcohol (PVA), used in the sensors depicted in Figs. 7.15 [158] and 7.16 [174], have been applied to the measurement of ambient humidity. Further examples of humidity sensors based on PVA are given in [209–211]. Other polymers useful for humidity measurements are Kapton [212, 213] and poly (2-hydroxyethyl methacrylate) (PHEMA) [214]. There are also polymers sensitive to various stimuli simultaneously, for instance polyetherurethane (PEUT) [215] (temperature and humidity), PEDOT:PSS (temperature, humidity, and pH), polydimethylsiloxane (PDMS) (pH and pressure) [216, 217], polymethyl methacrylate (PMMA) (humidity and gas sensing) [218], or cellulose (humidity and chemical sensing) [219–223].

Due to their high sensitivity and selectivity, long-term reliability and stability, low-temperature operation, and capability for integration of arrays on a massive scale, nanostructures constitute a powerful sensing platform for many different applications, including biosensing, electrochemical sensing, and gas detection, among others [224–229]. Many examples of RFID sensors based on nanostructured metallic oxides, for instance SnO_2 [230], silicon nanowires [155, 231–233], and CNTs [234–246] have been reported in the literature. Nevertheless, due to their very rich mechanical, electrical, and dielectric properties, CNTs constitute one of the most exhaustively used nanostructures in RFID sensing. Both conductance and capacitance changes have been exploited in CNT-based RFID sensors. Applications mainly include humidity, temperature, and mechanical strain sensing, as well as gas detection (in this case, selectivity to specific substances can be achieved by doping or by using functional groups). Further details on the underlying physics behind the transduction mechanisms and the specific processes for the preparation of the sensing structures are out of the scope of this book, but can be found in [227, 247].

Another "smart" material that deserves special attention is graphene, a two-dimensional sheet of pure carbon, arranged in a hexagonal lattice, and equivalent to an unraveled single-wall CNT. This material, discovered by Andre Geim and Konstantin Novoselov in 2004 [248], exhibits extraordinary chemical, physical, and electrical properties, and, similar to CNTs, it can be functionalized or doped in diverse ways, in order to enhance its sensing properties. The reported sensing applications of graphene and its derivatives, i.e. graphene oxide (GO) and reduced graphene oxide (rGO), are innumerable [249] and are out of the scope of this chapter, exclusively devoted to RFID-based sensors. Thus, accordingly, let us simply mention some examples of RFID sensors based on graphene. For example, in [250], an RFID-sensor system that uses Pt-decorated reduced graphene oxide (Pt-rGO), spin-coated in the tag antenna, was applied to the detection of hydrogen gas. In [251, 252], the authors report an RFID GO sensor for the monitoring and discrimination of breath anomalies, utilizing the moisture emitted during inhalations and exhalations, and its effect on the DC resistance of a GO-based electrode. In [253], the dependence of the dielectric constant of GO with the humidity level was used to implement a battery-free RFID humidity sensor by coating a printed graphene antenna with a GO layer. An interesting survey of the up-to-date progress in graphene-based RF sensing platforms, including RFID sensors, can be found in [254] (see also the references therein for further details on the reported applications).

Cadmium sulfide (CdS), barium strontium titanate (BST), and water absorbent material (cork) have been applied to light [255], temperature [256], and humidity [257] sensing, respectively, by means of RFID systems. Further references relative to "smart" materials for sensing can be found in various review papers devoted to RFID sensors [178, 179, 258, 259]. To end this section, let us emphasize that the discussed "smart" materials (or at least most of them) can also be applied to most of the dielectric-characterization sensors reported in the previous chapters, in order to broaden their range of applications and extend them, e.g. to temperature, humidity, chemical and biosensing, among others. For that purpose, the procedure is to coat the sensitive elements of the sensors (resonators or transmission line sections, typically) with the stimulus-dependent "smart" materials, in a similar way to the reported RFID sensors.

7.4 Applications

The applications of RFID sensors are reviewed in various papers and references therein [19, 20, 179, 180, 259]. There are many fields, or areas, where RFID technology may provide interesting sensing solutions, including healthcare, food and agriculture, smart packaging, heavy machinery industry (e.g. motion control), chemical and civil engineering (e.g. gas sensing and structural health monitoring), automotive industry, space, wearables and implants, localization, and activity monitoring. This section is not intended to provide illustrative examples of such applications (which can be found in the above cited references and in further references to be listed next), but to succinctly discuss the advantageous aspects and limitations of RFID sensing technology in the considered fields, as well as to highlight some specific target applications.

7.4.1 Healthcare, Wearables, and Implants

With the continuous growing of the percentage of elder people in the world population, the states must increasingly allocate huge amounts of money to the public health systems. The budget includes not only treatments against pathologies or diseases but also prevention, monitoring, and tracking systems. Within the framework of the "smart health" system paradigm, RFID sensing technology may play a crucial role, as far as a continuous monitoring of persons, health parameters, and environmental variables at low cost and wirelessly is a fundamental requirement [260–262].[13] Wearable, implanted, and environmental low-cost RFID sensors have the potential to make the "smart health" concept a reality, or at least contribute to obtain relevant medical information autonomously, thereby assisting the physicians, and other people involved in the healthcare chain, to take the convenient actions or appropriate assistance procedures, even remotely. RFID sensors are thus key enabling devices for IoT in healthcare (also called Internet of Medical Things – IoMT – or Health IoT – HIoT), a fast-growing market [263, 264]. In a typical scenario, an IoMT device (i.e. an RFID sensor) collects certain vital data (e.g. the blood pressure) of the patient, who is not physically present in the healthcare facility, thereby eliminating the need for patients to move to specialized data collecting places, such as hospitals or medical centers. The collected patient data are then forwarded to a software application, where healthcare professionals and/or patients can view it, and such data is then analyzed in order to recommend treatments or generate alerts. For example, an RFID sensor that detects an unusually high or low blood pressure may generate an alert, so that healthcare professionals can intervene.

Key aspects for the deployment of RFID technology in healthcare are sensor cost and performance. Battery-free and chipless-RFID sensing tags can be implemented at very low cost, but the typical sensitivities, data storage capacities, and read distances cannot compete against those offered by chip-based active or semi-passive tags, much more expensive. On the other hand, it must be taken into account that implantable and wearable RFID sensing tags must operate inside or in proximity to the human body. Therefore, further issues related to the frequency of operation,[14] tag size, dielectric influence of the human body [265, 266], and safety aspects[15] must be considered; and, obviously, special attention should be put on tag materials and fabrication techniques, as far as implantable and wearable sensors should be biocompatible, conformal, and robust against wearing, among other restrictions. It is thus clear that the implementation of low-cost reliable RFID systems in an IoMT scenario is an actual challenge. Temperature, humidity, chemical sensors, etc., able to continuously monitor vital variables such as body temperature, electrolyte and glucose content in blood and urine, arterial pressure, and cardiac rhythm are of actual interest. Examples of wearable and implant sensors, several of them devoted to smart health, can be found in [193, 267–278]. Monitoring other health aspects, such as depression and mood, or Parkinson's disease, are also of the highest interest, but not absent of certain difficulties due to the lack of clear quantitative variables indicative of the degree of development/incidence of such diseases in patients. The implementation of ingestible sensors is another topic of interest within the framework of IoMT [279–281].

RFID sensors can also be of interest for macroscopic motion sensing in order to identify behaviors that can be precursor of dangerous anomalies [282–284]. Motion sensing can be achieved by means of arrays of RFID sensors conveniently distributed at home, in clothes, or in personal belongings. Thus, human body movements in proximity to the tags may introduce scattering and shadowing effects, thereby altering the communication link between a fixed reader and the tags. Such changes in the signals received by a combination of wearable tags and ambient tags can be used to monitor human activity. Such monitoring becomes important, e.g. in disabled and elderly people, especially

---

13 Smart health pursues the continuous (remote) monitoring of health of patients, with the ultimate objective of providing medical data to patients and physicians at anytime and anywhere.

14 The size of the tag antenna and resonant elements decreases with the operation frequency. However, in biological tissues, RF/microwave losses are lower at low operating frequencies. Thus, a trade-off is necessary, and the final operating frequency and tag size depend on the specific implant and its application.

15 Due to safety aspects, the read range for in-body devices is limited to a few centimeters. The reason is the maximum authorized energy absorption, which limits the power delivered by the reader, and prevents from tag detection at higher distances.

in domestic environments, in order to detect unexpected body movements, accidental falls, or long inactivity periods (which might be indicative of potential dangerous events, e.g. lipothymia, ictus, epilepsy seizure, or even death).

7.4.2 Food, Smart Packaging, and Agriculture

Our society is increasingly evolving toward schemes based on quality control, safety legislation, and regulation standards. Food is not absent of this trend, in part motivated by consumer demands, by health security reasons, and by the fact that our daily lives force us to store food and consume it over time. Food safety and management systems should be able to control food safety hazards and ensure compliance with food safety regulations. For example, ensuring the preservation of the cold chain is an important aspect, but there are many other safety rules (e.g. keep cleaning, and separate cook and raw products). Within this context, it follows that RFID sensors may play an important role in food quality/safety and smart packaging [285–288], since these sensors constitute an enabling technology for smart tagging of food products, necessary to provide information indicative of their freshness, cleanness, purity, and other aspects of interest for the consumers and suppliers.

Since food products have, in general, a moderate or low prize, sensor cost must be small. Moreover, the considered sensors must be able to operate autonomously for months or even years. Thus, the convenient RFID sensor solutions for the food industry are battery-less devices [128, 169]. The sensors can be attached directly either to the food product (in case it is possible) or to the food package. The most usual working principle for sensing is based on the dielectric properties of the food or its environment, but temperature sensors are also needed in some cases (e.g. for monitoring the cold chain [289–291]). Note that changes in the food, e.g. freshness, chemical changes due to ripening and bacterial growth, might be manifested as changes in the dielectric constant, which can be detected by means of RFID permittivity sensors. Obviously, smart sensing films can be added in order to enhance the sensitivity, or to use other input variables, such as humidity or pH, or the presence of certain gases, among others, for sensing (food spoilage monitoring relies sometimes on the measurement of volatiles or analytes in food packaging). Indeed, many of the sensors (devoted to temperature, humidity, or gas sensing) reported so far, including those quoted in the reference list, can be potentially used for food monitoring, as reported in [288]. Nevertheless, there are many examples of RFID sensors specifically devoted to food sensing and smart packaging in the literature (some of them can be found in [159, 292–302]). Important aspects for the future deployment of RFID technology in food quality assessment are recyclability and biocompatibility. Sustainability issues, especially critical in mass-production products, force the progressive implementation of "green" products. The biocompatibility of the sensors is another challenge, as far as they might be in direct contact with the food inside the packages (i.e. the materials used for sensor fabrication should not be toxic or generate any dangerous effect on the tagged/monitored food). Further challenges of RFID technology in food sensing and smart packaging are pointed out in [288].

IoT has also penetrated in the agriculture sector, giving rise to the so-called smart agriculture [303–307]. By introducing intelligence and sensing capabilities, smart agriculture provides means to automate most aspects of the plant growing process, to save human efforts, and to improve the efficiency, yield, and quality of crops and harvests. RFID identification and sensing constitutes also a key enabling technology in smart agriculture, and a new series of applications have emerged in this field. Such applications are focused on aspects as diverse as measuring environmental variables of interest for farmers, monitoring soil conditions and plant growth, and assessing the quality of the harvests, among others. Significant research efforts have been dedicated in two main areas: the development of RFID sensors for leaf sensing, a new technology useful for monitoring water content and plant health (examples of leaf RFID sensors are given in [308–310]); and the implementation of RFID sensors able to provide information on soil conditions. A huge quantity of soil moisture and temperature RFID sensors have been reported (it is impossible to cite all the published papers on this topic, but an exhaustive list can be found in [112, 311–315]).

7.4.3 Civil Engineering: Structural Health Monitoring (SHM)

Civil engineering is another area where RFID sensing may play a key role. There are many infrastructures, such as buildings, bridges, railways, and pipelines, that are subjected to continuous stress and/or hostile environmental conditions. Hence, such infrastructures can suffer structural damage, which can be irreversible and even catastrophic in some cases. Thus, there is a continuous need to test the health of infrastructures, a topic known as structural health testing, or monitoring (SHM) [316]. Visual inspection is, of course, an option for structural health testing, but this depends on human action and consequently cannot offer the necessary reliability that the security of infrastructures demand. Moreover, not all the parts of the infrastructures (and in some cases none of them) are always accessible. There are several nondestructive approaches to monitor structural health, which provide good resolution, sensitivity, and reliability, including ultrasonic techniques [317] and eddy-current-based techniques [318]. Fiber-optic methods for SHM have also been reported [319]. Nevertheless, there is a growing demand to implement low-cost wireless systems for SHM. In this regard, battery-free RFID sensors may constitute a convenient solution, as far as the presence of batteries would limit the lifetime of the sensor networks, or generate maintenance problems (note that dense networks with multitude of tagging sensors are typically needed in SHM systems, especially in large infrastructures) [320].

Chip-based far-field RFID SHM sensors typically provide long read distances and data storage capacity [114, 130, 320–325], but such sensors are costly as compared with their chipless-RFID counterparts. Moreover, the ICs may be subjected to extreme environments. For these main reasons, there has been an intense research activity to the implementation of chipless-based RFID SHM solutions in recent years [147, 154, 164, 165, 172, 173, 245, 326–337]. Among these chipless-RFID solutions, time-domain and frequency-domain devices have been proposed for identification and sensing, and the reported sensors have been applied to monitoring deformation, strain, mechanical defects, corrosion, humidity, and cracks. Many of these structural health problems affect the permittivity and conductivity of the surrounding medium of the tag (which can be disposed as a "sensitive skin" of the tagged infrastructure [164], or embedded in it [147]). Thus, from these variations, the structural health can be monitored (see, for example, the structure of Fig. 7.14 and related text in Section 7.2.2.2 [164]). Deformation and strain typically cause changes in the electrical length of planar antennas, resonant elements, or transmission lines, and this provides a path for sensing. Corrosion, on the other hand, generates thin oxide films on metals and other metallic-based elements, and this causes changes in the conductivity and permittivity of the materials. Since corrosion is often caused by water absorption, preventive actions are potentially possible by measuring the moisture level of infrastructures by means of humidity sensors (examples are given in Section 7.2.2.2, see Figs. 7.15 and 7.16). Obviously, the sensing tags can be functionalized with smart materials in order to enhance their sensitivities to structural damage or deterioration, as reported in Section 7.3.

7.4.4 Automotive Industry, Smart Cities, and Space

Automotive industry, smart cities, and space are sectors where RFID identification and sensing have experienced an appreciable growth and have further application potential. Security and reliability issues, the paradigm of the intelligent and autonomous car, as well as the need to automate and improve manufacturing and logistic processes have contributed to the penetration of RFID in automotive industry. Nevertheless, there are not many papers devoted to RFID identification and sensing in the automotive field. Examples are tire pressure monitoring [338–340], license plate RF identification system [341], intelligent parking [342], as well as child detection in vehicles [343, 344].

Obviously, it is also well-known that RFID technology can be used for vehicle control and management, e.g. toll payment is a canonical application of this technology (although this is achieved by means of expensive active tags). Within the framework of the "smart cities" concept, localization, and activity monitoring are other aspects that can be benefited from RFID technology. However, in such applications, the tags are not true sensing elements, acting as transducers, but collect information relative to indoor and outdoor activities, including transportation, movement flow of citizens, and localization of items [345–351].

In space platforms, weight reduction is a critical aspect, with direct impact on cost. Thus, it is desirable to eliminate as much hardware and wire connections as possible. Another important aspect in space environments is energy saving, especially for long-term missions. Within this scenario, it follows that battery-free and wireless sensors are desirable, and RFID technology thereby constitutes an interesting solution [352, 353]. Furthermore, chipless-RFID identification and sensing technology minimize risks, since the tags do not include ICs that may jeopardize the sensor functionality as consequence of the extreme ambient conditions that may be present in space environments (however, chipless-RFID technology may be adopted for short-range communications only).

7.5 Commercial Solutions, Limitations, and Future Prospects

RFID sensors exhibit very interesting features that make them very attractive for IoT applications, including low power consumption, low cost, small size, wireless connectivity (without line-of-sight requirement in some cases), and robustness against harsh conditions. Due to their high potential and to the fact that there are still many aspects that need further improvement (e.g. read range, sensor sensitivity, and accuracy), RFID sensors have been the subject of an intensive research activity in the academia (universities and research centers). This is clear, as revealed by the huge number of published journal and conference papers on the topic (the cited references of this chapter constitute only a representative subset of all of them). Nevertheless, there are also commercial RFID sensors available in the market, most of them based on chipped tags, passive or semi-passive (i.e. battery assisted). As pointed out in a pair of recently published review papers [19, 20], several companies, such as *Texas Instruments, NXP, On Semiconductors*, and *Farsens* have produced RFID sensors devoted to the measurement of different types of variables (e.g. temperature, humidity, pressure, and force/strain) of interest for sectors as diverse as agriculture, health, logistics, industrial applications, etc. Some of these sensors operate in the near field (HF band) and are able to provide sensor data in NFC-enabled smartphones, an interesting solution in applications where the requirement of a centralized reader collecting data of multiple sensor nodes does not apply. However, most commercially available RFID sensors operate in the far field (UHF band), with a cost exceeding, in general, the one of NFC-based sensors (see [19, 20] for further details). Indeed, within the framework of the IoT paradigm, far-field sensors, able to provide their data to the reader over distances of various meters, are the ideal solution. As it can be appreciated, and pointed out before in this chapter, the necessary performance in many IoT applications (mainly the read range) demands solutions that do not satisfy cost requirements, at least in situations where hundreds or thousands of sensor nodes are a due. Chipless-RFID sensors are cheap, but they cannot compete against their chip-based counterparts in terms of read range and data storage capacity. Thus, the lack of available RFID sensors satisfying simultaneously the cost and performance requirements that many IoT applications need slows down the further penetration of RFID sensing technology in the market.

Probably the most challenging aspect in RFID sensor technology is the development of low-cost (below 1US$) sensors with high performance in terms of read range, sensor characteristics (accuracy, sensitivity, etc.), and data capacity. Even though the chipless-RFID approach has raised an increasing interest among researchers, the intense activity worldwide carried out in recent years has not yet generated solutions able to satisfy these demands. Another important aspect that further boosts up the challenge is the growing need to implement eco-friendly devices, i.e. battery-free and chipless sensor tags implemented with organic or biodegradable materials. As it has been discussed, these materials typically exhibit electromagnetic performances not comparable with those of the materials used for the implementation of ordinary tags. Thus, this makes even more complex to obtain the performance that would be desirable in many IoT applications. Within this context, it is clear that further investigations on materials and processes, as well as on data processing algorithms focused on improving tag performance and read range, are of primarily interest and constitute a future research prospect. Among the applications discussed in the previous section, it is believed that healthcare, agriculture and food industries, and IoT-related smart-cities applications might be benefited from the envisaged future improvements of RFID sensing technology.

References

1 M. Wollschlaeger, T. Sauter, and J. Jasperneite, "The future of industrial communication: Automation networks in the era of the internet of things and industry 4.0," *IEEE Ind. Electron. Mag.*, vol., 11, pp. 17–27, 2017.

2 E. Sisinni, A. Saifullah, S. Han, U. Jennehag, and M. Gidlund, "Industrial internet of things: challenges, opportunities, and directions," *IEEE Trans. Ind. Inform.*, vol. 14, pp. 4724–4734, 2018.

3 J. Davies and C. Fortuna, Ed, *The Internet of Things: From Data to Insight*, John Wiley, Hoboken, NJ, USA, 2020.

4 Q. F. Hassan, Ed., *Internet of Things A to Z*, John Wiley, Hoboken, NJ, USA, 2018.

5 T. J. Kaźmierski and S. Beeby, Ed, *Energy Harvesting Systems: Principles, Modeling and Applications*, Springer-Verlag, New York, USA, 2011.

6 K. Finkenzeller, *RFID Handbook: Radio-Frequency Identification: Fundamentals and Applications*, 2nd ed., John Wiley, Hoboken, NJ, USA, 2004.

7 V. D. Hunt, A. Puglia, and M. Puglia, *RFID: A Guide to Radio Frequency Identification*, John Wiley, Hoboken, NJ, USA, 2007.

8 K. Finkenzeller, *RFID Handbook: Fundamentals and Applications in Contactless Smart Cards, Radio Frequency Identification and Near-Field Communications*, 3rd ed., John Wiley, Hoboken, NJ, USA, 2010.

9 M. Chen and S. Chen, *RFID Technologies for Internet of Things*, Springer, Cham, Switzerland, 2016.

10 R. Want, "Enabling ubiquitous sensing with RFID," *Computer*, vol. 37, pp. 84–86, Apr. 2004.

11 A. Rida, L. Yang, and M. Tentzeris, *RFID-enabled Sensor Design and Applications*, Artech House, Norwood, MA, USA, 2010.

12 G. Marrocco, "Pervasive electromagnetics: sensing paradigms by passive RFID technology," *IEEE Wireless Commun.*, vol. 17, no. 6, pp. 10–17, Dec. 2010.

13 L. Roselli, Ed., *Green RFID Systems*, Cambridge University Press, Cambridge, United Kingdom, 2014.

14 E. Perret, *Radio Frequency Identification and Sensors: From RFID to Chipless RFID*, John Wiley, Hoboken, NJ, USA, 2014.

15 S. Dey, J. K. Saha, and N. C. Karmakar, "Smart sensing: chipless RFID solutions for the internet of everything," *IEEE Microw. Mag.*, vol. 16, no. 10, pp. 26–39, Nov. 2015.

16 M. Forouzandeh and N. C. Karmakar, "Chipless RFID tags and sensors: a review on time-domain techniques," *Wirel. Power Transf.*, vol. 2, no. 2, pp. 62–77, Oct. 2015.

17 N. C. Karmakar, E. M. Amin, and J. K. Saha, *Chipless RFID Sensors*, John Wiley, Hoboken, NJ, USA, 2016.

18 A. Ramos, A. Lazaro, D. Girbau, and R. Villarino, *RFID and Wireless Sensors Using Ultra-Wideband Technology*, ISTE Press/Elsevier, London, UK, 2016.

19 L. Cui, Z. Zhang, N. Gao, Z. Meng, and Z. Li, "Radio frequency identification and sensing techniques and their applications - a review of the state-of-the-art," *Sensors*, vol. 19, paper 4012, 2019.

20 F. Costa, S. Genovesi, M. Borgese, A. Michel, F. A. Dicandia, and G. Manara, "A review of RFID sensors, the new frontier of Internet of Things," *Sensors*, vol. 21, paper 3138, 2021.

21 F. Martín, C. Herrojo, J. Mata-Contreras, and F. Paredes, *Time-Domain Signature Barcodes for Chipless-RFID and Sensing Applications*, Springer, Cham, Switzerland, 2020.

22 N. Khalid, R. Mirzavand, and A. K. Iyer, "A survey on battery-less RFID-based wireless sensors," *Micromachines*, vol. 12, paper 819, 2021.

23 S. Preradovic and N. C. Karmakar, "Chipless RFID: bar code of the future," *IEEE Microw. Mag.*, vol. 11, pp. 87–97, 2010.

24 S. Preradovic and N. C. Karmakar, *Multiresonator-Based Chipless RFID: Barcode of the Future*, Springer, 2011.

25 N. C. Karmakar, R. Koswatta, P. Kalansuriya, and R. E-Azim, *Chipless RFID Reader Architecture*, Artech House, 2013.

26 R. Rezaiesarlak and M. Manteghi, *Chipless RFID: Design Procedure and Detection Techniques*, Springer, 2015.

27 N. C. Karmakar, M. Zomorrodi, and C. Divarathne, *Advanced Chipless RFID*, John Wiley, Hoboken, NJ, USA, 2016.

28 S. Tedjini, N. Karmakar, E. Perret, A. Vena, R. Koswatta, and R. E-Azim, "Hold the chips: chipless technology, an alternative technique for RFID," *IEEE Microw. Mag.*, vol. 14, no. 5, pp. 56–65, Jul. 2013.

29 N. C. Karmakar, "Tag, you're it radar cross section of chipless RFID tags," *IEEE Microw. Mag.*, vol. 17, no. 7, pp. 64–74, Jul. 2016.

30 B. Shao, Q. Chen, Y. Amin, R. Liu, and L. R. Zheng, "Chipless RFID tags fabricated by fully printing of metallic inks," *Ann. Telecommun.*, vol. 68, no. 7, pp. 401–413, Aug. 2013.

31 A. Vena, E. Perret, and S. Tedjini, "Design rules for chipless RFID tags based on multiple scatterers," *Ann. Telecommun.*, vol. 68, no. 7, pp. 361–374, Aug. 2013.

32 A. Vena, E. Perret, and S. Tedjini, *Chipless RFID Based on RF Encoding Particle: Realization, Coding and Reading System*, ISTE Press – Elsevier, 2016.

33 O. Rance, E. Perret, R. Siragusa, and P. Lemaitre-Auger, *RCS Synthesis for Chipless RFID: Theory and Design*, Elsevier, 2017.

34 C. Herrojo, F. Paredes, J. Mata-Contreras, E. Ramon, A. Núñez, and F. Martín, "Time-domain signature barcodes: near-field chipless-RFID systems with high data capacity," *IEEE Microw. Mag.*, vol. 20, no. 12, pp. 87–101, Dec. 2019.

35 C. Herrojo, F. Paredes, J. Mata-Contreras, and F. Martín, "Chipless-RFID: a review and recent developments," *Sensors*, vol. 19, page 3385, 2019.

36 J. Zhang, S. C. Periaswamy, S. Mao, and J. Patton, "Standards for passive UHF RFID," *GetMobile Mob. Comput. Commun.*, vol. 23, pp. 10–15, 2020.

37 V. Chawla and D. S. Ha, "An overview of passive RFID," *IEEE Communications Mag.*, vol. 45, no. 9, pp. 11–17, Sep. 2007.

38 W. Yoon, S. Chung, S. Lee, and Y. Moon, "Design and implementation of an active RFID system for fast tag collection," *7th IEEE International Conference on Computer and Information Technology (CIT 2007)*, Aizu-Wakamatsu, Japan, Oct. 2007, pp. 961–966.

39 V. Pillai, H. Heinrich, D. Dieska, P. V. Kikitin, R. Martinez, and K. V. S. Rao, "An ultra-low-power long range battery/passive RFID tag for UHF and microwave bands with a current consumption of 700 nA at 1.5 V," *IEEE Trans. Circuits Syst. I: Reg. Papers*, vol. 54, no. 7, pp. 1500–1512, Jul. 2007.

40 S. Kim, J. H. Cho, H. S. Kim, H. Kim, H. B. Kang, and S. K. Hong, "An EPC Gen-2 compatible passive/semi-active UHF RFID transponder with embedded FeRAM and temperature sensor," *2017 IEEE Asian Solid-State Circuits Conference*, Jeju, Nov. 2007, pp. 135–138.

41 W. Che, W. Chen, D. Meng, X. Wang, X. Tan, N. Yan, and H. Min, "Power management unit for battery assisted passive RFID tag," *Electron. Lett.*, vol. 46, no. 8, pp. 589–590, Apr. 2010.

42 V. Duong, N. X. Hieu, H. Lee, and J. Lee, "A battery-assisted passive EPC Gen-2 RFID sensor tag IC with efficient battery power management and RF energy harvesting," *IEEE Trans. Ind. Electron.*, vol. 63, no. 11, pp. 7112–7123, Nov. 2016.

43 C. S. Hartmann, "A global SAW ID tag with large data capacity," *2002 IEEE Ultrasonics Symposium*, Munich, Germany, Oct. 2002, vol. 1, pp. 65–69.

44 N. Saldanha and D. C. Malocha, "Design parameters for SAW multi-tone frequency coded reflectors," *2007 IEEE Ultrasonics Symposium Proceedings*, New York, NY, USA, Oct. 2007, pp. 2087–2090.

45 S. Harma, V. P. Plessky, C. S. Hartmann, and W. Steichen, "Z-path SAW RFID tag," *IEEE Trans. Ultrasonics, Ferroelectric Freq. Control*, vol. 55, pp. 208–213, 2008.

46 T. Han, W. Wang, H. Wu, and Y. Shui, "Reflection and scattering characteristics of reflectors in SAW tags," *IEEE Trans. Ultrason. Ferroelectr. Freq. Control*, vol. 55, no. 6, pp. 1387–1390, Jun. 2008.

47 S. Harma, V. P. Plessky, X. Li, and P. Hartogh, "Feasibility of ultra-wideband SAW RFID tags meeting FCC rules," *IEEE Trans. Ultrasonics, Ferroelectric Freq. Control*, vol. 56, pp. 812–820, 2009.

48 C. Hartmann, P. Hartmann, P. Brown, J. Bellamy, L. Claiborne, and W. Bonner, "Anti-collision methods for global SAW RFID tag systems," *2004 IEEE Ultrasonics Symposium*, Montreal, QC, Canada, Aug. 2004, vol. 2, pp. 805–808.

49 L. Wei, H. Tao, and S. Yongan, "Surface acoustic wave based radio frequency identification tags," *2008 IEEE International Conference on e-Business Engineering*, Xian, China, Oct. 2008, pp. 563–567.

50 A. Chamarti and K. Varahramyan, "Transmission delay line based ID generation circuit for RFID applications," *IEEE Microw. Wireless Compon. Lett.*, vol. 16, pp. 588–590, 2006.

51 M. Schüßler, C. Damm, and R. Jakoby, "Periodically LC loaded lines for RFID backscatter applications," *Proc. Metamaterials 2007*, Rome, Italy, Oct. 2007, pp. 103–106.

52 M. Schüßler, C. Damm, M. Maasch, and R. Jakoby, "Performance evaluation of left-handed delay lines for RFID backscatter applications," *2008 IEEE MTT-S Int. Microw. Symp. Digest*, Atlanta, GA, USA, Jun. 2008, pp. 177–180.

53 B. Shao, Q. Chen, Y. Amin, D. S. Mendoza, R. Liu, and L. R. Zheng, "An ultra-low-cost RFID tag with 1.67 Gbps data rate by ink-jet printing on paper substrate," *2010 IEEE Asian Solid State-Circuits Conf.*, Beijing, China, Nov. 2010.

54 F. J. Herraiz-Martínez, F. Paredes, G. Zamora, F. Martín, and J. Bonache, "Printed magnetoinductive-wave (MIW) delay lines for chipless RFID applications," *IEEE Trans. Ant. Propag.*, vol. 60, pp. 5075–5082, Nov. 2012.

55 S. Tedjini, E. Perret, A. Vena, and D. Kaddout, "Mastering the electromagnetic signature of chipless RFID tags," in: IGI Global. (ed) *Chipless and Conventional Radiofrequency Identification*, IGI Global, 2012, pp. 146–174.

56 L. Zhang, S. Rodriguez, H. Tenhunen, and L. R. Zheng, "An innovative fully printable RFID technology based on high speed time-domain reflections," *Conference on High Density Microsystem Design and Packaging and Component Failure Analysis, 2006 (HDP'06)*, Shanghai, China, Jun. 2006, pp. 166–170.

57 L. Zheng, S. Rodriguez, L. Zhang, B. Shao, and L. R. Zheng, "Design and implementation of a fully reconfigurable chipless RFID tag using Inkjet printing technology," *2008 IEEE International Symposium on Circuits and Systems (ISCAS)*, Seattle, WA, USA, May 2008, pp. 1524–1527.

58 C. Mandel, M. Schussler, M. Maasch, and R. Jakoby, "A novel passive phase modulator based on LH delay lines for chipless microwave RFID applications," *2009 IEEE MTT-S International Microwave Workshop on Wireless Sensing, Local Positioning, and RFID*, Cavtat, Croatia, Sept. 2009.

59 R. Nair, E. Perret, and S. Tedjini, "Temporal multi-frequency encoding technique for chipless RFID applications," *2012 IEEE MTT-S Int. Microw. Symp. Digest*, Montreal, QC, Canada, Jun. 2012.

60 I. Jalaly and I. D. Robertson, "RF barcodes using multiple frequency bands," *2005 IEEE MTT-S Int. Microw. Symp. Digest*, Long Beach, CA, USA, Jun. 2005, pp. 139–142.

61 S. Preradovic, I. Balbin, N. C. Karmakar, and G. F. Swiegers, "Multiresonator-based chipless RFID system for low-cost item tracking," *IEEE Trans. Microw. Theory Techn.*, vol. 57, pp. 1411–1419, 2009.

62 S. Preradovic and N. C. Karmakar, "Design of chipless RFID tag for operation on flexible laminates," *IEEE Anten. Wireless Propag. Lett.*, vol. 9, pp. 207–210, 2010.

63 J. McVay, A. Hoorfar, and N. Engheta, "Space-filling curve RFID tags," *2006 IEEE Radio and Wireless Symp.*, San Diego, CA, USA, Oct. 2006, pp. 199–202.

64 I. Jalaly and D. Robertson, "Capacitively-tuned split microstrip resonators for RFID barcodes," *2005 European Microwave Conference*, Paris, France, Oct. 2005, vol. 2, pp. 4–7.

65 H. S. Jang, W. G. Lim, K. S. Oh, S. M. Moon, and J. W. Yu, "Design of low-cost chipless system using printable chipless tag with electromagnetic code," *IEEE Microw. Wireless Compon. Lett.*, vol. 20, pp. 640–642, 2010.

66 A. Vena, E. Perret, and S. Tedjini, "A fully printable chipless RFID tag with detuning correction technique," *IEEE Microw. Wireless Compon. Lett.*, vol. 22, pp. 209–211, 2012.

67 A. Vena, E. Perret, and S. Tedjini, "Design of compact and auto-compensated single-layer chipless RFID tag," *IEEE Trans. Microw. Theory Techn.*, vol. 60, pp. 2913–2924, Sep. 2012.

68 A. Vena, E. Perret, and S. Tedjini, "High-capacity chipless RFID tag insensitive to the polarization," *IEEE Trans. Ant. Propag.*, vol. 60, pp. 4509–4515, Oct. 2012.

69 D. Girbau, J. Lorenzo, A. Lazaro, C. Ferrater, and R. Villarino, "Frequency-coded chipless RFID tag based on dual-band resonators," *IEEE Ant. Wireless Propag. Lett.*, vol. 11, pp. 126–128, 2012.

70 M. M. Khan, F. A. Tahir, M. F. Farooqui, A. Shamim, and H. M. Cheema, "3.56-bits/cm^2 compact inkjet printed and application specific chipless RFID tag," *IEEE Ant. Wireless Propag. Lett.*, vol. 15, pp. 1109–1112, 2016.

71 R. Rezaiesarlak and M. Manteghi, "Complex-natural-resonance-based design of chipless RFID tag for high-density data," *IEEE Trans. Ant. Propag.*, vol. 62, pp. 898–904, Feb. 2014.

72 M. S. Bhuiyan and N. Karmakar, "A spectrally efficient chipless RFID tag based on split-wheel resonator," *2014 International Workshop on Antenna Technology: Small Antennas, Novel EM Structures and Materials, and Applications* (iWAT), Sydney, NSW, Australia, Mar. 2014.

73 C. M. Nijas, U. Deepak, P. V. Vinesh, R. Sujith, S. Mridula, K. Vasudevan, and P. Mohanan, "Low-cost multiple-bit encoded chipless RFID tag using stepped impedance resonator," *IEEE Trans. Ant. Propag.*, vol. 62, no. 9, pp. 4762–4770, Sep. 2014.

74 J. Machac and M. Polivka, "Influence of mutual coupling on performance of small scatterers for chipless RFID tags," *2014 24th International Conference Radioelektronika*, Bratislava, Slovakia, Apr. 2014.

75 M. Svanda, J. Machac, M. Polivka, and J. Havlicek, "A comparison of two ways to reducing the mutual coupling of chipless RFID tag scatterers," *2016 21st International Conference on Microwave, Radar and Wireless Communications (MIKON)*, Krakow, Poland, May 2016.

76 C. Herrojo, J. Naqui, F. Paredes, and F. Martín, "Spectral signature barcodes based on S-shaped Split ring resonators (S-SRR)," *EPJ Appl. Metamaterials*, vol. 3, pp. 1–6, Jun. 2016.

77 A. Vena, E. Perret, and S. Tedjini, "Chipless RFID tag using hybrid coding technique," *IEEE Trans. Microw. Theory Techn.*, vol. 59, pp. 3356–3364, Dec. 2011.

78 A. Vena, E. Perret, and S. Tedjini, "A compact chipless RFID tag using polarization diversity for encoding and sensing," *2012 IEEE International Conference on RFID (RFID)*, Orlando, FL, USA, Apr. 2012, pp. 191–197.

79 M. A. Islam and N. C. Karmakar, "A novel compact printable dual-polarized chipless RFID system," *IEEE Trans. Microw. Theory Techn.*, vol. 60, pp. 2142–2151, Jul. 2012.

80 I. Balbin and N. C. Karmakar, "Phase-encoded chipless RFID transponder for large scale low cost applications," *IEEE Microw. Wireless. Compon. Lett.*, vol. 19, pp. 509–511, 2009.

81 S. Genovesi, F. Costa, A. Monorchio, and G. Manara, "Chipless RFID tag exploiting multifrequency delta-phase quantization encoding," *IEEE Ant. Wireless Propag. Lett.*, vol. 15, pp. 738–741, 2015.

82 O. Rance, R. Siragusa, P. Lemaitre-Auger, and E. Perret, "RCS magnitude coding for chipless RFID based on depolarizing tag," *2015 IEEE MTT-S Int. Microw. Symp. Digest*, Phoenix, AZ, USA, May 2015.

83 O. Rance, R. Siragusa, P. Lemaître-Auger, and E. Perret, "Toward RCS magnitude level coding for chipless RFID," *IEEE Trans. Microw. Theory Techn.*, vol. 64, pp. 2315–2325, Jul. 2016.

84 C. Herrojo, J. Naqui, F. Paredes, and F. Martín, "Spectral signature barcodes implemented by multi-state multi-resonator circuits for chipless RFID tags," *2016 IEEE MTT-S Int. Microw. Symp. Digest (IMS'16)*, San Francisco, USA, May 2016.

85 C. Herrojo, F. Paredes, J. Mata-Contreras, S. Zuffanelli, and F. Martín, "Multi-state multi-resonator spectral signature barcodes implemented by means of S-shaped Split Ring Resonators (S-SRR)," *IEEE Trans. Microw. Theory Techn.*, vol. 65, no. 7, pp. 2341–2352, Jul. 2017.

86 S. Gupta, B. Nikfal, and C. Caloz, "Chipless RFID system based on group delay engineered dispersive delay structures," *IEEE Ant. Wireless Propag. Lett.*, vol. 10, pp. 1366–1368, 2011.

87 R. Nair, E. Perret, and S. Tedjini, "Chipless RFID based on group delay encoding," *2011 IEEE Int. Conf. RFID-Technol. Appl.*, Sitges, Spain, Sept. 2011, pp. 214–218.

88 C. Feng, W. Zhang, L. Li, L. Han, X. Chen, and R. Ma, "Angle-based chipless RFID tag with high capacity and insensitivity to polarization," *IEEE Trans. Ant. Propag.*, vol. 63, no. 4, pp. 1789–1797, Apr. 2015.

89 A. El-Awamry, M. Khaliel, A. Fawky, M. El-Hadidy, and T. Kaiser, "Novel notch modulation algorithm for enhancing the chipless RFID tags coding capacity," *2015 IEEE Int. Conf. RFID*, San Diego, CA, USA, Apr. 2015, pp. 25–31.

90 A. Vena, A. A. Babar, L. Sydanheimo, M. M. Tentzeris, and L. Ukkonen, "A novel near-transparent ASK-reconfigurable inkjet-printed chipless RFID tag," *IEEE Ant. Wireless Propag. Lett.*, vol. 12, pp. 753–756, 2013.

91 M. Mohaisen, H. Yoon, and K. Chang, "Radio transmission performance of EPC global Gen-2 RFID system," *2008 10th International Conference on Advanced Communication Technology*, Gangwon, South Korea, Feb. 2008, pp. 1423–1428.

92 C. Herrojo, J. Mata-Contreras, F. Paredes, and F. Martín, "Near-field chipless RFID encoders with sequential bit reading and high data capacity," *2017 IEEE MTT-S Int. Microw. Symp. Digest (IMS'17)*, Honolulu, HI, USA, Jun. 2017.

93 C. Herrojo, J. Mata-Contreras, F. Paredes, and F. Martín, "Microwave encoders for chipless RFID and angular velocity sensors based on S-shaped split ring resonators (S-SRRs)," *IEEE Sensors J.*, vol. 17, pp. 4805–4813, Aug. 2017.

94 C. Herrojo, J. Mata-Contreras, F. Paredes, A. Núñez, E. Ramón, and F. Martín, "Near-field chipless-RFID tags with sequential bit reading implemented in plastic substrates," *Int. J. Magnetism. Magnetic Mat.*, vol. 459 pp. 322–327, 2018.

95 C. Herrojo, J. Mata-Contreras, F. Paredes, and F. Martín, "High data density and capacity in chipless radiofrequency identification (chipless-RFID) tags based on double-chains of S-shaped split ring resonators (S-SRRs)," *EPJ Appl. Metamat.*, vol. 4, article 8, 6 pages, Oct. 2017.

96 C. Herrojo, J. Mata-Contreras, F. Paredes, and F. Martín, "Near-field chipless RFID system with high data capacity for security and authentication applications," *IEEE Trans. Microw. Theory Techn.*, vol. 65, pp. 5298–5308, Dec. 2017.

97 C. Herrojo, J. Mata-Contreras, F. Paredes, A. Núñez, E. Ramon, and F. Martín, "Near-field chipless-RFID system with erasable/programmable 40-bit tags inkjet printed on paper substrates," *IEEE Microw. Wireless Compon. Lett.*, vol. 28, pp. 272–274, Mar. 2018.

98 C. Herrojo, J. Mata-Contreras, F. Paredes, A. Núñez, E. Ramon, and F. Martín, "Very low-cost 80-bit chipless-RFID tags inkjet printed on ordinary paper," *Technologies*, vol. 6, no. 2, p. 52, May 2018.

99 J. Havlíček, C. Herrojo, F. Paredes, J. Mata-Contreras, and F. Martín, "Enhancing the per-unit-length data density in near-field chipless-RFID systems with sequential bit reading," *IEEE Ant. Wireless Propag. Lett.*, vol. 18, pp. 89–92, Jan. 2019.

100 C. Herrojo, F. Paredes, and F. Martín, "Double-stub loaded microstrip line reader for very high data density microwave encoders," *IEEE Trans. Microw. Theory and Techn.*, vol.67, pp. 3527–3536, Sep. 2019.

101 C. Herrojo, F. Paredes, and F. Martín, "3D-printed high data-density electromagnetic encoders based on permittivity contrast for motion control and chipless-RFID," *IEEE Trans. Microw. Theory Techn.*, vol. 68, no. 5, pp. 1839–1850, May 2020.

102 F. Paredes, C. Herrojo, R. Escudé, E. Ramon, and F. Martín, "High data density near-field chipless-RFID tags with synchronous reading," *IEEE J. RFID*, vol. 4, no. 4, pp. 517–524, Dec. 2020.

103 C. Herrojo, F. Paredes, and F. Martín "Synchronism and direction detection in high resolution/high-density electromagnetic encoders," *IEEE Sensors J.*, vol. 21, no. 3, pp. 2873–2882, Feb. 2021.

104 M. Moras, C. Martínez-Domingo, C. Herrojo, F. Paredes, L. Terés, F. Martin, and E. Ramon, "Programmable organic chipless-RFID tags inkjet printed on paper substrates," *Appl. Sci.*, vol. 11, paper 7832, 2021.

105 W. Dargie and C. Poellabauer, *Fundamentals of Wireless Sensor Networks: Theory and Practice*, John Wiley, Hoboken, NJ, USA, 2010.

106 Z. Xiao, X. Tan, X. Chen, S. Chen, Z. Zhang, H. Zhang, J. Wang, Y. Huang, P. Zhang, L. Zheng, and H. Min, "An implantable RFID sensor tag toward continuous glucose monitoring," *IEEE Journal of Biomedical and Health Informatics*, vol. 19, no. 3, pp. 910–919, May 2015.

107 A. Wickramasinghe and D. C. Ranasinghe, "Ambulatory monitoring using passive computational RFID sensors," *IEEE Sensors J.*, vol. 15, no. 10, pp. 5859–5869, Oct. 2015.

108 E. Moradi, L. Sydänheimo, G. S. Bova, and L. Ukkonen, "Measurement of wireless power transfer to deep-tissue RFID-based implants using wireless repeater node," *IEEE Ant. Wireless Propag. Lett.*, vol. 16, pp. 2171–2174, 2017.

109 J. J. Baek, S. W. Kim, K. H. Park, M. J. Jeong, and Y. T. Kim, "Design and performance evaluation of 13.56-MHz passive RFID for E-skin sensor application," *IEEE Microw. Wireless Compon. Lett.*, vol. 28, no. 12, pp. 1074–1076, Dec. 2018.

110 E. Smits, J. Schram, M. Nagelkerke, R. H. L. Kusters, G. V. Heck, V. V. Acht, M. M. Koetse, J. V. D. Brand, G. H. Gelinck, and H. F. M. Schoo, "Development of printed RFID sensor tags for smart food packaging," *14th International Meeting on Chemical Sensors*, Nuremberg, Germany, May 2012, pp. 20–23.

111 K. H. Eom, K. H. Hyun, S. Lin, and J. W. Kim, "The meat freshness monitoring system using the smart RFID tag," *Int. J. Distrib. Sens. Netw.*, vol. 10, paper 591812, 2014.

112 S. F. Pichorim, N. J. Gomes, and J.C. Batchelor, "Two solutions of soil moisture sensing with RFID for landslide monitoring," *Sensors*, vol. 18, no. 2, paper 452, 2018.

113 G. S. Lorite, T. Selkälä, T. Sipola, J. Palenzuela, E. Jubete, A. Viñuales, G. Cabañero, H. J. Grande, J. Tuominen, S. Uusitalo, L. Hakalahti, K. Kordas, and G. Toth, "Novel, smart and RFID assisted critical temperature indicator for supply chain monitoring," *J. Food Eng.*, vol. 193, pp. 20–28, 2017.

114 E. DiGiampaolo, A. DiCarlofelice, and A. Gregori, "An RFID-enabled wireless strain gauge sensor for static and dynamic structural monitoring," *IEEE Sensors J.*, vol. 17, no. 2, pp. 286–294, Jan. 2017.

115 W. D. Leon-Salas and C. Halmen, "A RFID sensor for corrosion monitoring in concrete," *IEEE Sensors J.*, vol. 16, no. 1, pp. 32–42, Jan. 2016.

116 D. Petrov, M. Schmidt, U. Hilleringmann, C. Hedayat, and T. Otto, "RFID based sensor platform for industry 4.0 application," *Smart Systems Integration; 13th International Conference and Exhibition on Integration Issues of Miniaturized Systems*, Barcelona, Spain, Apr. 2019.

117 C. Kollegger, P. Greiner, C. Steffan, M. Wiessflecker, H. Froehlich, T. Kautzsch, G. Holweg, and B. Deutschmann, "A system-on-chip NFC bicycle tire pressure measurement system," *2017 IEEE 60th International Midwest Symposium on Circuits and Systems (MWSCAS)*, Boston, MA, USA, Aug. 2017, pp. 60–63.

118 Y. Huo, Y. Lu, W. Cheng, and T. Jing, "Vehicle road distance measurement and maintenance in RFID systems on roads," *2014 International Conference on Connected Vehicles and Expo* (ICCVE), Vienna, Austria, Nov. 2014, pp. 30–36.

119 T. Wang, Y. He, Q. Luo, F. Deng, and C. Zhang, "Self-powered RFID sensor tag for fault diagnosis and prognosis of transformer winding," *IEEE Sensors J.*, vol. 17, no. 19, pp. 6418–6430, Oct. 2017.

120 P. Escobedo, M. M. Erenas, N. Lopez-Ruiz, M. A. Carvajal, S. Gonzalez-Chocano, I. de Orbe-Payá, L. F. Capitán-Valley, A. J. Palma, and A. Martínez-Olmos, "Flexible passive near field communication tag for multigas sensing," *Anal. Chem.*, vol. 89, pp. 1697–1703, 2017.

121 Z. Li, P. Dharap, S. Nagarajaiah, E. V. Barrera, and J. D. Kim, "Carbon nanotube film sensors," *Adv. Mater.*, vol. 16, no. 7, pp. 640–643, Apr. 2004.

122 S. Kim, Y. Kawahara, A. Georgiadis, A. Collado, and M. M. Tentzeris, "Low-cost inkjet-printed fully passive RFID tags for calibration-free capacitive/haptic sensor applications," *IEEE Sensors J.*, vol. 15, no. 6, pp. 3135–3145, Jun. 2015.

123 H. Wegleiter, B. Schweighofer, C. Deinhammer, G. Holler, and P. Fulmek, "Automatic antenna tuning unit to improve RFID system performance," *IEEE Trans. Instrum. Measur.*, vol. 60, no. 8, pp. 2797–2803, Aug. 2011.

124 M. C. Caccami and G. Marrocco, "Electromagnetic modeling of self-tuning RFID sensor antennas in linear and nonlinear regimes," *IEEE Trans. Ant. Propag.*, vol. 66, no. 6, pp. 2779–2787, Jun. 2018.

125 K. Rasilainen, J. Ilvonen, A. Lehtovuori, J. Hannula, and V. Viikari, "On design and evaluation of harmonic transponders," *IEEE Trans. Ant. Propag.*, vol. 63, no. 1, pp. 15–23, Jan. 2015.

126 F. Alimenti and L. Roselli, "Theory of zero-power RFID sensors based on harmonic generation and orthogonally polarized antennas," *Prog. Electromagn. Res.*, vol. 134, pp. 337–357, 2013.

127 A. Sharif, J. Ouyang, F. Yang, H. T. Chattha, M. A. Imran, A. Alomainy, and Q. H. Abbasi, "Low-cost inkjet-printed UHF RFID tag-based system for Internet of Things applications using characteristic modes," *IEEE Internet of Things J.*, vol. 6, no. 2, pp. 3962–3975, Apr. 2019.

128 R. A. Potyrailo, N. Nagraj, Z. Tang, F. J. Mondello, C. Surman, and W. Morris, "Battery-free radio frequency identification (RFID) sensors for food quality and safety," *J. Agric. Food Chem.*, vol. 60, no. 35, pp. 8535–8543, Sep. 2012.

129 S. Manzari, A. Catini, G. Pomarico, C. D. Natale, and G. Marrocco, "Development of an UHF RFID chemical sensor array for battery-less ambient sensing," *IEEE Sens. J.*, vol. 14, pp. 3616–3623, 2014.

130 S. Caizzone and E. DiGiampaolo, "Wireless passive RFID crack width sensor for structural health monitoring," *IEEE Sensors J.*, vol. 15, no. 12, pp. 6767–6774, Dec. 2015.

131 S. Genovesi, F. Costa, M. Borgese, A. F. Dicandia, G. Manara, S. Tedjini, and E. Perret, "Enhanced chipless RFID tags for sensors design," *2016 IEEE Int. Symp. Ant. Propag. (APSURSI)*, Fajardo, PR, USA, Jun. 2016, pp. 1275–1276.

132 S. Kim, C. Mariotti, F. Alimenti, P. Mezzanotte, A. Georgiadis, A. Collado, L. Roselli and M.M. Tentzeris, "No battery required: perpetual RFID-enabled wireless sensors for cognitive intelligence applications," *IEEE Microw. Mag.*, vol. 14, no. 5, pp. 66–77, July-Aug. 2013.

133 A. Vena, E. Perret, D. Kaddour, and T. Baron, "Toward a reliable chipless RFID humidity sensor tag based on silicon nanowires," *IEEE Trans. Microw. Theory Tech.*, vol. 64, no. 9. pp. 2977–2985, 2016.

134 E. M. Amin, J. K. Saha, and N. C. Karmakar, "Smart sensing materials for low-cost chipless RFID sensor," *IEEE Sensors J.*, vol. 14, pp. 2198–2207, 2014.

135 S. Genovesi, F. Costa, M. Borgese, F. A. Dicandia, A. Monorchio, and G. Manara, "Chipless RFID sensor for rotation monitoring," *2017 IEEE International Conference on RFID Technology Application (RFID-TA)*, Warsaw, Poland, Sep. 2017, pp. 233–236.

136 B. Ando and S. Baglio, "Inkjet-printed sensors: a useful approach for low cost, rapid prototyping," *IEEE Instrum. Meas. Mag.*, vol. 14, pp. 36–40, 2011.

137 C. Mandel, B. Kubina, M. Schüßler, and R. Jakoby, "Passive chipless wireless sensor for two-dimensional displacement measurement," *2011 41st European Microwave Conference*, Manchester, UK, Oct. 2011, pp. 79–82.

138 V. P. Plessky and L. M. Reindl, "Review on SAW RFID tags," *IEEE Trans. Ultrason. Ferroelectr. Freq. Control.*, vol. 57, pp. 654–668, 2010.

139 A. Pohl, G. Ostermayer, L. Reindl, and F. Seifert, "Monitoring the tire pressure at cars using passive SAW sensors," *1997 IEEE Ultrasonics Symposium Proceedings. An International Symposium* (Cat. No.97CH36118), Toronto, ON, Canada, Oct. 1997, vol. 1, pp. 471–474.

140 F. Seifert, W. E Bulst, and C. Ruppel, "Mechanical sensors based on surface acoustic waves," *Sens. Actuators Phys.*, vol. 44, pp. 231–239, 1994.

141 A. Pohl, "A review of wireless SAW sensors," *IEEE Trans. Ultrason. Ferroelectr. Freq. Control.*, vol. 47, pp. 317–332, 2000.

142 D. Puccio, D. C. Malocha, D. Gallagher, and J. Hines, "SAW sensors using orthogonal frequency coding," *2004 IEEE International Frequency Control Symposium and Exposition*, Montreal, QC, Canada, Aug. 2004, pp. 307–310.

143 E. Cantatore, T. C. T. Geuns, G. H. Gelinck, E. van Veenendaal, A. F. A. Gruijthuijsen, L. Schrijnemakers, S. Drews, and D. M. de Leeuw, "A 13.56-MHz RFID system based on organic transponders," *IEEE J. Solid-State Circ.*, vol. 42, no. 1, pp. 84–92, Jan. 2007.

144 S. Conti, L. Pimpolari, G. Calabrese, R. Worsley, S. Majee, D. K. Polyushkin, M. Paur, S. Pace, D. H. Keum, F. Fabbri, and G. Iannaccone, "Low-voltage 2D materials-based printed field-effect transistors for integrated digital and analog electronics on paper," *Nat. Commun.*, vol. 11, no. 1, pp. 1–9, Jul. 2020.

145 K. Myny, "The development of flexible integrated circuits based on thin-film transistors," *Nat. Electron.*, vol. 1, no. 1, pp. 30–39, Jan. 2018.

146 Y. Khan, A. Thielens, S. Muin, J. Ting, C. Baumbauer, and A. C. Arias, "A new frontier of printed electronics: flexible hybrid electronics," *Adv. Mater.*, vol. 32, no. 15, paper 1905279, Apr. 2020.

147 A. Ramos, D. Girbau, A. Lazaro, and R. Villarino, "Wireless concrete mixture composition sensor based on time-coded UWB RFID," *IEEE Microw. Wireless Compon. Lett.*, vol. 25, no. 10, pp. 681–683, Oct. 2015.

148 D. Girbau, Á. Ramos, A. Lazaro, S. Rima, and R. Villarino, "Passive wireless temperature sensor based on time-coded UWB chipless RFID tags," *IEEE Trans. Microw. Theory Techn.*, vol. 60, no. 11, pp. 3623–3632, Nov. 2012.

149 G. Kaiser, *A Friendly Guide to Wavelets*, Birkhauser, Boston, MA, 1994.

150 A. Lazaro, D. Girbau, and R. Villarino, "Wavelet-based breast tumor localization technique of microwave imaging using UWB," *Prog. Electromagn. Res.*, vol. 94, pp. 264–280, 2009.

151 A. Lazaro, A. Ramos, D. Girbau, and R. Villarino, "Chipless UWB RFID tag detection using continuous wavelet transform," *IEEE Ant. Wireless Propag. Lett.*, vol. 10, pp. 520–523, 2011.

152 A. Ramos, D. Girbau, A. Lazaro, and R. Villarino, "Permittivity sensor using chipless time-coded UWB RFID," *2014 XXXIth URSI General Assembly and Scientific Symposium (URSI GASS)*, Beijing, China, Aug. 2014.

153 S. Shrestha, M. Balachandran, M. Agarwal, V. V. Phoha, and K. Varahramyan, "A chipless RFID sensor system for cyber centric monitoring applications," *IEEE Trans. Microw. Theory Techn.*, vol. 57, no. 5, pp. 1303–1309, May 2009.

154 P. Kalansuriya, R. Bhattacharyya, S. Sarma, and N. Karmakar, "Towards chipless RFID-based sensing for pervasive surface crack detection," *2012 IEEE International Conference on RFID-Technologies and Applications (RFID-TA)*, Nice, France, Nov. 2012, pp. 46–51.

155 R. S. Nair, E. Perret, S. Tedjini, and T. Baron, "A group-delay-based chipless RFID humidity tag sensor using silicon nanowires," *IEEE Ant. Wireless Propag. Lett.*, vol. 12, pp. 729–732, 2013.

156 A. Ramos, A. Lazaro, R. Villarino, and D. Girbau, "Sensing of thermal thresholds using UWB RFID passive tags," *2014 IEEE Sensors*, Valencia, Spain, Nov. 2014, pp. 1503–1506.

157 Y. Wang, C. H. Quan, F. X. Liu, X. Y. Zhang, and J. C. Lee, "A new chipless RFID permittivity sensor system," *IEEE Access*, vol. 9, pp. 35027–35033, 2021.

158 E. M. Amin, M. S. Bhuiyan, N. C. Karmakar, and B. Winther-Jensen, "Development of a low cost printable chipless RFID humidity sensor," *IEEE Sensors J.*, vol. 14, no. 1, pp. 140–149, Jan. 2014.

159 Y. Feng, L. Xie, Q. Chen, and L. Zheng, "Low-cost printed chipless RFID humidity sensor tag for intelligent packaging," *IEEE Sensors J.*, vol. 15, no. 6, pp. 3201–3208, Jun. 2015.

160 M. M. Forouzandeh and N. Karmakar, "Towards the improvement of frequency-domain chipless RFID readers," *2018 IEEE Wireless Power Transfer Conference (WPTC)*, Montreal, QC, Canada, Feb. 2018.

161 N. Javed, A. Habib, Y. Amin, J. Loo, A. Akram, and H. Tenhunen, "Directly printable moisture sensor tag for intelligent packaging," *IEEE Sensors J.*, vol. 16, no. 16, pp. 6147–6148, Aug. 2016.

162 M. Borgese, F. A. Dicandia, F. Costa, S. Genovesi, and G. Manara, "An inkjet printed chipless RFID sensor for wireless humidity monitoring," *IEEE Sensors J.*, vol. 17, no. 15, pp. 4699–4707, Aug. 2017.

163 H. E. Matbouly, K. Zannas, Y. Duroc, and S. Tedjini, "Chipless wireless temperature sensor based on C-like scatterer for standard RFID reader," *2017 XXXIInd General Assembly and Scientific*

Symposium of the International Union of Radio Science (URSI GASS), Montreal, QC, Canada, Aug. 2017.

164 A. Lázaro, R. Villarino, F. Costa, S. Genovesi, A. Gentile, L. Buoncristiani, and D. Girbau, "Chipless dielectric constant sensor for structural health testing," *IEEE Sensors J.*, vol. 18, no. 13, pp. 5576–5585, Jul. 2018.

165 A. M. J. Marindra and G. Y. Tian, "Chipless RFID sensor tag for metal crack detection and characterization," *IEEE Trans. Microw. Theory Techn.*, vol. 66, no. 5, pp. 2452–2462, May 2018.

166 J. A. Satti, A. Habib, H. Anam, S. Zeb, Y. Amin, J. Loo, and H. Tenhunen, "Miniaturized humidity and temperature sensing RFID enabled tags," *Int J. RF Microw. Comput. Aided Eng.*, vol. 28, no. 1, pp. e21151, Jan. 2018.

167 N. Javed, A. Habib, T. Noor, Y. Amin, and H. Tenhunen, "RFID enabled chipless humidity sensor," *The Nucleus*, vol. 56, no. 1, pp. 27–30, 2019.

168 W. M. Abdulkawi and A. F. A. Sheta, "Two-bit chipless RFID for temperature and humidity sensing," *2019 IEEE Asia-Pacific Microwave Conference (APMC)*, Singapore, Dec. 2019, pp. 1443–1445.

169 R. Raju, G. E. Bridges, and S. Bhadra, "Wireless passive sensors for food quality monitoring: improving the safety of food products," *IEEE Ant. Propag. Mag.*, vol. 62, no. 5, pp. 76–89, Oct. 2020.

170 W. M. Abdulkawi and A. F. A. Sheta, "Chipless RFID sensors based on multistate coupled line resonators," *Sens. Act. A: Phys.*, vol. 309, paper 112025, Jul. 2020.

171 D. P. Mishra, T. K. Das, and S. K. Behera, "Design of a 3-bit chipless RFID tag using circular split-ring resonators for retail and healthcare applications," *2020 National Conference on Communications* (NCC), Kharagpur, India, Feb. 2020.

172 N. Javed, M. A. Azam, and Y. Amin, "Chipless RFID multisensor for temperature sensing and crack monitoring in an IoT environment," *IEEE Sensors Lett.*, vol. 5, no. 6, paper 6001404, Jun. 2021.

173 S. Dey, R. Bhattacharyya, S. E. Sarma, and N. C. Karmakar, "A novel "smart skin" sensor for chipless RFID-based structural health monitoring applications," *IEEE Internet of Things Journal*, vol. 8, no. 5, pp. 3955–3971, Mar. 2021.

174 J. Yeo, J. I. Lee, and Y. Kwon, "Humidity-sensing chipless RFID tag with enhanced sensitivity using an interdigital capacitor structure," *Sensors*, vol. 21, paper 6550, 2021.

175 F. Costa, A. Gentile, S. Genovesi, L. Buoncristiani, A. Lazaro, R. Villarino, and D. Girbau, "A depolarizing chipless RF label for dielectric permittivity sensing," *IEEE Microw. Wireless Compon. Lett.*, vol. 28, no. 5, pp. 371–373, May 2018.

176 A. Vena, E. Perret, and S. Tedjni, "A depolarizing chipless RFID tag for robust detection and its FCC compliant UWB reading system," *IEEE Trans. Microw. Theory Techn.*, vol. 61, no. 8, pp. 2982–2994, Aug. 2013.

177 S. Genovesi, F. Costa, F. A. Dicandia, M. Borgese, and G. Manara, "Orientation-insensitive and normalization-free reading chipless RFID system based on circular polarization interrogation," *IEEE Trans. Ant. Propag.*, vol. 68, no. 3, pp. 2370–2378, Mar. 2020.

178 F. Costa, S. Genovesi, and A. Monorchio, "Normalization-free chipless RFIDs by using dual-polarized interrogation," *IEEE Trans. Microw. Theory Techn.*, vol. 64, no. 1, pp. 310–318, Jan. 2016.

179 K. M. Gee, P. Anandarajah, and D. Collins, "A review of chipless remote sensing solutions based on RFID technology," *Sensors*, vol. 19, paper 4829, 2019.

180 V. Mulloni and M. Donelli, "Chipless RFID sensors for the Internet of Things: challenges and opportunities," *Sensors*, vol. 20, paper 2135, 2020.

181 F. Babaeian and N. C. Karmakar "Time and frequency domains analysis of chipless RFID back-scattered tag reflection," *IoT*, vol. 1, pp. 109–127, 2020.

182 A. Blayo, and B. Pineaux, "Printing processes and their potential for RFID printing," *2005 Joint Conference on Smart Objects and Ambient Intelligence*, Grenoble, France, Oct. 2005, pp 27–30.

183 K. Weigelt, M. Hambsch, G. Karacs, T. Zillger, and A. C. Hubler, "Labeling the world: tagging mass products with printing processes," *IEEE Pervasive Computing*, vol. 9, no. 2, pp. 59–63, Mar. 2010.

184 T. Kellomäki, J. Virkki, S. Merilampi, and L. Ukkonen, "Towards washable wearable antennas: a comparison of coating materials for screen-printed textile-based UHF RFID tags," *Int. J. Antennas Propag.*, vol. 2012, article ID 476570, 2012.

185 M. I. Maksud, M. S. Yusof, and M. M. A. Jamil, "An investigation into printing processes and feasibility study for RFID tag antennas," *Applied Mechanics and Materials*, vol. 315, pp. 468–71, Apr. 2013.

186 A. Vena, E. Perret, S. Tedjini, G. E. P. Tourtollet, A. Delattre, F. Garet, and Y. Boutant, "Design of chipless RFID tags printed on paper by flexography," *IEEE Trans. Ant. Propag.*, vol. 61, no. 12, pp. 5868–5877, Dec. 2013.

187 S. Khan, L. Lorenzelli, and R. S. Dahiya, "Technologies for printing sensors and electronics over large flexible substrates: a review," *IEEE Sens. J.*, vol. 15, pp. 3164–3185, 2015.

188 M. Caironi and Y. Y. Noh, *Large Area and Flexible Electronics*, Wiley-VCH Verlag, Weinheim, Germany, 2015.

189 P. C. Joshi, T. Kuruganti, and C. E. Duty, in: Srivatsan, T. S., Sudarshan, T. S. (eds) *Additive Manufacturing: Innovations, Advances, and Applications*, CRC Press, Boca Raton, FL, USA, 2016.

190 T. Athauda and N. C. Karmakar, "Screen printed chipless RFID resonator design for remote sensing applications," *2018 IEEE Asia-Pacific Microwave Conference (APMC)*, Kyoto, Japan, Nov. 2018, pp. 1321–1323.

191 T. H. Phung, A. N. Gafurov, I. Kim, S. Y. Kim, K. M. Kim, and T. M. Lee, "IoT device fabrication using roll-to-roll printing process," *Sci. Rep.*, vol. 11, paper 19982, 2021.

192 J. G. D. Hester, J. Kimionis, and M. M. Tentzeris, "Printed motes for IoT wireless networks: State of the Art, challenges, and outlooks," *IEEE Trans. Microw. Theory Tech.*, vol. 65, pp. 1819–1830, 2017.

193 V. Sanchez-Romaguera, M. A. Ziai, D. Oyeka, S. Barbosa, J. S. R. Wheeler, J. C. Batchelor, E. A. Parker, and S. G. Yeates, "Towards inkjet-printed low cost passive UHF RFID skin mounted tattoo paper tags based on silver nanoparticle inks," *J. Mater. Chem. C*, vol. 1, pp. 6395–6402, 2013.

194 S. Kim, A. Georgiadis, and M. M. Tentzeris, "Design of inkjet-printed RFID-based sensor on paper: single- and dual-tag sensor topologies," *Sensors*, vol. 18, no. 6 paper 1958, 2018.

195 A. Kamyshny, J. Steinke, and S. Magdass "Metal-based inkjet inks for printed electronics," *The Open Applied Physics Journal*, vol. 4, pp. 19–36, 2011.

196 K. R. R. Venkata, K. A. Venkata, P. S. Karthik, and S. P. Singh, "Conductive silver inks and their applications in printed and flexible electronics," *RSC Adv.*, vol. 5, no. 95, pp. 77760–77790, 2015.

197 X. Huang, T. Leng, M. Zhu, X. Zhang, J. Chen, K. Chang, M. Aqeeli, A. K. Geim, K. S. Novoselov, and Z. Hu, "Highly flexible and conductive printed graphene for wireless wearable communications applications," *Sci. Rep.*, vol. 5, no. 1, paper 18298, 2016.

198 T. Leng, X. Huang, K. Chang, J. Chen, M. A. Abdalla, and Z. Hu, "Graphene nanoflakes printed flexible meandered-line dipole antenna on paper substrate for low-cost RFID and sensing applications," *IEEE Ant. Wireless Propag. Lett.*, vol. 15, pp. 1565–1568, 2016.

199 K. Pan, T. Leng, Y. Jiang, Y. Fang, X. Zhou, M. A. Abdalla, H. Ouslimani, and Z. Hu, "Graphene printed UWB monopole antenna for wireless communication applications," *2019 IEEE International Symposium on Antennas and Propagation and USNC-URSI Radio Science Meeting*, Atlanta, GA, USA, Jul. 2019, pp. 1739–1740.

200 T. Leng, K. Pan, and Z. Hu, "Printed graphene radio frequency and sensing applications for Internet of Things," in: Harun S. W. (ed) *Handbook of Graphene, Volume 8: Technology and Innovations*, Chapter 2, Wiley, Hoboken, NJ, USA, 2019.

201 G. Orecchini, L. Yang, A. Rida, F. Alimenti, M. M. Tentzeris, and L. Roselli, "Green technologies and RFID: present and future," *Applied Computational Electromagnetics Society Journal*, vol. 25, pp. 230–238, May 2010.

202 X. Huang, T. Leng, X. Zhang, J. Chen, K. Chang, A. Geim, K. Novoselov, and Z. Hu, "Binder-free highly conductive graphene laminate for low cost printed radio frequency applications," *Appl. Phys. Lett.*, vol. 106, no. 20, p. 203105, 2015.

203 T. Leng, *Printed Graphene Antennas and Systems for RFID Applications*, PhD Dissertation, The University of Manchester, UK, 2018.

204 S. Genovesi, F. Costa, M. Borgese, A. Monorchio, and G. Manara, "Chipless RFID tag exploiting cross polarization for angular rotation sensing," *2016 IEEE International Conference on Wireless for Space and Extreme Environments (WiSEE)*, Aachen, Germany, Sept. 2016, pp. 158–160.

205 G. Harsányi, "Polymer films in sensor applications: a review of present uses and future possibilities," *Sens. Rev.*, vol. 20, pp. 98–105, 2000.

206 G. Pawlikowski, "Effects of polymer material variations on high frequency dielectric properties," *MRS Online Proceedings Library (OPL)*, Vol. 1156, pp. D02–D05, 2009.

207 E. M. Amin and N. C. Karmakar, "Development of a chipless RFID temperature sensor using cascaded spiral resonators," *2011 IEEE Sensors*, Limerick, Ireland, Oct. 2011, pp. 554–557.

208 K. C. Kao, *Dielectric Phenomena in Solids*, 1st ed., Academic Press, San Diego, CA, USA, 2004.

209 Y. T. Chen and H. L. Kao, "Humidity sensors made on polyvinyl-alcohol film coated SAW devices," *Electron. Lett.*, vol. 42, paper 948, 2006.

210 E. M. Amin, N. C. Karmakar, and B. Winther-Jensen, "Polyvinyl-Alcohol (PVA)-based RF humidity sensor in microwave frequency," *Prog. Electromagn. Res. B.*, vol. 54, pp. 149–166, 2013.

211 E. M. Amin, S. Bhuiyan, N. C. Karmakar, and B. Winther-Jensen, "A novel EM barcode for humidity sensing," *2013 IEEE International Conference on RFID* (RFID), Penang, Malaysia, Apr.-May 2013, pp. 82–87.

212 J. Virtanen, L. Ukkonen, T. Björninen, and L. Sydänheimo, "Printed humidity sensor for UHF RFID systems," *2010 IEEE Sensors Applications Symposium* (SAS), Limerick, Ireland, Feb. 2010, pp. 269–272.

213 J. Virtanen, L. Ukkonen, T. Bjorninen, A. Z. Elsherbeni, and L. Sydänheimo, "Inkjet-printed humidity sensor for passive UHF RFID systems," *IEEE Trans. Instrum. Measur.*, vol. 60, no. 8, pp. 2768–2777, Aug. 2011.

214 A. S. G. Reddy, B. B. Narakathu, M. Z. Atashbar, M. Rebros, E. Rebrosova, and M. K. Joyce, "Fully printed flexible humidity sensor," *Procedia Engineering*, vol. 25, pp. 120–123, 2011.

215 R. A. Potyrailo and C. Surman, "A passive radio-frequency identification (RFID) gas sensor with self-correction against fluctuations of ambient temperature," *Sens. Act. B: Chem.*, vol 185, pp. 587–593, 2013.

216 S. Manzari, C. Occhiuzzi, S. Nawale, A. Catini, C. di Natale, and G. Marrocco, "Polymer-doped UHF RFID tag for wireless-sensing of humidity," *2012 IEEE International Conference on RFID (RFID)*, Orlando, FL, USA, Apr. 2012, pp. 124–129.

217 C. Bali, A. Brandlmaier, A. Ganster, O. Raab, J. Zapf, and A. Hübler, "Fully inkjet-printed flexible temperature sensors based on carbon and PEDOT: PSS," *Mater. Today Proc.*, vol. 3, pp. 739–745, 2016.

218 H. D. Zhang, C. C. Tang, Y. Z. Long, J. C. Zhang, R. Huang, J. J. Li, and C. Z. Gu, "High-sensitivity gas sensors based on arranged polyaniline/PMMA composite fibers," *Sens. Act. A: Phys*, vol. 219, pp. 123–127, 2014.

219 G. A. Eyebe, B. Bideau, N. Boubekeur, E. Loranger, and F. Domingue, "Environmentally-friendly cellulose nanofibre sheets for humidity sensing in microwave frequencies," *Sens. Act. B: Chem.*, vol. 245, pp. 484–492, 2017.

220 S. Yun and J. Kim, "Multi-walled carbon nanotubes–cellulose paper for a chemical vapor sensor," *Sens. Act. B: Chem.*, vol. 150, pp. 308–313, 2010.

221 R. A. Potyrail and A. W. G. Morris, "Multianalyte chemical identification and quantitation using a single radio frequency identification sensor," *ACS Anal. Chem.*, vol. 79, pp. 45–51, 2007.

222 F. Alimenti, C. Mariotti, V. Palazzi, M. Virili, G. Orecchini, P. Mezzanotte, and L. Roselli, "Communication and sensing circuits on cellulose," *J. Low Power Electron. Appl.*, vol. 5, pp. 151–164, 2015.

223 J. Fan, S. Zhang, F. Li, and J. Shi, "Cellulose-based sensors for metal ions detection," *Cellulose*, vol. 27, pp. 5477–5507, 2020.

224 Y. F. Sun, S. B. Liu, F. L. Meng, J. Y. Liu, Z. Jin, L. T. Kong, and J. H. Liu, "Metal oxide nanostructures and their gas sensing properties: a review," *Sensors*, vol. 12, no. 3, pp. 2610–2631, 2012.

225 R. Viter and I. Iatsunskyi, "Metal oxide nanostructures in sensing," in: Zenkina O. V. (ed) *Micro and Nano Technologies, Nanomaterials Design for Sensing Applications*, Elsevier, 2019, pp. 41–91.

226 T. Ghafouri, R. Tahmasebi, N. Manavizadeh, and E. Nadimi, "Gas sensing properties of silicon nanowires with different cross-sectional shapes toward ammonia: a first-principles study," *Journal of Nanoparticle Research*, vol. 21, paper 157, 2019.

227 I. V. Zaporotskova, N. P. Boroznina, Y. N. Parkhomenko, and L. V. Kozhitov, "Carbon nanotubes: sensor properties. a review," *Mod. Electron. Mater.*, vol. 2, pp. 95–105, 2016.

228 L. Camilli, and M. Passacantando, "Advances on sensors based on carbon nanotubes," *Chemosensors*, vol. 6, p. 62, 2018.

229 H. Guerin, H. Le Poche, R. Pohle, M. Fernández-Bolaños, J. Dijon, and A. M. Ionescu, "Carbon nanotube resistors as gas sensors: towards selective analyte detection with various metal-nanotube interfaces," *European Solid-State Device Research Conference*, Bucharest, Romania, Sep. 2013, pp. 326–329.

230 M. D. Balachandran, S. Shrestha, M. Agarwal, Y. Lvov, and K. Varahramyan, "SnO_2 capacitive sensor integrated with microstrip patch antenna for passive wireless detection of ethylene gas," *Elect. Lett.*, vol. 44, pp. 464–466, Mar. 2008.

231 R. Nair, E. Perret, S. Tedjini, and T. Barron, "A humidity sensor for passive chipless RFID applications," *2012 IEEE International Conference on RFID-Technologies and Applications (RFID-TA)*, Nice, France, Nov. 2012, pp. 29–33.

232 A. Vena, E. Perret, S. Tedjini, D. Kaddour, A. Potie, and T. Barron, "A compact chipless RFID tag with environment sensing capability," *2012 IEEE/MTT-S Int. Microw. Symp. Digest*, Montreal, QC, Canada, Jun. 2012.

233 S. Ahoulou, E. Perret, and J. M. Nedelec, "Functionalization and cCharacterization of sSilicon nNanowires for sSensing aApplications: aA rReview," *Nanomaterials*, vol. 11, paper 999, 2021.

234 L. Yang, R. Zhang, D. Staiculescu, C. P. Wong, and M. M. Tentzeris, "A novel conformal RFID-enabled module utilizing inkjet-printed antennas and carbon nanotubes for gas-detection applications," *IEEE Ant. Wirel. Propag. Lett.*, vol. 8, pp. 653–656, 2009.

235 L. Yang, D. Staiculescu, R. Zhang, C. P. Wong, and M. M. Tentzeris, "A novel "green" fully-integrated ultrasensitive RFID-enabled gas sensor utilizing inkjet-printed antennas and carbon nanotubes," *2009 IEEE Antennas and Propagation Society International Symposium*, North Charleston, SC, USA, Jun. 2009.

236 C. Occhiuzzi, A. Rida, G. Marrocco, and M. M. Tentzeris, "Passive ammonia sensor: RFID tag integrating carbon nanotubes," *2011 IEEE International Symposium on Antennas and Propagation (APSURSI)*, Spokane, WA, USA, Jul. 2011, pp. 1413–1416.

237 C. Occhiuzzi, A. Rida, G. Marrocco, and M. M. Tentzeris, "CNT-based RFID passive gas sensor," *2011 IEEE MTT-S Int. Microw. Symp.*, Baltimore, MD, USA, Jun. 2011.

238 C. Occhiuzzi, A. Rida, G. Marrocco, and M. Tentzeris, "RFID passive gas sensor integrating carbon nanotubes," *IEEE Trans. Microw. Theory Techn.*, vol. 59, no. 10, pp. 2674–2684, Oct. 2011.

239 R. Baccarelli, G. Orecchini, F. Alimenti, and L. Roselli, "Feasibility study of a fully organic, CNT based, harmonic RFID gas sensor," *2012 IEEE International Conference on RFID-Technologies and Applications* (RFID-TA 2012), Nice, France, Nov. 2012, pp. 419–422.

240 A. Vena, L. Sydänheimo, M. M. Tentzeris, and L. Ukkonen, "A novel inkjet printed carbon nanotube-based chipless RFID sensor for gas detection," *2013 European Microwave Conference*, Nuremberg, Germany, Oct. 2013, pp. 9–12.

241 P. Gou, N. D. Kraut, I. M. Feigel, H. Bai, G. J. Morga, Y. Chen, Y. Tang, K. Bocan, J. Stachel, L. Berger, M. Mickle, E. Sejdic, and A. Star, "Carbon nanotube chemiresistor for wireless pH sensing", *Sci. Rep.*, vol. 4, paper 4468, 2014.

242 I. Gammoudi, B. Aissa, M. Nedil, and M. M. Abdallah, "CNT-RFID passive tag antenna for gas sensing in underground mine," *2015 IEEE International Symposium on Antennas and Propagation & USNC/URSI National Radio Science Meeting*, Vancouver, BC, Canada, Jul. 2015, pp. 1758–1759.

243 A. A. Kutty, T. Björninen, L. Sydänheimo, and L. Ukkonen, "A novel carbon nanotube loaded passive UHF RFID sensor tag with built-in reference for wireless gas sensing," *2016 IEEE MTT-S Int. Microw. Symp. (IMS'16)*, San Francisco, CA, USA, May 2016.

244 N. N. Le, E. Fribourg-Blanc, H. C. T. Phan, D. M. T. Dang, and C. M. Dang, "A RFID-based wireless NH_3 gas detector using spin coated carbon nanotubes as sensitive layer," *Int. J. Nanotechnology*, vol. 15, no. 1–3, pp. 3–13, 2018.

245 S. H. Min, H. J. Kim, Y. J. Quan, H. S. Kim, J. H. Lyu, G. Y. Lee, and S. H. Ahn, "Stretchable chipless RFID multi-strain sensors using direct printing of aerosolised nanocomposite," *Sens. Act. A: Phys.*, vol. 313, paper 112224, 2020.

246 N. Javed, M. A. Azam, I. Qazi, Y. Amin, and H. Tenhunen, "A novel multi-parameter chipless RFID sensor for green networks," *AEU - International Journal of Electronics and Communications*, vol. 128, paper 153512, 2021.

247 Y. Wang and J. T. W. Yeow, "A review of carbon nanotubes-based gas sensors," *J. Sens.*, vol. 2009, pp. 1–24, 2009.

248 K. S. Novoselov, G. K. Geim, S. V. Morozov, D. Jiang, Y. Zhang, S. V. Dubonos, I. V. Grigorieva, and A. A. Firsov, "Electric field in atomically thin carbon films," *Science*, vol. 306, pp. 666–669, 2004.

249 A. Nag, A. Mitra, S. C. Mukhopadhyay, "Graphene and its sensor-based applications: a review," *Sens. Act. A: Phys.*, vol. 270, pp. 177–194, 2018.

250 J. S. Lee, J. Oh, J. Jun, and J. Jang, "Wireless hydrogen smart sensor based on pt/graphene-immobilized radio-frequency identification tag," *ACS Nano*, vol. 9, pp. 7783–7790, 2015.

251 M. C. Caccami, M. Y. S. Mulla, C. Occhiuzzi, C. Di Natale, and G. Marrocco, "Design and experimentation of a batteryless on-skin RFID graphene-oxide sensor for the monitoring and discrimination of breath anomalies," *IEEE Sensors J.*, vol. 18, no. 21, pp. 8893–8901, Nov. 2018.

252 M. C. Caccami, C. Miozzi, M. Y. S. Mulla, C. Di Natale, and G. Marrocco, "An epidermal graphene oxide-based RFID sensor for the wireless analysis of human breath," *2017 IEEE International Conference on RFID Technology & Application (RFID-TA)*, Warsaw, Poland, Sept. 2017, pp. 191–195.

253 X. Huang, T. Leng, T. Georgiou, J. Abraham, R. R. Nair, K. S. Novoselov, and Z. Hu, "Graphene oxide dielectric permittivity at GHz and its applications for wireless humidity sensing," *Sci. Rep.*, vol. 8, no. 1, pp. 1–7, 2018.

254 H. J. Lee, "Recent progress in radio-frequency sensing platforms with graphene/graphene oxide for wireless health care system," *Appl. Sci.*, vol. 11, no. 5, paper 2291, 2021.

255 E. M. Amin, R. Bhattacharyya, S. Sarma, and N. C. Karmakar, "Chipless RFID tag for light sensing," *2014 IEEE Antennas and Propagation Society International Symposium (APSURSI)*, Memphis, TN, USA, Jul. 2014, pp. 1308–1309.

256 C. Mandel, H. Maune, M. Maasch, M. Sazegar, M. Schüßler, and R. Jakoby, "Passive wireless temperature sensing with BST-based chipless transponder," *2011 German Microwave Conference*, Darmstadt, Germany, Mar. 2011.

257 R. Gonçalves, P. Pinho, B. B. Carvalho, and M. M. Tentzeris, "Humidity passive sensors based on UHF RFID using cork dielectric slabs," *2015 9th European Conference on Antennas and Propagation (EuCAP)*, Lisbon, Portugal, Apr. 2015.

258 R. Singh, E. Singh, and H. Singh Nalwa, "Inkjet printed nanomaterial based flexible radio frequency identification (RFID) tag sensors for the internet of nano things," *RSC Adv.*, vol. 7, pp. 48597–48630, 2017.

259 P. Mezzanotte, V. Palazzi, F. Alimenti, and L. Roselli, "Innovative RFID sensors for Internet of Things applications," *IEEE J. Microw.*, vol. 1, no. 1, pp. 55–65, Jan. 2021.

260 V. Vippalapalli and S. Ananthula, "Internet of things (IoT) based smart health care system," *2016 International Conference on Signal Processing, Communication, Power and Embedded System (SCOPES)*, Paralakhemundi, India, Oct. 2016, pp. 1229–1233.

261 K. Ullah, M. A. Shah, and S. Zhang, "Effective ways to use Internet of Things in the field of medical and smart health care," *2016 International Conference on Intelligent Systems Engineering (ICISE)*, Islamabad, Pakistan, Jan. 2016, pp. 372–379.

262 M. Haddara and A. Staaby, "RFID applications and adoptions in healthcare: a review on patient safety," *Procedia Computer Science*, vol. 138, pp. 80–88, 2018.

263 S. Vishnu, S. R. J. Ramson, and R. Jegan, "Internet of Medical Things (IoMT) - an overview," *2020 5th International Conference on Devices, Circuits and Systems* (ICDCS), Coimbatore, India, Mar. 2020, pp. 101–104.

264 B. Pradhan, S. Bhattacharyya, and K. Pal, "IoT-based applications in healthcare devices," *Journal of Healthcare Engineering*, vol. 2021, paper 6632599, 18 pages, 2021.

265 S. Gabriel, R. W. Lau, and C. Gabriel, "The dielectric properties of biological tissues: II. Measurements in the frequency range 10 Hz to 20 GHz," *Phys. Med. Biol.*, vol. 41, pp. 2251–2269, 1996.

266 S. Gabriel, R. W. Lau, and C. Gabriel, "The dielectric properties of biological tissues: III. Parametric models for the dielectric spectrum of tissues," *Phys. Med. Biol.*, vol. 41, pp. 2271–2293, 1996.

267 A. Vena, E. Moradi, K. Koski, A. A. Babar, L. Sydanheimo, L. Ukkonen, and M. M. Tentzeris, "Design and realization of stretchable sewn chipless RFID tags and sensors for wearable applications," *2013 IEEE International Conference on RFID* (RFID), Orlando, FL, USA, Apr. 2013–May 2013, pp. 176–183.

268 J. G. D. Hester and M. M. Tentzeris, "Inkjet-printed flexible mm-wave Van-Atta reflectarrays: a solution for ultralong-range dense multitag and multisensing chipless RFID implementations for IoT dmart skins," *IEEE Trans. Microw. Theory Techn.*, vol. 64, no. 12, pp. 4763–4773, Dec. 2016.

269 L. Corchia, G. Monti, E. De Benedetto, and L. Tarricone, "Low-cost chipless sensor tags for wearable user interfaces," *IEEE Sensors J.*, vol. 19, no. 21, pp. 10046–10053, Nov. 2019.

270 S. Amendola, L. Bianchi, and G. Marrocco, "Movement detection of human body segments: passive radio-frequency identification and machine-learning technologies," *IEEE Ant. Propag. Mag.*, vol. 57, no. 3, pp. 23–37, Jun. 2015.

271 L. Corchia, G. Monti, E. De Benedetto, A. Cataldo, L. Angrisani, P. Arpaia, and L. Tarricone, "Fully-textile, wearable chipless tags for identification and tracking applications," *Sensors*, vol. 20, paper 429, 2020.

272 T. Andriamiharivolamena, A. Vena, E. Perret, P. Lemaitre-Auger, and S. Tedjini, "Chipless identification applied to human body," *2014 IEEE RFID Technology and Applications Conference (RFID-TA)*, Tampere, Finland, Sept. 2014, pp. 241–245.

273 L. Corchia, G. Monti, and L. Tarricone, "A frequency signature RFID chipless tag for wearable applications," *Sensors*, vol. 19, paper 494, 2019.

274 L. Corchia, G. Monti, E. D. Benedetto, and L. Tarricone, "A chipless humidity sensor for wearable applications," *2019 IEEE International Conference on RFID Technology and Applications (RFID-TA)*, Pisa, Italy, Sept. 2019, pp. 174–177.

275 J. Kim, Z. Wang, and W. S. Kim, "Stretchable RFID for wireless strain sensing with silver nano ink," *IEEE Sens. J.*, vol. 14, pp. 4395–4401, 2014.

276 T. Ativanichayaphong, J. Wang, W. Huang, S. Rao, H. F. Tibbals, S. Tang, S. J. Spechler, H. Stephanou, and J. Chiao, "Development of an implanted RFID impedance sensor for detecting gastroesophageal reflux," *2007 IEEE International Conference on RFID*, Grapevine, TX, USA, Mar. 2007, pp. 127–133.

277 C. M. Boutry, H. Chandrahalim, P. Streit, M. Schinhammer, A. C. Hänzi, and C. Hierold, "Towards biodegradable wireless implants," *Philos. Trans. R. Soc. A Math. Phys. Eng. Sci.*, vol. 370, pp. 2418–2432, 2012.

278 C. Occhiuzzi, G. Contri, and G. Marrocco, "Design of implanted RFID tags for passive sensing of human body: the STENTag," *IEEE Trans. Ant. Propag.*, vol. 60, no. 7, pp. 3146–3154, Jul. 2012.

279 H. Rajagopalan and Y. Rahmat-Samii, "Ingestible RFID bio-capsule tag design for medical monitoring," *2010 IEEE Antennas and Propagation Society International Symposium*, Toronto, ON, Canada, Jul. 2010.

280 P. R. Chai, J. Castillo-Mancilla, E. Buffkin, C. Darling, R. K. Rosen, K. J. Horvath, E. D. Boudreaux, G. K. Robbins, P. L. Hibberd, and E. W. Boyer, "Utilizing an ingestible biosensor to assess real-time medication adherence," *J. Med. Toxicol.*, vol. 11, pp. 439–444, 2015.

281 A. J. Healey, P. Fathi, and N. C. Karmakar, "RFID sensors in medical applications," *IEEE Journal of Radio Frequency Identification*, vol. 4, no. 3, pp. 212–221, Sep. 2020.

282 S. Amendola, R. Lodato, S. Manzari, C. Occhiuzzi, and G. Marrocco, "RFID technology for IoT-based personal healthcare in smart spaces," *IEEE Internet Things J.*, vol. 1, pp. 144–152, 2014.

283 R. Krigslund, S. Dosen, P. Popovski, J. L. Dideriksen, G. F. Pedersen, and D. Farina, "A novel technology for motion capture using passive UHF RFID tags," *IEEE Trans. Biomed. Eng.*, vol. 60, pp. 1453–1457, 2013.

284 S. Manzari, C. Occhiuzzi, and G. Marrocco, "Feasibility of body-centric systems using passive textile RFID tags," *IEEE Antennas Propag. Mag.*, vol. 54, pp. 49–62, 2012.

285 P. Kumar, H. W. Reinitz, J. Simunovic, K. P. Sandeep, and P. D. Franzon, "Overview of RFID technology and its applications in the food industry," *J. Food Sci.*, vol. 74, no. 8, pp. 101–106, Oct. 2009.

286 K. B. Biji, C. N. Ravishankar, C. O. Mohan, and T. K. Srinivasa Gopal, "Smart packaging systems for food applications: aA review," *J. Food Sci. Technol.*, vol. 52, no. 10, pp. 6125–6135, Oct. 2015.

287 F. Bibi, C. Guillaume, N. Gontard, and B. Sorli, "A review: RFID technology having sensing aptitudes for food industry and their contribution to tracking and monitoring of food products," *Trends Food Sci. Technol.*, vol. 62, pp. 91–103, Apr. 2017.

288 P. Fathi, N. C. Karmakar, M. Bhattacharya, and S. Bhattacharya, "Potential chipless RFID sensors for food packaging applications: a review," *IEEE Sensors J.*, vol. 20, no. 17, pp. 9618–9636, Sept. 2020.

289 B. Yan and D. Lee, "Application of RFID in cold chain temperature monitoring system," *2009 ISECS International Colloquium on Computing, Communication, Control, and Management*, Sanya, China, Aug. 2009, pp. 258–261.

290 E. Abad, F. Palacio, M. Nuin, A. González de Zárate, A. Juarros, J. M. Gómez, and S. Marco, "RFID smart tag for traceability and cold chain monitoring of foods: demonstration in an intercontinental fresh fish logistic chain," *Journal of Food Engineering*, vol. 93, no. 4, pp. 394–399, 2009.

291 F. Vivaldi, B. Melai, A. Bonini, N. Poma, P. Salvo, A. Kirchhain, S. Tintori, A. Bigongiari, F. Bertuccelli, G. Isola, and F. Di Francesco, "A temperature-sensitive RFID tag for the identification of cold chain failures," *Sens. Act. A: Phys.*, vol. 313, paper 112182, 2020.

292 H. Tao, M. A. Brenckle, M. Yang, J. Zhang, M. Liu, S. M. Siebert, R. D. Averitt, M. S. Mannoor, M. C. McAlpine, J. A. Rogers, D. L. Kaplan, and F. G. Omenetto, "Silk-based conformal, adhesive, edible food sensors," *Adv. Mater.*, vol. 24, pp. 1067–1072, 2012.

293 S. D. Nguyen, T. T. Pham, E. F. Blanc, N. L. Nguyen, C. M. Dang, and S. Tedjini, "Approach for quality detection of food by RFID-based wireless sensor tag," *IET Elect. Lett.*, vol. 49, pp.1588–1589, 2013.

294 S. Y. Wu, C. Yang, W. Hsu, and L. Lin, "3D-printed microelectronics for integrated circuitry and passive wireless sensors," *Microsyst Nanoeng.*, vol. 1, no. 1, pp. 1–9, 2015.

295 S. Bhadra, D. J. Thomson, and G. E. Bridges, "Near field chipless tag for food quality monitoring," *Proc. 16th Int. Symp. Antenna Technol. Appl. Electromagn. (ANTEM)*, Victoria, BC, Canada, Jul. 2014.

296 S. Karuppuswami, L. L. Matta, E. C. Alocilja, and P. Chahal, "A wireless RFID compatible sensor tag using gold nanoparticle markers for pathogen detection in the liquid food supply chain," *IEEE Sensors Lett.*, vol. 2, no. 2, Art. no. 3500704, Jun. 2018.

297 S. Karuppuswami, A. Kaur, H. Arangali, and P. P. Chahal, "A hybrid magnetoelastic wireless sensor for detection of food adulteration," *IEEE Sensors J.*, vol. 17, no. 6, pp. 1706–1714, Mar. 2017.

298 M. Yuan, R. Ghannam, P. Karadimas, and H. Heidari, "Flexible RFID patch for food spoilage monitoring," *2018 IEEE Asia Pacific Conference on Postgraduate Research in Microelectronics and Electronics (PrimeAsia)*, Chengdu, China, Oct. 2018, pp. 68–71.

299 T. Athauda and N. C. Karmakar, "Review of RFID-based sensing in monitoring physical stimuli in smart packaging for food-freshness applications," *Wireless Power Transfer*, vol. 6, pp. 161–174, 2019.

300 B. Saggin, Y. Belaizi, A. Vena, B. Sorli, V. Guillard, and I. Dedieu, "A flexible biopolymer based UHF RFID-sensor for food quality monitoring," *2019 IEEE International Conference on RFID Technology and Applications (RFID-TA)*, Pisa, Italy, Sept. 2019, pp. 484–487.

301 S. Karuppuswami, S. Mondal, D. Kumar, and P. Chahal, "RFID coupled passive digital ammonia sensor for quality control of packaged food," *IEEE Sensors J.*, vol. 20, no. 9, pp. 4679–4687, May 2020.

302 H. Zhou, S. Li, S. Chen, Q. Zhang, W. Liu, and X. Guo, "Enabling low cost flexible smart packaging system with Internet-of-Things connectivity via flexible hybrid integration of silicon RFID chip and printed polymer sensors," *IEEE Sensors J.*, vol. 20, no. 9, pp. 5004–5011, May 2020.

303 L. Ruiz-Garcia and L. Lunadei, "The role of RFID in agriculture: Applications, limitations and challenges," *Computers and Electronics in Agriculture*, vol. 79, no.1, pp. 42–50, 2011.

304 G. Zhao, H. Yu, G. Wang, Y. Sui, and L. Zhang, "Applied research of IOT and RFID technology in agricultural product traceability system," in: Li D., Chen Y. (eds) *Computer and Computing Technologies in Agriculture VIII. (CCTA 2014), IFIP Advances in Information and Communication Technology*, vol. 452, Springer.

305 T. Wasson, T. Choudhury, S. Sharma, and P. Kumar, "Integration of RFID and sensor in agriculture using IOT," *2017 International Conference on Smart Technologies For Smart Nation (SmartTechCon)*, Bengaluru, India, Aug. 2017, pp. 217–222.

306 L. García, L. Parra, J. M. Jimenez, J. Lloret, and P. Lorenz, "IoT-based smart irrigation systems: an overview on the recent trends on sensors and IoT systems for irrigation in precision agriculture," *Sensors*, vol. 20, no. 4, paper 1042, 2020.

307 R. Rayhana, G. Xiao, and Z. Liu, "RFID sensing technologies for smart agriculture," *IEEE Instrum. Measur. Mag.*, vol. 24, no. 3, pp. 50–60, May 2021.

308 V. Palazzi, F. Gelati, U. Vaglioni, F. Alimenti, P. Mezzanotte, and L. Roselli, "Leaf-compatible autonomous RFID-based wireless temperature sensors for precision agriculture," *2019 IEEE Topical Conference on Wireless Sensors and Sensor Networks* (WiSNet), Orlando, FL, USA, Jan. 2019.

309 S. N. Daskalakis, G. Goussetis, S. D. Assimonis, M. M. Tentzeris, and A. Georgiadis, "A UW backscatter-morse-leaf sensor for low-power agricultural wireless sensor networks," *IEEE Sensors J.*, vol. 18, pp. 7889–7898, 2018.

310 S. Dey, E. M. Amin, and N. C. Karmakar, "Paper based chipless RFID leaf wetness detector for plant health monitoring," *IEEE Access*, vol. 8, pp. 191986–191996, 2020.

311 S. Kim, T. Le, M. M. Tentzeris, A. Harrabi, A. Collado, and A. Georgiadis, "An RFID-enabled inkjet-printed soil moisture sensor on paper for "smart" agricultural applications," *2014 IEEE SENSORS*, Valencia, Spain, Nov. 2014, pp. 1507–1510.

312 J. Wang, L. Chang, S. Aggarwal, O. Abari, and S. Keshav, "Soil moisture sensing with commodity RFID systems," *18th International Conference on Mobile Systems, Applications, and Services*, Toronto, ON, Canada, Jun. 2020, pp. 273–285.

313 M. Boada, A. Lázaro, R. Villarino, and D. Girbau, "Battery-less soil moisture measurement system based on a NFC device with energy harvesting capability," *IEEE Sensors J.*, vol. 18, no. 13, pp. 5541–5549, Jul. 2018.

314 N. S. S. M. Da Fonseca, R. C. S. Freire, A. Batista, G. Fontgalland, and S. Tedjini, "A passive capacitive soil moisture and environment temperature UHF RFID based sensor for low cost agricultural applications," *2017 SBMO/IEEE MTT-S International Microwave and Optoelectronics Conference (IMOC)*, Aguas de Lindoia, Brazil, Aug. 2017.

315 F. Deng, P. Zuo, K. Wen, and X. Wu, "Novel soil environment monitoring system based on RFID sensor and LoRa," *Computers and Electronics in Agriculture*, vol. 169, paper 105169, 2020.

316 W. Ostachowicz and J. A. Güemes, Edts., *New Trends in Structural Health Monitoring*, Springer, 2013.

317 F. Hernandez-Valle, A. R. Clough, and R. S. Edwards, "Stress corrosion cracking detection using non-contact ultrasonic techniques," *Corros. Sci.*, vol. 78, pp. 335–342, 2014.

318 G. Y. Tian, Y. He, I. Adewale, and A. Simm, "Research on spectral response of pulsed eddy current and NDE applications," *Sens. Act. A: Phys.*, vol. 189, pp. 313–320, 2013.

319 B. Glišić and D. Inaudi, *Fibre Optic Methods for Structural Health Monitoring*, John Wiley, Hoboken, NJ, USA, 2007.

320 J. Zhang, G. Y. Tian, A. M. J. Marindra, A. I. Sunny, and A. B. Zhao, "A review of passive RFID tag antenna-based sensors and systems for structural health monitoring applications," *Sensors*, vol. 17, paper 265, 2017.

321 M. Donelli and F. Viani, "Remote inspection of the structural integrity of engineering structures and materials with passive MST probes," *IEEE Trans. Geosci. Remote Sens.*, vol. 55, pp. 6756–6766, 2017.

322 M. Donelli and D. Franceschini, "Experiments with a modulated scattering system for through-wall identification," *IEEE Ant. Wireless. Propag. Lett.*, vol. 9, pp. 20–23, 2010.

323 M. Donelli, "An RFID-based sensor for masonry crack monitoring," *Sensors*, vol.18, paper 4485, 2018.

324 C. Occhiuzzi, C. Paggi, and G. Marrocco, "Passive RFID strain-sensor based on meander-line antennas," *IEEE Trans. Ant. Propag.*, vol. 59, no. 12, pp. 4836–4840, Dec. 2011.

325 X. Yi, C. Cho, J. Cooper, Y. Wang, M. M. Tentzeris, and R. T. Leon, "Passive wireless antenna sensor for strain and crack sensing—Electromagnetic modeling, simulation, and testing," *Smart Mater. Struct.*, vol. 22, paper 085009, 2013.

326 A. Vena, M. Tedjini, T. Björninen, L. Sydänheimo, L. Ukkonen, and M. M. Tentzeris, "A novel inkjet-printed wireless chipless strain and crack sensor on flexible laminates," *2014 IEEE Antennas and Propagation Society International Symposium* (APS-URSI), Memphis, TN, USA, Jul. 2014, pp. 1294–1295.

327 S. Dey, P. Kalansuriya, and N. C. Karmakar, "Chipless RFID based high resolution crack sensing through SWB technology," *2014 IEEE International Microwave and RF Conference (IMaRC)*, Bangalore, India, Dec. 2014, pp. 330–333.

328 A. Ramos, A. Lazaro, and D. Girbau, "Time-coded chipless sensors to detect quality of materials in civil engineering," *2015 9th European Conference on Antennas and Propagation (EuCAP)*, Lisbon, Portugal, Apr. 2015.

329 A. M. J. Marindra, R. Sutthaweekul, and G. Y. Tian, "Depolarizing chipless RFID sensor tag for characterization of metal cracks based on dual resonance features," *2018 10th International Conference on Information Technology and Electrical Engineering* (ICITEE), Bali, Indonesia, Jul. 2018, pp. 73–78.

330 K. Brinker and R. Zoughi, "Embedded chipless RFID measurement methodology for microwave materials characterization," *2018 IEEE International Instrumentation and Measurement Technology Conference* (I2MTC), Houston, TX, USA, May 2018.

331 A. M. J. Marindra and G. Y. Tian, "Multiresonance chipless RFID sensor tag for metal defect characterization using principal component analysis," *IEEE Sensors J.*, vol. 19, no. 18, pp. 8037–8046, Sep. 2019.

332 M. M. Li, G. C. Wan, M. Yang, and M. S. Tong, "A chipless RFID sensor for metal crack detection based on notch characteristics," *2019 Photonics & Electromagnetics Research Symposium - Fall (PIERS - Fall)*, Xiamen, China, Dec. 2019, pp. 217–221.

333 A. M. J. Marindra and G. Y. Tian "Chipless RFID sensor for corrosion characterization based on frequency selective surface and feature fusion," *Smart Mater. Struct.*, vol. 29, paper 125010, 2020.

334 S. Deif and M. Daneshmand, "Multiresonant chipless RFID array system for coating defect detection and corrosion prediction," *IEEE Tran. Ind. Electron.*, vol. 67, no. 10, pp. 8868–8877, Oct. 2020.

335 G. Wan, W. Kang, C. Wang, W. Li, M. Li, L. Xie, and L. Chen, "Separating strain sensor based on dual-resonant circular patch antenna with chipless RFID tag," *Smart Mater. Struct.*, vol. 30, paper 015007, 2020.

336 W. Chompoosawat, A. Boonpoonga, P. Akkaraekthalin, L. Bannawat, and T. Lertwiriyaprapa, "Single-layer chipless RFID sensor for metal crack detection," *2021 9th International Electrical Engineering Congress (iEECON)*, Pattaya, Thailand, Mar. 2021, pp. 575–578.

337 K. Mc Gee, P. Anandarajah, and D. Collins, "Proof of concept novel configurable chipless RFID strain sensor," *Sensors*, vol. 21, paper 6224, 2021.

338 M. C. Caccami, S. Amendola, and C. Occhiuzzi, "Method and system for reading RFID tags embedded into tires on conveyors," *2019 IEEE International Conference on RFID Technology and Applications (RFID-TA)*, Pisa, Italy, Sep. 2019, pp. 141–144.

339 J. Grosinger, L. W. Mayer, C. F. Mecklenbrauker, and A. L. Scholtz, "Input impedance measurement of a dipole antenna mounted on a car tire," *2009 International Symposium on Antennas and Propagation (ISAP 2009)*, Bangkok, Thailand, Oct. 2009.

340 S. Shao, A. Kiourti, B. Burkholder, and J. Volakis, "Broadband and flexible textile RFID tags for tires," *2014 IEEE Antennas and Propagation Society International Symposium (APS-URSI)*, Memphis, TN, USA, Jul. 2014, p. 1507.

341 Z. Liang, J. Ouyang, F. Yang, and L. Zhou, "Design of license plate RFID tag antenna using characteristic mode pattern synthesis," *IEEE Trans. Ant. Propag.*, vol. 65, no. 10, pp. 4964–4970, Oct. 2017.

342 A. T. Mobashsher, A. J. Pretorius, and A. M. Abbosh, "Low-profile vertical polarized slotted antenna for on-road RFID-enabled intelligent parking," *IEEE Trans. Ant. Propag.*, vol. 68, no. 1, pp. 527–532, Jan. 2020.

343 J. K. Lee and D. R. Spach, *Wireless System to Detect Presence of Child in a Baby Car Seat*. U.S. Patent 773,322,8 B2, 8 Jun. 2010.

344 A. Leschke, F. Weinert, M. Obermeier, S. Kubica, and V. Bonaiuto, "Method for classification of frontal collision events in passenger cars based on measurement of local component-specific decelerations," *Int. J. Automot. Technol.*, vol. 21, pp. 785–794, 2020.

345 M. Buettner, R. Prasad, M. Philipose, and D. Wetherall, "Recognizing daily activities with RFID-based sensors," *11th International Conference on Ubiquitous Computing*, New York, NY, USA, Sep. 2009, pp. 51–60.

346 A. Motroni, A. Buffi, and P. Nepa, "A survey on indoor vehicle localization through RFID technology," *IEEE Access*, vol. 9, pp. 17921–17942, 2021.

347 H. Mora-Mora, V. Gilart-Iglesias, D. Gil, and A. Sirvent-Llamas, "A computational architecture based on RFID sensors for traceability in smart cities," *Sensors*, vol. 15, pp. 13591–13626, 2015.

348 A. Buffi, P. Nepa, and F. Lombardini, "A phase-based technique for localization of UHF-RFID tags moving on a conveyor belt: performance analysis and test-case measurements," *IEEE Sensors J.*, vol. 15, no. 1, pp. 387–396, Jan. 2015.

349 F. Bernardini, A. Buffi, D. Fontanelli, D. Macii, V. Magnago, M. Marracci, A. Motroni, P. Nepa, and B. Tellini, "Robot-based indoor positioning of UHF-RFID tags: the SAR method with multiple trajectories," *IEEE Trans. Instrum. Meas.*, vol. 70, pp. 1–15, 2021.

350 V. Magnago, L. Palopoli, A. Buffi, B. Tellini, A. Motroni, P. Nepa, D. Macii, and D. Fontanelli, "Ranging-free UHF-RFID robot positioning through phase measurements of passive tags," *IEEE Trans. Instrum. Meas.*, vol. 69, pp. 2408–2418, 2020.

351 A. Buffi, A. Motroni, P. Nepa, B. Tellini, and R. Cioni, "A SAR-based measurement method for passive-tag positioning with a flying UHF-RFID reader," *IEEE Trans. Instrum. Meas.*, vol. 68, pp. 845–853, 2019.

352 S. Moscato, R. Moro, M. Bozzi, L. Perregrini, S. Sakouhi, F. Dhawadi, A. Gharsallah, P. Savazzi, A. Vizziello, and P. Gamba, "Chipless RFID for space applications," *2014 IEEE International Conference on Wireless for Space and Extreme Environments* (WiSEE), Noordwijk, Netherlands, Oct. 2014.

353 C. Qi, J. D. Griffin, and G. D. Durgin, "Low-power and compact microwave RFID reader for sensing applications in space," *2018 IEEE International Conference on RFID Technology Application (RFID-TA)*, Macau, China, Sep. 2018.

8

Comparative Analysis and Concluding Remarks

The objectives of this succinct chapter are to compare qualitatively (and briefly) the planar microwave sensors presented in the previous chapters and to highlight their many relevant aspects (as discussed in Chapter 1, providing a quantitative comparison between sensors involving different input and output variables is meaningless). RFID sensors are excluded of such comparison since many RFID sensing tags, discussed in Chapter 7, cannot be considered to be true planar sensors (e.g. chip-based tags), and chipless-RFID sensor tags (i.e. full planar) typically (although not exclusively) exploit frequency variation as the sensing mechanism.

In Section 1.4 of Chapter 1, planar microwave sensors were qualitatively compared with other sensing approaches, including not only nonplanar microwave sensors, but also sensors based on different technologies (such as optics or acoustics, among others). It was pointed out in that chapter that planar microwave sensors, regardless of their working principle, exhibit numerous advantageous aspects, which can be summarized as follows:

- Low cost, small size, and low profile.
- Possibility of sensor implementation not only on rigid substrates (e.g. commercial microwave substrates or general-purpose printed circuit boards – PCB – such as *FR4*), but also on flexible substrates, including polymers, paper, and organic substrates.
- Compatibility with both subtractive fabrication techniques (e.g. photo-etching or drilling) and additive processes (e.g. screen-printing or inkjet-printing).
- Compatibility with other technologies, e.g. microfluidics, textiles, and micromachining.
- Potential for conformal, wearable sensors, and implants.
- Potential for "green" sensors, implemented with recyclable or biodegradable materials.
- Possibility of integrating the associated electronics, needed for signal generation and processing, in the same substrate.
- Easy sample preparation and low-sized samples.
- High versatility for the measurement of many different types of variables, including permittivity and related variables (e.g. material composition or characterization of liquid samples), motion variables, as well as many physical, chemical, and biological parameters by using smart or functional materials, reagents, or bio-receptors.
- Potential for the implementation of wireless sensors and RFID sensor systems based on planar sensing tags.
- Operation under different modes, i.e. contact/contactless, reflection-mode/transmission-mode, single-ended/differential, invasive/noninvasive, and intrusive/nonintrusive.
- Possibility of operation over a wide band of frequencies, from UHF up to millimeter-wave frequencies.

Planar Microwave Sensors, First Edition. Ferran Martín, Paris Vélez, Jonathan Muñoz-Enano, and Lijuan Su.
© 2023 John Wiley & Sons, Inc. Published 2023 by John Wiley & Sons, Inc.

The previous characteristics are very interesting for the deployment of the so-called "Smart World," with sensors present everywhere, in everything, in everyone, as discussed before in this book. Nevertheless, it should also be mentioned that planar microwave sensors exhibit limited performance in certain applications, where other sensor types are preferred (sometimes with the penalty of higher cost or other drawbacks). For example, for the accurate measurement of the complex permittivity of materials, resonant cavities are preferred over planar sensors (though resonant cavities are bulky and expensive). Electromagnetic encoders, to cite another example, cannot compete against their optical counterparts in terms of sensor resolution (though optical encoders are less tolerant to the effects of pollution, dirt, or grease, present in many industrial scenarios). Let us also mention that for biosensing, optical methods are, in general, preferred, as such methods exhibit very good sensitivity and accuracy. On the other hand, planar microwave sensors present intrinsic (and well-known) drawbacks, including air-gap effects (in solid samples), or substrate absorption (in liquid samples). These limitative aspects have been discussed throughout the book, and solutions to palliate their effects have been proposed.

In this chapter, the comparison is restricted to microwave sensors implemented in planar technology, and grouped by their working principle, in coherence with the chapters of the book. Table 8.1 reports such comparison, where different aspects of interest for planar microwave sensors (the subject of comparison) are included in the first column.

The previous table offers a bird's eye overview of the most relevant characteristics of the different planar microwave sensors. However, let us next dedicate some words to clarify some aspects that need further discussion. Concerning size and cost, it has been pointed out in the previous list of advantages of planar microwave sensors that their dimensions and cost are very competitive, as compared to other sensors. However, it is obvious that there are differences between the various planar microwave sensor approaches. Thus, the terms "high/low" and "large/small" referred to cost and size, respectively, in Table 8.1, should be understood as relative to the type of sensors included in the table. Thus, among them, differential sensors are considered to be the most expensive and largest sensors, because such sensors are based on a sensor pair. By contrast, frequency-variation sensors exhibit a small size as far as the sensing element is an electrically small planar resonator (the size is inversely proportional to its resonance frequency). The electromagnetic module of frequency-variation sensors should be low cost, in coherence with their small size. However, the cost of these sensors has been termed as "moderate" because wideband signals are needed to retrieve the resonance frequency, the canonical output variable of these sensors, and, for that purpose, wideband (i.e. moderately costly) voltage-controlled oscillators (VCOs) are needed. In this regard, single-frequency sensors, such as phase-variation and coupling-modulation sensors constitute the most interesting approach, since a harmonic signal suffices for sensing. Note, however, that the size of phase-variation sensors has been considered to be moderate in the table, in coherence with the reported implementations of Chapter 3. In some cases, however, the size could be considered to be large (e.g. in highly sensitive phase-variation sensors based on meandered lines), but the size can be reduced, maintaining the sensitivity, by increasing the operational frequency, as discussed in Chapter 3. In reflective-mode phase-variation sensors based on step-impedance 90° lines terminated with quarter- or half-wavelength open-ended lines, or electrically small resonators (the sensing elements), the overall sensor size (electromagnetic module) might be relatively considerable, but the sensing area is determined by the dimensions of the resonator (either semi-lumped or distributed) terminating the step-impedance line. Sensor size can thus be very small if the sensor is solely implemented by means of the sensing resonator. Thus, in reality, the size of phase-variation sensors can vary significantly among the different implementations, and the authors have opted to assign the term "moderate," as a qualitative average, for the size of these sensors. As can be seen in Table 8.1, the size and cost of coupling-modulation sensors is considered small and low,

Table 8.1 Qualitative comparison of planar microwave sensors.

	Frequency-variation	Phase-variation	Coupling-modulation	Frequency-splitting	Differential-mode
Size	Small	Moderate	Small	Moderate	Large
Cost	Moderate	Low	Low	Moderate	High
Tolerance against ambient factors	Limited	Limited	Good	Good	Very good
Frequency span	Wideband	Single frequency	Single-frequency	Wideband	—
Robustness against noise and EMI	Good	Good	Limited	Good	—
Sensitivity	Good	Very good	Moderate	Moderate	Good
Resolution	Good	Good	Good	Limited	Good
Design and fabrication complexity	Simple	Moderate	Simple	Complex	Moderate
Canonical measurand	Permittivity	Permittivity	Displacement	Permittivity	Permittivity
Canonical output variable	Notch or peak frequency	Phase (reflection or transmission coefficient)	Magnitude (transmission or reflection coefficient)	Difference between split frequencies	Cross-mode transmission or reflection coefficient

respectively, since these sensors operate at a single-frequency, and many implementations, e.g. those exploiting electromagnetic symmetry, are based on resonator-loaded lines. However, in other implementations, e.g. electromagnetic encoders, the size of the rotor may be significant, depending on the required number of pulses per revolution (intimately related to sensor resolution). By contrast, the size of the stator, typically a resonator-loaded line, can be made relatively small.

Another aspect included in the comparative analysis refers to sensor tolerance against the effects of ambient factors, or cross-sensitivities to ambient factors, such as temperature or humidity (or even atmospheric pressure). As discussed exhaustively in previous chapters, as far as such environmental factors do not change at the typical scale of the sensors, they are "seen" as common-mode stimuli by the sensors, and their effects are cancelled by using differential-mode or quasi-differential-mode sensor schemes. Thus, besides differential sensors, which intrinsically exhibit small cross-sensitivities to ambient factors, frequency-splitting sensors and coupling-modulation sensors based on electromagnetic symmetry perturbation are robust against the effects caused by variations in environmental conditions.

Concerning the effects of noise and electromagnetic interference (EMI), it is well known that magnitude measurements are less robust than frequency or phase measurements. For that reason, coupling-modulation sensors are those devices exhibiting the poorest performance with regard to this aspect. Nevertheless, it should be mentioned that there are phase-variation sensors where phase-to-magnitude (voltage) transformation is applied, e.g. by including phase-to-magnitude converters based on hybrid couplers. Thus, in such cases, the robustness of the sensors against the effects of noise and EMI is also limited. Note that for differential-mode sensors, the corresponding box in the table has been left blank, since the differential output variable in such sensors can be a frequency, a phase, or a magnitude (or a magnitude combined with a frequency, e.g. the frequency position and magnitude of the maximum value of the cross-mode transmission coefficient, as reported in various examples in Chapter 6).

In the present book, the main parameter indicative of sensor performance is considered to be the sensitivity. The sensitivity can vary significantly among different implementations within each sensor type of Table 8.1. Thus, the qualitative comparison reported in the table refers to the maximum achievable sensitivity within each group. For this reason, the highest sensitivity is attributed to phase-variation sensors. As it has been demonstrated in Chapter 3, reflective-mode phase-variation sensors based on a cascade of high/low 90° line sections terminated with a distributed or a semi-lumped resonator exhibit an unprecedented sensitivity. However, such high sensitivity is achieved at the expense of a limited input dynamic range. Thus, these highly sensitive sensors are of special interest in applications where the measurement of small perturbations with regard to a reference value is necessary (for example, the detection of tiny defects in samples, or the measurement of small concentrations of solute in diluted solutions). Obviously, relaxing sensitivity optimization improves the input dynamic range, since better linearity is achieved. Thus, depending on the application, a trade-off is necessary. Examples of frequency-variation sensors and differential-mode sensors with good sensitivity have also been reported in this book. In frequency-variation sensors, the sensitivity can be improved by merely increasing the resonance frequency of the sensing resonator. In these sensors, the relevant figure of merit is the relative sensitivity, i.e. the sensitivity normalized to the resonance frequency of the bare resonator, and it has been demonstrated that in sensors based on slotted resonators, the maximum achievable relative sensitivity scarcely depends on the geometry of the resonator, but on the dielectric constant of the substrate. In coupling-modulation sensors, the sensitivity is not an easily controllable parameter since the output variable is typically the magnitude of the transmission coefficient at a certain frequency, and it depends on losses. In these sensors, and frequency-splitting sensors, the sensitivity is considered to be moderate, as compared to the sensitivities achievable in the other sensor types.

Sensor resolution is another important performance parameter, in part related to the sensitivity. The achievable resolution with planar microwave sensors is, in general, good. Particularly, very good resolutions have been demonstrated in differential-mode sensors devoted to electrolyte concentration measurements in DI water (see Chapter 6). The resolution achieved in electromagnetic encoders (specifically in rotary encoders), a type of coupling-modulation sensors, is also reasonably good, with 0.3° angle resolution in one of the implementations reported in Chapter 4. Probably the poorest sensors concerning resolution are those based on frequency splitting. This is in part due to sensitivity degradation at small perturbations (nevertheless, Chapter 5 reports solutions to this issue).

Concerning design and fabrication complexity, frequency-variation and coupling-modulation sensors are considered to be the most favorable devices, since, in their simplest implementations, such sensors consist of a transmission line loaded with a resonant element. Some phase-variation sensors consisting merely in a sensing transmission line (straight or meandered) can be even more easily designed and fabricated. However, phase-variation sensors based on artificial lines are not exempt of certain design difficulties. Moreover, detuning in such sensors may be critical, and, for this reason, accurate fabrication processes are necessary in order to achieve the required performance. This applies also to differential sensors, where good balance between the sensing elements of the sensor pair is necessary, especially in highly sensitive devices. The more complex devices in terms of design and fabrication are considered to be the frequency-splitting sensors, as far as these devices require good balance between the sensing elements (similar to differential-mode sensors), but, additionally, special care must be focused on their design. The reason is that strategies to mitigate the effects of inter-resonator coupling (which tends to degrade the sensitivity at small perturbations, as indicated in the preceding paragraph) are needed.

The last two rows of Table 8.1 indicate the canonical input and output variables of the considered sensor types. Note that planar microwave sensors are essentially permittivity sensors, i.e. devices sensitive to the dielectric characteristics of the surrounding medium. However, the permittivity correlates with many other variables, and for this reason, these sensors can be applied to the measurement of multitude of variables, as pointed out in the preceding list of advantages, where versatility is a key aspect. Nevertheless, in coupling-modulation sensors, the natural input variable is displacement, either linear or angular, and for this reason, such sensors are mostly devoted to motion sensing (displacement and velocities). As for the output variables, besides the canonical one, indicated in Table 8.1 for each sensor class, an additional output variable is needed for the measurement of magnitudes involving two variables, e.g. the complex permittivity of samples. Thus, frequency-variation sensors typically utilize the magnitude of the notch, or peak, of the resonance as second output variable. Similarly, in phase-variation sensors, the phase and the magnitude of the transmission or reflection coefficient at the operating frequency are the considered output variables when the complex permittivity should be inferred. In frequency-splitting sensors, the pair of output variables are the difference between the split resonance frequencies and magnitudes (notch or peaks). Finally, in various differential-mode sensors, mostly reported in Chapter 6, the output variable is the cross-mode transmission or reflection coefficient, and retrieving the complex permittivity of material is possible by considering the magnitude and frequency position of the maximum of the cross-mode transmission or reflection coefficient.

To conclude this chapter, and the whole book, let us mention that, in the humble opinion of the authors, planar microwave sensors are envisaged to play an increasingly important role in next future (many reported examples throughout the book are indicative of this trend). Nevertheless, further efforts are still needed in order to bring many of the research results (presented in this book and in other sources) to the market. It should also be mentioned that, despite their inherent advantages and their potential for the deployment of the so-called Internet of Things (IoT) and the "Smart World," planar microwave sensors are not the panacea. Many limitations and challenges need to be addressed in next future, especially related to aspects such as wireless connectivity and sustainability (see Chapter 7 for further details). Moreover, other sensor technologies, such as optics, are also very competitive, and are expected to dominate the market in certain areas (indeed, the trend in future smart systems, with high levels of complexity, is the coexistence of various sensing technologies). This book has been conceived as a comprehensive review on the specific topic of planar microwave sensors, with an increasing research activity in recent years. It is the authors' hope that the book constitutes an inspiring source for many researchers and professionals involved in microwave and sensor technologies and related topics.

Index

Printed and bound by CPI Group (UK) Ltd, Croydon, CR0 4YY

16/04/2025

14658424-0005